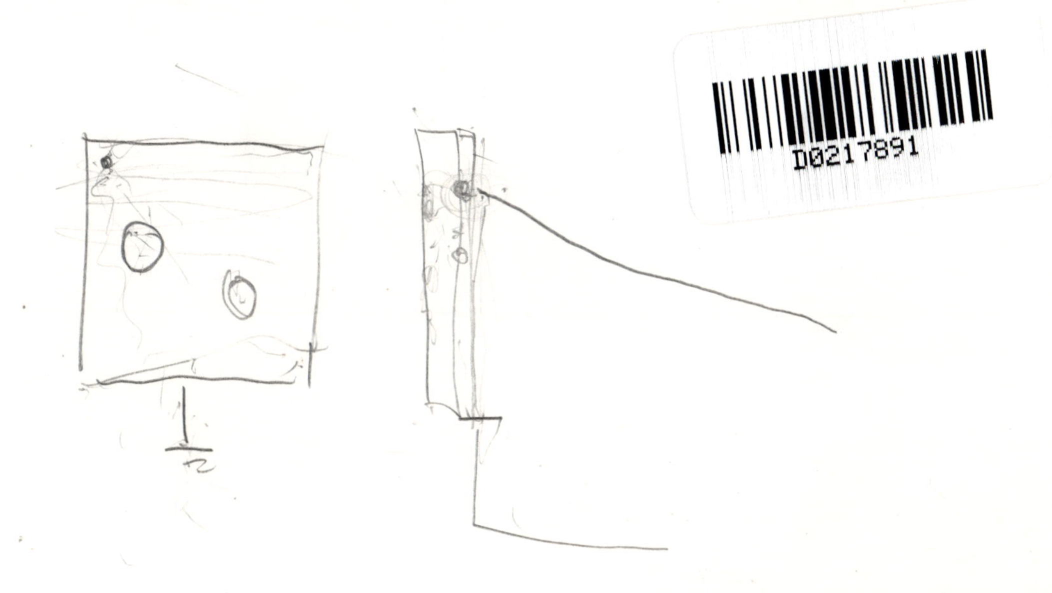

Laser Engineering

Kelin Kuhn
University of Washington

Library of Congress Cataloging-in-Publication Data
Kuhn, Kelin J.
Laser engineering / Kelin J. Kuhn
p. cm.
Includes index.
ISBN 0-02-366921-7 (hardcover)
1. Lasers—Design and construction. 2. Nonlinear optics.
I. Title.
TA1675.K84 1998 97-53211
CIP

Acquisition Editor: Eric Svendsen
Editor-in-Chief: Marcia Horton
Production Manager: Bayani Mendoza de Leon
Editor-in-Chief: Jerome Grant
Director of Production and Manufacturing: David W. Riccardi
Manufacturing Manager: Trudy Pisciotti
Full Service Coordinator: Donna Sullivan
Composition/Production Service: ETP Harrison
Editorial Assistant: Andrea Au
Creative Director: Paula Maylahn
Art Director: Jayne Conte
Cover Designer: Bruce Kenselaar

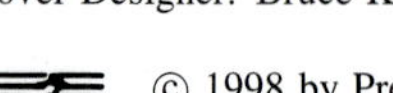

Printed in the United States of America
10 9 8 7 6 5 4 3 2

ISBN 0-02-366921-7

Prentice-Hall International (UK) Limited, *London*
Prentice-Hall of Australia Pty. Limited, *Sydney*
Prentice-Hall Canada, Inc., *Toronto*
Prentice-Hall Hispanoamericana, S.A., *Mexico*
Prentice-Hall of India Private Limited, *New Delhi*
Prentice-Hall of Japan, Inc., *Tokyo*
Simon & Schuster Asia Pte. Ltd., *Singapore*
Editora Prentice-Hall do Brasil, Ltda., *Rio de Janeiro*

To my parents

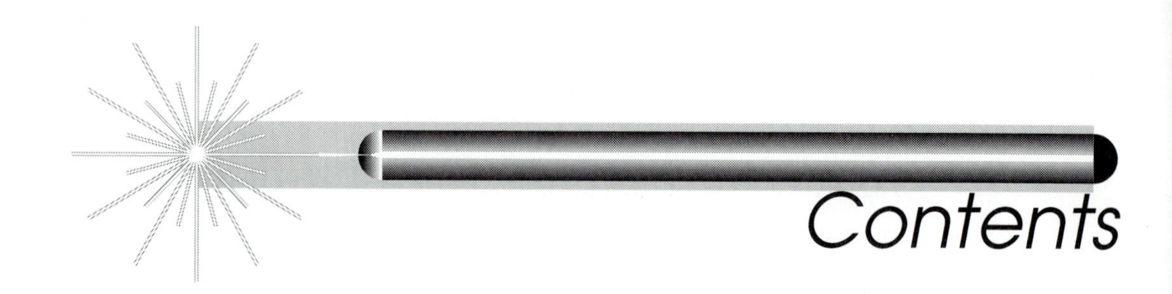

Contents

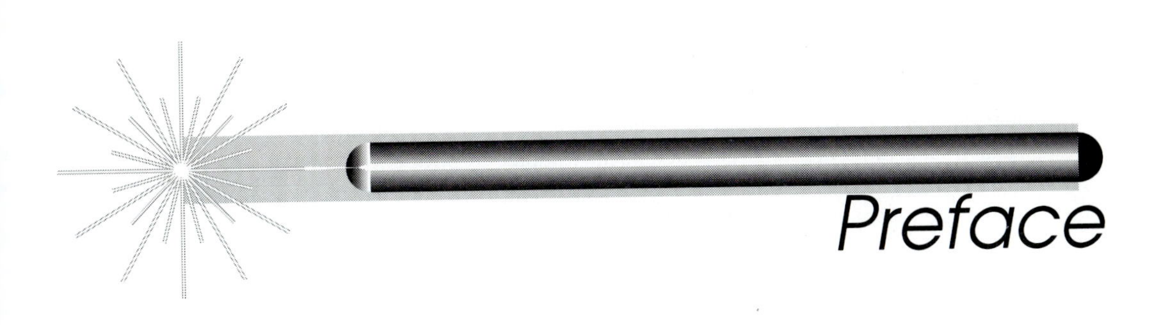

Preface

This text was developed for junior and senior electrical engineering students planning to work for companies that make or use lasers. Thus, the first goal of the text is to introduce students to those fundamental principles of lasers that are immediately relevant to the lasers that the students will encounter in the workplace.

This text is also intended for a junior or senior engineering class with some design content. This content could be in the form of design problems given during the class or an end-of-quarter design project. Thus, the second goal of this text is to provide a resource to assist students with design activities.

In order to meet these two goals, the text is organized in two parts. Part I (Chapters 1–8) covers laser fundamentals. Part II (Chapters 9–12) covers the design details for various laser systems. The exercises in Part I are primarily engineering science and the exercises in Part II are primarily engineering design.

ORGANIZATION

Chapters 1–6 of the book contain the fundamental theory of lasers.

Chapter 1 presents a broad and qualitative picture of the essential concepts underlying laser systems. All the topics presented in overview in Chapter 1 are covered in more detail later in the book. Copious forward references are provided in Chapter 1 to alert students to the location of the more detailed material.

Chapters 2–4 contain the fundamental material on gain, energy bands, transverse modes, longitudinal modes, Fabry-Perot etalons, and Gaussian beam propagation essential in any laser application; and constitute a basic introduction to engineering laser science.

Chapters 5 and 6 present solutions to the laser rate equations. Rate equations are often new to students taking the course, and so a detailed introduction to creating and solving steady-state rate equations is provided in Chapter 5. Modern computational tools such as *Mathematica* permit a more visual and dynamic approach to teaching the transient solutions to the rate equations. Thus, *Mathematica* examples are presented and compared to analytical solutions in Chapter 6.

Chapter 7 is an introduction to nonlinear optics (which is becoming increasingly important in modern commercial lasers) and Chapter 8 provides some background on essential support technologies (multilayer dielectric films, birefringent crystals, and photodetectors).

Chapters 9–12 are stand-alone chapters that highlight the design and applications of various laser systems. Design project ideas can be found in the applications sections of Chapters 9–12 and in the Appendix.

TECHNICAL BACKGROUND

The text assumes that students have a sophomore engineering background including a year of freshman physics, a quarter of freshman chemistry, calculus (including differential equations), and basic circuit theory. It is helpful if the students have also had basic engineering electromagnetism (including the vector form of Maxwell's equations), some practical optics, and basic electronics.

PEDAGOGY

There are two common teaching pedagogies. In deductive teaching, the principles are presented first and applications follow later. In inductive teaching, the applications are presented first and are followed by general principles.

The physical organization of the material in the book follows the deductive approach. Thus, instructors using a deductive pedagogy will be able to follow the material in the order presented.

However, there is an increasing body of research that suggests that inductive teaching promotes deeper learning and longer retention of information.[1] This is especially true when inductive techniques can be coupled with an active learning environment such as that offered by a design activity.[2] Therefore, the book has been organized to permit instructors to present the material inductively. Specifically, Chapters 9–12 are independent of each other and heavily cross-referenced to permit their presentation at any point in the course. An instructor could use the Chapter 9–12 material to motivate the Chapter 1–8 material. As a specific example, the HeNe material in Chapter 9 could be used as an introduction for the gain and energy state material in Chapter 2.

---

[1] R. Felder and L. Silverman, *Engineering Education* 78:674 (1988); W. J. McKeachie, ed, *Learning, Cognition, and College Teaching* (San Francisco, CA: Jossey-Bass, 1980).

[2] W. J. McKeachie, *Teaching Tips: A Guidebook for the Beginning College Teacher*, 8th ed. (Lexington, MA: D. C. Heath and Co., 1986).

SCHEDULING

A one-quarter, three-credit class with an included design project can typically cover the material in Chapters 1–6 in detail, and use selections from the material in the remaining chapters. In this format, the available time generally does not permit both laboratories and a design activity. For a one-quarter, three-credit class with an included design activity, an effective schedule is:

Weeks 1–5: Chapters 1–4
Week 6: Oral student design proposal presentations
Weeks 7–9: Chapters 5–6, with selections from others
Weeks 7–9: Design activity, open lab
Week 10: Project Open House

A one-quarter, four-credit class with an included design project can typically cover the material in Chapters 1–7 in detail and use selections from the material in the remaining chapters. In this format, it is also possible (with careful scheduling!) to run both laboratories and a design activity (e.g., the laboratories can be run the first five weeks, and the design activity the last three). For a one-quarter, four-credit class with an included design activity and no laboratories, a common schedule is:

Weeks 1–4: Chapters 1–4, then midterm exam
Week 5: Begin Chapter 5
Week 6: Oral student design proposal presentations
Weeks 7–9: Chapters 6–7, with selections from others
Week 7–9: Student design activity, open lab
Week 10: Project Open House

For a one-quarter, four-credit class with both an included design activity and laboratories, a common schedule is:

Weeks 1–5: Chapters 1–4
Weeks 1–5: HeNe and argon-ion laboratories in lab session
Week 6: Oral student design proposal presentations
Weeks 7–9: Chapters 5–6, with selections from others
Weeks 7–9: Student design activity in lab session
Week 10: Project Open House

A one semester, three- or four-credit class provides the flexibility to cover the majority of the material and include either a significant design experience, or a mix of design and organized laboratories. Schedules for semester courses can be constructed by various combinations of the quarter schedules given above.

ACKNOWLEDGMENTS

This is a complex book, and I would like to thank the many individuals and organizations that have contributed to its completion.

To begin with, I would like to acknowledge the support of the National Science Foundation Presidential Young Investigator award program in the completion of this textbook. In addition, many of the laboratory experiments were performed using equipment purchased under a National Science Foundation Instrumentation and Laboratory Improvement grant. This textbook would not exist if it were not for the help of the National Science Foundation.

In addition, I would like to thank Prof. Thomas Plant of Oregon State University. Prof. Plant's help and good advice during the latter stages of the book development were critical in its completion.

I would also like to thank the numerous University of Washington and Oregon State University students who took my Laser Engineering class from 1987 to 1997. Their enthusiasm for lasers and their original and clever design projects inspired this book.

In addition, I would like to thank the many individuals who were graduate students while this book was being completed, particularly Carrie Cornish, Brian Read, Chris Doughty, David Dawson, Marc Daoura, Kyle Johnston, Chuck Jung, and Kevin Welton. Their advice and contributions were invaluable.

I would also like to thank my colleagues, Prof. Bruce Darling and Prof. Randy Babbitt, who both made serious contributions to the development of this text.

This book would not have been possible without the overwhelming support of the contributing laser companies. Specifically, I would like to thank:

Melles Griot: Lisa M. Tsufura, Nina Richards, Peter M. Allard

Spectra-Physics Lasers, Inc.: Bruce B. Craig, David L. Wright, Steven M. Jarrett, James Clark, James Kafka

SDL, Inc.: Sean Ogarrio

Coherent, Inc., Laser Group: Mark M. Gitin, Anne Gilbert, John Nightengale, Murray Reed

XMR, Inc.: Michael Simile, Yale H. Sun, Jenn Ying Liu, Leonard Goldfine

Synrad, Inc.: Peter Laakmann, Dave Clarke, Yong Fang Zhang

Crystal Technology, Inc.: John Fowler

EG&G, Inc.: Jim Reed

Lawrence Livermore National Laboratories: Steve Wampler

Additionally, I would like to thank my Department Chair, Prof. Greg Zick, who was highly supportive of this project during its development.

Finally, I would like to thank Prof. Thomas Sigmon and Prof. John Arthur, who have demonstrated to their students and colleagues the highest levels of technical competence and professional ethics.

Kelin J. Kuhn

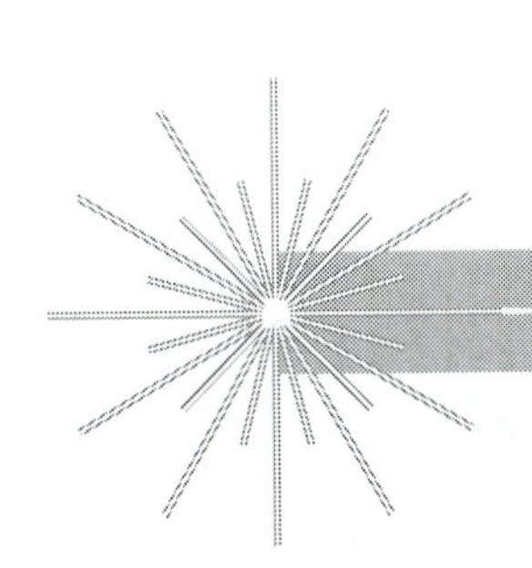

Part I

Laser Fundamentals

1 Introduction to Lasers

Objectives

- To calculate the transition wavelength between two states from the transition energy (and to calculate the transition energy from the transition wavelength).
- To calculate the population ratio between two states in thermal equilibrium.
- To understand the differences between energy, average power, peak power, energy density, average power density, and peak power density (as they apply to lasers); and to calculate these quantities given the appropriate parameters.
- To understand the concept of linewidth and to convert a linewidth given in frequency (Hz) to a linewidth given in length (meters).
- To draw a rough picture of the longitudinal mode spectrum of a laser (given the frequency parameters for the laser) and to distinguish between the linewidths of each of the modes and the linewidth of the gain curve.
- To calculate the frequency spacing between longitudinal modes.

1.1 A BRIEF HISTORY

As with most scientific fields, it is very difficult to say, "On this day the idea of a laser was conceived." Roots of the laser concept lie as far back as 1940, when Valentin A. Fabrikant speculated on the possibility of "molecular amplification" in his doctoral thesis.[1] However, it is traditional to place the origin of the maser and laser in the late 1940s. A number of

[1]V. A. Fabrikant, referenced by Jeff Hecht, *Laser Pioneers*, revised ed. (San Diego, CA: Academic Press, 1992), pp. 5–6, as being cited in Mario Bertolotti, *Masers and Lasers: A Historical Approach* (Bristol, England: Adam Hilger Ltd., 1983).

excellent reviews of the history of this era have been published[2,3,4] and only the major points are summarized here.

For a variety of reasons, it is easier to obtain amplification of electromagnetic waves in the microwave spectrum. Thus, the first devices to demonstrate *gain* (see the Appendix for definitions of common laser terms and Section 2.2 for more information on gain) by *stimulated emission* (see Section 1.4) were microwave devices termed masers (**M**icrowave **A**mplification by the **S**timulated **E**mission of **R**adiation).

The idea of using stimulated emission as a means of amplifying electromagnetic radiation in the microwave spectrum seems to have been independently conceived of by Charles H. Townes at Columbia University, Joseph Weber at the University of Maryland,[5] and Alexander M. Prokhorov and Nikolai G. Basov at the Lebedev Physics Institute (Moscow).[6] The first maser (a 24-GHz ammonia device) was operated by James P. Gordon, Townes, and Herbert J. Zeiger at Columbia in 1954.[7]

The successful operation of the ammonia maser immediately generated discussion as to whether these principles could be applied to visible wavelengths. At the time, there were a number of issues that suggested that it would be quite difficult to construct a visible maser. The three major issues were: 1) increased pumping requirements as the wavelength decreases, 2) creating a single mode cavity (as is traditional for masers) at visible wavelengths, and 3) locating materials possessing visible transitions with a sufficiently high *quantum efficiency* (see Section 2.1).

In 1951, Fabrikant filed a patent entitled "A method for the amplification of electromagnetic radiation (ultraviolet, visible, infrared and radio) distinguished by the fact that the amplified radiation is passed through a medium which, by means of auxiliary radiation of other means, generates excess concentrations, in comparison with the equilibrium concentration of atoms, other particles, or systems at upper energy states corresponding to the excited states." The patent was filed on June 18, 1951, but not granted until 1959.[8] For a variety of reasons, this patent had little impact on either the Soviet Union or the international development of lasers.

In 1954, Robert H. Dicke developed the idea of using a short excitation pulse to produce a *population inversion* (see Section 1.4). This inversion would then generate an intense burst of amplified spontaneous emission.[9] This idea plus the idea of using a *Fabry-Perot etalon* (see Section 3.2) as a *resonant cavity* (see Chapter 4) appear

---

[2]Jeff Hecht, *Laser Pioneers*, revised ed. (San Diego, CA: Academic Press, 1992).

[3]Joan L. Bromberg, *The Laser in America 1950–1970* (Cambridge, MA: MIT Press, 1991).

[4]William Broad, *Star Warriors* (New York: Simon and Schuster, 1985).

[5]J. Weber, *Trans. IRE Prof. Group on Electron Dev.* PGED-3:1 (1953).

[6]N. G. Basov and A. M. Prokhorov, *JETP* 27:431 (1954).

[7]J. P. Gordon, H. J. Zeiger, and C. H. Townes, *Phys. Rev.* 95:282 (1954).

[8]V. A. Fabrikant referenced by Jeff Hecht, *Laser Pioneers*, revised ed. (San Diego, CA: Academic Press, 1992), pp. 12–13 as being cited in Mario Bertolotti, *Masers and Lasers: A Historical Approach* (Bristol, England: Adam Hilger Ltd., 1983).

[9]R .H. Dicke, *Phys. Rev.* 93:99 (1954).

in his 1958 patent entitled "Molecular Amplification and Generation Systems and Methods."[10]

In 1957, Gordon Gould conceived of the idea of using a Fabry-Perot cavity as part of a laser structure. He documented his ideas in his laboratory notebook under the title of "laser" (**L**ight **A**mplification by **S**timulated **E**mission of **R**adiation). He had his notebook notarized on November 13, 1957, by a candy-store clerk. Gould then attempted to acquire a patent on these ideas. After a great deal of legal fuss, he obtained four patents: one in 1977 on optically pumped laser amplifiers,[11] one in 1979 on a broad range of laser applications,[12] one in 1987 on electrical discharge pumped lasers,[13] and one in 1988 on Brewster angle windows for lasers.[14] Many U.S. laser companies still pay royalties under license agreements on these patents.

In 1958, Schawlow and Townes wrote a seminal paper entitled "Infrared and Optical Masers,"[15] discussing the various aspects of constructing an optical maser. In this paper a number of questions regarding the practicality of lasers were discussed. The required pumping power was calculated (10 mW for a 1-cm cube) and shown to be practical, the question of using a multimode cavity was discussed, and a number of schemes for mode selection (including a long cavity Fabry-Perot) were presented. The possibility for *three- and four-state* (see Section 2.1) solid-state lasers, *linewidth* (see Section 2.1), and *tunability* (see Chapter 11) were also briefly mentioned.

The appearance of this paper caused a great deal of interest in the scientific community. A number of laser programs were initiated, mostly distinguished by the choice of the laser material. Peter P. Sorokin and Mirek Stevenson at IBM focused on calcium fluoride doped with a rare earth, Theodore Maiman at Hughes pursued ruby, and Ali Javan at Bell Laboratories worked with a helium-neon mixture. (Ruby as a laser material is discussed in Section 11.3 and helium-neon as a laser material is discussed in Section 9.1.)

On May 16, 1960, Maiman and his coworkers achieved laser action in a "pink" (low chromium concentration) ruby rod. Maiman submitted his paper to *Physical Review Letters*, where it was rejected. Maiman then sent a paper to *Nature*[16] and arranged a press conference to discuss the development. The result was received with skepticism, because in Schawlow's paper, ruby had been rejected as a laser material due to a presumed low quantum efficiency.

As a result of the Hughes publicity, a group from Bell Laboratories reproduced Maiman's results and submitted a paper to *Physical Review Letters*. This paper was accepted

---

[10] R. H. Dicke, "Molecular Amplification and Generation Systems and Methods," U.S. Patent #2,581,652, Sept. 9, 1958.

[11] U.S. Patent #4,053,845, August 16, 1977.

[12] U.S. Patent #4,161,436, July 17, 1979.

[13] U.S. Patent #4,704,583, November 3, 1987.

[14] U.S. Patent #4,746,201, May 24, 1988.

[15] A. L. Schawlow and C. H. Townes, *Phys. Rev.* 112:1940 (1958).

[16] T. H. Maiman, *Nature* 187:493 (1960).

and published on October 1, 1960.[17] Soon after, the Bell Laboratories group published another paper reporting laser action in "red" ruby.[18]

This flurry of *Physical Review Letter* papers from the Bell Laboratories group caused much confusion as to who was responsible for the first demonstration of laser action. Many scientists assumed that the Bell Laboratories group had obtained laser action prior to Maiman at Hughes. Although Maiman published in *Physical Review* somewhat later,[19] uncertainty remained for several years.

The majority of early laser schemes were complex. Maiman's demonstration of laser action in a simple and elegant experiment dramatically altered the direction of laser development. Soon after Maiman's demonstration of laser action in ruby, Sorokin and Stevenson of IBM switched to a flashlamp pumped rod design for their uranium-doped calcium fluoride laser. This laser lased on its first try in November 1960.[20] A few weeks later, Sorokin and Stevenson obtained laser action in samarium-doped calcium fluoride.[21] Although doped calcium fluoride is not a commonly used laser material today, these experiments were the first demonstrations of laser action in a four-state material.

The floodgates were now open to laser development. In rapid succession, a host of new materials were found to lase. On December 12, 1960, Ali Javan, William R. Bennett, Jr., and Donald Herriott obtained laser action in a helium-neon gas mixture.[22] In 1963, C. Kumar N. Patel obtained laser action in carbon dioxide.[23] In 1964, Joseph Geusic, H. M. Marcos, and Le Grand Van Uitert obtained laser action in Nd:YAG;[24] and William Bridges obtained laser action in argon-ion.[25] (Carbon dioxide lasers are discussed in Section 12.1, Nd:YAG lasers are discussed in Chapter 10, and argon-ion lasers in Section 9.2.)

In re-reading Schawlow and Townes' paper today, there is a clear feeling that visible maser (laser) construction was expected to be very difficult. To everyone's surprise, it turned out to be quite straightforward. Today, lasers have been demonstrated in solid, liquid, gas, and plasma materials at virtually every wavelength (see Figures 1.1 and 1.2).

1.2 THE LASER MARKET

Lasers are used in a number of commercial and research applications. Some of these (such as laser light shows) are dramatic, but involve a relatively small number of lasers. Others (such as laser marking of aluminum) are less well-known, but constitute a significant dollar value of the laser market.

---

[17] R. J. Collins, D. F. Nelson, A. L. Schawlow, W. Bond, C. G. B. Garrett, and W. Kaiser, *Phys. Rev. Lett.* 5:303 (1960).

[18] A. L. Schawlow and G. E. Devlin, *Phys. Rev. Lett.* 6:96 (1961).

[19] T. H. Maiman, R. H. Hoskins, J. D'Haenens, C. K. Asawa, and V. Evtuhov, *Phys. Rev.* 123:1151 (1961).

[20] P. P. Sorokin and M. J. Stevenson, *Phys. Rev. Lett.* 5:557 (1960).

[21] P. P. Sorokin and M. J. Stevenson, *IBM Jour. of Res. and Dev.* 5:56 (1961).

[22] A. Javan, W. R. Bennett, Jr., and D. R. Herriott, *Phys. Rev. Lett.* 6:106 (1961).

[23] C. K. N. Patel, *Phys. Rev.* A136:1187 (1964).

[24] J. E. Geusic, H. M. Marcos, and L. G. Van Uitert, *Appl. Phys. Lett.* 4:182 (1964).

[25] W. B. Bridges, *Appl. Phys. Lett.* 4:128 (1964); and erratum 5:39 (1964).

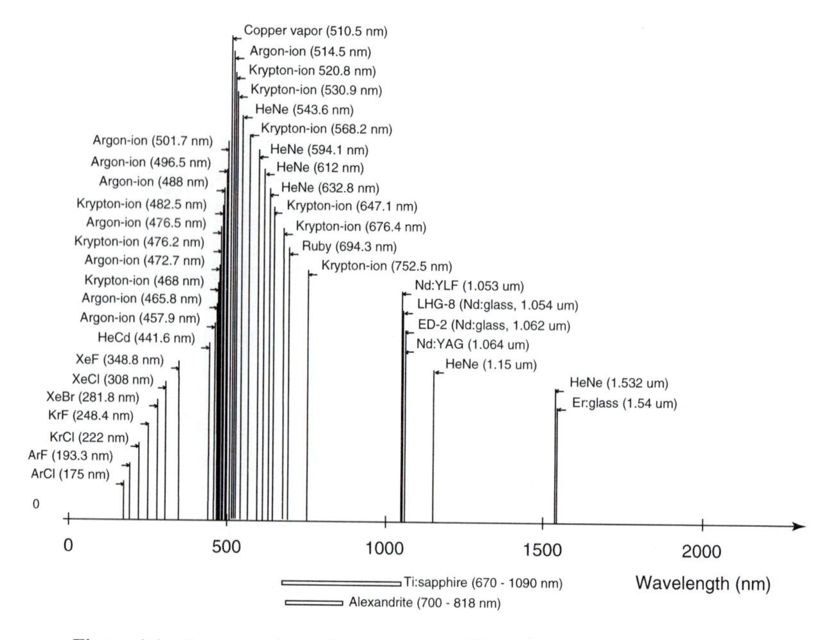

Figure 1.1 Summary chart of commonly used lasers organized by wavelength.

Figure 1.3 illustrates worldwide laser sales in (in millions of dollars) from 1985-1993 by laser type.[26] Figure 1.4 contains the same information as Figure 1.3, but sorted by laser application. It is interesting to see materials processing (laser machining) heading the list, followed closely by medicine and then by research and development. Laser communications are also increasing dramatically. Notice that laser applications most commonly cited as dominating the commercial laser market (inspection/measurement, laser printing, and barcode reading) are actually a relatively small dollar-value component of the market as a whole.

Carbon dioxide (CO_2) lasers tend to consistently head the list in dollar volume of sales. This is largely due to strong industrial applications in materials processing (such as laser marking, cutting, and welding). The large economic market tends to disguise the relatively small number of systems sold. In 1993, 4,352 CO_2 laser systems were sold, with 2,780 in laser machining applications, 1,200 in medicine, and 342 in research and development. (Carbon dioxide lasers are discussed in Section 12.1.)

Solid-state lasers have also made a consistent and strong showing over the years, typically second only to CO_2 lasers. Solid-state lasers also find strong markets in laser machining and in medical applications. Again, the relatively high cost of these systems tends to disguise the small volumes. In 1993, 6,320 solid-state laser systems were sold, with 2,430 in laser machining applications, 1,260 in medicine, and 1,815 in research and development. (Solid-state lasers are discussed in more detail in Chapter 10.)

---

[26]This particular group of dates was selected for two reasons. First, it covers the period when the laser community switched from a primary focus of defense-related research and development to one of commercial application. Second, *Laser Focus / Laser Focus World* maintained the same type of data base for laser sales over these years. In 1994, *Laser Focus / Laser Focus World* changed their method of organizing the data, and the results can not be easily compared with the earlier numbers.

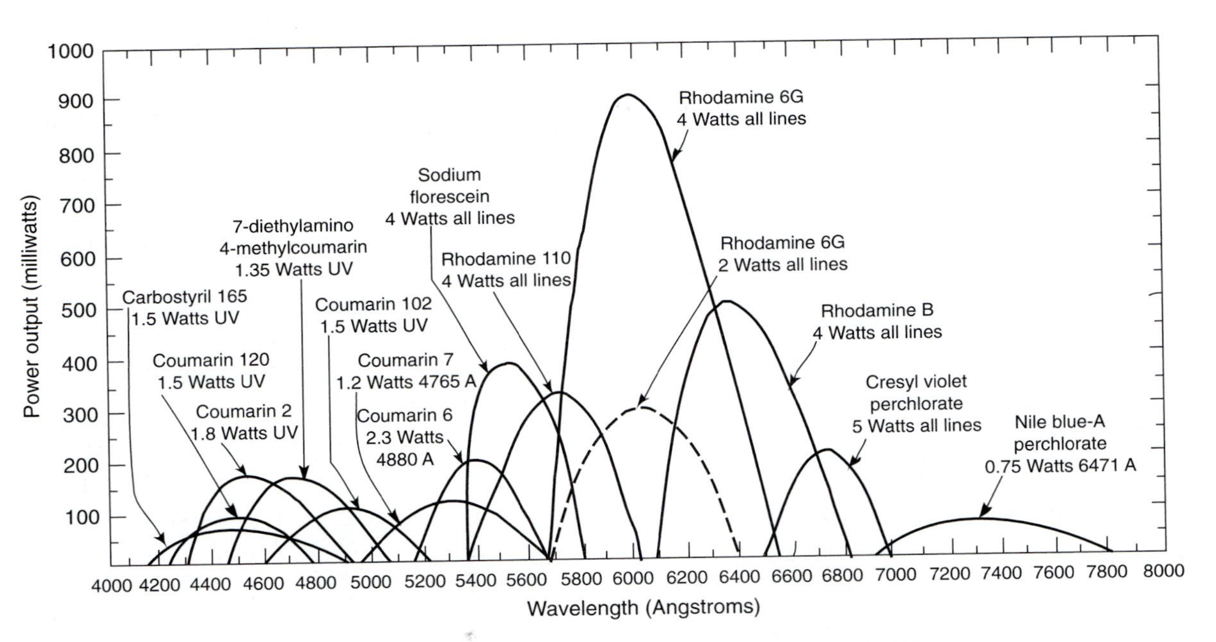

Figure 1.2 Summary chart of common laser dyes organized by wavelength. (Courtesy of Coherent, Inc., Laser Group)

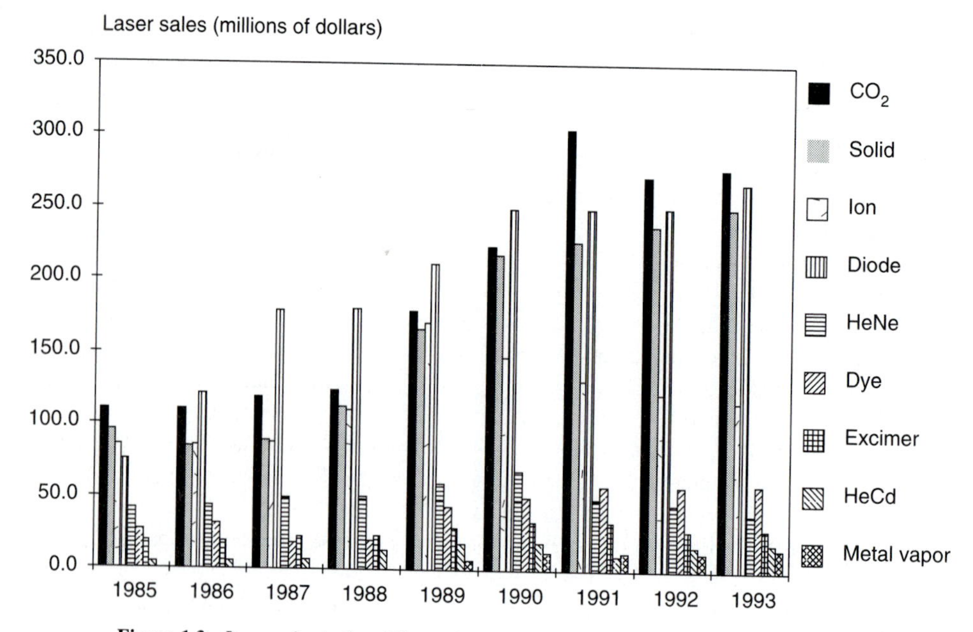

Figure 1.3 Laser sales in (in millions of dollars) between 1985 and 1993 sorted by laser type. (Data from *Laser Focus / Laser Focus World* 1985–1993, © PennWell Publishing Company. Used by permission.)

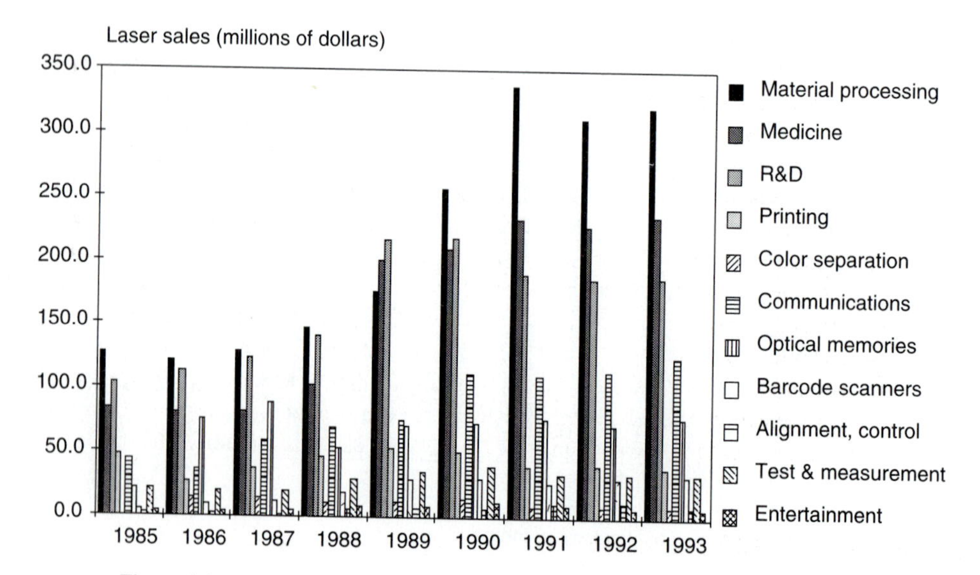

Figure 1.4 Laser sales in (in millions of dollars) between 1985 and 1993 sorted by application. (Data from *Laser Focus / Laser Focus World* 1985–1993, © PennWell Publishing Company. Used by permission.)

It is interesting to compare the steady, solid growth of CO_2 and solid-state lasers with the decline of ion lasers. Ion laser sales peaked in roughly 1989 and have been dropping ever since. Notice that ion lasers find the largest economic market in research and development, with secondary applications in medicine and in laser printing. Roughly 17,156 ion lasers were sold in 1993, with 5,838 in medical application, 2,588 in research and development, and 5,580 for laser printing and color separation. (Ion lasers are discussed in more detail in Section 9.2.)

Semiconductor diode lasers have shown the most dramatic change over this period. From 1985 to 1992, dollar sales of diode lasers have virtually quadrupled. Unit sales have gone up a factor of twenty, from 2,216,230 in 1985 to 40,022,050 in 1993. Primary applications of diode lasers are in optical memory products (such as audio and digital compact disks) and in laser printing. Notice that the relatively low cost of these units disguises the true differences in magnitude between the laser diode market and any other laser market. In 1993, 40,022,050 laser diodes were sold; 31,500,000 laser diodes specifically for optical memory applications and 7,500,000 specifically for laser printing. (Semiconductor diode lasers are discussed in more detail in Section 12.3.)

An interesting evolution is appearing in the laser market with regard to HeNe lasers and red semiconductor diode lasers. When red semiconductor laser diodes first appeared, it was widely claimed that these diodes would essentially eliminate HeNe lasers as a viable product. However, this has not turned out to be the case. Although HeNe lasers have lost some market share in the barcode scanning arena (compare the peak of 240,000 HeNe units sold for barcode scanning in 1990 to 170,000 in 1993), many test and alignment applications have returned to HeNe lasers due to their better beam quality (compare 57,000 HeNe units sold for alignment test and measurement in 1990 to the 71,000 sold in 1991). (HeNe lasers are discussed in more detail in Section 9.1.)

Excimer lasers have also undergone a change in market over this period. Excimer lasers were originally developed for the research and development market. However, as the research and development market has declined, excimer lasers have moved to more commercial markets. In 1993, 500 excimer systems were sold; 250 for research and development, 180 for medical applications and 70 for laser machining. This compares with 1990 when 560 systems were sold; 390 for research and development, 80 for medical applications and 90 for laser machining. (Excimer lasers are discussed in more detail in Section 12.2.)

At one time lasers were called "a solution looking for a problem." However, a very healthy recent trend is the reduction in the number of lasers used by the research and development community, and an increase in commercial lasers used for materials processing and medicine. Prior to roughly 1989, the laser industry was largely driven by research and development applications. Today, the increasing number of commercial applications for lasers is providing a sound foundation for the industry as a whole.

1.3 ENERGY STATES IN ATOMS

Lasers can be constructed from solids, liquids, gases, or plasmas. In all cases laser action occurs because photons are emitted as the system transitions between two *energy states*. In some cases (such as CO_2 lasers) these two energy states are two vibrational modes of a molecule (see Section 12.1). In other cases (such as semiconductor diode lasers) these two states are the conduction and valence bands of a semiconductor (see Section 12.3). However, in most cases, these two states are *electronic states* in an atom or ion.

To describe electronic states further, recall that the atom can be modeled as a central nucleus of protons and neutrons surrounded by a cloud of electrons. These electrons can only occupy certain well-defined energy states. These states are organized into groups (often called the main shells of the atom) characterized by the principal quantum number n. Within the main shells, the electrons are further divided into electron orbitals. The simplest orbitals (called **s**-orbitals) are spherical orbitals that contain two electrons of opposite spins. The more complex **p**-orbitals are dumbbell-shaped (with a dumbbell in each of the three Cartesian directions) and contain six total electrons. The orbitals after **s** and **p** possess increasingly complex shapes and follow the sequence: **d**, **f**, **g**, **h**, **i**, **k**, **l**, ... (no **j**!).

The electronic structure of an atom can be expressed by a shorthand notation that describes the principal quantum number, the electron orbital letter, and the number of electrons in each orbital. For example, argon can be expressed as

$$\text{argon} = 1\mathbf{s}^2 2\mathbf{s}^2 2\mathbf{p}^6 3\mathbf{s}^2 3\mathbf{p}^6 \tag{1.1}$$

and krypton as

$$\text{krypton} = 1\mathbf{s}^2 2\mathbf{s}^2 2\mathbf{p}^6 3\mathbf{s}^2 3\mathbf{p}^6 3\mathbf{d}^{10} 4\mathbf{s}^2 4\mathbf{p}^6. \tag{1.2}$$

In real atoms, the various electron orbitals are divided yet again into sublevels whose energies are determined by more complex interactions within the atom (see Figure 1.5).

Most lasers use transitions between electronic states in the vicinity of the outermost shell of the atom. For example, in the argon-ion laser, laser action occurs when electrons transition from the 4**p** to the 4**s** states (see Figure 1.5). These states are significantly higher in energy than the ground state configuration given in Equation (1.1).

Example 1.1

The rare-earth atom neodymium is often used as a dopant in solid-state laser systems. What is the electronic configuration of neodymium?

Solution. The solution can be found in the electronic configuration tables in a chemistry textbook and is found to be

$$\text{neodymium} = 1\mathbf{s}^2 2\mathbf{s}^2 2\mathbf{p}^6 3\mathbf{s}^2 3\mathbf{p}^6 3\mathbf{d}^{10} 4\mathbf{s}^2 4\mathbf{p}^6 4\mathbf{d}^{10} 5\mathbf{s}^2 5\mathbf{p}^6 6\mathbf{s}^2 4\mathbf{f}^4. \tag{1.3}$$

1.4 BASIC STIMULATED EMISSION

1.4.1 Transitions Between Laser States

The word laser is an acronym for **L**ight **A**mplification by **S**timulated **E**mission of **R**adiation. It is one of the few acronyms that is now accepted as a word. Thus, the verb forms "to lase" and "lasing" are accepted as real words. (Check a dictionary!)

But, what do the words "light amplification by stimulated emission of radiation" mean?

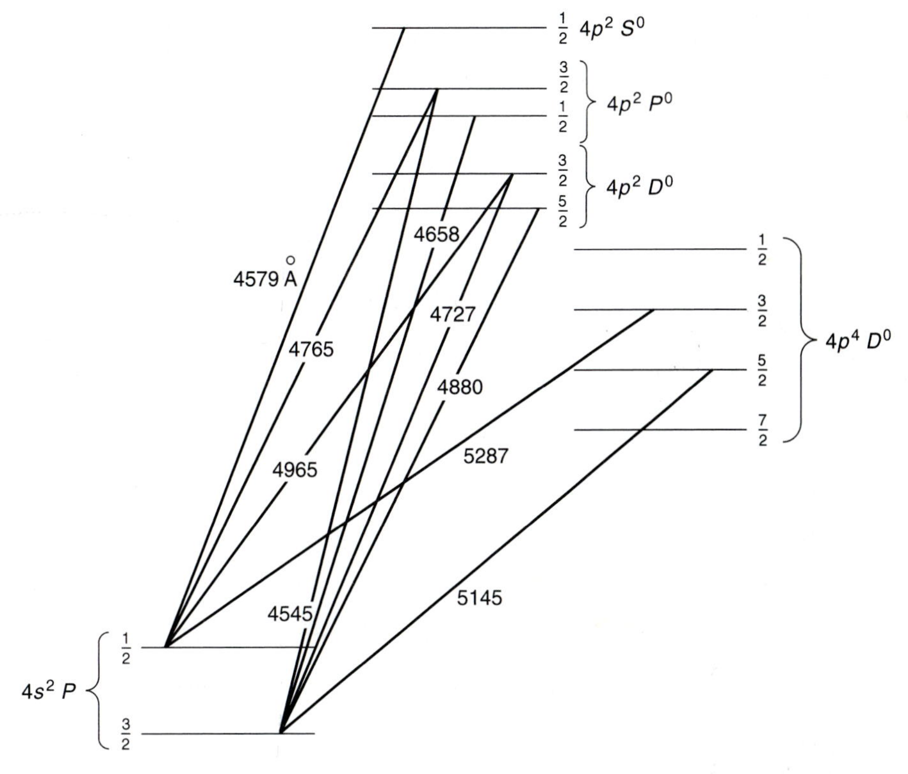

Figure 1.5 Laser states in the argon-ion laser. (Reprinted with permission from W. B. Bridges, *Appl. Phys. Lett.* 4:128, 1964, with a correction noted in *Appl. Phys. Lett.* 5:39, 1964. Copyright 1964 American Institute of Physics.)

To answer this, recall how photons are created in an atom. When an electron spontaneously decays from one energy state E_2 to a lower energy state E_1, it emits a photon of the energy

$$\boxed{E = h\nu = E_2 - E_1} \tag{1.4}$$

where h is Planck's constant and ν is the frequency of the laser light. This process is called *spontaneous emission* (see Figure 1.6).

Spontaneous emission does not occur instantaneously. Instead, the electrons reside in the upper energy state for a certain period of time before they spontaneously decay to the lower energy state. The time constant for the spontaneous emission process is termed the *spontaneous lifetime*.

However, it is also possible to force a transition from one state to another by means of a photon. In other words, a photon of the energy $h\nu$ can force or *stimulate* an electron to transfer between states 2 and 1, yielding another photon of the energy $h\nu = E_2 - E_1$. This *stimulated emission* process results in two photons of the energy $h\nu$. Furthermore, these two photons will be in phase, of the same polarization, and heading in the same direction (see Figure 1.7).

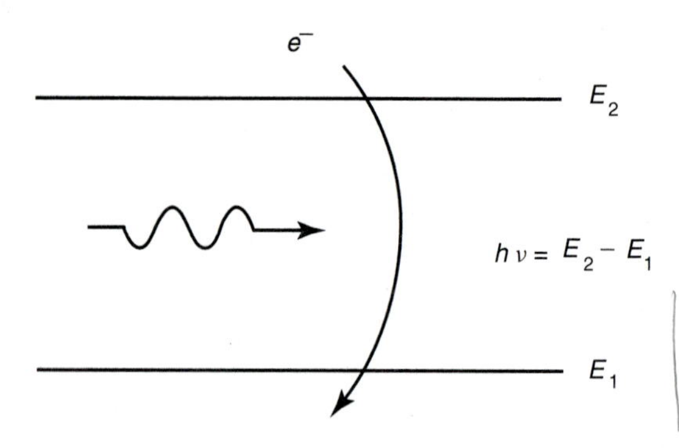

Figure 1.6 When an electron spontaneously decays from one atomic state to another, it *spontaneously* emits a photon of the energy $h\nu = E_2 - E_1$.

Thus, ideal laser light is formed of groups of photons where all the photons are at exactly the same frequency (wavelength) and all the photons are in phase. Speaking more intuitively, laser light is composed of photons that are all "one color" and all "marching together."

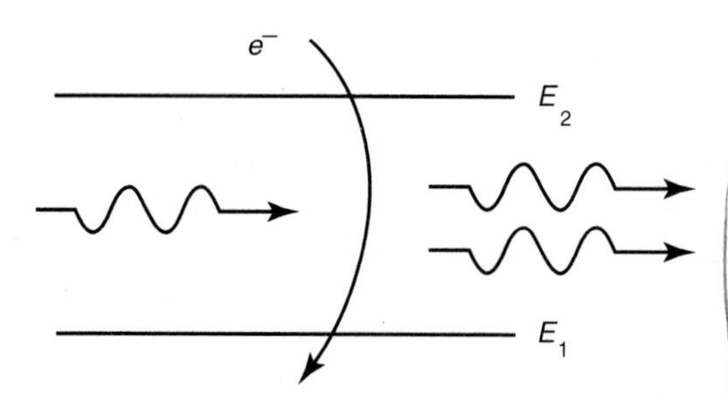

Figure 1.7 A photon of the energy $h\nu$ can force or *stimulate* an electron to transfer between states 2 and 1. This *stimulated emission* process results in two photons of the energy $h\nu$ that are of the same polarization and in phase.

Example 1.2

Consider two energy states, $E_2 = 2.50$ eV and $E_1 = 1.0$ eV. What is the energy of the photon that is spontaneously emitted between these states? What is the wavelength of the photon?

Solution. The energy is given by Equation (1.4) as

$$E = h\nu = E_2 - E_1 = 2.50 \text{ eV} - 1.0 \text{ eV} = 1.50 \text{ eV}. \tag{1.5}$$

The wavelength is determined by rewriting Equation (1.4) as

$$\lambda = \frac{hc_o}{E_2 - E_1} = \frac{6.626 \cdot 10^{-34} \text{ J-sec} \cdot 3 \cdot 10^8 \text{ m/sec}}{(2.5 \text{ J/coul} - 1.0 \text{ J/coul})(1.602 \cdot 10^{-19} \text{ coul})} = 827.216 \text{ nm} \tag{1.6}$$

or, in the "easy-to-remember" form

$$\lambda(\mu\text{m}) = \frac{hc_o}{E_2 - E_1} = \frac{1.24 \text{ (eV} \cdot \mu\text{m)}}{E \text{ (eV)}} = 0.827 \ \mu\text{m}. \tag{1.7}$$

Example 1.3

In laser engineering, problems often arise in the casual use of significant figures. To illustrate the difficulty, perform two calculations for the wavelength of the laser transition between the two energy states, $E_2 = 2.50$ eV and $E_1 = 1.0$ eV. In the first calculation, use the exact values for h and c_o. In the second calculation, use the approximate values $6.6 \cdot 10^{-34}$ J-sec and $3 \cdot 10^8$ m/sec.

Solution. Repeating the calculation from Example 1.2:

$$\lambda = \frac{hc_o}{E_2 - E_1} = \frac{6.6260755 \cdot 10^{-34} \text{ J-sec} \cdot 299,792,458 \text{ m/sec}}{(2.5 \text{ J/coul} - 1.0 \text{ J/coul})(1.60217733 \cdot 10^{-19} \text{ coul})} = 826.562 \text{ nm} \quad (1.8)$$

$$\lambda = \frac{hc_o}{E_2 - E_1} = \frac{6.6 \cdot 10^{-34} \text{ J-sec} \cdot 3 \cdot 10^{8} \text{ m/sec}}{(2.5 \text{ J/coul} - 1.0 \text{ J/coul})(1.602 \cdot 10^{-19} \text{ coul})} = 823.97 \text{ nm.} \quad (1.9)$$

So, the calculation using the approximate values for h and c_o differs by approximately 25.9 angstroms from the calculation using the more precise values. This wavelength difference is significantly larger than the linewidth of a typical laser transition. In fact, it is on the order of the difference between one distinct laser transition and another!

The use of precision and significant figures has been well-defined in sciences such as chemistry and biology. However, the established definitions from the natural sciences are not commonly used in laser engineering. The guidelines used in this text for significant figures are given in the Appendix.

1.4.2 Population Inversion

It is only possible to achieve gain in a laser if the population in the upper laser state is greater than the population in the lower laser state. This condition is termed a *population inversion.* (The physics underlying this process is discussed in more detail in Section 2.2.)

In *thermal equilibrium,* the population ratio between two states is governed by the Boltzman equation

$$\boxed{\frac{N_2}{N_1} = e^{-(E_2 - E_1)/k_B T}} \quad (1.10)$$

where N_2 is the population in the upper state, N_1 is the population in the lower state, k_B is Boltzman's constant, and T is the temperature.

Notice that the negative sign in the exponent suggests that a population inversion is only permitted under the conditions of "negative temperature"! This result was very disturbing to early laser researchers, as negative temperatures are not physically realizable. However, the Boltzman equation only describes conditions of *thermal equilibrium.* Lasers are not operated in thermal equilibrium. Instead, the upper state is populated by *pumping* it via some nonequilibrium process. A pulse of light, an electrical spark or a chemical reaction can all be used to populate the upper laser state.

Example 1.4

Consider two energy states, $E_2 = 2.10$ eV and $E_1 = 1.0$ eV. Assume that there are $1.0 \cdot 10^{16}$ electrons/cm^3 in E_1. At a temperature of 1000.0 K, how many electrons are in state E_2?

Solution. This uses the Boltzman equation [Equation (1.10)] as

$$N_2 = N_1 \cdot e^{-(E_2 - E_1)/k_B T} \quad (1.11)$$

$$N_2 = \left(1.0 \cdot 10^{16} \text{ electrons/cm}^3\right) \cdot e^{-(2.10 \text{ eV} - 1.0 \text{ eV})/(k_B \cdot 1000.0 \text{ K})} \quad (1.12)$$

$$N_2 = 2.85942 \cdot 10^{10} \text{ electrons/cm}^3. \quad (1.13)$$

$K_B = 1.38 \times 10^{-23}$ J/K

Example 1.5

Consider two energy states, $E_2 = 2.10$ eV and $E_1 = 1.0$ eV. Assume that there are $1.0 \cdot 10^{16}$ electrons/cm^3 in E_2 and $1.0 \cdot 10^{15}$ electrons/cm^3 in E_1. What temperature is required to create this population distribution in thermal equilibrium?

Solution. The Boltzman equation [Equation (1.10)] can be rewritten as

$$T = \frac{-(E_2 - E_1)}{k_B \cdot \ln(N_2/N_1)} \tag{1.14}$$

$$T = \frac{-(2.10 \text{ eV} - 1.0 \text{ eV})}{k_B \cdot \ln\left(1.0 \cdot 10^{16} \text{ electrons/cm}^3 / \, 1.0 \cdot 10^{15} \text{ electrons/cm}^3\right)} \tag{1.15}$$

$$T = -5543.72 \text{ K} \tag{1.16}$$

Notice the negative temperature! This situation will not occur in thermal equilibrium, as negative temperatures are not physically realizable. However, similar population ratios are quite common in real laser systems under nonequilibrium pumping conditions.

1.5 POWER AND ENERGY

Lasers can operate either in a *continuous wave* (*cw*) or a *pulsed* mode. Pulsed operation is occasionally performed to reduce the heating of the laser (common for semiconductor diode lasers). However, in most cases, pulsed operation is combined with techniques such as *Q-switching* (which concentrates the laser energy into the pulse) and *mode-locking* (which shortens the width of the pulse in time). Q-switched, mode-locked lasers have the ability to concentrate very high *peak power* densities due to the relatively short length of the pulse. (For more information on Q-switching and mode-locking, see Sections 6.2 and 6.4.)

Some care must be taken in discussing the properties of pulsed versus cw lasers, as the use of watts (to describe peak power) can be easily confused with the use of watts (to describe *average power*). In this text, the following definitions will be used.

Average power: The power (in watts) of a continuous wave laser, or the energy per pulse (J) times the pulse repetition rate (Hz) for pulsed lasers as

$$P_{\text{avg}} = E_{\text{pulse}} \cdot R_{\text{reprate}}. \tag{1.17}$$

Peak power: The energy per pulse (J) divided by the temporal length of the pulse (seconds) as

$$P_{\text{peak}} = \frac{E_{\text{pulse}}}{t_{\text{pulse}}}. \tag{1.18}$$

Average energy density: The energy (joules) per unit area (cm^2 or m^2).

Average power density: The average power (watts) per unit area (cm^2 or m^2).

Peak power density: The peak power (watts) per unit area (cm^2 or m^2).

Example 1.6

Compute the average and peak power, as well as the average and peak power densities, for a Nd:glass laser with an output energy of 9.0 J/pulse, a repetition rate of 2.0 Hz, and a pulsewidth of 20.0 ns. Assume that the laser is focused into a spot 0.250 mm in radius and that the intensity is constant across the spot.

Solution. Equations (1.17) and (1.18) may be evaluated as

$$P_{\text{avg}} = E_{\text{pulse}} \cdot R_{\text{reprate}} = 9.0 \text{ joule/pulse} \cdot 2.0 \text{ Hz} = 18.0 \text{ watts} \tag{1.19}$$

$$P_{\text{peak}} = \frac{E_{\text{pulse}}}{t_{\text{pulse}}} = \frac{9.0 \text{ joule/pulse}}{20.0 \text{ nsec/pulse}} = 4.50 \cdot 10^{8} \text{ watt} = 450.0 \text{ MW} \tag{1.20}$$

$$E_{\text{den}} = \frac{9.0 \text{ joule/pulse}}{\pi r^2} = 4583.66 \text{ joule/cm}^2 \tag{1.21}$$

$$P_{\text{avg-den}} = \frac{P_{\text{avg}}}{\pi r^2} = 9.16733 \cdot 10^{3} \text{ watt/cm}^2 = 9.16733 \text{ kW/cm}^2 \tag{1.22}$$

$$P_{\text{peak-den}} = \frac{P_{\text{peak}}}{\pi r^2} = 2.29183 \cdot 10^{11} \text{ watt/cm}^2 = 229.183 \text{ GW/cm}^2. \tag{1.23}$$

1.6 MONOCHROMATICITY, COHERENCY, AND LINEWIDTH

The property of having a group of photons at exactly one frequency is referred to as *monochromaticity*. The property of having a group of photons with the same relative phase is referred to as *coherency*. Thus, lasers are often termed *monochromatic* and *coherent* sources of light (see Figure 1.8). (Notice that this is a redundant definition for an ideal laser, because a perfectly coherent source of light must be monochromatic. However, notice also that a perfectly monochromatic source of light, such as spontaneous emission, need not be coherent!)

Typically, the "color" of a laser beam is characterized by the center wavelength of the laser line. In actual practice, both wavelength λ (in angstroms or nm) and frequency ν (in

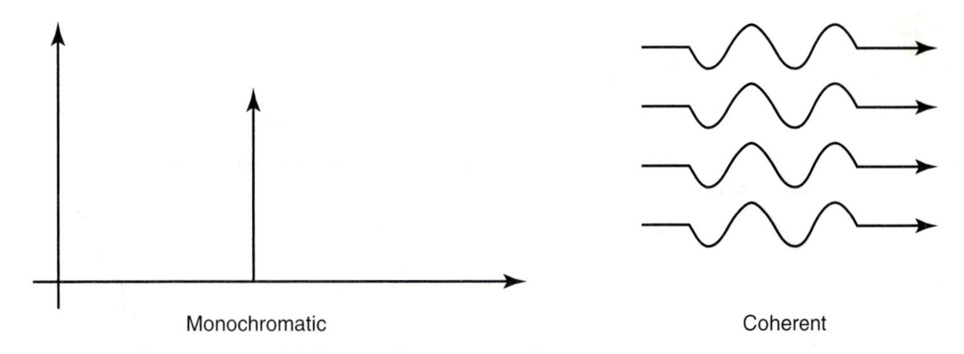

Figure 1.8 An ideal monochromatic source of light has a group of photons with exactly one frequency. An ideal coherent source of light has a group of photons with the same relative phase.

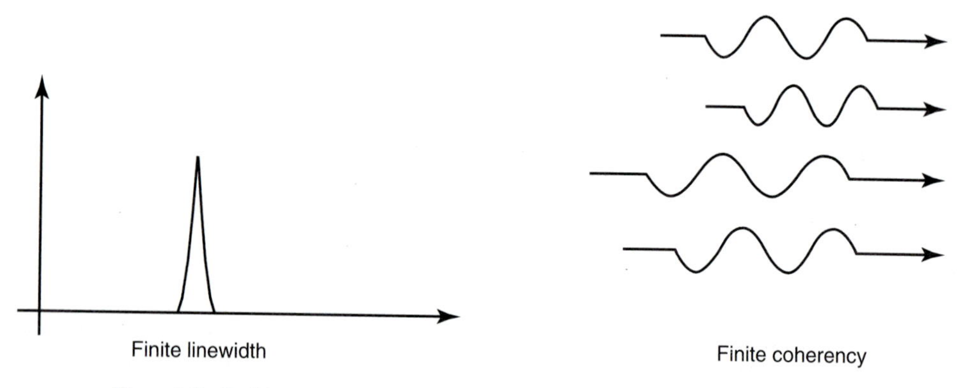

Figure 1.9 Real laser sources are neither perfectly monochromatic nor perfectly coherent. A real laser source will have a finite linewidth and finite coherency.

Hz) are used. Thus, the center wavelength λ of a laser line could be expressed in terms of frequency ν as

$$\lambda = \frac{c_o}{\nu}. \tag{1.24}$$

Of course, real lasers are neither perfectly monochromatic nor perfectly coherent (see Figure 1.9). However, when characterizing a real laser system, it is generally assumed that the laser beam was initially in phase and the incoherence of the laser arises only from the lack of monochromaticity of the source. (This is a reasonable assumption for conventional lasers with feedback, but may not be sufficiently accurate for unusual laser systems.) Thus, coherency and monochromaticity are generally assumed to measure the same parameter.

The monochromaticity of a laser beam is described by its *wavelength linewidth* $\Delta\lambda$ (in angstroms or nm) or its *frequency linewidth* $\Delta\nu$ (in Hz). The two quantities are related as

$$\Delta\nu = \nu_1 - \nu_2 = \frac{c_o}{\lambda_1} - \frac{c_o}{\lambda_2} = c_o\left(\frac{\lambda_2 - \lambda_1}{\lambda_1\lambda_2}\right) \tag{1.25}$$

which (assuming that λ_1 and λ_2 are much larger than $\lambda_2 - \lambda_1$) can be approximated by

$$\boxed{\Delta\nu = \nu_1 - \nu_2 \approx c_o\left(\frac{\Delta\lambda}{\lambda^2}\right).} \tag{1.26}$$

The fundamental linewidth for an ideal laser line is extremely small.[27] In practice, various *broadening mechanisms* will increase this fundamental linewidth in real lasers (see Section 2.1).

The monochromaticity of a laser beam can also be described in terms of coherency. Notice that the length of time it takes for two oscillations differing in frequency by $\Delta\nu$ to get out of phase by a full cycle is approximately $1/\Delta\nu$. Thus, the *coherence time* $\delta\tau$ is

[27] The fundamental linewidth is given by the Schawlow-Townes equation as $\Delta\nu = (h\nu/2\pi\tau_{\text{cav}}^2 P_{\text{out}})$ where $\Delta\nu$ is the linewidth, τ_{cav} is the cavity decay time and P_{out} is the output laser power.

given by

$$\delta\tau = \frac{1}{\Delta\nu} \tag{1.27}$$

and the *coherence length* l_c is expressed as

$$l_c = c_o \cdot \delta\tau = \frac{c_o}{\Delta\nu}. \tag{1.28}$$

Example 1.7

Consider an argon-ion laser with a center wavelength of 514.5 nm. Calculate the energy of the laser transition in eV, the frequency of the transition in Hz, and the spectroscopic wavenumber ($1/\lambda$) in cm^{-1}.

Solution. The transition energy is determined by rewriting Equation (1.4) as

$$E_2 - E_1 = \frac{hc_o}{\lambda} = \frac{hc_o}{514.5 \cdot 10^{-9}\ \text{m}} = 2.4098\ \text{eV}. \tag{1.29}$$

The optical frequency is determined by rewriting Equation (1.24) as

$$\nu = \frac{c_o}{\lambda} = \frac{c_o}{514.5 \cdot 10^{-9}\ \text{m}} = 5.827 \cdot 10^{14}\ \text{Hz} = 582.7\ \text{THz}. \tag{1.30}$$

The spectroscopic wavenumber is determined by simply inverting the wavelength as

$$w_n = \frac{1}{\lambda} = \frac{1}{514.5 \cdot 10^{-9}\ \text{m}} = 19436\ \text{cm}^{-1}. \tag{1.31}$$

Example 1.8

Consider a semiconductor laser diode with a laser linewidth of 2.0 angstroms and a center operating wavelength of 870 nm. What is the frequency linewidth $\Delta\nu$ in Hertz?

Solution. This uses Equation (1.26). Substituting in the relevant parameters gives

$$\Delta\nu \approx c_o \left(\frac{\Delta\lambda}{\lambda^2}\right) = c_o \cdot \left(\frac{2.0 \cdot 10^{-10}\ \text{m}}{\left(870 \cdot 10^{-9}\ \text{m}\right)^2}\right) = 79.2\ \text{GHz}. \tag{1.32}$$

Example 1.9

Consider a HeNe laser with a linewidth of 1.50 GHz and a center operating wavelength of 632.8 nm. What is the laser linewidth $\Delta\lambda$ in angstroms?

Solution. This uses Equation (1.26) reorganized as

$$\Delta\lambda \approx \lambda^2 \left(\frac{\Delta\nu}{c_o}\right). \tag{1.33}$$

Substituting in the relevant parameters gives

$$\Delta\lambda \approx \lambda^2 \left(\frac{\Delta\nu}{c_o}\right) = \left(632.8 \cdot 10^{-9}\ \text{m}\right)^2 \left(\frac{1.5 \cdot 10^9\ \text{Hz}}{c_o}\right) = 0.002004\ \text{nm}. \tag{1.34}$$

Example 1.10

Consider a Nd:YAG laser with a laser linewidth of 4.50 angstroms and a center operating wavelength of 1.064 μm. What is its coherence length?

Solution. This uses Equation (1.26) to determine the linewidth in Hz. Substituting in the relevant parameters gives

$$\Delta\nu \approx c_o\left(\frac{\Delta\lambda}{\lambda^2}\right) = c_o \cdot \left(\frac{4.5 \cdot 10^{-10}\ \text{m}}{\left(1.064 \cdot 10^{-6}\ \text{m}\right)^2}\right) = 119.17\ \text{GHz}. \tag{1.35}$$

The coherence length is then determined by Equation (1.28) as

$$l_c = \frac{c_o}{\Delta\nu} = \frac{c_o}{119.16\ \text{GHz}} = 2.516\ \text{mm}. \tag{1.36}$$

Example 1.11

The uninitiated observer might confuse the center frequency of the laser line ν (in Hz) with the laser linewidth $\Delta\nu$ (also in Hz). However, the two numbers differ significantly in magnitude. To demonstrate this, calculate the center frequency ν (in Hz) and the laser linewidth $\Delta\nu$ (in Hz) for a semiconductor diode laser with a center operating wavelength of 760 nm and a laser linewidth $\Delta\lambda$ of 3.0 angstroms.

Solution. The center wavelength can be converted to frequency units by using equation (1.24). For the semiconductor diode laser of Example 1.1, this yields

$$\nu = \frac{c_o}{\lambda} = \frac{c_o}{760 \cdot 10^{-9}\ \text{m}} = 3.945 \cdot 10^{14}\ \text{Hz} = 394.5\ \text{THz}. \tag{1.37}$$

The laser linewidth uses Equation (1.26). Substituting in the relevant parameters gives

$$\Delta\nu \approx c_o\left(\frac{\Delta\lambda}{\lambda^2}\right) = c_o \cdot \left(\frac{3.0 \cdot 10^{-10}\ \text{m}}{\left(760 \cdot 10^{-9}\ \text{m}\right)^2}\right) = 155.7\ \text{GHz}. \tag{1.38}$$

Notice that the center frequency ν will be on the order of many THz. The laser linewidth $\Delta\nu$ of most lasers is on the order of MHz to GHz. Thus, the two numbers differ by many orders of magnitude!

1.7 SPATIAL COHERENCE AND LASER SPECKLE

Spatial coherence is another important property of laser systems. To understand spatial coherence, imagine a beam of constant intensity propagating in the z direction. Consider any two points P_1 and P_2 on the beam chosen in a plane perpendicular to the z-axis (see Figure 1.10). If the relative phase difference measured between the two points P_1 and P_2 is constant in time, then the beam is termed spatially coherent.

Notice that spatial coherence does not require photon coherence. Thus spatial coherence is not a direct result of the laser action, but is actually a property of the electromagnetic modes of the laser. (For example, spatially coherent light can be obtained by shining sunlight through a pinhole.) However, lasers offer the highest intensity sources of spatially coherent light and thus are used in applications (such as holography and interferometry) that require spatially coherent sources (see Section 9.1 for more information on these applications).

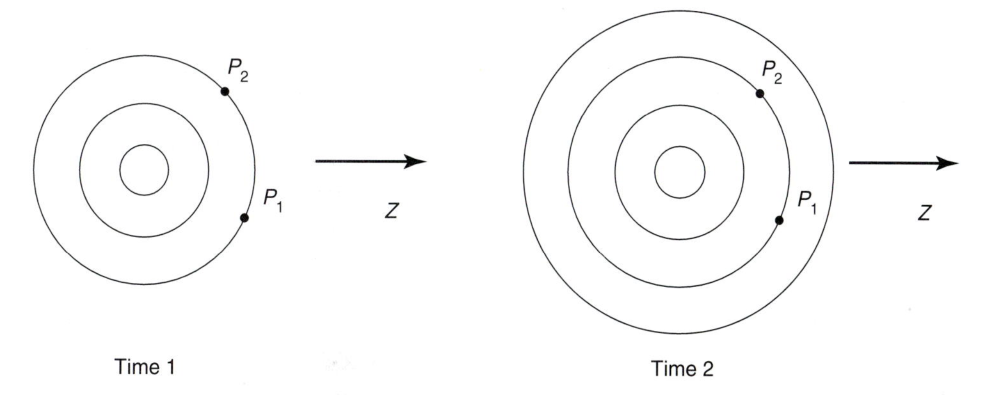

Figure 1.10 Imagine a beam of constant intensity propagating in the z direction. Consider any two points P_1 and P_2 chosen in a plane perpendicular to the z axis on this beam. If the relative phase difference measured between the two points P_1 and P_2 is constant in time, then the beam is termed spatially coherent.

Laser speckle is one of the consequences of the high spatial coherency of a light beam. When a laser shines on a rough surface (such as a wall) each small irregularity reflects the light from a slightly different location. This creates an interference pattern formed of bright (*constructive*) and dark (*destructive*) regions. Notice that you cannot focus on the speckle pattern. Indeed, you can only see the speckle pattern if you focus your eyes on some intermediate point. (Eyeglass wearers will discover that they can see the speckle pattern better if they remove their glasses!) Interestingly enough, if you move your head from side to side, the speckle pattern will also move. If it moves in the opposite direction of your head, you are near-sighted; if it moves in the same direction, you are far-sighted.

1.8 THE GENERIC LASER

Most lasers are constructed of three important elements, a *gain material*, a *pumping source* and a *resonant cavity* (see Figure 1.11).

The *gain material* is the location of the energy states which participate in stimulated emission. The material can be solid (Nd:YAG, ruby, GGG, GSGG, alexandrite, emerald, Cr:sapphire, Ti:sapphire, AlGaAs/GaAs, etc.; see Chapters 10 and 11), liquid (dye, chelate, etc.), gas (krypton, argon, nitrogen, helium-neon, CO_2, KrF, XeCl, etc.; see Sections 9.1, 9.2, 12.1, and 12.2), or plasma (x-ray, free-electron, etc.).

The *pumping source* provides the energy to set up the energy states so that stimulated emission can occur. Lasers can be optically pumped using lamps or other lasers (most solid-state lasers, see Chapters 10 and 11), electrically pumped using a pn-junction (semiconductor diodes, see Section 12.3) or an electric discharge (most gas lasers, see Sections 9.1 and 9.2), or pumped by a chemical reaction (HF, iodine, etc.). Lasers have even been pumped by such dramatic methods as jet engines (gasdynamic CO_2 lasers, see Section 12.1) or atomic bombs (x-ray lasers).

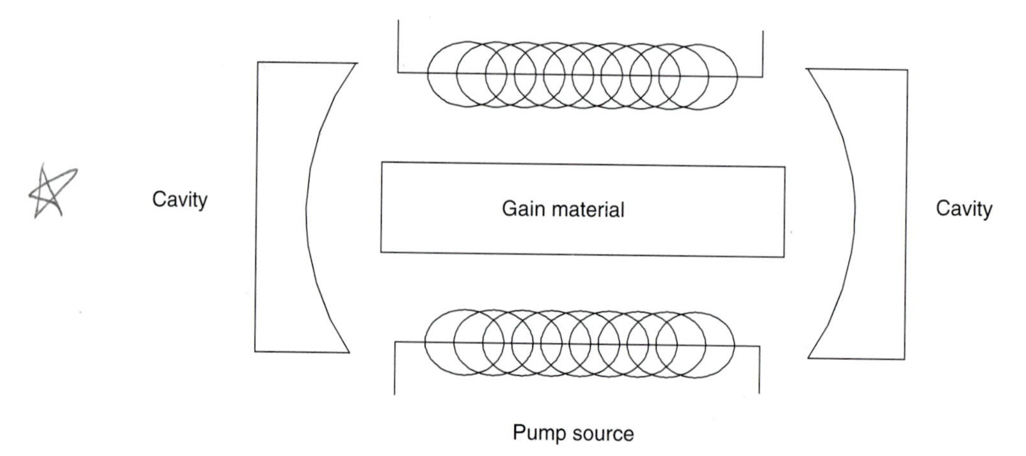

Figure 1.11 Most lasers are constructed from three important elements: a *gain material*, a *pumping source*, and a *resonant cavity*.

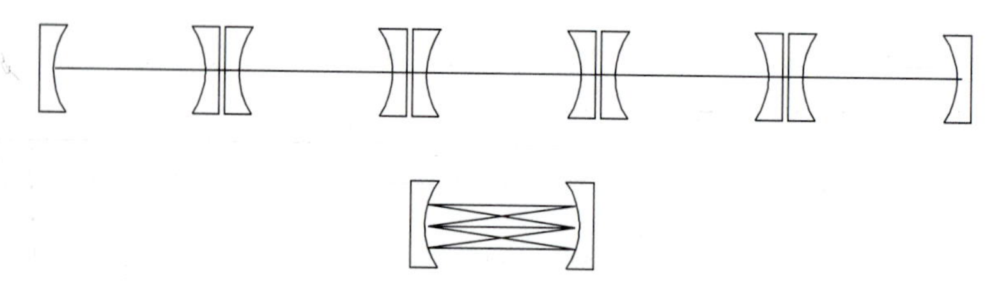

Figure 1.12 The laser path can be considered to be "folded up" between the resonant cavity mirrors. The functions of the folded resonant cavity are to physically shorten the laser and tailor the profile of the electromagnetic mode.

The *resonant cavity* provides a regenerative path for the photons. In essence, the functions of the resonant cavity are to 1) physically shorten the laser, and 2) tailor the profile of the electromagnetic mode. (General resonant cavity design is discussed in Chapter 4.)

In some sense, the laser is simply "folded up" between the two resonant cavity mirrors (see Figure 1.12). Although the resonant cavity is a key part of most commercial lasers, there are many lasers for which the resonant cavity is not necessary. It is certainly possible to make a laser long enough so that a reasonable intensity of light beam emerges without a resonant cavity (most x-ray lasers are made this way). However, such lasers tend to have poor output-beam quality.

1.9 TRANSVERSE AND LONGITUDINAL MODES

The output spot of the laser beam is termed the *transverse electromagnetic mode (TEM)*. In most commercial lasers the transverse electromagnetic mode is $TEM_{0,0}$ (see Figure 1.13), which is a round mode with a *Gaussian* profile in cross-section. (The $TEM_{0,0}$ mode is discussed in more detail in Section 4.3.)

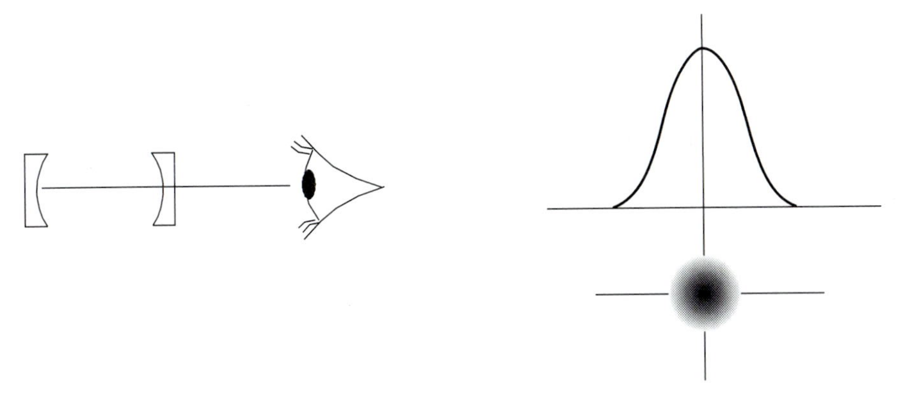

Figure 1.13 In most commercial lasers the transverse mode is $TEM_{0,0}$, which is a round mode with a Gaussian profile in cross-section.

However, it is possible to operate on a wide variety of other *transverse mode* configurations (see Section 4.2). In these configurations, the output spot will have a much more peculiar shape (see Figure 1.14).

A laser can only lase at those wavelengths for which an integral multiple of half-wavelengths precisely fit into the cavity (see Figure 1.15). The set of possible integral multiples of the cavity length is termed the set of *longitudinal electromagnetic modes* of the cavity (or simply the *longitudinal modes*). The frequencies of these modes are given by

$$\boxed{\nu = \frac{p \cdot c_o}{2nL} \quad \text{where } p = 1, 2, 3, \ldots} \tag{1.39}$$

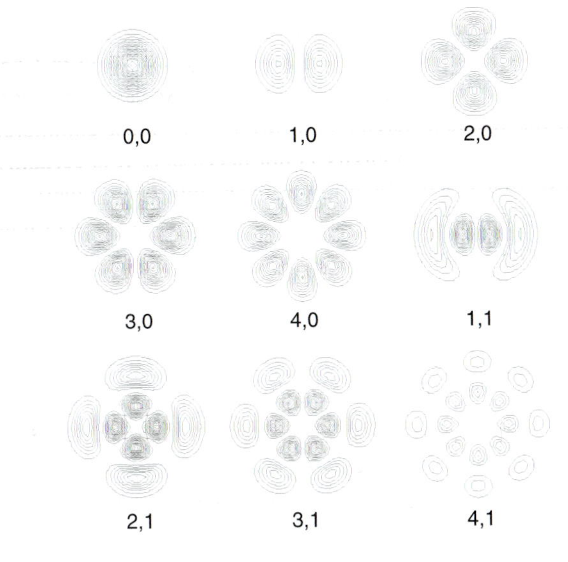

Figure 1.14 It is possible to operate a laser with a wide variety of other transverse electromagnetic modes.

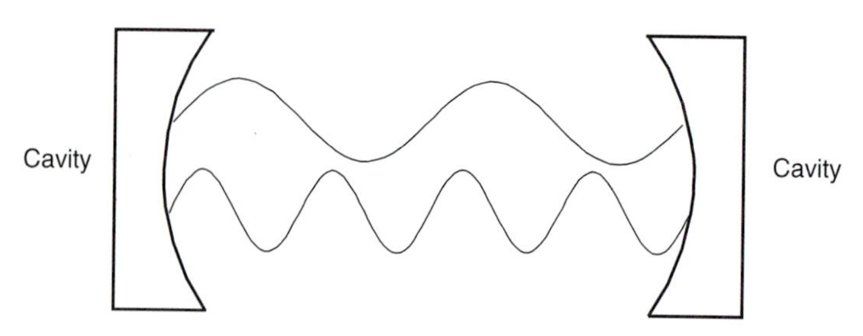

Figure 1.15 A laser can only lase at those wavelengths (longitudinal modes) for which an integral multiple of wavelengths fit into the cavity.

where n is the index of refraction in the cavity and L is the cavity length. Notice that p will be a very large number in a typical laser system (see Section 3.1).

The *longitudinal mode spacing* is frequently of interest in designing or using a laser system. The longitudinal mode spacing is given as

$$\boxed{\Delta\nu_{\text{FSR}} = \frac{c_o}{2nL}.} \tag{1.40}$$

Example 1.12

An argon-ion laser with a 1.0 meter resonant cavity is lasing with a center wavelength (in vacuum) at 488 nm. What is the mode number p for the mode closest to 488 nm? (Assume that the index of refraction for the cavity is 1.0.)

Solution. This uses Equation (1.39) reorganized as

$$p = \frac{2nL}{\lambda} = \frac{2 \cdot 1.0 \cdot 1.0 \text{ m}}{488.0 \cdot 10^{-9} \text{ m}} = 4098360.65 \rightarrow 4098361 \tag{1.41}$$

which may be verified by using equation (1.39),

$$\lambda = \frac{2nL}{p} = \frac{2 \cdot 1.0 \cdot 1.0 \text{ m}}{4098361} = 487.999959 \cdot 10^{-9} \text{ m}. \tag{1.42}$$

Example 1.13

A HeNe laser with a 30.0 cm resonant cavity is lasing with a center wavelength (in vacuum) at 632.8 nm. What is the longitudinal mode spacing (in Hz)? (Assume that the index of refraction for the cavity is 1.0)

Solution. This uses Equation (1.40) as

$$\Delta\nu_{FSR} = \frac{c_o}{2nL} = \frac{c_o}{2 \cdot 1.0 \cdot 0.30 \text{ m}} = 499.654 \text{ MHz}. \tag{1.43}$$

1.10 THE GAIN PROFILE

A particular laser material does not have gain at all frequencies. The function that describes the frequency dependence of the gain is termed the *gain profile* $g(\nu)$ (see Figure 1.16). (The gain profile is discussed in more detail in Section 2.1.)

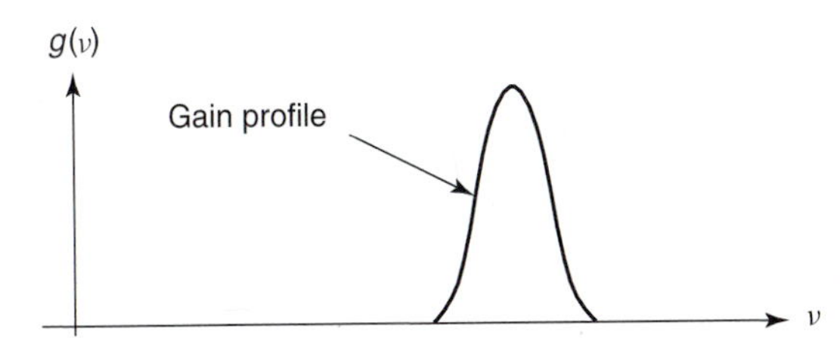

Figure 1.16 The function that describes the frequency dependence of the gain is termed the gain profile $g(\nu)$.

There are an infinite number of integral multiples of cavity length. However, only a finite number will fit into the gain profile of the laser gain material. Thus, the actual output of the laser is the intersection of the set of possible longitudinal modes with the gain profile (see Figure 1.17).

This may create opportunities for confusion. There is a center wavelength of the gain profile and its corresponding frequency, there is a center wavelength for each longitudinal mode and their corresponding frequencies, there is a linewidth for the gain profile, a linewidth for each longitudinal mode, and a longitudinal mode separation (see Figure 1.18). Furthermore, the wavelength may be given as the vacuum wavelength or the wavelength in the material (vacuum wavelength divided by the index of refraction, $\lambda = \lambda_o/n$). Additionally, all of these quantities can be measured in either frequency (Hz) or length (nm) units, and several of them have the same magnitude. To add to the confusion, there is no standard notation distinguishing the various quantities!

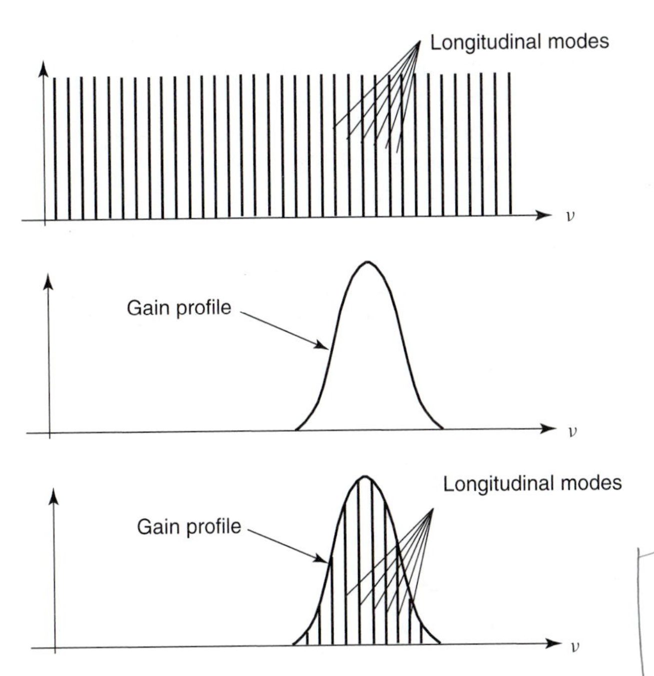

Figure 1.17 The actual output of a laser is the intersection of the set of possible longitudinal modes with the gain profile.

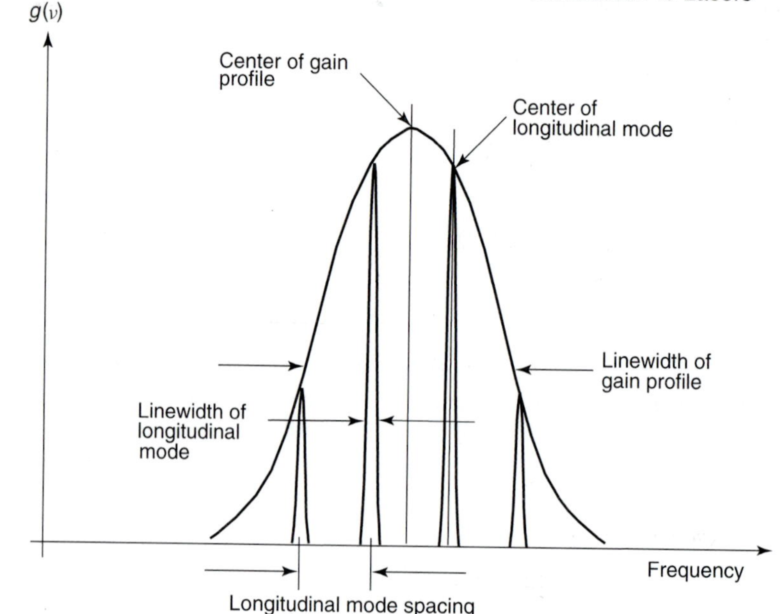

Figure 1.18 There is a center wavelength for the gain profile and its corresponding frequency, a center wavelength for each longitudinal mode and their corresponding frequencies, a linewidth for the gain profile, a linewidth for each longitudinal mode, and a longitudinal mode separation.

1.11 LASER SAFETY

Laser safety is an extremely important part of laser engineering—not only in using lasers, but in designing lasers and laser systems to be used by others. The Appendix presents a discussion of laser safety for laser courses with an included laboratory. For more information on laser safety issues, the reader is strongly encouraged to obtain a copy of the ANSI publication ANSI Z136.1-1993 (American National Standard for the Safe Use of Lasers).[28] Additionally there are a number of excellent references that are specifically directed toward the issues of laser safety.[29,30,31,32,33,34]

[28]Published by the Laser Institute of America.

[29]D. H. Sliney and M. L. Wolbarsht, *Safety with Lasers and Other Optical Sources* (New York: Plenum Publishing Co., 1980).

[30]L. Goldman, R. J. Rockwell Jr., and P. Hornby, "Laser Laboratory Design and Personnel Protection from High Energy Lasers," in *Handbook of Laboratory Safety*, ed N. V. Steere, 2d ed. (Cleveland, OH: Chemical Rubber Company, 1971).

[31]R. J. Rockwell, Jr., *Lasers and Applications* 5,(5): 97–103; and 5,(9): 93–99 (1986).

[32]"Laser Health Hazards Control," U.S. Dept. of the Air Force Manual, AFM-161-32, 1973 or latest edition.

[33]"Controls to Hazards to Health from Laser Radiation," U.S. Dept. of the Army Technical Bulletin, TB-MED-524, June 1985 or latest edition.

[34]"Standard Radiation Safety of Laser Products, Equipment Classification, Requirements and Users Guide," IEC-825 (Geneva, Switzerland, 1990).

SYMBOLS USED IN THE CHAPTER

s,p,d,f: Notation for electron orbitals

E: Photon energy (joules or eV)

E_2 and E_1: Upper and lower laser state energies (joules or eV)

ν: Nonspecific frequency (Hz)

λ: Nonspecific wavelength in a material (cm or m)

N_2 and N_1: Upper and lower laser state populations (electrons/cm^3)

T: Temperature (degrees K)

E_{pulse}: Energy per pulse (joules)

R_{reprate}: Repetition rate (Hz)

P_{avg}: Average power (watts)

P_{peak}: Peak power (energy/pulse width) (watts)

t_{pulse}: Pulse width (seconds)

E_{den}: Energy per unit area (joules/cm^2)

$P_{\text{peak-den}}$: Peak power per unit area (watts/cm^2)

$P_{\text{avg-den}}$: Average power per unit area (watts/cm^2)

ν_g, λ_g: Frequency and wavelength at the center of the gain curve

ν_o, λ_o: Frequency and wavelength at the center of a particular longitudinal mode (Hz, and m or nm)

$\Delta\nu_g$, $\Delta\lambda_g$: Frequency and wavelength linewidths for the gain curve (Hz, and m or nm)

$\Delta\nu$, $\Delta\lambda$: Frequency and wavelength linewidth of a particular longitudinal mode (Hz, and m or nm)

$\Delta\nu_{\text{FSR}}$, $\Delta\lambda_{\text{FSR}}$: Frequency and wavelength spacings between longitudinal modes (Hz, and m or nm)

τ_{cav}: Cavity decay time (seconds)

P_{out}: Output laser power (watts)

$\delta\tau$: Coherence time (seconds)

l_c: Coherence length (cm or m)

w_n: Spectroscopic wavenumber ($1/\lambda$) (cm^{-1})

P_2 and P_1: Locations on a wavefront

L: Length of a resonant cavity (cm or m)

p: Integer describing the "number" of a particular longitudinal mode

n: Real index of refraction (unitless)

EXERCISES

Basics

1.1 **(a)** Describe briefly the difference between a penlight laser and a penlight flashlight.
(b) Propose a simple test to differentiate a penlight laser from a penlight flashlight.

1.2 The United Federation of Planets starship Voyager is armed with phasors and photon torpedoes. Explain (citing fundamental physical principles as appropriate) why neither phasors nor photon torpedoes are advanced versions of lasers. (For those who are not Trekkies, phasors and photon torpedoes are fictional weapons from *Star Trek*. A phasor can be set to "stun" or "kill": an individual shot by a phasor on "stun" falls over in a faint; on "kill" the person is vaporized. Phasors usually appear orange on the TV screen and can be seen to visually travel from the source to target. Photon torpedoes are round objects shot from the starship. They impact on the target and it explodes. They also can be seen to visually travel from the source to target.)

1.3 Construct a logarithmic wavelength-frequency scale for electromagnetic radiation. Mark both frequency and wavelength on the x-axis of the scale in appropriate units (nm, μm, cm, m, etc.) and (Hz, kHz, GHz, etc.). On the scale indicate the following regions: **a.** visible light (include red, yellow and blue), **b.** AM radio, **c.** FM radio, **d.** x-rays, **e.** gamma rays, **f.** S- and X-band traffic radar, and **g.** TV.

1.4 Complete the following table for various types of electromagnetic radiation.

	Center wavelength, λ	Center frequency, ν
Red light	650 nm	
Green light		571 THz
Blue light	450 nm	
Yellow light		517 THz
AM radio	560.3 m–186.8 m	
FM radio		88 MHz–108 MHz
X-rays	0.1–100 Å	
Gamma rays		> $3 \cdot 10^{19}$ Hz
Near infrared	800 nm–5 μm	
S-band radar		2 GHz–4 GHz
TV channels 7–13	1.72 m–1.38 m	

Transitions between laser states

1.5 There are a number of ways of specifying the separation between two laser states. These include the energy difference $E_2 - E_1$ in joules or eV, the wavelength λ (in angstroms, nm or microns), the frequency ν (in Hz) and the spectroscopic wavenumber ($1/\lambda$ in cm^{-1}). Determine appropriate conversion equations for all quantities and complete the following table.

(a) General lasers

	E (eV)	E (J)	λ (nm)	ν (Hz)	$1/\lambda$ (cm^{-1})
XeCl	4.025				
Coumarin-2		$4.414 \cdot 10^{-19}$			
Argon-ion			514.5		
Ti:sapphire				$3.944 \cdot 10^{14}$	
Nd:YAG					9398.5

(b) Gas and excimer lasers

	E (eV)	E (J)	λ (nm)	ν (Hz)	$1/\lambda$ (cm^{-1})
KrF	4.979				
XeF		$5.695 \cdot 10^{-19}$			
Argon-ion			488.0		
HeNe-IR				$2.607 \cdot 10^{14}$	
CO_2					943.396

(c) Dye lasers

	E (eV)	E (J)	λ (nm)	ν (Hz)	$1/\lambda$ (cm^{-1})
Coumarin-102	2.53				
Coumarin-7		$4.169 \cdot 10^{-19}$			
Rhodamine 110			570.0		
Cresyl violet				$4.409 \cdot 10^{14}$	
Nile blue-A					13698.6

(d) Solid-state and semiconductor diode lasers

	E (eV)	E (J)	λ (nm)	ν (Hz)	$1/\lambda$ (cm^{-1})
AlGaAs	1.589				
Nd:YLF		$1.886 \cdot 10^{-19}$			
Nd:glass (LHG-8)			1054		
Nd:glass (LG-670)				$2.826 \cdot 10^{14}$	
Er:YAG					3401.36

(e) More solid-state and semiconductor diode lasers

	E (eV)	E (J)	λ (nm)	ν (Hz)	$1/\lambda$ (cm^{-1})
AlGaAs	1.851				
Ruby		$2.861 \cdot 10^{-19}$			
Alexandrite			760		
Cr:LiSAF				$3.568 \cdot 10^{14}$	
Er:glass					6493.51

Population inversion

1.6 Two energy levels E_2 and E_1 are separated by an energy gap $E_2 - E_1$. Evaluate the ratio N_2/N_1 (where N_2 is the population in level E_2 and N_1 is the population in level E_1) for the following cases.

$T(K)$	Energy gap	N_2/N_1
300	6328 Å	
300	11 GHz	
300	6 eV	
1000	6328 Å	
1000	11 GHz	
1000	6 eV	

1.7 Two energy levels E_2 and E_1 are separated by an energy gap $E_2 - E_1$. Evaluate the temperature that gives the specific population ratio N_2/N_1 (where N_2 is the population in level E_2 and N_1 is the population in level E_1) for the following cases.

N_2/N_1	Energy gap	$T(K)$
0.01	6328 Å	
0.01	11 GHz	
0.01	6 eV	
0.1	6328 Å	
0.1	11 GHz	
0.1	6 eV	

Power and energy

1.8 **(a)** A Nd:YAG laser has an energy of 0.4 joules/pulse, a repetition rate of 30 Hz, a pulsewidth of 20 nsec and has been focused into a spot of radius 0.5 mm. Calculate the average and peak power of the laser. Also calculate the energy density and the peak power density at the focus.

(b) A Nd:glass laser has an energy of 10 joules/pulse, a repetition rate of 1 Hz, a pulsewidth of 20 nanoseconds, and has been focused into a spot of radius 100 μm. Calculate the average and peak power of the laser. Also calculate the energy density and the peak power density at the focus.

(c) A pulsed dye laser has an energy of 0.1 joules/pulse, a repetition rate of 100 Hz, a pulsewidth of 50 femtoseconds and has been focused into a spot of radius 20 μm. Calculate the average and peak power of the laser. Also calculate the energy density and the peak power density at the focus.

Linewidth

1.9 Gain linewidth for lasers ($\Delta\nu_g$, $\Delta\lambda_g$) are often measured in angstroms if they are large and in Hz if they are small (see Figure 1.19). Complete the following table.

	wavelength, λ	$\Delta\nu_g$(Hz)	$\Delta\lambda_g$ (Å)
Nd:YAG	1.064 μm	300 GHz	
Nd:glass	1.055 μm		24.13 Å
HeNe	632.8 nm	1500 MHz	
GaAs - diode	780.0 nm		81.17 Å

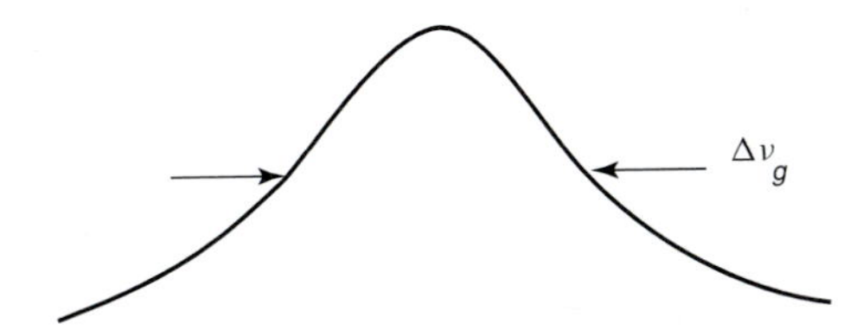

Figure 1.19 Gain linewidth for lasers ($\delta\nu_g$, $\delta\lambda_g$).

1.10 A laser spectrum is composed of a number of longitudinal modes. These modes are separated by the longitudinal mode spacing ($\Delta\nu_{\text{FSR}}$, $\Delta\lambda_{\text{SFR}}$) of the laser cavity. The total linewidth of the gain curve ($\Delta\nu_g$, $\Delta\lambda_g$) is the envelope of these modes (see Figure 1.20). Assuming that the index of refraction is equal to air ($n = 1$) in all cases, complete the following tables.

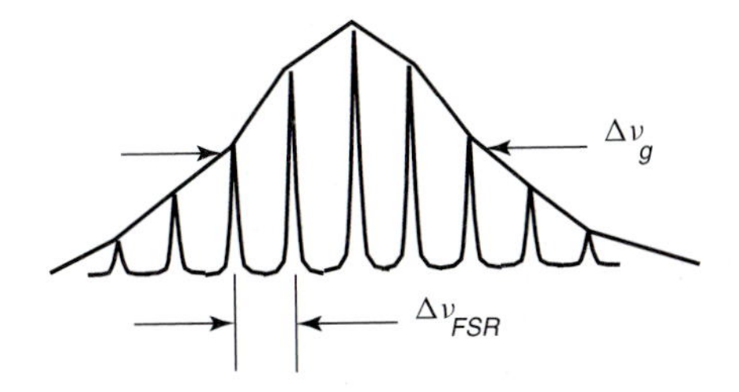

Figure 1.20 Laser modes are separated by the longitudinal mode spacing ($\Delta\nu_{\text{FSR}}$, $\Delta\lambda_{\text{SFR}}$) of the laser cavity.

(a) Major commercial HeNe lines

	λ (nm)	L (cm)	$\Delta\nu_{FSR}$ (Hz)	$\Delta\lambda_{FSR}$ (Å)
Green	543.5	22.4		
Yellow	594.1	26.4		
Orange	612.0	35.7		
Red	632.8	25.1		
Near IR	1150.0	20.3		
Fiber-optic IR	1523.0	24.8		
IR	3390.0	34.9		

(b) Major commercial argon-ion laser lines

	λ (nm)	L (cm)	$\Delta\nu_{FSR}$ (Hz)	$\Delta\lambda_{FSR}$ (Å)
Blue	465.8	65.5		
Blue	472.7	71.1		
Blue-green	476.5	89.1		
Blue-green	488.0	73.9		
Blue-green	496.5	78.7		
Green	501.7	88.4		
Emerald green	514.5	93.5		

(c) Major solid-state laser materials

	λ (nm)	L (cm)	$\Delta\nu_{FSR}$ (Hz)	$\Delta\lambda_{FSR}$ (Å)
Ruby	694.3	68.2		
Ti:sapphire	760	75.2		
Cr:LiSAF	840	87.4		
Nd:YLF	1053.0	59.4		
Nd:glass (LHG-5)	1053.0	63.8		
Nd:glass (ED-2)	1062.3	79.1		
Nd:YAG	1064.0	85.9		

Review of basic optics—Lenses

1.11 The following are simple review lens problems.

(a) Consider a convex glass lens where the radius of the front surface is 25 cm and the radius of the back surface is 40 cm. Assume that the index of refraction is 1.5. What is the focal length of the lens? Repeat the calculation assuming the lens is concave.

(b) Consider a plano-convex lens, where the radius of the back surface is 50 cm. Assume that the index of refraction is 1.5. What is the focal length of the lens? Repeat the calculation assuming the lens is plano-concave.

(c) Consider a meniscus convex glass lens where the radius of the front surface is 25 cm and the radius of the back surface is −40 cm. Assume that the index of refraction is 1.5. What is the focal length of the lens?

(d) Consider a meniscus concave glass lens where the radius of the front surface is 50 cm and the radius of the back surface is −30 cm. Assume that the index of refraction is 1.5. What is the focal length of the lens?

(e) Consider a convex lens with a focal length of 50 cm. If a point source is placed 65 cm in front of the lens, where is the image located? If the point source is moved to a distance 15 cm from the front of the lens, where is the image now located?

(f) Consider a concave lens with a focal length of −50 cm. If a point source is placed 65 cm in front of the lens, where is the image located? If the point source is moved to a distance 15 cm from the front of the lens, where is the image now located?

Review of basic optics—Telescope problems

1.12 Consider a telescope similar to Figure A.4. Assume that the front lens is convex with a focal length of 50 cm. Assume the back lens is convex with a focal length of 70 cm. Assume that the lenses are separated by 1 meter and that the image is located a distance of 30 cm from the front of first lens. Calculate the location of the object. Is this a real or virtual object?

1.13 Consider a telescope similar to Figure A.5. Assume that the front lens is concave with a focal length of -20 cm. Assume the back lens is convex with a focal length of 60 cm. Assume that the lenses are separated by 70 cm and that the image is located a distance of 25 cm from the front of first lens. Calculate the location of the object. Is this a real or virtual object?

1.14 Consider a telescope similar to Figure A.4. Assume that the front lens is convex with a focal length of 40 cm. Assume the back lens is convex with a focal length of 80 cm. Assume the input beam is collimated.

(a) Plot the location of the focus spot as the distance separating the lenses is varied from 0 to 250 cm.

(b) Discuss the plot in terms of the laboratory construction of a small telescope to be used as a beam expander.

1.15 Consider a telescope similar to Figure A.5. Assume that the front lens is concave with a focal length of −30 cm. Assume the back lens is convex with a focal length of 100 cm. Assume the input beam is collimated.

(a) Plot the location of the focus spot as the distance separating the lenses is varied from 0 to 250 cm.

(b) Discuss the graph in terms of laboratory construction of a small telescope to be used as a beam expander.

Review of basic optics—Reflection and refraction

1.16 Consider the situation where a laser beam is traveling from one material to another material. Assume the incident angle is 15 degrees from the normal. In each of the three cases, calculate the reflected and transmitted angle (as measured from the normal) and create a sketch (roughly to scale) of the incident, reflected and refracted rays.

(a) A laser beam traveling in ethanol ($n = 1.361$) that enters a fused silica block ($n = 1.458$).

(b) A laser beam traveling in gallium arsenide ($n = 3.655$) that enters air ($n = 1$).

(c) A laser beam traveling in lithium niobate ($n = 2.094$) that enters water ($n = 1.33$).

1.17 Two polar bears are on display at the Point Defiance Park Zoo (see Figure 1.21). Their habitat contains a large pool with a window. (They spend a great deal of time playing in the pool.) The window into the habitat is oriented so that visitors can see both the underwater and above-water scene. For example, when the bears are mock-fighting in the water, the visitors can see both the bears' heads (above water) and the bears' bodies (below water). However, when standing in the viewing area, the polar bears' heads (as viewed above the water) are offset from their bodies (as viewed below the water). When children come into the viewing area, they always think there are four bears because the heads are offset significantly from the bodies. Calculate the distance that the heads are apparently offset from the bodies.

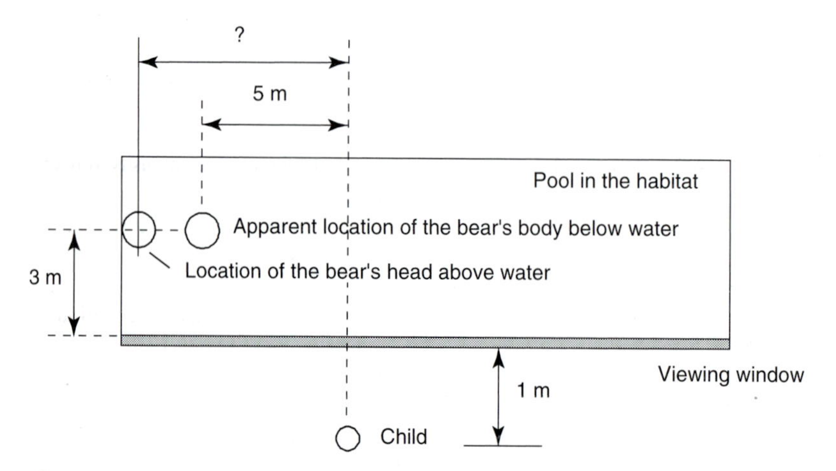

Figure 1.21 Two polar bears on display at the Point Defiance Park Zoo.

Brewster's angle

1.18 It is quite common to insert optical components inside a laser cavity at Brewster's angle to minimize the reflection of the p or || polarization component (see Figure 1.22).

(a) Calculate the length of the major axis ($2a$) of the beam image on the optical surface, assuming that the optical element is a piece of quartz (index of refraction = 1.458). Assume that the beam diameter is 10 mm.

(b) Calculate the length of the major axis ($2a$) of the beam image on the optical surface, assuming that the optical element is a piece of lithium niobate (index of refraction = 2.094). Assume that the beam diameter is 10 mm.

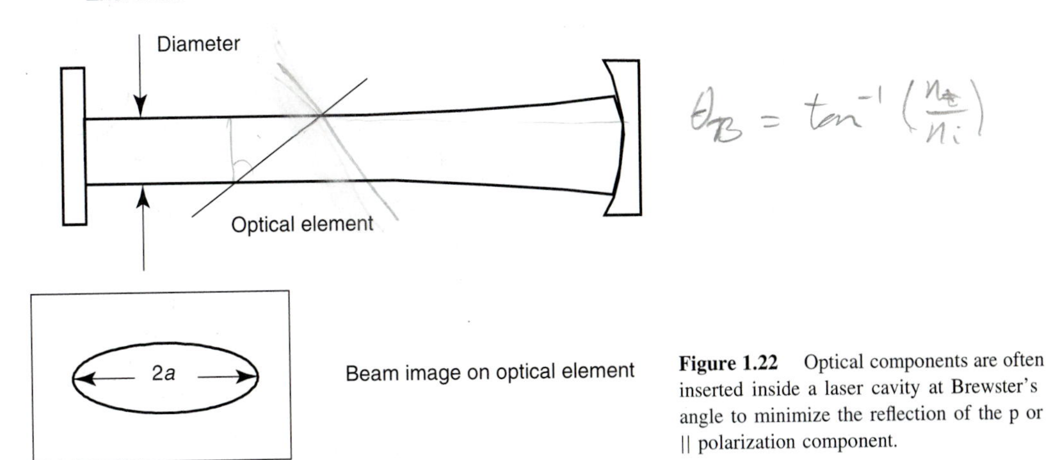

Figure 1.22 Optical components are often inserted inside a laser cavity at Brewster's angle to minimize the reflection of the p or || polarization component.

(c) Calculate the length of the major axis ($2a$) of the beam image on the optical surface, assuming that the optical element is a piece of GaAs (index of refraction = 3.655). Assume that the beam diameter is 10 mm.

Review of basic optics—Fresnel equations

1.19 Consider the method of separating real diamonds from fake diamonds by dropping them in olive oil. The idea is that real diamonds can still be seen and fake diamonds "vanish."

- **(a)** Calculate the intensity reflection at normal incidence for diamonds ($n = 2.41$) in olive oil ($n = 1.4679$).
- **(b)** Calculate the intensity reflection at normal incidence for lead glass ($n = 1.6$) in olive oil.
- **(c)** YAG is also used as simulated diamond. Calculate the intensity reflection at normal incidence for YAG ($n = 1.82$) in olive oil.
- **(d)** Do you think that this separation method works? Explain why or why not.

1.20 Consider a glass microscope slide suspended in the center of an aquarium full of water. Assume a randomly polarized laser beam is traveling through the aquarium and intersects the microscope slide. Calculate the parallel and perpendicular polarization reflectances of the laser beam from the microscope slide assuming the slide is at 25 degrees to the laser beam (25 degrees from normal). Assume the microscope slide has an index of refraction of $n = 1.55$.

1.21 Use of computer program (e.g., Mathematica, Maple, MatLab, MathCad, Lotus 123, Excel) to generate a plot of the reflection and transmission coefficients for parallel and perpendicular polarizations (a plot similar to Figure A.8). However, assume that the material is YAG with an index of refraction $n = 1.82$.

1.22 Use a computer program (e.g., Mathematica, Maple, MatLab, MathCad, Lotus 123, Excel) to generate a plot of the reflection and transmission coefficients for parallel and perpendicular polarizations. However, instead of having the x-axis represent θ_{in}, have the x-axis represent index of refraction (n). Plot the reflectance and transmittance for both polarizations at $\theta_{in} = 25$ degrees assuming an index of refraction range of 1.0–4.0. Assume $n_i = 1.0$.

2

Energy States and Gain

Objectives

- To distinguish between three-state and four-state lasers.
- To analyze simple multiple-state laser systems where states may be "stealing" gain from each other due to shared states or cascade situations.
- To describe the difference between a homogeneously and an inhomogeneously broadened laser.
- To compute and plot the gain profile $g(\nu)$ for a Lorentzian or Gaussian line.
- To compute the gain $\gamma(\nu)$ of a laser from fundamental parameters such as A_{21}, n, λ_o, N_2, N_1, g_2, g_1, and $g(\nu)$.
- To follow the derivation for the blackbody radiation density inside the cavity using both the classical approach and the Planck approach.
- To compute and plot the radiation density outside a blackbody for a useful blackbody such as a noble gas discharge lamp.
- To write the Einstein detailed balance equation and explain the significance of the terms.
- To compute the cross-section $\sigma_{21}(\nu)$, either for a whole gain curve or at the peak of a gain curve.
- To compute the gain of a laser from the cross-section $\sigma_{21}(\nu)$ and N_2, N_1, g_2, and g_1.
- To compute the single pass or round trip gain of a laser given the mirror reflectances and either fundamental parameters (such as A_{21}, n, γ_o, N_2, N_1, g_2, and g_1), or the cross-section $\sigma(\nu)$ and pumping parameters (such as N_2, N_1, g_2, and g_1).

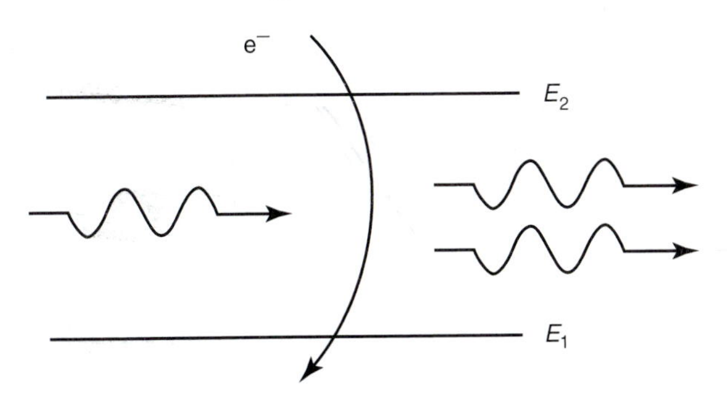

Figure 2.1 Stimulated emission between two laser states.

2.1 ENERGY STATES

2.1.1 Laser States

In Chapter 1, the stimulated emission process was described as occurring between two ideal energy states, E_2 (the upper laser state) and E_1 (the lower laser state). (See Figure 2.1.) However, it is rare for real lasers to only possess two laser states. Only semiconductor lasers (see Section 12.3) and vibronic lasers such as Ti:sapphire (see Chapter 11) have a two-state energy-band structure. Most laser materials possess a multiplicity of interacting states, some serving as laser states and others as pump states.

Three-state lasers. Most laser systems incorporate a *pump state* (or states) in addition to the upper and lower laser states. For example, *three-state* laser systems (see Figure 2.2) consist of a *ground state* (0), an *upper laser state* (2) and a *pumping state* or *states* (3).

The general process is that a *pump source* with a broad energy spectrum provides population to a system of *pump states*. These states then decay to provide population to the upper laser state. The upper laser state is chosen to have a long transition lifetime so that the population waits in the upper laser state (without decaying to some other state) until stimulated by a photon to transition to the lower state.

In an optimal three-state configuration, the transition lifetime from state 3 to state 2 should be very short and the transition lifetime from 2 to 0 very long. The long lifetime of the upper state is essential so that stimulated emission can occur before the upper-state population is depleted by recombination processes. In the optimal three-state configuration,

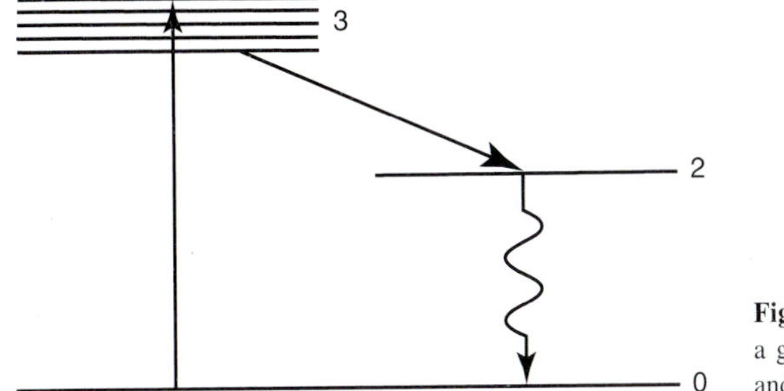

Figure 2.2 A three-state laser consists of a ground state (0), an upper laser state (2), and a pumping state or states (3).

population pumped to state 3 will rapidly decay to state 2, and then wait there until stimulated to state 0. (Such long-lived states are often termed *metastable states*). An important property of an efficient laser is that the majority of the population pumped to the pump bands should end up at the upper laser state. The percentage of the population pumped to the pump bands that makes it to the upper laser state is often termed the *quantum efficiency*.

The difficulty with three-state systems is that (given a short lifetime in state 3) slightly more than one-half the population in state 0 must be pumped to state 2 before any population inversion (or gain!) occurs. This means that very few practical lasers are three-state lasers. The one exception is ruby, due to the high quantum efficiency as well as the very wide pump bands (which overlap well with a broadband pumping source).

Four-state lasers. The most common laser configuration is a *four-state laser*. A four state laser system (see Figure 2.3) consists of a *ground state* 0, a *pumping state* (or band of states) 3, an *upper laser state* 2, and a *lower laser state* 1. The difference between a three-state and a four-state laser is that in a four-state laser the lower laser state (state 1) is lifted above the ground state. The idea is to locate a lower state that is depopulated and is several $k_B T$ from the ground state (so as to minimize thermal transitions from the ground state to the lower laser state). Thus, any population that ends up at the upper laser state contributes to a population inversion.

In an optimal four-state configuration, the transition lifetime from state 3 to state 2 is very short, the transition lifetime from 2 to 1 is very long (metastable state) and the transition lifetime from state 1 to state 0 is very short. Thus, population pumped to state 3 will rapidly decay to state 2 and then wait there until stimulated to state 1. Population stimulated to state 1 by laser action will then rapidly decay to state 0. This means that it is relatively easy to sustain a population inversion so long as state 1 is sufficiently far from state 0 to minimize thermal population.

2.1.2 Multiple-State Laser Systems

The stylized illustrations of three- and four-state lasers (see Figures 2.2 and 2.3) only include one pump band and one pair of laser states. However, real lasers may have several possible

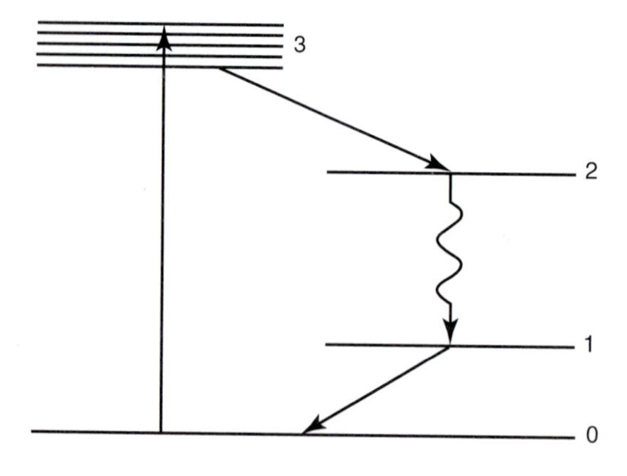

Figure 2.3 A four-state laser consists of a ground state 0, a pumping state (or band of states) 3, an upper laser state 2, and a lower laser state 1.

pump and laser state combinations. These additional states may be beneficial or detrimental depending on the circumstances.

Sharing an upper laser state. One common situation is for two laser transitions to share an upper laser state, as illustrated in Figure 2.4. If the laser is intended to only operate on one laser transition, then the other transition represents a *parasitic transition*. If this parasitic transition is allowed to operate, it will "steal" upper-state population from the primary laser line.

Although it is impossible to stop spontaneous emission on the undesired transition, it is possible to reduce the number of stimulated transitions. One method for reducing the number of stimulated transitions is to minimize the number of photons in the cavity at the undesired wavelength. (Wavelength-selective absorbers placed in the cavity are one way of minimizing photons at undesirable wavelengths. However, it is often difficult to fabricate an optical element that absorbs well at one wavelength while simultaneously transmitting at a neighboring wavelength.)

Laser action occurs when the *round-trip gain* between the end mirrors of the laser is greater than 1. If the round-trip gain for the undesired transition can be kept to less than 1, then the undesired transition will not lase. (There may be some stimulated radiation emitted from the laser at that transition. However, without a *regenerative* (closed) path, there will be no laser action and the depletion in upper-state population will be minimized.) The most common way to reduce round-trip gain on the undesired transition is to fabricate end mirrors that are highly reflective at the laser wavelength of interest and poorly reflective at the undesired wavelength. Such mirrors are a critical aspect of modern laser design and are fabricated using multilayer dielectric film technology. (Details on multilayer dielectric mirrors are discussed in Section 8.2.)

Sharing a lower laser state. Another common situation is for two laser transitions to share a lower laser state, as illustrated in Figure 2.5. Since laser action is driven by the population difference between the upper and lower states, sharing a lower state has some of the same consequences as sharing an upper state. Unless the lower state has a very

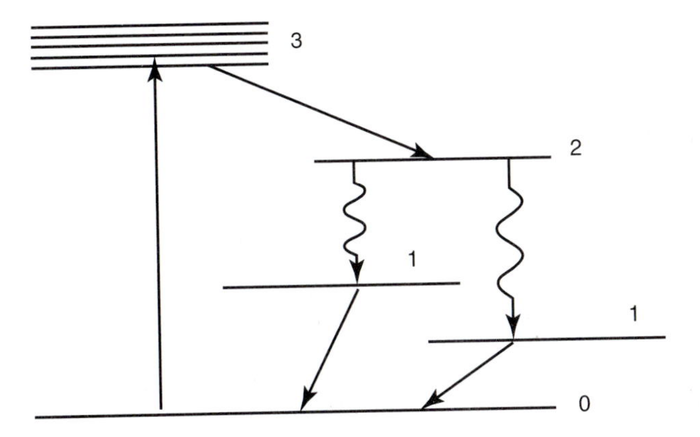

Figure 2.4 One common situation is for two laser transitions to share an upper state.

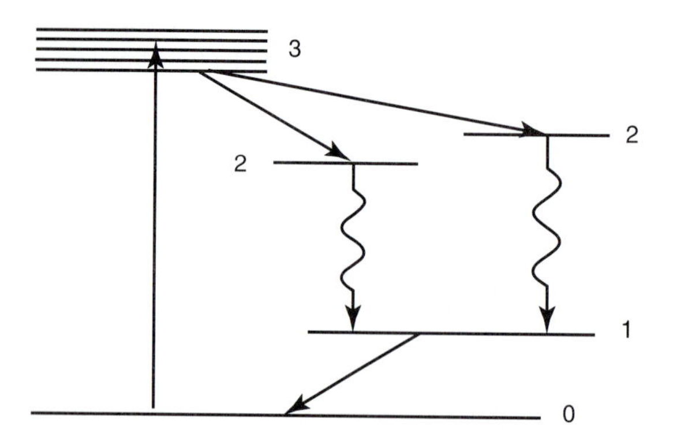

Figure 2.5 Another common situation is for two laser transitions to share a lower state.

rapid spontaneous lifetime, electrons collecting in the lower state from laser action on one line will reduce the gain of the other line.

As with the situation of sharing an upper laser state, reducing the loss from the undesired transition requires minimizing the number of undesired photons in the cavity (for example, with selective absorbers) or reducing the round-trip gain at the undesired wavelength (for example, with selectively reflecting mirrors).

More complex situations. There are a number of other possible situations that can arise in real laser materials. For example, an undesired laser line may have a lower state that is higher in energy than the lower state of a desired laser line (see Figure 2.6). Although the two transitions do not share a lower laser state, electrons in the lower state of the undesired transition may decay downward to populate the lower state of the desired transition. This has the same consequences as sharing a lower laser state.

Another common situation is for the laser material to have multiple sets of pump bands and laser lines (see Figure 2.7). This is usually not a problem when operation is desired on the higher energy system of pump bands and laser lines. However, it can be a problem when operation is desired on the lower energy system. Laser lines operating from the higher energy system may populate the lower laser states of the lower energy system. This is similar to the situation of sharing a lower laser state, except that the population originates in a different system of pump bands.

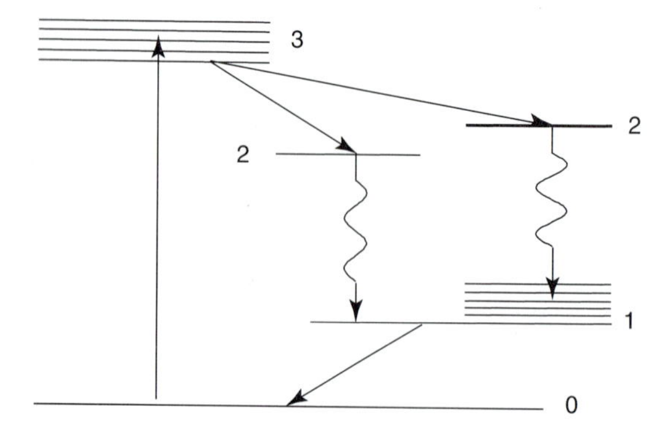

Figure 2.6 Although two transitions may not share a lower laser state, electrons in the lower state of an undesired transition may decay downward to populate the lower state of a desired transition.

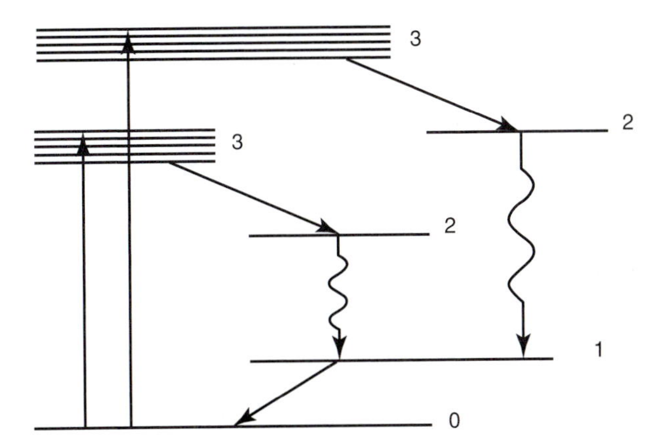

Figure 2.7 Another common situation is for the laser material to have multiple sets of pump bands and laser lines.

2.1.3 Linewidth and the Uncertainty Principle

In Chapter 1, the transition energy $E_2 - E_1$ between the two states ($|2\rangle$ and $|1\rangle$) was presented as having an infinitely narrow linewidth. Of course, there is some finite linewidth to any transition in a real system! However, there are also some fundamental considerations that limit the linewidth in an ideal system. Such linewidth considerations arise from a Fourier-transform limited property called the Heisenberg Uncertainty Principle.[1]

Consider a wavepacket described by a square time distribution (where $\delta t = 2b$ is the time width of the wavepacket) as illustrated in Figure 2.8.

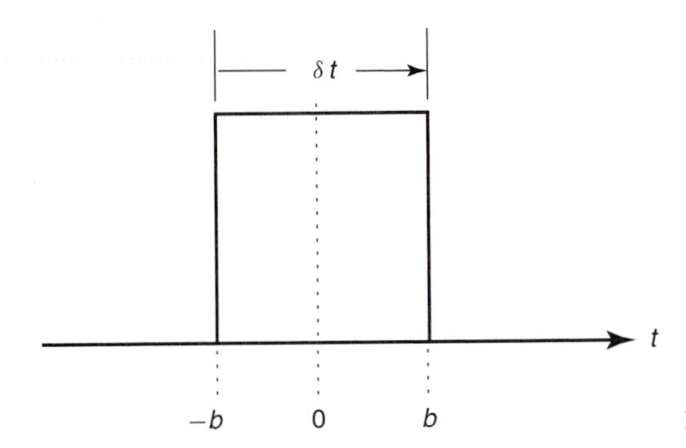

Figure 2.8 A square wavepacket in time.

The Fourier transform of this wavepacket is given by[2,3]

$$X(\omega) = \int_{-\infty}^{+\infty} x(t) e^{-j\omega t} dt = \int_{-b}^{+b} e^{-j\omega t} dt = 2b \frac{\sin(b\omega)}{b\omega} = 2b \cdot \text{sinc}(b\omega). \tag{2.1}$$

[1]The Heisenberg uncertainty principle is discussed in all treatments of quantum mechanics at any level. For an interesting view of the principle from its developer, consider W. Heisenberg, *The Physical Principles of the Quantum Theory*, Dover Publications, 1930. For an excellent overview and placement of the principle in the body of quantum theory, consider J. Gribbin, *Schrödinger's Kittens and the Search for Reality* (Boston, MA: Little Brown and Company, 1995).

[2]C. L. Liu and Jane W. S. Liu, *Linear System Analysis* (New York: McGraw Hill, 1975).

[3]Ronald N. Bracewell, *The Fourier Transform and Its Applications*, 2d ed. (New York: McGraw Hill, 1978).

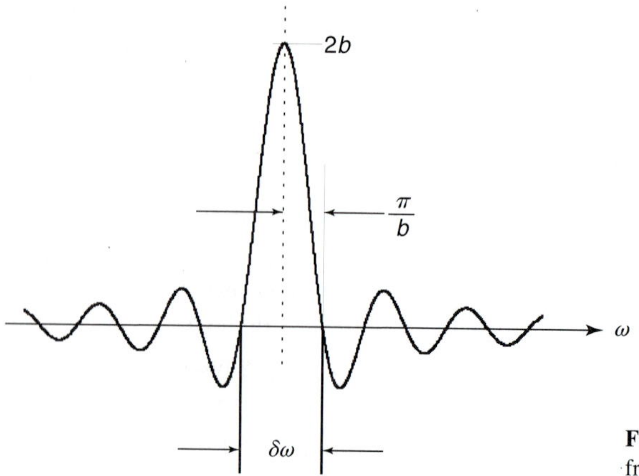

Figure 2.9 A sinc wavepacket in frequency.

(The term $\sin(x)/x$ is commonly seen in laser engineering and is called a $\mathrm{sinc}(x)$ function.) Thus, the Fourier transform of the "top hat" time distribution given in Figure 2.8 is the $\mathrm{sinc}(b\omega)$ frequency function given in Figure 2.9.

Notice that the width $2b$ of the time wavepacket corresponds to the width (at first zero crossing) of $2\pi/b$ for the frequency packet. So, a relationship exists between δt and $\delta\omega$ on the order of

$$\delta t \cdot \delta\omega \sim 2b \cdot \frac{2\pi}{b} = 4\pi. \tag{2.2}$$

If the waves are electromagnetic waves, then there is a lower limit on the energy given by $\hbar\omega$ resulting in

$$\delta t \cdot \delta E \sim \hbar \cdot 2b \cdot \frac{2\pi}{b} = 4\pi\hbar \tag{2.3}$$

and

$$\delta E \sim \frac{4\pi\hbar}{\delta t}. \tag{2.4}$$

Essentially, this suggests that a laser transition cannot simultaneously have an instantaneous spontaneous lifetime ($\delta t \to 0$) and be sharply defined ($\delta E \to 0$). An instantaneous spontaneous lifetime suggests that $\delta t \to 0$. However, if $\delta t \to 0$ then $\delta E \to \infty$.

Heisenberg postulated that this uncertainty principle applies to all quanta and that it establishes a limit for how precisely δE or δt can be specified. (Note that the usual form of the uncertainty relation is derived in a more elaborate way and results in a slightly different definition for $\delta t \cdot \delta E$ than indicated above. This more elaborate derivation gives the conventional formulation $\delta E \cdot \delta t \geq \hbar$ for the uncertainty principle.[4,5])

---

[4] Stephen Gasiorowicz, *Quantum Physics* (New York: John Wiley and Sons, 1975), p. 36.

[5] In many elementary physics references, the 4π is simply explained away by assuming that $4\pi \approx 1$. For a more detailed and complete treatment on this issue, see Sakurai, *Modern Quantum Mechanics* (Redwood City, CA: Addison-Wesley Pub. Co., Inc., 1985), Chapter 1.

Now, consider the picture of an electron in the sharply defined state E_1. Sharply defined implies that $\delta E_1 = 0$. If $\delta E_1 = 0$ then $\delta t \rightarrow \infty$ and spontaneous emission cannot occur. But it does. Thus, the states E_n cannot be sharply defined, but must possess some finite linewidth.

2.1.4 Broadening of Fundamental Linewidths

Although the Heisenberg Uncertainty Principle describes the fundamental limit to the narrowness of a laser transition, in actual practice transitions usually possess much greater linewidths determined by various *broadening* processes in the material. The actual shape of real energy transitions is described by the gain profile $g(\nu)$, which is a *normalized function*, meaning that

$$\int_0^\infty g(\nu)d\nu = 1. \tag{2.5}$$

The form of $g(\nu)$ is a function of the type of broadening. There are two major types of broadening in laser systems, *homogeneous* and *inhomogeneous*.

Homogeneous broadening. Homogeneous broadening mechanisms are those that operate on all atoms in the system equally.[6] *Collisional broadening* in a gas (where each atom is essentially the same as all other atoms) is an example of homogeneous broadening.

If the number of photons in the cavity is small, the upper-state population will not be significantly affected by the laser action. This regime of operation is often termed the *unsaturated* or *small signal* gain regime (discussed in more detail in Section 2.2). As the number of photons in the cavity increases, the upper-state population is affected by the laser action. If the intensity in the cavity is sufficiently high, the upper-state population is depleted by the laser process and the gain begins to drop. This process is called *gain saturation* or *saturation* and is discussed in more detail in Chapter 5.

A major characteristic of homogeneous broadening is that when the transition is saturated the *entire* lineshape will decrease proportionally (see Figure 2.10).

The gain profile $g(\nu)$ for homogeneous broadening is the normalized *Lorentzian* function[7]

$$g(\nu) = \frac{\Delta\nu}{2\pi\left((\nu_o - \nu)^2 + (\Delta\nu/2)^2\right)} \tag{2.6}$$

where $\Delta\nu$ is the frequency linewidth at *full-width half-maximum* (*FWHM*). Here, ν_o is the center frequency, and the peak value for the Lorentzian curve is given by

$$g(\nu_o) = \frac{2}{\pi\Delta\nu}. \tag{2.7}$$

---

[6]Examples of homogeneous broadening include lifetime broadening (a Fourier transform broadening determined by the radiative lifetime), collisional broadening, phonon broadening and dipolar broadening.

[7]For a derivation of this using a Fourier transform approach, see Joseph T. Verdeyen, *Laser Electronics*, 2d ed. (Englewood Cliffs, NJ: Prentice-Hall, 1989), pp. 164–169.

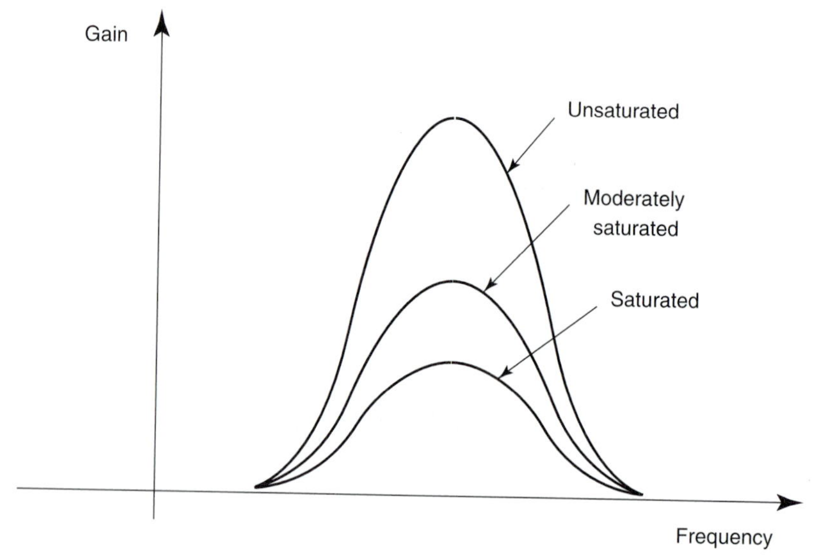

Figure 2.10 A major characteristic of homogeneous broadening is that the entire lineshape will decrease proportionally when saturated.

Inhomogeneous broadening. Inhomogeneous broadening mechanisms are those that operate on some groups of atoms in the system differently than on other groups of atoms.[8] Each specific group of atoms will have a Lorentzian homogeneous profile and the total profile will be the sum of the Lorentzian profiles of the various groups. For example, neon is broadened by a mechanism called *isotope broadening*.[9] Neon is composed of two isotopes. One isotope has an atomic mass of 20 and the other has an atomic mass of 22. The total lineshape is the sum of the individual lineshapes of the two isotopes.

A major characteristic of inhomogeneous broadening is that parts of the transition can be saturated *independently* of other parts of the transition. In effect, the transition is the sum of numerous homogeneous transitions (see Figure 2.11).

Lineshapes for inhomogeneous broadening. The gain profile $g(\nu)$ for inhomogeneous broadening is composed of the sum of numerous homogeneous lineshapes. Such a function is called a Voight function and is generally not analytic.[10]

However, if the inhomogeneous linewidth is much larger than the homogeneous linewidth, then it is possible to take the homogeneous part and replace it with a delta function. Under these circumstances the gain profile for inhomogeneous broadening reduces

[8] Examples of inhomogeneous broadening include Doppler broadening, strain broadening, and isotope broadening.

[9] A. Szöke and A. Javan, *Phys. Rev. Lett.* 10:521 (1963).

[10] For an excellent description of the Voight function and its application to laser engineering check A. Siegman, *Lasers* (Mill Valley, CA: University Science Books, 1986), pp. 173–5, and associated references.

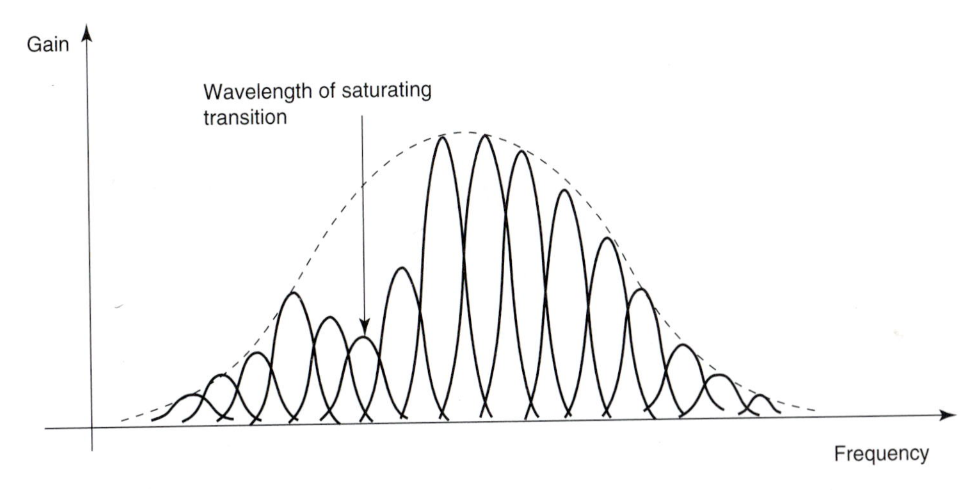

Figure 2.11 A major characteristic of inhomogeneous broadening is that parts of the transition can be saturated independently of other parts of the transition.

to a normalized Gaussian profile[11]

$$g(\nu) = \left(\frac{4\ln 2}{\Delta\nu^2\pi}\right)^{1/2} \exp\left(-4\ln 2\left(\frac{(\nu-\nu_0)}{\Delta\nu}\right)^2\right) \tag{2.8}$$

where $\Delta\nu$ is the frequency linewidth at full-width half-maximum (FWHM). Here ν_o is the center frequency and the peak value for the Gaussian curve is given by

$$g(\nu_o) = \frac{2}{\Delta\nu}\left(\frac{\ln 2}{\pi}\right)^{1/2}. \tag{2.9}$$

Thus, the gain profile $g(\nu)$ for purely homogeneous broadening is a normalized Lorentzian function, the gain profile $g(\nu)$ for purely inhomogeneous broadening is a normalized Gaussian function, and the gain profile $g(\nu)$ for broadening mechanisms in between homogeneous and inhomogeneous is an in-between nonanalytic function termed a Voight function (see Figure 2.12).

2.2 GAIN

2.2.1 Basics of Gain

When a laser is initially turned on, the pump source begins to excite electrons into the upper laser state (see Section 2.1 for more information on laser states). The increasing number of electrons in the upper laser state will cause an increase in the number of spontaneously emitted photons. Eventually, a photon will be emitted whose direction of travel is in line

[11] Amnon Yariv, *Quantum Electronics*, 4th ed. (New York: John Wiley and Sons, 1975), p. 167.

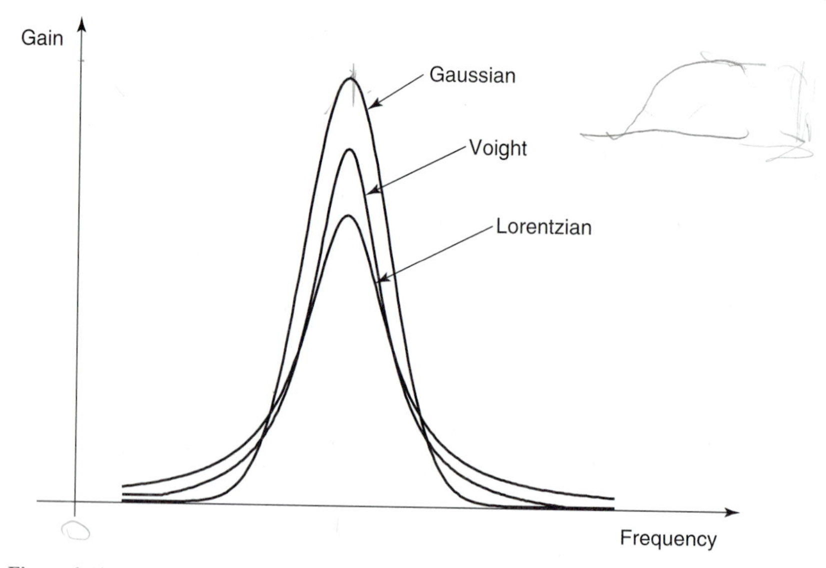

Figure 2.12 The line-shape function for purely homogeneous broadening is a normalized Lorentzian function: the line-shape function for purely inhomogeneous broadening is a normalized Gaussian function, and the line-shape function for broadening mechanisms "in between" homogeneous and inhomogeneous is an "in-between" nonanalytic function termed a Voight function.

with the resonant cavity mirrors. As this photon bounces back and forth between the mirrors, it stimulates the gain media to emit additional photons at the same wavelength, in phase, and heading in the same direction (see Figure 2.13).

The large number of photons involved in laser action permits the photons to be treated collectively. Depending on the circumstances, these photons can be envisioned to form a classical electromagnetic wave, an idealized ray, or a short pulse.

This collective behavior also allows the unsaturated gain to be modeled by a simple exponential equation. For example, consider a resonant cavity of length L with an electromagnetic wave of an intensity I_{start} beginning at one end of the resonant cavity (see

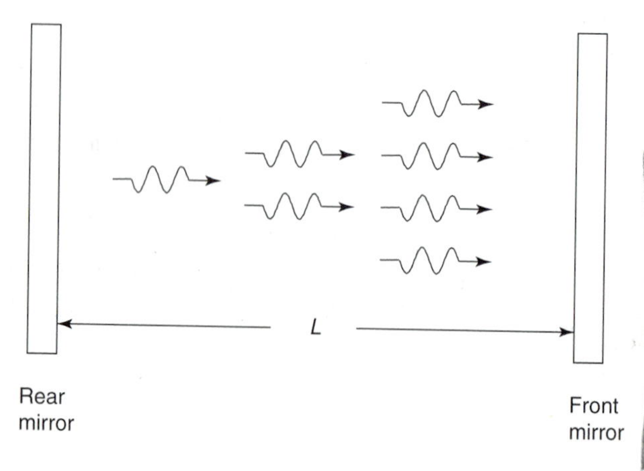

Figure 2.13 Schematic of photons traveling in a laser resonant cavity of length L. The first photon begins at the rear mirror. As it travels down the resonant cavity, it stimulates electron transitions from the upper to the lower state. These electron transitions contribute additional photons at the same wavelength, in phase, at the same polarization and heading in the same direction.

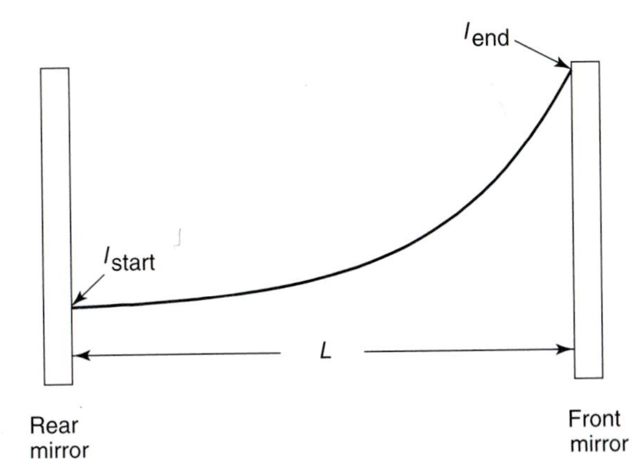

Figure 2.14 Schematic of a laser beam traveling in the resonant cavity. The laser beam begins at the rear mirror with an intensity I_{start}. Just before it is incident on the the front mirror, it has an intensity $I_{end} = I_{start}e^{\gamma(\nu)L}$.

Figure 2.14). The intensity I_{end} just before the ray is incident on the front mirror is expressed as

$$I_{end} = I_{start}e^{\gamma(\nu)L} \tag{2.10}$$

where $\gamma(\nu)$ is the gain coefficient and is given as

$$\gamma(\nu) = g(\nu)\left(\frac{A_{21}\lambda_o^2}{8\pi n^2}\right)(N_2 - N_1\left(\frac{g_2}{g_1}\right)) \tag{2.11}$$

where $g(\nu)$ is the gain profile (see Section 2.1), A_{21} is the Einstein coefficient for the transition (see Section 2.2.2), λ_o is the free-space wavelength of the laser transition, n is the index of refraction, N_2 and N_1 are the electron population densities for levels 2 and 1, respectively (which are determined by the pumping of the laser), and g_2 and g_1 are the *degeneracies* (the number of states at the same energy) for levels 2 and 1 (a property of the gain material).

Once the beam is incident on the front mirror of the resonant cavity, a certain amount of the light is reflected back into the resonant cavity and a certain amount is transmitted. The transmitted light,

$$I_t = T_F \cdot I_{start}e^{\gamma(\nu)L} = (1 - R_F) \cdot I_{start}e^{\gamma(\nu)L} \tag{2.12}$$

is the useful external light from the laser. The reflected light,

$$I_r = R_F \cdot I_{start}e^{\gamma(\nu)L} \tag{2.13}$$

remains inside the resonant cavity to stimulate more transitions (see Figure 2.15). Thus, a round-trip pass of the resonant cavity results in a final intensity (inside the resonant cavity) of

$$I_{rt} = I_{start} \cdot R_B R_F \cdot e^{2\gamma(\nu)L} \tag{2.14}$$

where R_F and R_B are the reflectances of the front and back mirrors.

The *multiplicative gain G* is often used to describe the gain process. G is simply given by the ratio of the ending versus starting intensities for either a *single pass* of the resonant cavity (starting at the rear mirror, exiting from the front),

$$G_{sp} = \frac{I_{end}}{I_{start}} = R_F e^{\gamma(\nu)L} \tag{2.15}$$

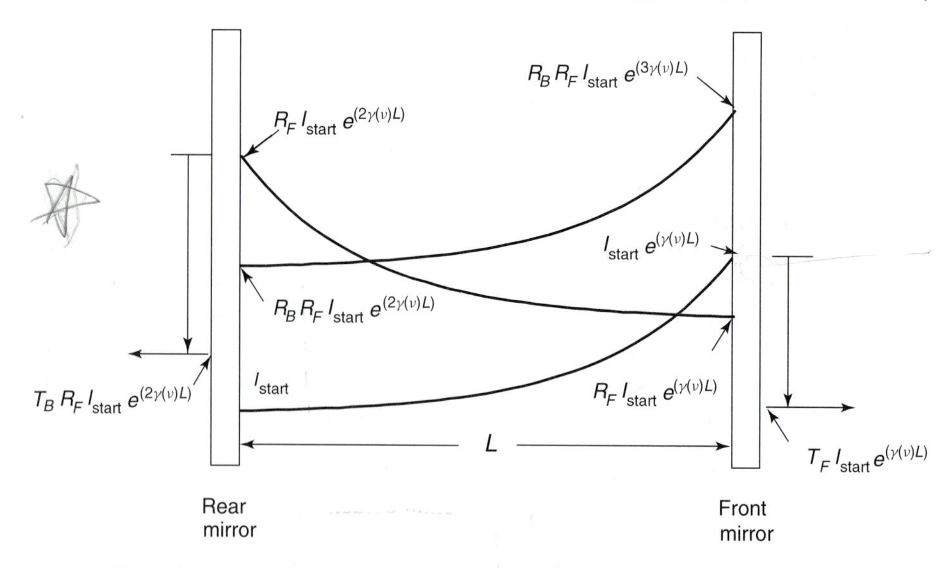

Figure 2.15 Schematic diagram of a laser beam after multiple reflections in the resonant cavity. The laser beam begins at the rear mirror with an intensity I_{start}. Just after it is incident on the front mirror, it has an intensity $R_F I_{start} e^{\gamma(\nu)L}$ inside the cavity and $T_F I_{start} e^{\gamma(\nu)L}$ outside the cavity. Just after it is incident on the rear mirror, it has an intensity $R_B R_F I_{start} e^{2\gamma(\nu)L}$ inside the cavity and $T_B R_F I_{start} e^{2\gamma(\nu)L}$ outside the cavity.

or a complete *round trip* of the resonant cavity,

$$G_{rt} = \frac{I_{end}}{I_{start}} = (R_B R_F)\, e^{2\gamma(\nu)L}. \tag{2.16}$$

The gain coefficient $\gamma(\nu)$ is often termed the *unsaturated* or *small signal* gain coefficient. The reason is that this exponential model for the gain process implicitly assumes that the upper-state population is not depleted. Of course, as the intensity in the cavity increases, the upper-state population *does* deplete and the gain begins to drop. This process is called *gain saturation* and is discussed in Chapter 5.

Example 2.1

Consider a HeNe laser system with a gain coefficient $\gamma = 0.010\ \text{cm}^{-1}$. Calculate the intensity of the laser beam after its first single pass between the resonant cavity mirrors, assuming that the starting beam was from the rear mirror. Calculate the intensity after 5 round trips. Assume that no saturation mechanisms exist, that the starting intensity is 1.0 pW, that the rear mirror is 100% reflecting, the front mirror is 98.5% reflecting and that the laser is 25.0 cm long.

Solution. The single pass multiplicative gain for this system is given as

$$G_{sp} = \frac{I_{end}}{I_{start}} = R_F e^{\gamma(\nu)L} = (0.985)\cdot e^{(0.010\ \text{cm}^{-1})\cdot(25.0\ \text{cm})} = 1.2648. \tag{2.17}$$

The intensity after the first pass is given by

$$I_{end} = I_{start} R_F e^{\gamma(\nu)L} = (1.0\cdot 10^{-12})\ \text{watt}\cdot(0.985)\cdot e^{(0.010\ \text{cm}^{-1})\cdot(25.0\ \text{cm})} \tag{2.18}$$

$$I_{end} = 1.0\cdot 10^{-12}\ \text{watt}\ \cdot(0.985)\cdot 1.284 = 1.2648\ \text{pW}. \tag{2.19}$$

The intensity after five round trips is given by

$$I_{\text{end}} = I_{\text{start}} (R_F R_B)^5 e^{2\cdot 5\cdot \gamma(\nu) L} = 1 \cdot 10^{-12} \text{ watt} \cdot (0.985 \cdot 1.0)^5 \cdot e^{(10)\cdot(0.010 \text{ cm}^{-1})\cdot(25.0 \text{ cm})} \quad (2.20)$$

$$I_{\text{end}} = 1 \cdot 10^{-12} \text{ watt } \cdot (0.985 \cdot 1.0)^5 \cdot (1.284)^{10} = 11.296 \text{ pW}. \quad (2.21)$$

2.2.2 Blackbody Radiation

The *blackbody radiation spectrum* is of great significance to the laser engineer. First, the historical development for the derivation of the blackbody radiation spectrum yields the *Einstein A and B coefficients.* These coefficients are critical for the calculation of laser gain. Second, blackbody radiation sources are routinely used as pump sources for laser systems. (See Sections 10.4 and 10.5 for an example of the use of blackbody noble-gas flashlamp sources as pumps for Nd:YAG lasers.)

Early in the 1900s, the great difference between the classical predictions for the blackbody radiation spectrum and the experimental evidence led a number of important scientists to study the problem of the blackbody radiation spectrum. Planck's approach in 1900 represented the first significant application of quantized energy states to a theoretical problem. Einstein's approach in 1917 was an alternative solution which led to the Einstein A and B coefficients. These two significant quantum mechanical developments are outlined in this section.

Quantum mechanics is a deep and widely diverse field. There are a number of outstanding quantum mechanics texts available for the reader who wishes to learn more. These range from delightful treatments for the layman,[12,13] through junior and senior college texts,[14,15,16] to advanced references.[17]

The classical approach. Blackbody radiation is the radiation produced by a hot object. As is well-known by potters who fire clay pots in a hot kiln, the color of the radiation from a hot dark object is not a function of the type of object, but only of the temperature. The peak wavelength emitted by a hot object is given by Wien's law as[18]

$$\lambda_{\text{peak}} = \frac{2.8978 \cdot 10^{-3} \text{ m} \cdot \text{K}}{T} \quad (2.22)$$

where T is the temperature in degrees K.

It turns out that a nearly ideal blackbody can be formed by making a small hole in a resonator. It is relatively easy to measure the radiation per unit wavelength $u(\lambda, T)\, d\lambda$

---

[12] John Gribbon, *Schrödinger's Kittens and the Search for Reality* (Boston, MA: Little, Brown and Company, 1995).

[13] P. C. W. Davis and J. R. Brown, *The Ghost in the Atom*, Canto ed. (Cambridge, MA: Cambridge University Press, 1993).

[14] R. H. Dicke and J. P. Wittke, *Introduction to Quantum Mechanics* (Reading, MA: Addison-Wesley, 1965).

[15] Stephen Gasiorowicz, *Quantum Physics* (New York: John Wiley and Sons, 1975).

[16] D. Park, *Introduction to Quantum Theory*, 2d ed. (New York: McGraw-Hill, 1974).

[17] J. J. Sakurai, *Modern Quantum Mechanics* (Redwood City, CA: Addison-Wesley, 1985).

[18] W. Wien and F. Harms, *Handbuch der Experimentallphysik*, Vol. 18, "Wellenoptik und Polarisation," Akadermische Verlagsgeselschaft, Leipzig, 1928.

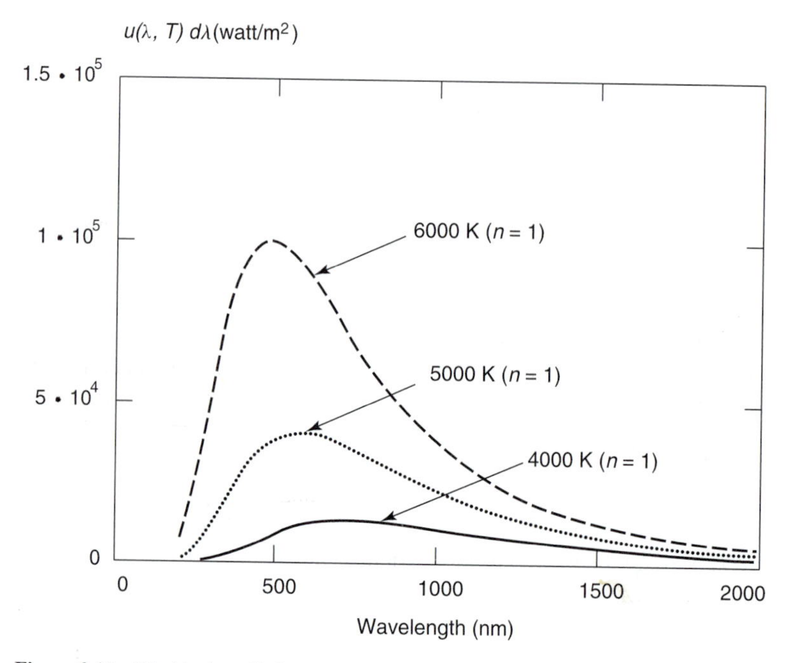

Figure 2.16 Blackbody radiation $u(\lambda, T)\, d\lambda$ as a function of wavelength for three different values of temperature. Notice that at higher temperatures the peak of the blackbody curve shifts downward in wavelength.

emitted from the hole as a function of wavelength, and this radiation takes the form illustrated in Figure 2.16.

Kirchhoff demonstrated that the emissive power *outside the cavity* per unit wavelength $u(\lambda, T)\, d\lambda$ is related to the blackbody energy density *inside the cavity* $\rho(\lambda, T)\, d\lambda$ by the simple relationship[19,20]

$$u(\lambda, T)\, d\lambda = \frac{\rho(\lambda, T)\, d\lambda \cdot c_o}{4}. \tag{2.23}$$

At first glance, it would seem to be straightforward to derive a classical expression for $\rho(\nu, T)$. The anticipated steps are as follows:

1. Since the blackbody radiation is an electromagnetic wave, basic electromagnetic theory can be used to determine the *electromagnetic mode density* in the resonator $N(\nu)\, d\nu$.
2. If electromagnetic modes correspond to the *degrees of freedom of a classical system*, then each mode should have the energy $k_B T$ where k_B is Boltzman's constant.[21]
3. Multiplying the density of modes by the energy per mode should yield the energy density $\rho(\nu)$.

---

[19] Eugene Hecht, *Optics*, 2d ed. (Reading, MA: Addison-Wesley, 1987), p. 540.

[20] Stephen Gasiorowicz, *Quantum Physics* (New York: John Wiley and Sons, 1975), p. 2.

[21] The equipartition theory predicts that the energy per degree of freedom should be $k_B T/2$. For an oscillator, the contribution of $k_B T/2$ from the kinetic energy is matched by a contribution of $k_B T/2$ from the potential energy.

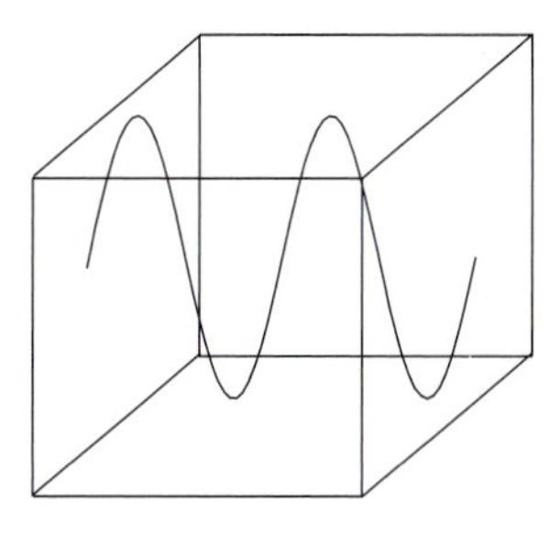

Figure 2.17 The frequency resonances of a rectangular electromagnetic resonator occur when each resonator dimension is an integral multiple of the half-wavelength for the Cartesian component in that direction.

Although the number of electromagnetic modes in a resonator is generally a function of the exact shape of the resonator, it is possible to derive a general expression for the mode density by assuming that the size of the resonator is much greater than the wavelength of the electromagnetic radiation. This derivation is given in some detail by Verdeyen[22] and Yariv,[23] so only the major points will be summarized here.

Recall that the magnitude of the resonant frequency of a rectangular electromagnetic resonator occurs when each resonator dimension is an integral multiple of the half-wavelength for the Cartesian component in that direction (see Figure 2.17). Mathematically this can be expressed in terms of frequency

$$|\nu| = \left(\nu_x^2 + \nu_y^2 + \nu_z^2\right)^{1/2} = \left(\left(\frac{m_x c_o}{2na_x}\right)^2 + \left(\frac{m_y c_o}{2na_y}\right)^2 + \left(\frac{m_z c_o}{2na_z}\right)^2\right)^{1/2} \tag{2.24}$$

where m_x, m_y, and m_z are integers, and a_x, a_y, and a_z are dimensions in the x, y, and z directions.

Now, consider a simple resonator with the dimensions $a_x = a_y = a_z = a$. If $m_x = m_y = m_z = m$, then the physical dimension a is related to the frequency projections ν_x, ν_y, and ν_z by

$$\nu_x = \nu_y = \nu_z = \frac{mc}{2a} = \frac{mc_o}{2na}. \tag{2.25}$$

Thus, the lowest-order frequency mode (where $m = 1$) has the projections

$$\nu_x = \nu_y = \nu_z = \frac{c_o}{2na} \tag{2.26}$$

and the minimum mode volume in frequency space is approximately given by

$$V_\nu = \nu_x \cdot \nu_y \cdot \nu_z = \left(\frac{c_o}{2na}\right)^3. \tag{2.27}$$

Now, consider the construction of frequency space with axes ν_x, ν_y, and ν_z as illustrated in Figure 2.18. To calculate the number of modes in frequency space, it is necessary to take the total available volume of frequency space and then divide by the volume of the minimum mode size in frequency space. Since frequency does not have a negative value, then the total available volume of frequency space only consists of that volume where

[22]Joseph T. Verdeyen, *Laser Electronics*, 2d ed. (Englewood Cliffs, NJ: Prentice-Hall, 1989), pp. 150–64.

[23]Amnon Yariv, *An Introduction to Theory and Applications of Quantum Mechanics* (New York: John Wiley and Sons, 1982), pp. 133–6.

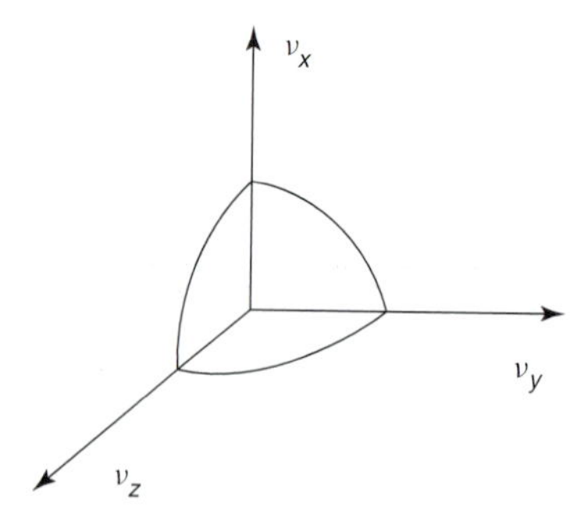

Figure 2.18 Frequency space with axes ν_x, ν_y, and ν_z. Since frequency does not have a negative value, then the total available volume of frequency space only consists of that volume where ν_x, ν_y, and ν_z are greater than zero.

ν_x, ν_y, and ν_z is greater than zero. This is 1/8 the total volume of a sphere or (assuming that $\nu = \nu_x = \nu_y = \nu_z$)

$$V_{\text{available}} = \frac{1}{8}\left(\frac{4}{3}\pi\nu^3\right) = \frac{\pi\nu^3}{6}. \tag{2.28}$$

The total number of modes in frequency space is then equal to the total mode volume divided by the volume per minimum mode, or

$$N = \frac{V_{\text{available}}}{V_\nu} = \frac{\pi\nu^3/6}{(c_o/2na)^3} = \frac{4\pi n^3\nu^3a^3}{3c_o^3}. \tag{2.29}$$

However, there are really two energy carrying modes per electromagnetic mode because both the tranverse electric (TE) and transverse magnetic (TM) polarizations can carry energy. (TE and TM notation is reviewed in Section A.5.) Thus, the final answer for the total number of modes is

$$N = 2\cdot\frac{4\pi n^3\nu^3a^3}{3c_o^3} = \frac{8\pi n^3\nu^3a^3}{3c_o^3}. \tag{2.30}$$

Now, in general it is more convenient to work with a mode density per unit volume in a small frequency interval $d\nu$. This is

$$\frac{1}{V}\cdot\frac{dN}{d\nu}\cdot d\nu = \frac{1}{a^3}\cdot\frac{3\cdot 8\pi n^3\nu^2a^3}{3c_o^3}\cdot d\nu = \frac{8\pi n^3\nu^2\,d\nu}{c_o^3} = \frac{8\pi n^3\,d\nu}{\lambda_o^2c_o} \tag{2.31}$$

(assuming that the index of refraction n does not vary with ν).

Now, in a classical electromagnetic system, each mode should have the energy k_BT (where k_B is Boltzman's constant). Thus, multiplying the energy per mode by the density of modes per unit volume should yield the energy density per unit volume in a small frequency interval $\rho(\nu)d\nu$ (as an aside, notice that the units of $\rho(\nu)d\nu$ are J/m^3 and the units of $\rho(\nu)$ are J-sec/m^3)

$$\rho(\nu)\,d\nu = k_BT\cdot\frac{N(\nu)d\nu}{a^3} = k_BT\left(\frac{8\pi n^3\nu^2}{c_o^3}\right)d\nu. \tag{2.32}$$

However, this expression goes to infinity at short wavelengths! (See Figure 2.19.)

Planck's approach to the blackbody radiation problem. After consideration of the mismatch between the classical theoretical expression, Equation (2.32), and experimental observations of the blackbody spectrum, Planck decided in 1900 that the classical

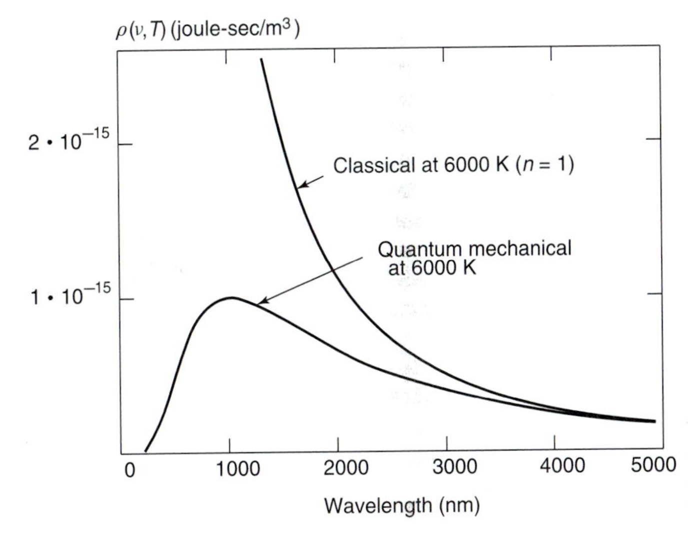

Figure 2.19 Comparison of the classical versus quantum mechanical expressions for $\rho(\nu, T)$. Notice that the classical result goes to infinity as the wavelength goes to zero.

assumption that each mode has the energy $k_B T$ must be incorrect.[24] Therefore, he constructed an averaging process for the energy that included the quantum mechanical limits on the mode energy.

In order to form this average, each state energy E_n must be weighted by the relative probability of occurrence and divided by the sum of the relative probabilities. The quantum states E_n, where $n = 1, 2, 3, \ldots$, have the energies $1h\nu$, $2h\nu$, $3h\nu$, ... The relative probability of each state is given by the Boltzman relations: $\exp(-h\nu/k_B T)$, $\exp(-2h\nu/k_B T)$, $\exp(-3h\nu/k_B T)$, ... Thus, the average state energy $\langle E \rangle$ is[25]

$$\langle E \rangle = \frac{h\nu e^{(-h\nu/k_B T)} + 2h\nu e^{(-2h\nu/k_B T)} + 3h\nu e^{(-3h\nu/k_B T)} + \cdots}{e^{(-h\nu/k_B T)} + e^{(-2h\nu/k_B T)} + e^{(-3h\nu/k_B T)} + \cdots} \tag{2.33}$$

It turns out that this series has an exact sum given as[26]

$$\langle E \rangle = \frac{h\nu}{e^{(h\nu/k_B T)} - 1} \tag{2.34}$$

which describes the average energy of the mode, assuming that the electromagnetic energy is quantized in packets of $h\nu$.

So, returning to the previous calculation, multiplying the average energy per mode $\langle E \rangle$ by the density of modes should yield the energy density per unit volume in a small frequency interval $\rho(\nu)d\nu$

$$\rho(\nu)\, d\nu = \frac{h\nu}{e^{(h\nu/k_B T)} - 1} \cdot \left(\frac{8\pi n^3 \nu^2}{c_o^3} \right) d\nu. \tag{2.35}$$

[24] M. Planck, *Verh. d. deutsch phys. Ges.*, 2:202 and 2:237 (1900); and *Ann. d. Physik* (4), 4:553 (1901).

[25] Stephen Gasiorowicz, *Quantum Physics* (New York: John Wiley and Sons, 1975) p. 6.

[26] M. Planck, *Verh. d. deutsch phys. Ges.*, 2:202 and 2:237 (1900); and *Ann. d. Physik* (4), 4:553 (1901).

Rewriting this slightly gives

$$\rho(\nu)\, d\nu = \frac{8h\pi n^3}{c_o^3}\left(\frac{\nu^3}{e^{(h\nu/k_B T)} - 1}\right) d\nu \tag{2.36}$$

which is the blackbody energy density inside the cavity per unit frequency.

Recall from Chapter 1 that

$$d\nu = \frac{c_o}{\lambda_o^2}\, d\lambda \tag{2.37}$$

so, Equation (2.36) can be rewritten as

$$\rho(\lambda, T)\, d\lambda = \frac{8h\pi n^3}{\lambda_o^3}\left(\frac{1}{e^{(h\nu/k_B T)} - 1}\right) \frac{c_o}{\lambda_o^2}\, d\lambda. \tag{2.38}$$

Now, remember from Equation (2.23) that

$$u(\lambda, T)\, d\lambda = \frac{\rho(\lambda, T)\, d\lambda \cdot c_o}{4}. \tag{2.39}$$

Thus, substituting in from Equation (2.38) gives

$$u(\lambda, T)\, d\lambda = \frac{c_o}{4}\left(\frac{8h\pi n^3}{\lambda_o^3}\left(\frac{1}{e^{(h\nu/k_B T)} - 1}\right) \frac{c_o}{\lambda_o^2}\, d\lambda\right) \tag{2.40}$$

and the radiation ($u(\lambda, T)d\lambda$) emitted by the blackbody outside the cavity per unit wavelength is can be written as[27]

$$u(\lambda, T)d\lambda = \left(\frac{2h\pi n^3 c_o^2}{\lambda_o^5}\left(\frac{1}{e^{(h\nu/k_B T)} - 1}\right) d\lambda\right). \tag{2.41}$$

Example 2.2

Calculate and plot $u(\lambda, T)\, d\lambda$ for the Sun. Assume that the peak of the solar spectrum is at 550 nm and that the index of refraction $n = 1$.

Solution. Wien's law given in Equation (2.22) can be used to find the temperature corresponding to a peak color of 550 nm as

$$T = \frac{2.8978 \cdot 10^{-3}\ \text{m} \cdot \text{K}}{550 \cdot 10^{-9}\ \text{m}} = 5268.727\ \text{K}. \tag{2.42}$$

This temperature can then be used in equation 2.41

$$u(\lambda, T)\, d\lambda = \left(\frac{2h\pi n^3 c_o^2}{\lambda_o^5}\left(\frac{1}{e^{(h\nu/k_B T)} - 1}\right) d\lambda\right) \tag{2.43}$$

to generate the plot in Figure 2.20.

[27] Jay Snell, "Radiometry and Photometry," in *Handbook of Optics*, eds W. G. Driscoll and W. Vaughan, sponsored by the Optical Society of America (New York: McGraw Hill, 1978), pp. 1–13.

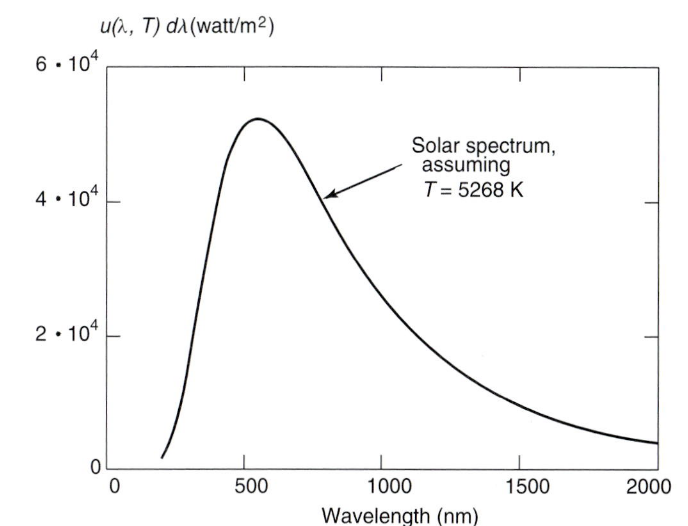

Figure 2.20 A plot of $u(\lambda, T)d\lambda$ for the solar spectrum.

Einstein's approach to the blackbody radiation problem. The next chronological step in solving the blackbody radiation problem was of immediate and practical interest to the laser field. In 1917, Einstein developed an alternative way of working the blackbody problem that also yielded the *spontaneous* and *stimulated emission rates*. These rates are critical in the calculation of laser gain and pumping requirements. Einstein's original paper, entitled "On the Quantum Theory of Radiation," provides an outstanding (and very readable) summary of his contributions to the blackbody problem.[28]

Einstein's basic idea was to identify three major transition rates (change in the number of atoms per unit time, dN/dt), the *spontaneous emission rate* (A_{21}), the *stimulated emission rate* (B_{21}), and the *induced absorption rate* (B_{12}).

In equilibrium there must be a *detailed balance* between processes transferring atoms from 1 to 2 and processes transferring atoms from 2 to 1 (see Figure 2.21). The detailed balance equation can be expressed as:

$$g_2 A_{21} e^{(-h\nu_2/k_B T)} + g_2 B_{21} e^{(-h\nu_2/k_B T)} \rho(\nu) = g_1 B_{12} e^{(-h\nu_1/k_B T)} \rho(\nu) \tag{2.44}$$

The first term ($g_2 A_{21} e^{(-h\nu_2/k_B T)}$), models *spontaneous* emission from the upper state. This term is the product of the degeneracy[29] of the upper state (g_2), the spontaneous rate at which electrons leave the upper state to go to the lower state (A_{21}), and the relative probability of being in the upper state ($e^{(-h\nu_2/k_B T)}$). The second term models *stimulated* emission from the upper state. The term incorporates the product of the degeneracy of the upper state (g_2), the stimulated rate at which electrons leave the upper state to go to the lower state (B_{21}), the relative probability of being in the upper state ($e^{(-h\nu_2/k_B T)}$), and the stimulating radiation field ($\rho(\nu)$). The two terms modeling the population *leaving* the upper state are equated to

[28] A. Einstein, "On the Quantum Theory of Radiation," *Physikalische Zeitschrift* 18:121 (1917), reprinted in *Laser Theory*, ed Frank S. Barnes (New York: IEEE Press, 1972), pp. 5–21; and in Elmsford, *The Old Quantum Theory* (New York: Pergamon, 1967), pp. 167–183.

[29] The degeneracy of a state is an integer describing the number of identical states at exactly the same energy. The use of g for the degeneracy should not be confused with the use of $g(\nu)$ for the gain profile.

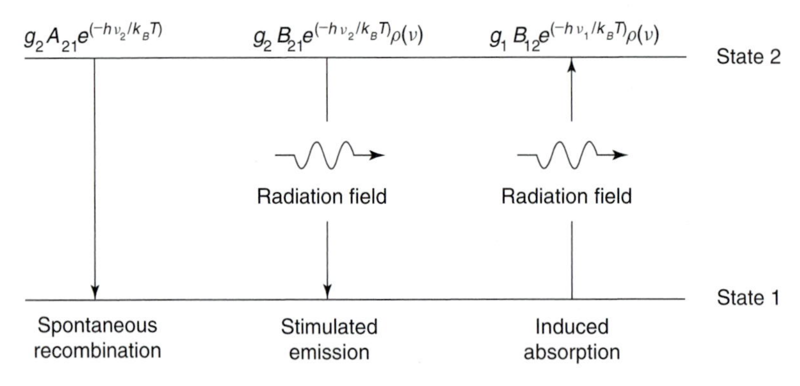

Figure 2.21 In equilibrium there must be a *detailed balance* between processes transferring atoms from 1 to 2 and processes transferring atoms from 2 to 1. This detailed balance includes the action of the stimulating radiation field $\rho(\nu)$.

a third term modeling population *entering* the upper state. The third term incorporates the product of the degeneracy of the lower state (g_1), the stimulated rate at which electrons go from the lower state to the upper state (B_{12}), the relative probability of being in the lower state ($e^{(-h\nu_1/k_B T)}$), and the stimulating radiation field ($\rho(\nu)$).

If we assume that $\rho(\nu)$ goes to infinity as T goes to infinity, then the boundary conditions at $T \to \infty$ are

$$g_2 B_{21} \rho(\nu) = g_1 B_{12} \rho(\nu). \tag{2.45}$$

Thus,

$$g_2 B_{21} = g_1 B_{12}. \tag{2.46}$$

With this, then $\rho(\nu)$ takes the form

$$\rho(\nu) = \frac{g_2 A_{21}}{g_1 B_{12} e^{((h\nu_2 - h\nu_1)/k_B T)} - g_2 B_{21}} = \frac{A_{21}}{B_{21}} \left(\frac{1}{e^{((h\nu_2 - h\nu_1)/k_B T)} - 1} \right). \tag{2.47}$$

Comparing this with the Planck formula, Equation (2.36) gives the additional result

$$\rho(\nu) d\nu = \frac{A_{21}}{B_{21}} \left(\frac{1}{e^{((h\nu_2 - h\nu_1)/k_B T)} - 1} \right) d\nu = \frac{8h\pi n^3}{c_o^3} \left(\frac{\nu^3}{e^{(h\nu/k_B T)} - 1} \right) d\nu \tag{2.48}$$

or

$$\frac{A_{21}}{B_{21}} = \frac{8h\pi n^3 \nu^3}{c_o^3}. \tag{2.49}$$

This derivation reproduces the Planck formula as well as relates the values for the various transition rates to known constants. In addition (since the emission rate from the upper state is the inverse of the lifetime of the state[30]) the derivation also yields the spontaneous lifetime for the upper state as $\tau_S = 1/A_{21}$.

[30]For a rigorous derivation of the inverse relation between the Einstein A_{21} coefficients and the spontaneous upper-state lifetime τ_S, see Joseph T. Verdeyen, *Laser Electronics*, 2d ed. (Englewood Cliffs, NJ: Prentice-Hall, 1989), pp. 160–164.

Example 2.3

Calculate the stimulated emission rate B_{21} for a transition at 632.8 nm, an average index of refraction of 1.0, and a spontaneous lifetime $\tau_S = 1/A_{21} = 0.30 \cdot 10^{-6}$ sec.

Solution. This uses Equation (2.49) with ν determined by the frequency of the transition. Evaluating for this case gives

$$\frac{A_{21}}{B_{21}} = \frac{8h\pi n^3 \nu^3}{c_o^3} \rightarrow B_{21} = \frac{A_{21} c_o^3}{8h\pi n^3 \nu^3} = \frac{\lambda_o^3}{8h\pi n^3 \tau_S} \tag{2.50}$$

$$B_{21} = \frac{\left(632.8 \cdot 10^{-9} \text{ m}\right)^3}{8h\pi \cdot 1.0^3 \cdot 0.30 \cdot 10^{-6} \text{ sec}} \tag{2.51}$$

$$B_{21} = 5.072 \cdot 10^{19} \text{m}^3/\text{J-sec}^2. \tag{2.52}$$

(Comment: Remember that the units of $B_{21}\rho(\nu)$ are 1/sec and the units of $\rho(\nu)$ are J-sec/m^3.)

2.2.3 Gain

In this section, the gain coefficient first presented in Section 2.2.1 will be derived using the Einstein A and B coefficients of Section 2.2.2. This derivation is not only interesting, it also yields alternative forms of the gain equation frequently used by vendors to specify the properties of laser materials.

Consider a laser resonator of length Δz experiencing gain. The change in intensity ΔI as a function of length can be calculated from the number of electrons in the upper state stimulated to fall to the lower state $N_2 B_{21} \rho(\nu)$, minus the number of electrons stimulated from the lower state to the upper state, $N_1 B_{12} \rho(\nu)$, multiplied by the energy of the transitional photon $h\nu$,

$$\frac{\Delta I}{S} \cdot \frac{1}{\Delta z} = (N_2 B_{21} \rho(\nu) - N_1 B_{12} \rho(\nu))\, h\nu \tag{2.53}$$

where N_2 and N_1 are population densities (atoms/m^3), $\rho(\nu)$ is the energy density per unit frequency of the laser beam (J-sec/m^3), and S is the cross-sectional area of the laser beam.

If the finite linewidths are included, the function $\rho(\nu)$ is replaced by $\rho \cdot g(\nu)$ where ρ is a constant describing the energy density (J/m^3) and $g(\nu)$ is the gain profile (1/Hz = sec).

$$\frac{\Delta I}{S} \cdot \frac{1}{\Delta z} = (N_2 g(\nu) B_{21} \rho - N_1 g(\nu) B_{12} \rho)\, h\nu \tag{2.54}$$

Now, recall that

$$g_2 B_{21} = g_1 B_{12} \tag{2.55}$$

so

$$\frac{\Delta I}{S} \cdot \frac{1}{\Delta z} = \left(N_2 g(\nu) B_{21} \rho - N_1 g(\nu) \left(\frac{g_2}{g_1}\right) B_{21} \rho\right) h\nu \tag{2.56}$$

or

$$\frac{\Delta I}{S} \cdot \frac{1}{\Delta z} = \left(g(\nu) B_{21} \rho \left(N_2 - N_1 \left(\frac{g_2}{g_1}\right)\right)\right) h\nu. \tag{2.57}$$

Since

$$\frac{A_{21}}{B_{21}} = \frac{8h\pi n^3 \nu^3}{c_o^3} \tag{2.58}$$

then

$$B_{21} = \frac{A_{21} c_o^3}{8h\pi n^3 \nu^3} \tag{2.59}$$

and

$$\frac{\Delta I}{S} \cdot \frac{1}{\Delta z} = \left(g(\nu) \frac{A_{21} c_o^3}{8h\pi n^3 \nu^3} \rho \left(N_2 - N_1 \left(\frac{g_2}{g_1} \right) \right) \right) h\nu. \tag{2.60}$$

The intensity per unit area of the laser beam (I/S = watts/m^2 = J/m^2-sec) is equal to the energy density ρ (J/m^3) times the velocity of the electromagnetic wave in the material c (m/sec),

$$\rho = \frac{I}{S \cdot c} = \frac{I \cdot n}{S \cdot c_o}, \tag{2.61}$$

Substituting into Equation (2.60) gives

$$\frac{\Delta I}{S} \cdot \frac{1}{\Delta z} = \left(\frac{g(\nu) \cdot n \cdot I}{S \cdot c_o} \left(\frac{A_{21} c_o^3}{8h\pi n^3 \nu^3} \right) \left(N_2 - N_1 \left(\frac{g_2}{g_1} \right) \right) \right) h\nu \tag{2.62}$$

or

$$\frac{\Delta I}{S} \cdot \frac{1}{\Delta z} = \left(\frac{g(\nu) I}{S} \left(\frac{A_{21} c_o^2}{8\pi n^2 \nu^2} \right) \left(N_2 - N_1 \left(\frac{g_2}{g_1} \right) \right) \right) \tag{2.63}$$

or, as more commonly written,

$$\frac{\Delta I}{\Delta z} = \left(g(\nu) \left(\frac{A_{21} \lambda_o^2}{8\pi n^2} \right) \left(N_2 - N_1 \left(\frac{g_2}{g_1} \right) \right) \right) I. \tag{2.64}$$

The gain coefficient $\gamma(\nu)$ (cm^{-1} or m^{-1}) is defined as

$$I_{\text{end}} = I_{\text{start}} e^{\gamma(\nu) z} \tag{2.65}$$

so

$$\frac{\Delta I_{\text{end}}}{\Delta z} \approx \frac{d I_{\text{end}}}{dz} = \gamma(\nu) I_{\text{start}} e^{\gamma(\nu) z} = \gamma(\nu)\, I_{\text{end}} \tag{2.66}$$

therefore

$$\boxed{\gamma(\nu) = g(\nu) \left(\frac{A_{21} \lambda_o^2}{8\pi n^2} \right) (N_2 - N_1 (g_2/g_1))\,.} \tag{2.67}$$

This is a very important formula! First, it clearly shows the need for a population inversion $N_2 > N_1(g_2/g_1)$ in order to obtain laser gain. Second, it provides a quantitative gain coefficient $\gamma(\nu)$ for the gain in an arbitrary laser material as a function of the gain profile $g(\nu)$, the Einstein coefficient for the transition A_{21} (which is a property of the gain material), the free-space wavelength of the laser transition λ_o, the index of refraction n, the electron population densities for levels 2 and 1, N_2 and N_1 (which are determined by the pumping of the laser), and the degeneracies for levels 2 and 1, g_2 and g_1 (again a property of the gain material).

Now, the spontaneous transition time (spontaneous lifetime) τ_s is equal to $1/A_{21}$. Thus the gain coefficient is often seen in the form

$$\gamma(\nu) = g(\nu)\left(\frac{\lambda_o^2}{\tau_s 8\pi n^2}\right)\left(N_2 - N_1\left(\frac{g_2}{g_1}\right)\right). \tag{2.68}$$

The *stimulated emission cross-section* σ_{21} is very commonly used in laser engineering to characterize laser materials[31]. The stimulated emission cross-section is defined by

$$\gamma(\nu) = \sigma_{21}(\nu) \cdot \Delta N. \tag{2.69}$$

The term $\Delta N = (N_2 - N_1(g_2/g_1))$ so

$$\sigma_{21}(\nu) = g(\nu)\left(\frac{A_{21}\lambda_o^2}{8\pi n^2}\right) = g(\nu)\left(\frac{\lambda_o^2}{\tau_s 8\pi n^2}\right) \tag{2.70}$$

and another form for $\gamma(\nu)$ is

$$\gamma(\nu) = \sigma_{21}(\nu)\left(N_2 - N_1\left(\frac{g_2}{g_1}\right)\right). \tag{2.71}$$

Still another form (one often used by vendors) is the stimulated emission cross-section at the center of the atomic transition. For a Lorentzian lineshape this is given as

$$\sigma_{21} = \frac{A_{21}\lambda_o^2}{4\pi^2 n^2 \Delta\nu} \tag{2.72}$$

and for a Gaussian lineshape as

$$\sigma_{21} = \frac{A_{21}\lambda_o^2}{4\pi n^2 \Delta\nu}\left(\frac{\ln 2}{\pi}\right)^{1/2} \tag{2.73}$$

where $\Delta\nu$ is the frequency linewidth of the transition (see Section 2.1 for more information on these types of lineshapes).

Example 2.4

Calculate the gain coefficient of a HeNe laser[32] assuming that $\sigma = 6.5 \cdot 10^{-13}$ cm^2 at the center of the gain curve, the population difference $N_2 - N_1 = 5 \cdot 10^9$ atoms/cm^3, the lower-state population density is negligible, and $g_1 = g_2 = 1$.

Solution. The gain coefficient at line center is given by Equation (2.71)

$$\gamma(\nu) = \sigma(\nu)\left(N_2 - N_1\left(\frac{g_2}{g_1}\right)\right) \tag{2.74}$$

where $N_2 - N_1 = 5 \cdot 10^9$ atoms/cm^3 and $g_1 = g_2 = 1$. Evaluating gives

$$\gamma(\nu) = 6.5 \cdot 10^{-13}\text{cm}^2 \cdot 5 \cdot 10^9 \text{ atoms/cm}^3 = 0.00325 \text{ cm}^{-1}.$$

---

[31] Walter Koechner, *Solid State Laser Engineering*, 3d ed. (Berlin, Germany: Springer-Verlag, 1992), p. 15.

[32] Data values from William Silvest, "Lasers," in *Handbook of Optics, Vol. I, Fundamental Techniques and Designs* (New York: McGraw Hill, 1995), p. 11.8, Table I.

SYMBOLS USED IN THE CHAPTER

E_n: General state energies (joules or eV)
δt: Change in time (sec)
$\delta \omega$: Change in frequency (rad/sec)
δE: Change in energy (joules or eV)
b: Width of a wavepacket in time (sec)
ω: Frequency (rad/sec)
$g(\nu)$: Function for the normalized gain profile (1/Hz = sec)
ν: General frequency (Hz)
ν_o: Center frequency of a transition (Hz)
$\Delta\nu$: Frequency linewidth of a transition at full width half maximum (FWHM) (Hz)
$\gamma(\nu)$: Gain coefficient as a function of frequency (cm^{-1} or m^{-1})
L: Laser resonator length (cm or m)
I_{start}, I_{end}, I_r, I_{rt}, I_t: Various light intensities (watts)
A_{12}, A_{21}: Spontaneous Einstein A coefficients (1/sec)
N_2, N_1: Upper and lower state populations (atoms/cm^3)
λ_o: Free space wavelength of light (m)
n: Real index of refraction (unitless)
g_2, g_1: Upper and lower state degeneracies (unitless)
R_F, R_B: Front and back mirror reflectances, between 0 and 1
G: General multiplicative gain (unitless)
G_{sp}, G_{rt}: Single pass and round trip gains (unitless)
λ_{peak}: Wavelength of the radiation peak from a hot object (cm or m)
T: Temperature (degrees K)
λ: General wavelength (cm or m)
$\mu(\lambda, T)\,d\lambda$: Spectral excitance outside the cavity in a small wavelength interval (watts/m^2)
$\rho(\lambda, T)$: Blackbody energy density inside the cavity in a small wavelength interval (J-sec/m^3)
ρ, $\rho(\nu)$, $\rho(\nu, T)$: Energy density inside a cavity in a small frequency interval (J-sec/m^3)
m_x, m_y, m_z: Mode integers along various Cartesian directions (unitless)
ν_x, ν_y, ν_z: Frequency projections along various Cartesian directions (Hz)
a_x, a_y, a_z: Resonator dimensions along various Cartesian directions (cm or m)
a: General resonator dimension (cm or m)
$V_{\text{available}}$, V_ν: Available and minimum mode volumes in frequency space (Hz^3)
N: Number of modes in frequency space (unitless)
V: Volume (cm^3 or m^3)
$\langle E \rangle$: Average energy (joules or eV)
B_{12}, B_{21}: Stimulated Einstein "B" coefficients (m^3/J-sec^2 or cm^3/J-sec^2)

τ_S : Spontaneous transition lifetime $\tau_S = 1/A_{21}$ (sec)
ΔI: Change in light intensity (watts or watts/cm^3)
Δz: Change in distance (cm or m)
S: Cross-sectional area of the laser beam (cm^2 or m^2)
$\sigma_{21}(\nu)$: Cross-section between states 2 and 1 (cm^2 or m^2)

EXERCISES

Energy states and lineshapes

2.1 Consider the relationships for Lorentzian and Gaussian lineshapes given in the text. Using a computer if possible, plot the Lorentzian against the Gaussian assuming that $\Delta\nu = 1.5$ GHz and that the laser is an argon-ion with a center linewidth of $\lambda_o = 514.5$ nm. Both plots should be normalized to the gain at line center of the Gaussian line. (This means that the gain at the center of the Gaussian line should be 1.) Your plot should have wavelength on the x-axis and normalized units on the y-axis. Discuss the similarities and differences between the lineshapes. Is there an easy way to distinguish between Lorentzian and Gaussian lines?

Gain—Basics of gain

2.2 Consider a laser system with a gain coefficient of $\gamma = 0.015$ cm^{-1}. Assume that a ray of the intensity 1 pW is injected into the cavity immediately in front of the rear mirror. Assume that the resonator is 50 cm long, that no additional rays are propagating in the resonator, and that no saturation mechanisms exist. Assume that both mirrors have a reflectivity of R = 100%.

(a) Calculate the intensity after six single passes (three round trips) of the resonator.

(b) Assume the front mirror is 80% reflecting and calculate the intensity after six single passes (three round trips) of the resonator.

(c) Assume the front mirror is 60% reflecting and calculate the intensity after six single passes (three round trips) of the resonator.

2.3 (design) Consider the gain of a laser. Is it always desirable to have high gain? Speculate on three circumstances for which it might be desirable to have a lower gain. Your answer should include a brief description of each of the three circumstances and any required supporting calculations.

2.4 Consider a laser system with a gain coefficient of $\gamma = 0.015$ cm^{-1}. Assume that a ray of the intensity 15 pW is injected into the cavity immediately in front of the rear mirror. Assume that the resonator is 50 cm long, that no additional rays are propagating in the resonator, and that no saturation mechanisms exist. Assume that both mirrors have a reflectivity of $R = 100\%$.

(a) Using a computer if possible, plot the intensity at each mirror of the resonator as a function of the number of single passes. Your plot should have the number of single passes (N) on the x-axis and intensity on the y-axis.

(b) Assume the front mirror is 80% reflecting and plot the intensity at each mirror of the resonator as a function of the number of single passes. Your plot should have the number of single passes (N) on the x-axis and intensity on the y-axis.

(c) Assume the front mirror is 60% reflecting and plot the intensity at each mirror of the resonator as a function of the number of single passes. Your plot should have the number of single passes (N) on the x-axis and intensity on the y-axis.

2.5 Consider a HeNe laser system with a gain coefficient of $\gamma = 0.01$ cm^{-1}. Assume that a single photon (at 632.8 nm) is injected into the cavity immediately in front of the rear mirror. Assume that the resonator is 50 cm long, that the pulse generated by this photon is the only one propagating in the resonator, and that no saturation mechanisms exist. Assume that both mirrors have a reflectivity of R = 100%.

(a) Calculate the energy of the pulse after 30 single passes of the resonator.

(b) Calculate the average power of the pulse after 30 single passes of the resonator.

Gain—Blackbody radiation

2.6 Assume a certain laser flashlamp has a peak spectral response at 400 nm. Determine the effective temperature of the lamp. Using a computer if possible, plot the blackbody response assuming that temperature. The x-axis of your plot should be wavelength over a range of 0 nm to 2 μm. The y-axis of your plot should be $u(\lambda, T)$ in units of watt/m^3. Assume the lamp is an ideal blackbody. What qualitative observations can you make about laser flashlamps from the plot?

2.7 (design) The derivation in Section 2.2.2 assumes a cubical resonator of the dimensions $a \times a \times a$. Choose some other shape of resonator and repeat the derivation for the classical expression for $\rho(\nu)d\nu$. Compare your result with the textbook answer in Equation (2.32). Can you speculate on a resonator shape that would make the classical expression more like the experimentally observed answer? Explain why or why not.

2.8 Assume you have a number of laser flashlamps with different spectral response peaks. Using a computer if possible, create a plot that relates the spectral response peak of the lamp to its temperature. Assume the lamps are ideal blackbodies. The x-axis of your plot should be wavelength over a range of 300 nm to 800 nm. The y-axis of your plot should be temperature in degrees kelvin. What qualitative observations can you make about laser flashlamps from the plot?

2.9 (design) Consider the average energy expression given in Equation (2.33). Say that you are a scientist living in the year 1900. Assume that your research area is blackbody radiation. Develop an alternative method for averaging the quantum mechanical state energies other than that given by Equation (2.33). Repeat the derivation for Equation (2.36) using your averaging technique. Your answer should include

(a) your proposed averaging method,

(b) a description of the technical justification for your method,

(c) your version for Equation (2.36), and

(d) a brief description of how your method compares with Planck's version.

2.10 Consider the classical [Equation (2.32)] and quantum mechanical [Equation (2.36)] expressions for energy density. Using a computer if possible, plot both expressions on the same graph as a function of frequency. The x-axis of your plot should be frequency over a range of 0 to 500 THz. The y-axis of your plot should be the energy density $\rho(\nu, T)$. At what value of wavelength do the expressions begin to diverge? Explain the underlying physics and mathematics for the divergence.

2.11 Consider the question of the emission from hot laser flashlamps. Using a computer if possible, plot the quantum mechanical $u(\lambda, T)$ (outside the flashlamp) for three different flashlamps. One flashlamp is at a temperature of 2000 K, the second at 3000 K, and the third at 4000 K. The x-axis of your plot should be wavelength over a range of 0 nm to 4 μm. The y-axis of your plot should be $u(\lambda, T)$ in units of watt/m^3. Assume the lamps are ideal blackbodies and $n = 1$. What qualitative observations can you make about laser flashlamps from the plot?

2.12 Calculate the stimulated emission rate for an argon-ion transition at $\lambda_o = 488$ nm, where $n = 1.5$, $d\nu = 10$ kHz, and the spontaneous lifetime is 200 μs.

2.13 Equations (2.7) and (2.9) describe the gain at the center of a Lorentzian or Gaussian line. Derive these equations.

2.14 (design) Consider the equations for the gain coefficient and cross-section [Equations (2.68) and (2.70)]. Assume that you want to determine the maximum possible realistic gain coefficient and its associated cross-section. Estimate the maximum (or minimum) realistic value for each term in the equation and justify your choice. (For example, $n = 1$ is the lowest reasonable index of refraction, since $n = 1$ is vacuum!) Using these terms, calculate the maximum realistic gain and maximum realistic cross-section. Compare these with the values for real lasers. Your answer should include three parts:

(a) the values for the various terms and a brief justification for each,
(b) the calculated maximum realistic gain and cross-section, and
(c) a brief discussion and comparison of these values with those of other real lasers.

Gain—Gain and lineshapes

2.15 Consider a HeNe laser with a laser wavelength of 632.8 nm and a linewidth of 1.5 GHz. Assume the length of the gain region is 30 cm long, the index of refraction of the cavity is 1.0, the spontaneous upper-state lifetime is 20 ms, the population in the upper-level laser state is $1 \cdot 10^{16}$ atoms/cm^3, and the population in the lower laser state is $5 \cdot 10^{14}$ atoms/cm^3. Assume that the degeneracy g_2 of the upper state is 3, and of the lower state g_1 is 5. Assume the line is Gaussian.

(a) Calculate the cross-section at line center.
(b) Calculate γ at line center.
(c) Calculate the single pass gain, G_{sp}.

2.16 Consider an argon-ion laser with a laser wavelength of 514.5 nm and a linewidth of 1.5 GHz. Assume the length of the gain region is 75 cm long, the index of refraction of the cavity is 1.0, the spontaneous upper-state lifetime is 300 μs, the population in the upper-level laser state is $1 \cdot 10^{16}$ atoms/cm^3, and the population in the lower laser state is $5 \cdot 10^{14}$ atoms/cm^3. Assume that the degeneracy g_2 of the upper state is 3, and of the lower state g_1 is 5. Assume the line is Lorentzian.

(a) Calculate the cross-section at line center.
(b) Calculate γ at line center.
(c) Calculate the single pass gain, G_{sp}.

$\lambda = 632.8$, $\Delta\nu = 1.5$ GHz, $L = 30$ cm $n = 1.0$

$t_s = 1/A_{21} = 20$ ms

$N_2 B_{21} \rho(\nu) = 1 \cdot 10^{16}$ atoms/cm3

$N_1 B_{12} \rho(\nu) = 5 \cdot 10^{14}$ atoms/cm3

$g_2 = 3$ $g_1 = 5$ $g_2 B_{21} = g_1 B_{12}$

3 The Fabry-Perot Etalon

Objectives

- To compute the change in frequency of a longitudinal mode as the cavity length of a laser is changed.
- To describe and sketch how an etalon is used to select out a single longitudinal mode from a laser.
- To sketch the approximate transmittance and reflectance spectrum of an etalon given its fabrication parameters.
- To calculate and plot the exact transmittance and reflectance spectrum of an etalon given its fabrication parameters.
- To compute the quality factor Q and the finesse F of an etalon from its fabrication parameters.
- To calculate the etalon coating reflectances, given the transmittance or reflectance spectrum of the etalon.

3.1 LONGITUDINAL MODES IN THE LASER RESONANT CAVITY

Recall from Chapter 1 that a laser can only lase on those longitudinal modes for which an integral multiple of half-wavelengths fit in the laser cavity. The operating frequency spectrum of the laser is the intersection between the set of possible longitudinal modes and the gain profile of the laser (see Figure 1.17 of Chapter 1). The frequencies of the longitudinal modes are a function of the laser cavity length, and small variations in the length of the cavity can result in large variations in the frequency location of the mode.

Example 3.1

Consider how much the cavity length of an argon-ion laser must change in order for a particular longitudinal mode to change by a frequency of one-half of the spacing between the modes (see Figure 3.1). Assume the argon-ion has a cavity length of 75 cm and is operating with a center wavelength of 514.5 nm.

Solution. The frequencies of the longitudinal modes are given by

$$\nu = \frac{p \cdot c_o}{2nL} \quad \text{where } p = 1, 2, 3, \ldots \tag{3.1}$$

Thus, the mode closest to the center wavelength is

$$p = \text{int}\left(\frac{2nL}{\lambda_o}\right) = \text{int}\left(\frac{2 \cdot 75 \text{ cm}}{514.50 \cdot 10^{-9} \text{ m}}\right) = 2915451 \tag{3.2}$$

where p is the mode number and int means to take the closest integer.

This may be checked by calculating the frequency of the mode

$$\nu = p \cdot \frac{c_o}{2nL} = p \cdot \frac{c_o}{2nL} = 582.687 \text{ THz} \tag{3.3}$$

and by calculating the wavelength of the mode

$$\lambda = \frac{c_o}{\nu} = \frac{2nL}{p} = 514.500158 \text{ nm}. \tag{3.4}$$

If an observer is tracking this longitudinal mode as the cavity length changes, then the mode number p will remain constant while the cavity length L changes.

If the cavity length has changed by an amount ΔL, then the new mode frequency is, from Equation (3.4),

$$\nu + \Delta\nu = p \cdot \frac{c_o}{2n(L + \Delta L)}. \tag{3.5}$$

Inverting this equation gives the change in length as a function of the change in frequency

$$(L + \Delta L) = p \cdot \frac{c_o}{2n(\nu + \Delta\nu)}. \tag{3.6}$$

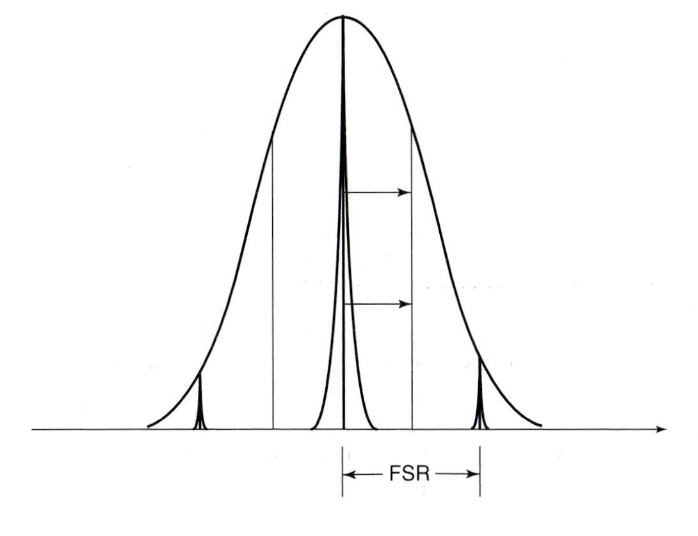

Figure 3.1 Consider how much the cavity length of an argon-ion laser must change in order for a particular longitudinal mode to move one-half of the free spectral range up in frequency.

Now, consider the frequency change of 1/2 of the original spacing between modes (1/2 of the free spectral range) as

$$\Delta\nu = \frac{1}{2}\left(\frac{c_o}{2nL}\right) \tag{3.7}$$

or, substituting this into Equation (3.6) and simplifying,

$$\Delta L = \frac{-L}{(2p+1)}. \tag{3.8}$$

Evaluating for the 75-cm argon-ion cavity gives

$$\Delta L = \frac{-L}{(2p+1)} = \frac{-75\text{cm}}{(2 \cdot 2915451 + 1)} = -128.625 \text{ nm}. \tag{3.9}$$

Notice that a very small change in the cavity length has resulted in a very large change in the position of a particular longitudinal mode. Thus, small changes in cavity length due to minor thermal or acoustical effects can result in very large changes in the longitudinal mode behavior of a laser. In practice, thermal variations as minor as cool air blowing across an operating laser can cause significant changes in longitudinal mode position.

As a consequence, many lasers are frequency-stabilized by using an active feedback system that changes the length of the cavity (using a *piezoelectric* crystal driving a mechanical stage) in order to keep the longitudinal mode at a constant frequency (see Section 9.2.4 for an example of a commercial frequency-stabilized argon-ion laser).

3.1.1 Using an Etalon for Single Longitudinal Mode Operation

An *etalon* is an optical element consisting of two plane parallel reflective surfaces. An etalon will transmit a comb of frequencies separated by the free spectral range ($c_o/2nL$). (See Figure 3.2.)

Etalons used as elements in laser cavities typically consist of two plane parallel optical surfaces coated on a glass or fused-silica substrate. This particular configuration is a particularly useful device in laser engineering. As an example, consider an argon-ion laser that is lasing on four longitudinal modes. An etalon can be selected with a mode spectrum possessing a different frequency spacing than that of the laser. This etalon can then be placed inside the argon-ion laser cavity. The laser will then lase on the one longitudinal mode that is transmitted through the etalon (see Figure 3.3). If the etalon spectrum doesn't quite match up with the laser spectrum, then the laser cavity length can be changed (or the etalon twisted) so that the modes match.

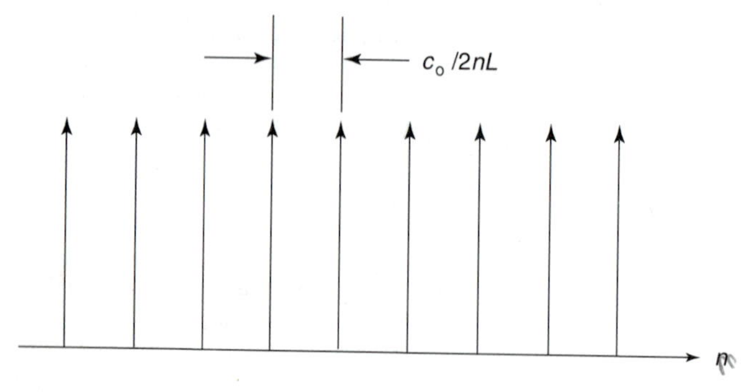

Figure 3.2 An etalon is a piece of glass with two plane parallel reflective surfaces. It will transmit a comb of frequencies separated by ($c_o/2nL$).

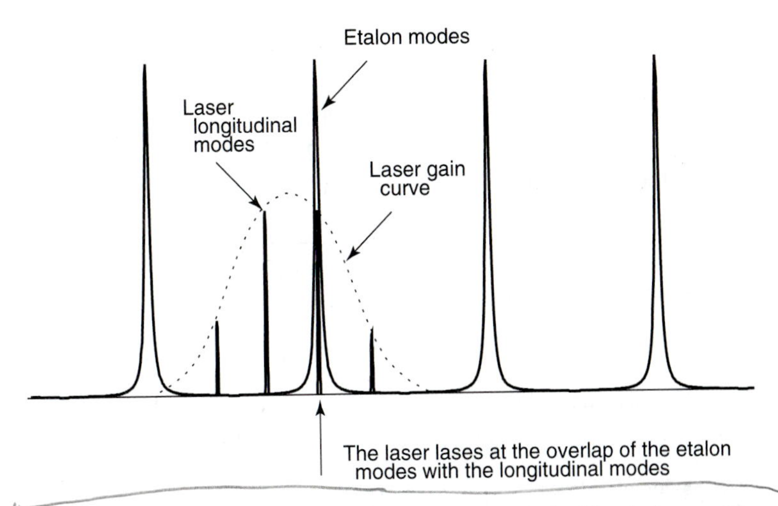

Figure 3.3 Single frequency operation of a laser can be obtained by placing an etalon inside the laser cavity. The laser will lase in the one longitudinal mode that represents the transmission overlap between the laser modes and the etalon modes.

The stable single longitudinal mode operation of a laser is critical for many applications. As an example, in order to create effective holograms, the average coherence length of the laser over the time of the hologram exposure must be significantly greater than the size of the object being imaged. (More generally, the larger the average coherence length, the higher the quality of the hologram.)

The average coherence length of a laser is given by

$$l_c = \frac{c_o}{\Delta\nu} \tag{3.10}$$

where $\Delta\nu$ is the average frequency bandwidth. If the laser is running on a single longitudinal mode that is drifting in frequency during the exposure, then $\Delta\nu$ is the total bandwidth covered by the drifting mode (typically several hundred GHz). However, if the single longitudinal mode can be frequency-stabilized, then $\Delta\nu$ is the linewidth of the single longitudinal mode (typically a much smaller number!).

3.2 QUANTITATIVE ANALYSIS OF A FABRY-PEROT ETALON

The Fabry-Perot etalon is an important element in laser engineering. As an independent optical element, it can be used as a very narrow linewidth filter to isolate a single longitudinal mode. In combination with a piezoelectric crystal to change its length, the etalon can also be used as a very precise instrument to measure the spectral character of laser lines.

3.2.1 Optical Path Relations in a Fabry-Perot Etalon

The ideal Fabry-Perot etalon consists of an ideal plane-parallel plate of glass of thickness L and of index n_{glass} immersed in a background of air at index n_{air}. When an optical beam of amplitude A_{in} strikes the first air-glass interface of the etalon, both an externally reflected beam B_1 and an internally transmitted beam C_1 are formed. The internally transmitted beam passes through the etalon and eventually strikes the glass-air interface at the other side of the etalon. When this occurs, both an internally reflected beam C_2 and an externally transmitted beam A_1 are formed. The beam continues to reflect back and forth inside the etalon producing a sequence of reflected beams B_1, B_2, ..., and transmitted beams A_1, A_2, ... The total reflected signal is the sum of all the reflected beams and the total transmitted signal is the sum of all transmitted beams (see Figure 3.4).

Now, consider the optical path length difference between two adjacent rays (see Figure 3.5). The path difference is given by

$$P_L = n_{\text{glass}} \left(\overline{AB} + \overline{BC}\right) - n_{\text{air}} \left(\overline{AD}\right) \tag{3.11}$$

where $\overline{AB} = \overline{BC} = L/\cos\theta_t$ and $\overline{AD} = \left(\overline{AC}\right)\sin\theta_i$.

Additionally, $\overline{AC} = 2L\tan\theta_t$, so substituting into the previous equation gives

$$P_L = n_{\text{glass}} \left(\frac{2L}{\cos\theta_t}\right) - n_{\text{air}} \left(2L\tan\theta_t \sin\theta_i\right). \tag{3.12}$$

Snell's law (Appendix A.5) can be applied to express $\sin\theta_i$ in terms of $\sin\theta_t$, yielding

$$P_L = n_{\text{glass}} \left(\frac{2L}{\cos\theta_t}\right) - n_{\text{air}} \left(2L\tan\theta_t \left(\frac{n_{\text{glass}}}{n_{\text{air}}}\sin\theta_t\right)\right) \tag{3.13}$$

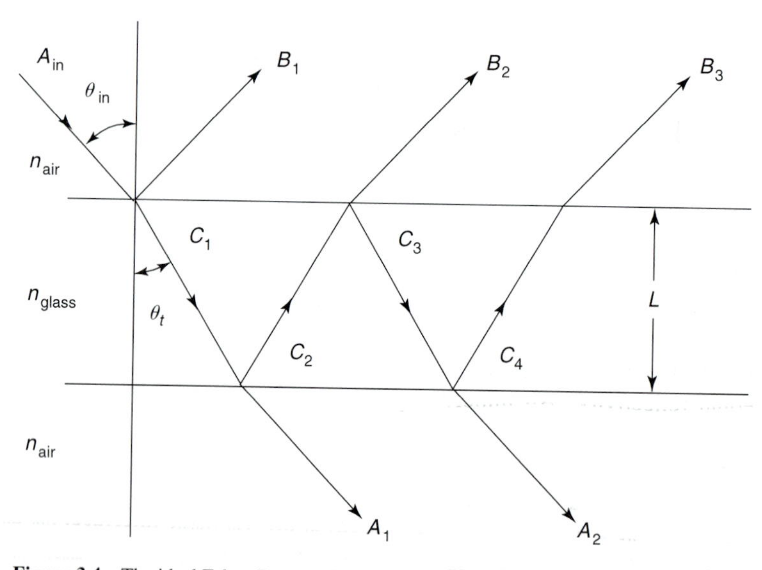

Figure 3.4 The ideal Fabry-Perot etalon consists of an ideal plane-parallel plate of glass of thickness L and of index n_{glass} immersed in a background of air at index n_{air}. When a beam of amplitude A_{in} strikes the first air-glass interface of the etalon, an externally reflected beam B_1 and an internally transmitted beam C_1 are formed. The internally transmitted beam passes through the etalon and strikes the glass-air interface at the other side of the etalon. An internally reflected beam C_2 and an externally transmitted beam A_1 are thereby formed.

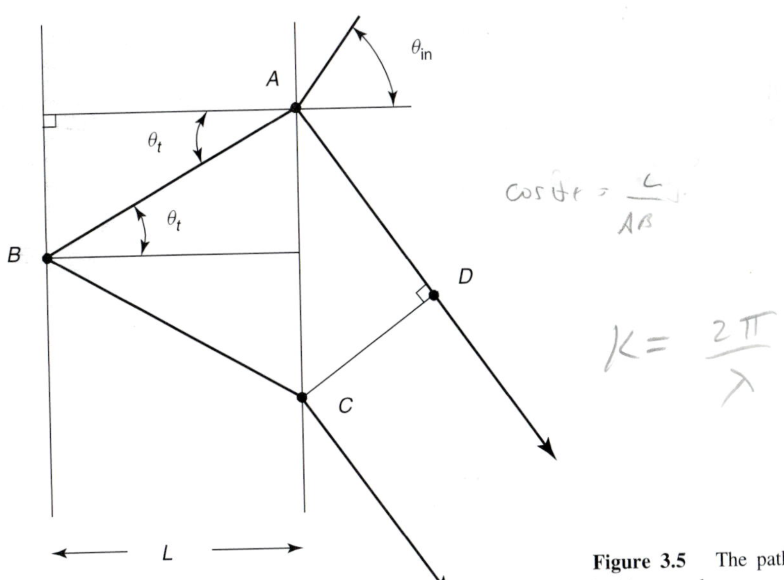

Figure 3.5 The path length difference inside an etalon.

or, simplifying,

$$P_L = n_{\text{glass}} \left(\frac{2L}{\cos \theta_t} \right) - n_{\text{glass}} \left(\frac{2L \sin^2 \theta_t}{\cos \theta_t} \right) = 2Ln_{\text{glass}} \cos \theta_t . \tag{3.14}$$

The phase difference (δ) is simply k_o times the optical path length difference. Thus, the phase difference is given by[1]

$$\delta = k_o P_L = \left(\frac{2\pi}{\lambda_o} \right) 2Ln_{\text{glass}} \cos \theta_t = \frac{4\pi \, Ln_{\text{glass}} \cos \theta_t}{\lambda_o} . \tag{3.15}$$

3.2.2 Reflection and Transmission Coefficients in a Fabry-Perot Etalon

A common etalon configuration is a glass or fused-silica substrate coated with two surfaces of equal reflectivity. An important property of this etalon is the ratio of the reflected and transmitted electric field. This ratio can be calculated by summing all of the reflected and transmitted waves as outlined in the following procedure. (The interested reader can find additional information in Yariv[2] and Verdeyen.[3])

Define r_{ag} as the reflection coefficient at the $n_{\text{air}}/n_{\text{glass}}$ interface, t_{ag} as the transmission coefficient from n_{air} to n_{glass}, $\overline{r_{ga}}$ as the reflection coefficient at the $n_{\text{glass}}/n_{\text{air}}$ interface, and $\overline{t_{ga}}$ as the transmission coefficient from n_{glass} to n_{air}. Notice that this definition does not include

[1] Amnon Yariv, *Optical Electronics*, 4th ed. (Philadelphia, Saunders College Publishing, 1991), p. 112.

[2] Amnon Yariv, *Optical Electronics*, 4th ed. (Philadelphia, PA: Saunders College Publishing, 1991), pp. 112–118.

[3] Joseph T. Verdeyen, *Laser Electronics*, 2d ed. (Englewood Cliffs, NJ: Prentice Hall, 1989), Chapter 6.

possible variations in reflection or transmission coefficients with angle or polarization. Also notice that $\left(\overline{r_{ga}}\right) = -\left(r_{ag}\right)$.

The complex amplitude of the total reflected wave is (see Figure 3.6)

$$A_r = \left(r_{ag}\right) A_{\text{in}} + \left(t_{ag}\right)\left(\overline{r_{ga}}\right)\left(\overline{t_{ga}}\right) A_{\text{in}} e^{i\delta} + \left(t_{ag}\right)\left(\overline{r_{ga}}\right)^3\left(\overline{t_{ga}}\right) A_{\text{in}} e^{2i\delta} + \cdots \quad (3.16)$$

and the complex amplitude of the transmitted wave is

$$A_t = \left(t_{ag}\right)\left(\overline{t_{ga}}\right) A_{\text{in}} + \left(t_{ag}\right)\left(\overline{r_{ga}}\right)^2\left(\overline{t_{ga}}\right) A_{\text{in}} e^{i\delta} + \left(t_{ag}\right)\left(\overline{r_{ga}}\right)^4\left(\overline{t_{ga}}\right) A_{\text{in}} e^{2i\delta} + \cdots \quad (3.17)$$

Simplifying the expression for the reflected wave leads to

$$A_r = A_{\text{in}}\left(\left(r_{ag}\right) + \left(t_{ag}\right)\left(\overline{t_{ga}}\right)\left(\overline{r_{ga}}\right) e^{i\delta}\left(1 + \left(\overline{r_{ga}}\right)^2 e^{i\delta} + \left(\overline{r_{ga}}\right)^4 e^{2i\delta} + \cdots\right)\right) \quad (3.18)$$

and similarly for the transmitted wave

$$A_t = A_{\text{in}}\left(t_{ag}\right)\left(\overline{t_{ga}}\right)\left(1 + \left(\overline{r_{ga}}\right)^2 e^{i\delta} + \left(\overline{r_{ga}}\right)^4 e^{2i\delta} + \cdots\right). \quad (3.19)$$

After some mathematical manipulation (and recognizing that $\left(\overline{r_{ga}}\right) = -\left(r_{ag}\right)$, and $\left(r_{ag}\right)^2 + \left(t_{ag}\right)\left(\overline{t_{ga}}\right) = R+T = 1$) Equations (3.18) and (3.19) can be reduced to an expression for the reflected wave as

$$\frac{A_r}{A_{\text{in}}} = \frac{(1 - e^{i\delta})\sqrt{R}}{(1 - Re^{i\delta})} \quad (3.20)$$

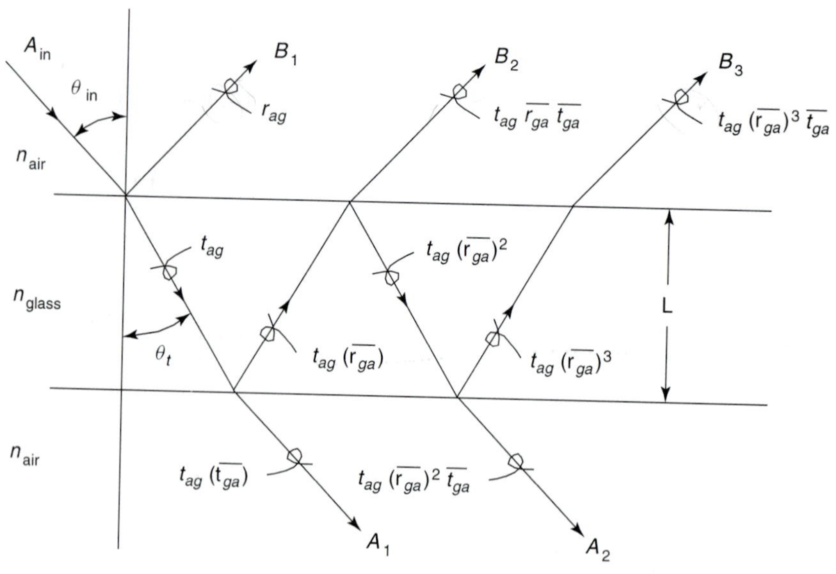

Figure 3.6 Detail of transmission and reflection coefficients inside an etalon. The term r_{ag} is defined as the reflection coefficient at the n_a/n_g interface, t_{ag} is the transmission coefficient from n_a to n_g, $\overline{r_{ga}}$ is the reflection coefficient at the n_g/n_a interface, and $\overline{t_{ga}}$ is the transmission coefficient from n_g to n_a. Notice that $\left(\overline{r_{ga}}\right) = -\left(r_{ag}\right)$.

and an expression for the transmitted wave as

$$\frac{A_t}{A_{\text{in}}} = \frac{T}{(1 - Re^{i\delta})} \tag{3.21}$$

where $(r_{ag})^2 = (\overline{r_{ga}})^2 = R$ and $(t_{ag})(\overline{t_{ga}}) = T$.

Example 3.2

Derive Equation (3.20).

Solution. Begin with Equation (3.18).

$$A_r = A_{\text{in}}\left((r_{ag}) + (t_{ag})(\overline{t_{ga}})(\overline{r_{ga}})e^{i\delta}\left(1 + (\overline{r_{ga}})^2 e^{i\delta} + (\overline{r_{ga}})^4 e^{2i\delta} + \cdots\right)\right). \tag{3.22}$$

If $(r_{ag})^2 = (\overline{r_{ga}})^2 = R$, then the previous series can be expressed as

$$\left(1 + Re^{i\delta} + (R)^2 e^{2i\delta} + \cdots\right) = \left(1 + \left(Re^{i\delta}\right)^1 + \left(Re^{i\delta}\right)^2 + \left(Re^{i\delta}\right)^3 \cdots\right) \tag{3.23}$$

which is the series for $1/(1-x)$ where $x = \left(Re^{i\delta}\right)$

$$\frac{1}{1-x} = 1 + x + x^2 + x^3 + \cdots \tag{3.24}$$

So, the equation for A_r can now be expressed as

$$A_r = A_{\text{in}}\left((r_{ag}) + (t_{ag})(\overline{t_{ga}})(\overline{r_{ga}})e^{i\delta}\left(\frac{1}{1 - \left(Re^{i\delta}\right)}\right)\right). \tag{3.25}$$

Taking advantage of the relations $(\overline{r_{ga}}) = -(r_{ag})$, $(r_{ag})^2 = (\overline{r_{ga}})^2 = R$, and $(t_{ag})(\overline{t_{ga}}) = T$ leads to

$$A_r = A_{\text{in}}(r_{ag})\left(1 - Te^{i\delta}\left(\frac{1}{1 - \left(Re^{i\delta}\right)}\right)\right) \tag{3.26}$$

or, after simplifying,

$$A_r = A_{\text{in}}\sqrt{R}\left(\left(\frac{1 - \left(Re^{i\delta}\right) - Te^{i\delta}}{1 - \left(Re^{i\delta}\right)}\right)\right) = A_{\text{in}}\sqrt{R}\left(\left(\frac{1 - (R+T)e^{i\delta}}{1 - \left(Re^{i\delta}\right)}\right)\right). \tag{3.27}$$

Recognizing that $R + T = 1$

$$A_r = A_{\text{in}}\sqrt{R}\left(\left(\frac{1 - e^{i\delta}}{1 - \left(Re^{i\delta}\right)}\right)\right) \tag{3.28}$$

and making the other substitutions of $(t_{ag})(\overline{t_{ga}}) = T$ and $1 - R = T$ gives

$$A_r = A_{\text{in}}(r_{ag})\left(1 + Te^{i\delta}\left(\frac{1}{1 - \left(Re^{i\delta}\right)}\right)\right). \tag{3.29}$$

Example 3.3

Key to the previous example is the relation $(\overline{r_{ga}}) = -(r_{ag})$. Verify this using the Fresnel equations (see the Appendix) for a lossless glass-air system with $\theta_{in} = 30$ degrees, and with $n_{\text{air}} = 1.0$ and $n_{\text{glass}} = 1.54$.

Solution. First calculate the transmitted angles in both the film and air using Snell's law as

$$\theta_{t-\text{film}} = \sin^{-1}\left(\frac{n_{\text{air}}}{n_{\text{glass}}} \cdot \sin\theta_i\right) = \sin^{-1}\left(\frac{1.0}{1.54} \cdot \sin\ 30\ \text{deg}\right) = 18.946\ \text{deg} \tag{3.30}$$

$$\theta_{t-\text{air}} = \sin^{-1}\left(\frac{n_{\text{glass}}}{n_{\text{air}}} \cdot \sin\theta_i\right) = \sin^{-1}\left(\frac{1.54}{1.0} \cdot \sin\ 18.946\ \text{deg}\right) = 30\ \text{deg} \tag{3.31}$$

Now calculate reflection coefficients for both polarizations for the glass to air interface as

$$r_\perp = \left(\frac{E_r}{E_i}\right)_\perp = \frac{n_{\text{air}}\cos\theta_i - n_{\text{glass}}\cos\theta_t}{n_{\text{air}}\cos\theta_i + n_{\text{glass}}\cos\theta_t} \tag{3.32}$$

$$r_\perp = \frac{1.0 \cdot \cos\ 30\ \text{deg} - 1.54 \cdot \cos\ 18.946\ \text{deg}}{1.0 \cdot \cos\ 30\ \text{deg} + 1.54 \cdot \cos\ 18.946\ \text{deg}} = -0.254 \tag{3.33}$$

$$r_\parallel = \left(\frac{E_r}{E_i}\right)_\parallel = \frac{n_{\text{glass}}\cos\theta_i - n_{\text{air}}\cos\theta_t}{n_{\text{air}}\cos\theta_t + n_{\text{glass}}\cos\theta_i} \tag{3.34}$$

$$r_\parallel = \left(\frac{E_r}{E_i}\right)_\parallel = \frac{1.54 \cdot \cos\ 30\ \text{deg} - 1.0 \cdot \cos\ 18.946\ \text{deg}}{1.0 \cdot \cos\ 18.946\ \text{deg} + 1.54 \cdot \cos\ 30\ \text{deg}} = 0.17 \tag{3.35}$$

and calculate reflection coefficients for both polarizations for the air to glass interface as

$$r_\perp = \left(\frac{E_r}{E_i}\right)_\perp = \frac{n_{\text{glass}}\cos\theta_i - n_{\text{air}}\cos\theta_t}{n_{\text{glass}}\cos\theta_i + n_{\text{air}}\cos\theta_t} \tag{3.36}$$

$$r_\perp = \frac{1.54 \cdot \cos\ 18.946\ \text{deg} - 1.0 \cdot \cos\ 30\ \text{deg}}{1.54 \cdot \cos\ 18.946\ \text{deg} + 1.0 \cdot \cos\ 30\ \text{deg}} = 0.254 \tag{3.37}$$

$$r_\parallel = \left(\frac{E_r}{E_i}\right)_\parallel = \frac{n_{\text{air}}\cos\theta_i - n_{\text{glass}}\cos\theta_t}{n_{\text{glass}}\cos\theta_t + n_{\text{air}}\cos\theta_i} \tag{3.38}$$

$$r_\parallel = \left(\frac{E_r}{E_i}\right)_\parallel = \frac{1.0 \cdot \cos\ 18.946\ \text{deg} - 1.54 \cdot \cos\ 30\ \text{deg}}{1.54 \cdot \cos\ 30\ \text{deg} + 1.0 \cdot \cos\ 18.946\ \text{deg}} = -0.17 \tag{3.39}$$

So, indeed, $(\overline{r_{ga}}) = -(r_{ag})$ for this case. (A proof can be easily constructed using this example as a basis.)

3.2.3 Calculating the Reflected and Transmitted Intensities for a Fabry-Perot Etalon with the Same Reflectances

In most laser applications, the primary quantity of interest is not the ratio of reflected and transmitted electric field values, but rather the ratio of reflected and transmitted intensities.

For reflectance, this ratio is given as

$$\boxed{R = \frac{I_r}{I_{\text{in}}} = \frac{A_r A_r^*}{A_{\text{in}} A_{\text{in}}^*} = \left(\frac{4R\sin^2(\delta/2)}{(1-R)^2 + 4R\sin^2(\delta/2)}\right)} \tag{3.40}$$

and for transmittance as

$$\boxed{T = \frac{I_t}{I_{\text{in}}} = \frac{A_t A_t^*}{A_{\text{in}} A_{\text{in}}^*} = \left(\frac{(1-R)^2}{(1-R)^2 + 4R\sin^2(\delta/2)}\right).} \tag{3.41}$$

Notice that the transmittance is unity whenever

$$\boxed{\delta = \frac{4\pi L n_{\text{glass}} \cos\theta_t}{\lambda_o} = \frac{2\omega n_{\text{glass}} L \cos\theta_t}{c_o} = 2m\pi \quad \text{where } m = \pm 1,\ \pm 2,\ \pm 3,\ \ldots} \tag{3.42}$$

Thus, the transmittance spectrum will consist of a *comb* of lines. Each line corresponds to a resonant *mode* of the etalon. Thus, the frequency of each line is given by

$$\nu = p \cdot \frac{c_o}{2n_{\text{glass}} L \cos\theta_t} \quad \text{where } p = \pm 1,\ \pm 2,\ \pm 3,\ \ldots \tag{3.43}$$

and the lines are separated in frequency by

$$\Delta\nu = \frac{c_o}{2n_{\text{glass}} L \cos\theta_t}. \tag{3.44}$$

Example 3.4

Derive Equations (3.40) and (3.41).

Solution. Begin with Equation (3.20).

$$\frac{A_r}{A_{\text{in}}} = \frac{(1-e^{i\delta})\sqrt{R}}{(1-Re^{i\delta})}. \tag{3.45}$$

Multiply Equation (3.20) by its complex conjugate

$$\frac{I_r}{I_{\text{in}}} = \frac{A_r A_r^*}{A_{\text{in}} A_{\text{in}}^*} = \frac{(1-e^{i\delta})\sqrt{R}}{(1-Re^{i\delta})} \cdot \frac{(1-e^{-i\delta})\sqrt{R}}{(1-Re^{-i\delta})} \tag{3.46}$$

$$\frac{I_r}{I_{\text{in}}} = \frac{A_r A_r^*}{A_{\text{in}} A_{\text{in}}^*} = \frac{(1-e^{i\delta} - e^{-i\delta} + 1)R}{(1-Re^{i\delta} - Re^{-i\delta} + R^2)} = \frac{(2-2\cos\delta)R}{(1-2R\cos\delta + R^2)} \tag{3.47}$$

and then use the identity $\cos\delta = 1 - 2\sin^2(\delta/2)$ to yield

$$\frac{I_r}{I_{\text{in}}} = \frac{A_r A_r^*}{A_{\text{in}} A_{\text{in}}^*} = \frac{(2-2\left[1-2\sin^2(\delta/2)\right])R}{(1-2R\left[1-2\sin^2(\delta/2)\right] + R^2)} = \frac{4R\sin^2(\delta/2)}{(1-R)^2 + 4R\sin^2(\delta/2)}. \tag{3.48}$$

Now consider Equation (3.21),

$$\frac{A_t}{A_{\text{in}}} = \frac{T}{(1-Re^{i\delta})}. \tag{3.49}$$

Multiply Equation (3.21) by its complex conjugate

$$\frac{I_t}{I_{\text{in}}} = \frac{A_t A_t^*}{A_{\text{in}} A_{\text{in}}^*} = \frac{T}{(1 - Re^{i\delta})} \cdot \frac{T}{(1 - Re^{-i\delta})} \tag{3.50}$$

$$\frac{A_t}{A_{\text{in}}} = \frac{A_t A_t^*}{A_{\text{in}} A_{\text{in}}^*} = \frac{T^2}{(1 - 2R\cos\delta + R^2)} = \frac{(1-R)^2}{(1 - 2R\cos\delta + R^2)}. \tag{3.51}$$

Then use the identity $\cos\delta = 1 - 2\sin^2(\delta/2)$ to yield

$$\frac{I_r}{I_{\text{in}}} = \frac{A_r A_r^*}{A_{\text{in}} A_{\text{in}}^*} = \frac{(1-R)^2}{(1 - 2R\left[1 - 2\sin^2(\delta/2)\right] + R^2)} = \frac{(1-R)^2}{(1-R)^2 + 4R\sin^2(\delta/2)}. \tag{3.52}$$

3.2.4 Calculating the Reflected and Transmitted Intensities for a Fabry-Perot Etalon with Different Reflectances

It is not uncommon to have an etalon system where the two reflectances are different. The calculations in the previous sections can be repeated for a system where the reflectances are different at each interface. In this case, the ratio of reflected intensities is given as

$$\boxed{\frac{I_r}{I_i} = \frac{(\sqrt{R_1} - \sqrt{R_2})^2 + 4\sqrt{R_1 R_2}\sin^2(\delta/2)}{(1 - \sqrt{R_1 R_2})^2 + 4\sqrt{R_1 R_2}\sin^2(\delta/2)}} \tag{3.53}$$

and the ratio of transmitted intensities is given as

$$\boxed{\frac{I_t}{I_i} = \frac{(1 - R_1)(1 - R_2)}{(1 - \sqrt{R_1 R_2})^2 + 4\sqrt{R_1 R_2}\sin^2(\delta/2)}}. \tag{3.54}$$

In the usual case, the laser beam is coming in normal to the surface of the etalon. Under these conditions, then the term

$$\boxed{\delta = \frac{4\pi L n_{\text{glass}} \cos\theta_t}{\lambda_o} = \frac{2\omega n_{\text{glass}} L \cos\theta_t}{c_o} = 2m\pi \quad \text{where } m = \pm 1, \pm 2, \pm 3, \ldots} \tag{3.55}$$

simplifies to

$$\delta = \frac{4\pi L n_{\text{glass}}}{\lambda_o} = \frac{2\omega n_{\text{glass}} L}{c_o} = 2m\pi \quad \text{where } m = \pm 1, \pm 2, \pm 3, \ldots \tag{3.56}$$

Thus, the ratio of reflected intensities takes the form

$$R = \frac{I_r}{I_i} = \frac{(\sqrt{R_1} - \sqrt{R_2})^2 + 4\sqrt{R_1 R_2}\sin^2(\omega n L/c_o)}{(1 - \sqrt{R_1 R_2})^2 + 4\sqrt{R_1 R_2}\sin^2(\omega n L/c_o)} \tag{3.57}$$

and the ratio of transmitted intensities is given by

$$T = \frac{I_t}{I_i} = \frac{(1 - R_1)(1 - R_2)}{(1 - \sqrt{R_1 R_2})^2 + 4\sqrt{R_1 R_2}\sin^2(\omega n L/c_o)}. \tag{3.58}$$

Under these circumstances, the maximum transmittance is

$$T_{\text{max}} = \frac{(1 - R_1)(1 - R_2)}{(1 - \sqrt{R_1 R_2})^2} \tag{3.59}$$

$\frac{1 - R_2 - R_1 + R_1R_2}{1 - \sqrt{R_1R_2} - \sqrt{R_1R_2} + R_1R_2 + 4\sqrt{R_1R_2}} = \frac{}{(1+\sqrt{R_1R_2})^2}$

and the minimum transmittance is

$$T_{\min} = \left(\frac{(1-\sqrt{R_1R_2})}{(1+\sqrt{R_1R_2})}\right)^2 T_{\max}. \tag{3.60}$$

3.2.5 Calculating the Q and the Finesse of a Fabry-Perot Etalon

Etalons are often characterized by their *quality factor Q*. This parameter is defined as the resonant frequency of the cavity ν_o divided by the frequency linewidth $\Delta\nu$ at FWHM (full-width half-maximum) or

$$Q = \frac{\nu_o}{\Delta\nu} = \frac{pc_o}{2n_{\text{glass}}L\Delta\nu} = \frac{2\pi n_{\text{glass}}L(R_1R_2)^{1/4}}{\lambda_o(1-\sqrt{R_1R_2})}. \tag{3.61}$$

Because Q is often very large, another measure of the line resonance is commonly used. This is termed the *finesse F* and is defined as the free spectral range divided by the frequency linewidth at FWHM or

$$F = \frac{\Delta\nu_{\text{FSR}}}{\Delta\nu} = \frac{c_o}{2n_{\text{glass}}L\Delta\nu} = \frac{\pi(R_1R_2)^{1/4}}{(1-\sqrt{R_1R_2})}. \tag{3.62}$$

3.3 ILLUSTRATIVE FABRY-PEROT ETALON CALCULATIONS

Example 3.5

Consider a Fabry-Perot etalon to be used as a line filter for an argon-ion laser operating at a center wavelength of 514.5 nm. Assume that the beam is entering the etalon at normal incidence ($\theta_t = 0$). Plot the transmittance spectra for an etalon thickness of 2 mm, an etalon index of refraction of 1.5, and reflectances (on both surfaces) of $R = 50\%$. Repeat for $R = 90\%$.

Solution. This uses Equations (3.40) and (3.41) as

$$\frac{I_r}{I_{\text{in}}} = \left(\frac{4R\sin^2(\delta/2)}{(1-R)^2 + 4R\sin^2(\delta/2)}\right) \tag{3.63}$$

and

$$\frac{I_t}{I_{\text{in}}} = \left(\frac{(1-R)^2}{(1-R)^2 + 4R\sin^2(\delta/2)}\right) \tag{3.64}$$

where

$$\delta = \frac{4\pi L n_{\text{glass}}}{\lambda_o}. \tag{3.65}$$

The spectra are plotted in Figure 3.7.

Example 3.6

Calculate the maximum transmittance $T_{\max}$, the minimum transmittance $T_{\min}$, the quality factor Q, and the finesse F for Example 3.5.

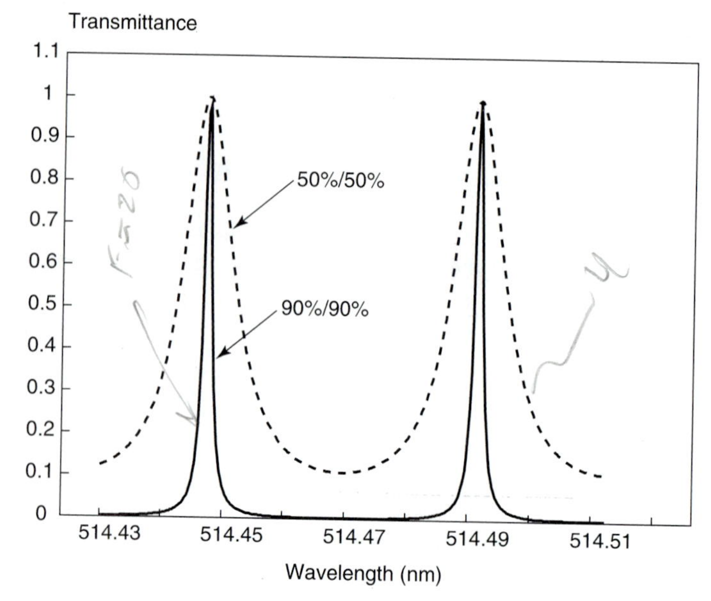

Figure 3.7 The transmittance spectra for two etalons with thicknesses of 2 mm, etalon indices of refraction of 1.5, and reflectances (on both surfaces) of $R = 50\%$ and $R = 90\%$. The laser is an argon-ion laser operating at a center wavelength of 514.5 nm and the beam is entering the etalons at normal incidence ($\theta_t = 0$).

Solution. This uses Equations (3.59) to (3.62). For the first case of 50% mirrors, this yields

$$T_{max} = \frac{(1 - R_1)(1 - R_2)}{(1 - \sqrt{R_1 R_2})^2} = \frac{(1 - 0.5)(1 - 0.5)}{(1 - \sqrt{0.5 \cdot 0.5})^2} = 1 \tag{3.66}$$

$$T_{min} = \left(\frac{(1 - \sqrt{R_1 R_2})}{(1 + \sqrt{R_1 R_2})}\right)^2 = \left(\frac{(1 - \sqrt{0.5 \cdot 0.5})}{(1 + \sqrt{0.5 \cdot 0.5})}\right)^2 \cdot 1.0 = 0.111 \tag{3.67}$$

$$Q = \frac{2\pi n_{glass} L (R_1 R_2)^{1/4}}{\lambda_o (1 - \sqrt{R_1 R_2})} = \frac{2 \cdot \pi \cdot 1.5 \cdot (2 \cdot 10^{-3}\ \text{m}) \cdot (0.5 \cdot 0.5)^{1/4}}{(5.145 \cdot 10^{-9}\ \text{m}) \cdot (1 - \sqrt{0.5 \cdot 0.5})} = 5.181 \cdot 10^4 \tag{3.68}$$

$$F = \frac{\pi (R_1 R_2)^{1/4}}{(1 - \sqrt{R_1 R_2})} = \frac{\pi \cdot (0.5 \cdot 0.5)^{1/4}}{(1 - \sqrt{0.5 \cdot 0.5})} = 4.443. \tag{3.69}$$

For the case of 90% mirrors, $T_{max} = 1$, $T_{min} = 0.003$, $Q = 3.476 \cdot 10^5$, and $F = 29.804$.

Example 3.7

Consider a Fabry-Perot etalon to be used as a line filter for an argon-ion laser operating at a center wavelength of 514.5 nm. Assume that the beam is entering the etalon at normal incidence ($\theta_t = 0$). Assume an etalon thickness of 2 mm and an etalon index of refraction of 1.5. Plot the transmittance spectrum assuming that one surface has a reflectance of 98%, while the other surface has a reflectance of 60%. Do this again, assuming one surface has a reflectance of 50% and the other surface has a reflectance of 30%.

Solution. This uses equations (3.53) and (3.54) as

$$\frac{I_r}{I_i} = \frac{(\sqrt{R_1} - \sqrt{R_2})^2 + 4\sqrt{R_1 R_2}\sin^2(\delta/2)}{(1 - \sqrt{R_1 R_2})^2 + 4\sqrt{R_1 R_2}\sin^2(\delta/2)} \tag{3.70}$$

and

$$\frac{I_t}{I_i} = \frac{(1 - R_1)(1 - R_2)}{(1 - \sqrt{R_1 R_2})^2 + 4\sqrt{R_1 R_2}\sin^2(\delta/2)} \tag{3.71}$$

where

$$\delta = \frac{4\pi L n_{\text{glass}}}{\lambda_o}. \tag{3.72}$$

The spectra are plotted in Figure 3.8.

Example 3.8

Calculate the maximum transmittance $T_{\max}$, the minimum transmittance $T_{\min}$, the quality factor Q, and the finesse F for Example 3.7.

Solution. This uses Equations (3.59) to (3.62).

$$T_{\max} = \frac{(1 - R_1)(1 - R_2)}{(1 - \sqrt{R_1 R_2})^2} = \frac{(1 - 0.6) \cdot (1 - 0.98)}{(1 - \sqrt{0.6 \cdot 0.98})^2} = 0.147 \tag{3.73}$$

$$T_{\min} = \left(\frac{(1 - \sqrt{R_1 R_2})}{(1 + \sqrt{R_1 R_2})}\right)^2 = \left(\frac{(1 - \sqrt{0.6 \cdot 0.98})}{(1 + \sqrt{0.6 \cdot 0.98})}\right)^2 \cdot 0.147 = 0.003 \tag{3.74}$$

$$Q = \frac{2\pi n_{\text{glass}} L (R_1 R_2)^{1/4}}{\lambda_o (1 - \sqrt{R_1 R_2})} = \frac{2 \cdot \pi \cdot 1.5 \cdot (2 \cdot 10^{-3}\ \text{m}) \cdot (0.6 \cdot 0.98)^{1/4}}{(5.145 \cdot 10^{-9}\ \text{m}) \cdot (1 - \sqrt{0.6 \cdot 0.98})} = 1.376 \cdot 10^5 \tag{3.75}$$

$$F = \frac{\pi (R_1 R_2)^{1/4}}{(1 - \sqrt{R_1 R_2})} = \frac{\pi \cdot (0.6 \cdot 0.98)^{1/4}}{(1 - \sqrt{0.6 \cdot 0.98})} = 11.797. \tag{3.76}$$

For the case of 50% and 30% mirrors, $T_{\max} = 0.932$, $T_{\min} = 0.182$, $Q = 3.721 \cdot 10^4$, and $F = 3.191$.

Example 3.9

In actual operation, there is no guarantee that the etalon transmittance peaks will exactly line up with the longitudinal modes of the laser line. Thus, the etalon may be twisted slightly (i.e. changing $\cos\theta_t$) to scan the transmittance peaks across the laser spectrum. Assume that a Fabry-Perot etalon is to be used with an argon-ion laser operating at a center wavelength of 514.5 nm. The etalon thickness is 2 mm, the etalon index of refraction is 1.5, and both surfaces possess 90% reflective coatings. Plot the transmittance spectrum assuming $\theta_t = 0$ radians, $\theta_t = 0.005$ radians, $\theta_t = 0.009$ radians, and $\theta_t = 0.02$ radians.

Solution. This uses equations (3.40) and (3.41).

$$\frac{I_r}{I_{\text{in}}} = \left(\frac{4R\sin^2(\delta/2)}{(1 - R)^2 + 4R\sin^2(\delta/2)}\right) \tag{3.77}$$

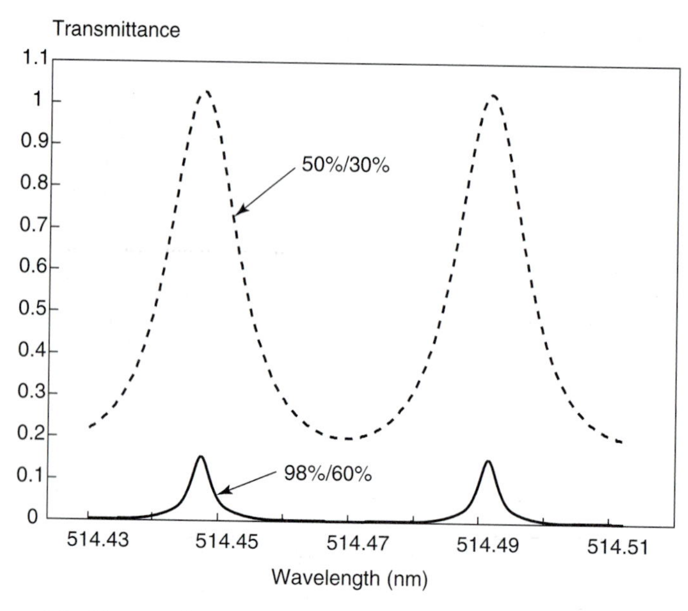

Figure 3.8 The transmittance spectra for two etalons with thicknesses of 2 mm and etalon indices of refraction of 1.5. The first case is one surface with a reflectance of 98%, while the other surface has a reflectance of 60%. The second case is one surface with a reflectance of 50%, while the other surface has a reflectance of 30%. The laser is an argon-ion laser operating at a center wavelength of 514.5 nm and the beam is entering the etalons at normal incidence ($\theta_t = 0$).

and

$$\frac{I_t}{I_{\text{in}}} = \left(\frac{(1-R)^2}{(1-R)^2 + 4R\sin^2(\delta/2)}\right) \tag{3.78}$$

where

$$\delta = \frac{4\pi L n_{\text{glass}} \cos\theta_t}{\lambda_o}. \tag{3.79}$$

The spectra are plotted in Figure 3.9.

Notice that increasing θ_t decreases $\cos\theta_t$. A decrease in the value of $\cos\theta_t$ increases the frequency of a single line, as given by

$$\nu = p \cdot \frac{c_o}{2n_{\text{glass}} L \cos\theta_t} \quad \text{where } p = \pm1,\ \pm 2,\ \pm 3,\ \ldots$$

However, frequency is inversely proportional to wavelength; thus as θ_t is increased, the wavelength decreases.

Also notice that $\theta_t = 0$ radians, $\theta_t = 0.005$ radians, and $\theta_t = 0.009$ radians all lie within the same free spectral range. However $\theta_t = 0.02$ radians has jumped by one free spectral range.

Example 3.10

An instrument called a scanning confocal interferometer is frequently used to measure the spectral character of laser lines. In essence, a scanning confocal interferometer is an etalon with an adjustable length L. In order to simulate the operation of the scanning confocal interferometer,

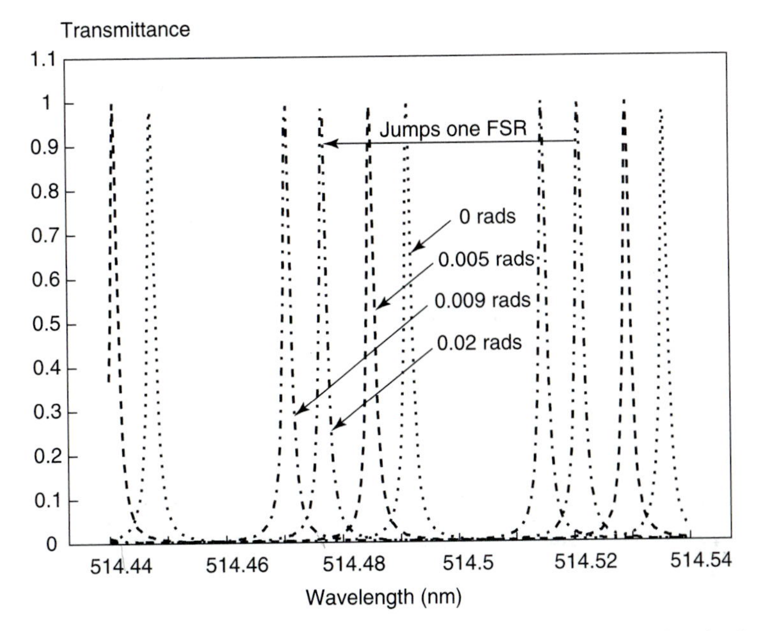

Figure 3.9 The transmittance of a Fabry-Perot etalon for several values of angle. In this case, a Fabry-Perot etalon is used with an argon-ion laser operating at a center wavelength of 514.5 nm. The etalon thickness is 2 mm, the etalon index of refraction is 1.5, and both surfaces have reflectances of 90%. The angle values are $\theta_t = 0$ radians, $\theta_t = 0.005$ radians, $\theta_t = 0.009$ radians, and $\theta_t = 0.02$ radians.

calculate the transmittance of a Fabry-Perot etalon for several values of length. Assume that the etalon is to be used with an argon-ion laser operating at a center wavelength of 514.5 nm. The etalon thickness is 15 mm, the etalon index of refraction is 1.0 and $\theta_t = 0$. Plot the transmittance spectrum assuming $L = 15$ mm + 0 nm, $L = 15$ mm + 1 nm, $L = 15$ mm + 10 nm, and $L = 15$ mm + 100 nm.

Solution. This uses Equations (3.40) and (3.41).

$$\frac{I_r}{I_{\text{in}}} = \left(\frac{4R\sin^2(\delta/2)}{(1-R)^2 + 4R\sin^2(\delta/2)}\right) \tag{3.80}$$

and

$$\frac{I_t}{I_{\text{in}}} = \left(\frac{(1-R)^2}{(1-R)^2 + 4R\sin^2(\delta/2)}\right) \tag{3.81}$$

where

$$\delta = \frac{4\pi L n_{\text{glass}} \cos\theta_t}{\lambda_o}. \tag{3.82}$$

The spectra are plotted in Figure 3.10.

Notice that increasing L decreases the frequency of a single line given by

$$\nu = p \cdot \frac{c_o}{2n_{\text{glass}} L \cos\theta_t} \qquad \text{where } p = \pm 1, \pm 2, \pm 3, \ldots \tag{3.83}$$

However, frequency is inversely proportional to wavelength, thus as L is increased, the wavelength increases.

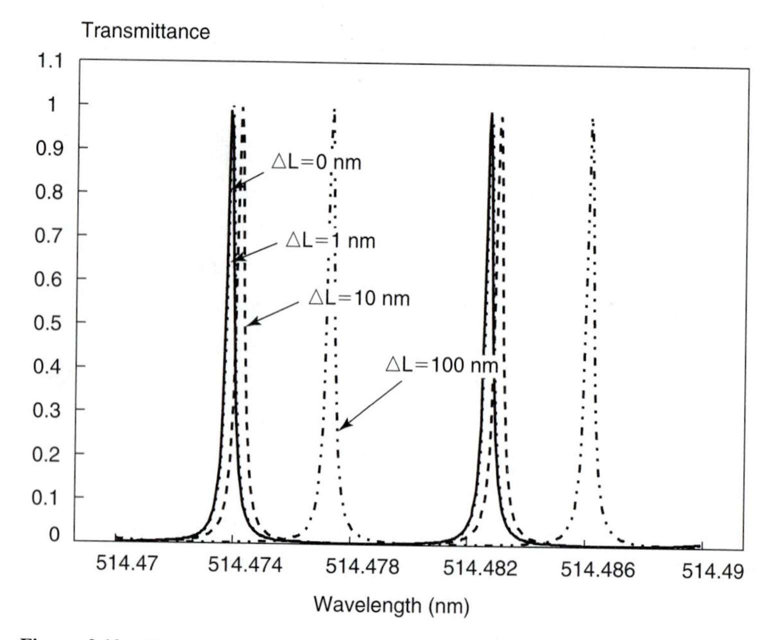

Figure 3.10 The transmittance of a Fabry-Perot etalon for several values of length. In this case, a Fabry-Perot etalon is used with an argon-ion laser operating at a center wavelength of 514.5 nm. The etalon thickness is 15 mm, the etalon index of refraction is 1.0 and $\theta_t = 0$. The length values are L = 15 mm + 0 nm, L = 15 mm + 1 nm, L = 15 mm + 10 nm, and L = 15 mm + 100 nm.

SYMBOLS USED IN THE CHAPTER

$\Delta\nu_{\text{FSR}}$: Free spectral range (Hz)

$\Delta\nu$: Frequency linewidth (Hz)

n: General real refractive index (unitless)

L: General length (cm or m)

p: Mode number (integer)

λ_o: Free space wavelength (cm or m)

λ: Material wavelength (cm or m)

ν: Frequency (Hz)

l_c: Coherence length (cm or m)

n_{air}: Real index of refraction for air, generally taken as 1.0 (unitless)

n_{lens}, n_{glass}, n_{film}: Real indices of refraction for various materials (unitless)

θ_i, θ_r, and θ_t: Incident, reflected, and transmitted angles. These angles lie in the plane of incidence and are measured from the surface normal (deg or rad)

P_L: Optical path difference between adjacent rays in an etalon (cm or m)

k_o: Analytic wavenumber in free space $2\pi/\lambda_o$ (m^{-1} or cm^{-1})

A_{in}: Incident ray in an etalon

A_1, B_1, and C_1: Reflected, transmitted and internal rays in an etalon

R, R_1, R_2: Intensity reflectivities of etalon surfaces (unitless)

δ: Phase difference between rays in an etalon (rad)

r_{ag}: Reflection coefficient at the $n_{\text{air}}/n_{\text{glass}}$ interface (unitless)

t_{ag}: Transmission coefficient from n_{air} to n_{glass} (unitless)

$\overline{r_{ga}}$: Reflection coefficient at the $n_{\text{glass}}/n_{\text{air}}$ interface (unitless)

$\overline{t_{ga}}$: Transmission coefficient from n_{glass} to n_{air} (unitless)

A_r: Complex amplitude of the total reflected wave in an etalon (volt/m)

A_t: Complex amplitude of the total transmitted wave in an etalon (volt/m)

r, t, $r_\perp$, $r_\parallel$, $t_\perp$, $t_\parallel$: Electric field reflection and transmission coefficients (unitless.)

I_r, I_t, and I_{in}: Reflected, transmitted, and incident intensities into an etalon (watts/cm^2)

R, T, A, $R_\perp$, $R_\parallel$, $T_\perp$, $T_\parallel$, $A_\perp$, $A_\parallel$: Intensity reflectance, transmittance, and absorptance coefficients (unitless)

$T_{\max}$: Maximum intensity transmittance of a Fabry-Perot etalon (unitless)

$T_{\min}$: Minimum intensity transmittance of a Fabry-Perot etalon (unitless)

Q: Quality factor of an etalon (unitless)

F: Finesse of an etalon (unitless)

EXERCISES

Longitudinal modes

3.1 Laser modes move in frequency as a cavity resonator heats up (see Figure 3.11). (Assume the cavity is 25 cm long, the laser is a HeNe at 632.8 nm, and the index of refraction is 1.)

(a) Derive an equation which describes how far a mode moves (in frequency) for a given change in length of the resonator cavity.

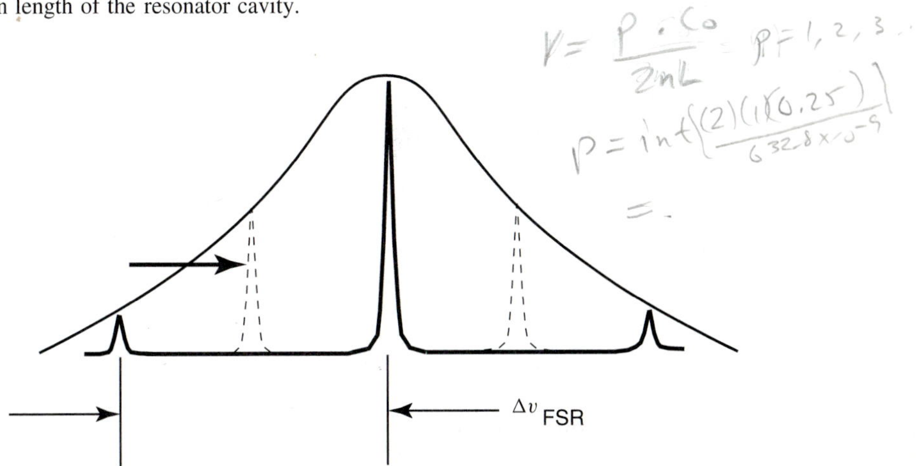

Figure 3.11 Laser modes move in frequency as a cavity resonator heats up.

(b) Once you have this equation, calculate how much the resonator cavity length must change for a mode to move 1/2 of the longitudinal mode spacing (the longitudinal mode spacing is also called the free spectral range or FSR).

3.2 From Exercise 3.1, it is clear that a laser is quite sensitive to changes in cavity length. Assume a laser cavity is made of quartz (with a thermal expansion coefficient of $(1/l)\,dl/dT$ per °C $= 5.5 \cdot 10^{-7}$), and calculate how much the modes move (in frequency) as the laser warms from 25 to 50°C. Assume the cavity is 25 cm long, the laser is HeNe at 632.8 nm, and the index of refraction is 1. Express your answer in terms of fractions of the free spectral range (for example, $0.35 \cdot$ FSR or $10.2 \cdot$ FSR).

3.3 (design) Consider the thermal drift of the longitudinal modes calculated in Exercise 3.2. Propose three laser resonator designs that would minimize this effect. Your answer should include a sketch of each of the three designs, as well as a concise description of how the designs work.

The Fabry-Perot etalon

3.4 Consider a soap film that is illuminated by white light. If the film is emerald green (wavelength = 510 nm), how thick is the film? Is this the only possible thickness? Assume the soap film has an index of approximately $n = 1.34$ and that you are viewing it at an angle of 45 degrees.

3.5 Consider the question of the maximum and minimum intensity transmittance, T_{max} and T_{min}, of an etalon with dissimilar mirrors.

(a) What does your intuition suggest about the maximum and minimum intensity transmittance? For example, what do you intuitively think the maximum and minimum intensity transmittance of an etalon with 65%–40% mirrors should be? What about 65%–90%?

(b) Develop a computer program that plots the maximum and minimum intensity transmittance, T_{max} and T_{min}, of an etalon with dissimilar mirrors. The y-axis should be transmittance T, the x-axis should be the reflectance R_2.

(c) Set the reflectance of one surface at $R_1 = 65\%$ and plot the maximum and minimum transmittances as a function of R_2.

(d) Set the reflectance of one surface at $R_1 = 90\%$ and plot the maximum and minimum transmittances as a function of R_2.

(e) Is your intuition correct? Why or why not?

3.6 Consider the finesse F of an etalon with two dissimilar mirrors, R_1 and R_2.

(a) What does your intuition suggest about the finesse as a function of the mirror reflectance? For example, what does your intuition suggest the finesse F of an etalon with 65%–40% mirrors should be? What about 65%–90%?

(b) Develop a computer program that plots the finesse F of an etalon with two dissimilar mirrors, R_1 and R_2. The y-axis should be finesse F, the x-axis should be reflectance R_2. Set up the program to plot two plots on the same graph. For one plot, set the reflectance of the constant surface at $R_1 = 65\%$. For the second plot, set the reflectance of the constant surface at $R_1 = 90\%$.

(c) Is your intuition correct? Why or why not?

3.7 Develop a computer program that plots the transmittance of the etalon as a function of wavelength. The transmittance T should be on the y-axis and wavelength (in angstroms or nm) on the x-axis. Set your program up so that 2 or 3 etalon peaks are visible in the plot.

(a) Consider an etalon 2.5 mm thick, fabricated of glass with an index $n = 1.58$. Assume $\lambda =$ 488 nm and $\theta_t = 0$ deg. Use your program to calculate the transmittance for the following three different reflectances. If possible, plot all three cases on the same plot (see Figure 3.7).

$$R_1 = R_2 = 50\%$$

$$R_1 = R_2 = 80\%$$

$$R_1 = R_2 = 95\%$$

(b) Consider an etalon 2.5 mm thick, fabricated of glass with an index $n = 1.58$. Assume $\lambda =$ 488 nm and $\theta_t = 0$ deg. Use your program to calculate the transmittance for the following three different reflectances. If possible, plot all three cases on the same plot (see Figure 3.8).

$$R_1 = 95\%, \quad R_2 = 10\%$$

$$R_1 = 95\%, \quad R_2 = 50\%$$

$$R_1 = 50\%, \quad R_2 = 10\%$$

(c) Etalons are very sensitive to small changes in length. Consider an etalon 5 mm thick. Assume that it will be used in a 488 nm laser, has an index of refraction of $n = 1.58$, and has both surfaces coated so that they have reflectances of 90%. Construct the plot so that all the transmittance spectra can be seen and compared (the plot should look similar to Figure 3.10). Generate spectra for $L = 5$ mm + 0 nm, $L = 5$ mm + 5 nm, $L = 5$ mm + 15 nm, and $L = 5$ mm + 30 nm. Put transmittance T on the y-axis and wavelength on the x-axis.

3.8 It is common to consider a laser resonator as an etalon. However, the finesse of a laser resonator is typically much smaller than the finesse of a typical etalon, and the free spectral range of the etalon is typically much larger than that of the cavity. To illustrate this, develop a computer program that plots the transmittance of a typical etalon as a function of wavelength. Transmittance T should be on the y-axis and wavelength (in angstroms or nm) on the x-axis. Plot the transmittance of an etalon 2.5 mm thick with an index of refraction of 1.5 and two surfaces each with reflectances of 90%. On the same plot, plot the transmittance of a laser resonator 15 cm long, with a front-mirror reflectance of 50%, a rear-mirror reflectance of 90%, and an index of refraction of 1.0. Assume $\lambda = 488$ nm and $\theta_t = 0$ deg. Construct the plot so that at least two of the etalon modes are visible.

3.9 It is rare for a mode of an etalon to exactly match the frequency of one of the modes of the laser. However, twisting the etalon slightly will change the optical path length and allow the modes to overlap.

(a) Consider an etalon 2.5 mm thick. Assume that it will be used in a 488 nm laser, has an index of refraction of $n = 1.58$, and has both surfaces coated so that they have reflectances of 90%. Use a computer program to generate several plots of the transmittance spectrum for this Fabry-Perot etalon. Construct the plot so that all the transmittance spectra can be seen and compared (the plot should look similar to Figure 3.9). Generate spectra for $\theta_t = 0$ radians, $\theta_t = 0.004$ radians, $\theta_t = 0.008$ radians, and $\theta_t = 0.012$ radians. Put transmittance (T) on the y-axis and wavelength on the x-axis.

(b) Considering the results of Exercise 3.8, generate an equation that predicts how many degrees an etalon must be twisted in order to move the etalon mode completely across the $\theta_t = 0$ radians free spectral range of the etalon. Solve the equation for the conditions given in Exercise 3.8. Is the answer consistent with your computer results?

3.10 An engineer has constructed a simple interferometer by taking an optical fiber and coating both ends with chromium. This creates a Fabry-Perot etalon. When the etalon is vibrated, the length of the fiber changes slightly. In the same spirit as Exercise 3.9, use a computer program to generate several plots of the transmittance spectrum of this fiber etalon as a function of the cavity length. Assume that the fiber is 35 cm long, that one mirror has a reflectance of 98%, that the other mirror has a reflectance of 90%, and that the laser is HeNe at $\lambda = 632.8$ nm. Generate plots for changes in length of 15 nm, 30 nm, and 45 nm. (i.e., plots for 35 cm + 0 nm, 35 cm + 15 nm, 35 cm + 30 nm, and 35 cm + 45 nm). Assume that the fiber has an index of refraction of 1.5 and that $\theta_t = 0$ degrees.

3.11 The fiber etalon (see Exercise 3.10) is being used to scan across a single longitudinal mode of a HeNe laser. However, the operator doesn't want to scan clear across to the next mode.

(a) Considering the results of Exercise 3.10, generate an equation that predicts the amount of movement (in frequency) of the etalon mode for a given change in length of the fiber.

(b) How much must the length of the etalon change to scan over the 20 MHz longitudinal mode of a HeNe laser?

(c) Assume that the maximum permitted scan range is the FSR of a 25 cm long HeNe laser. What is the maximum change in length to avoid scanning over two longitudinal modes?

4

Transverse Mode Properties

Objectives

- To compute the beam waist $w(z)$ of a laser as a function of position using both the exact and the hyperbolic angular approximation.
- To compute the beam radius $R(z)$ as a function of position.
- To perform computations of the beam waist $w(z)$ and the beam radius $R(z)$ for any combination of input parameters.
- To describe the differences between focusing a Gaussian beam and focusing a conventional beam.
- To determine the ray matrix of a simple optical system.
- To set up a ray matrix expression for a complex optical system and solve it using a computer.
- To write the *ABCD* law and describe what it means.
- To compute the stability of a laser resonator.
- To set up the matrix analysis for computing the beam waist size and location inside an arbitrary stable resonator.
- To compute the minimum beam waist size and location inside a resonator for a simple resonator with either one or two curved mirrors.

4.1 INTRODUCTION

The output spot of a laser beam is termed the transverse electromagnetic, $TEM_{x,y}$, mode. At least one (and frequently more) $TEM_{x,y}$ modes are present in an operating laser resonator. In most commercial lasers, the transverse electromagnetic mode is $TEM_{0,0}$ (which is a round mode with a Gaussian profile in cross-section). However, many other modes are possible. (See Figure 4.1 and 4.2.)

The laser resonator structure is responsible for both the cross-sectional appearance of the spot (see Section 4.2) and the mode propagation properties of the laser beam in space (see Section 4.3). Many commercially interesting properties of lasers (such as the tightly focused high-energy laser beams used for laser welding or the propagation of laser beams over long distances for surveying) are fundamentally determined by the properties of the transverse electromagnetic mode.

4.2 $TEM_{x,y}$ TRANSVERSE MODES

Transverse modes are identified by the convention $TEM_{x,y}$ where x and y refer to the subscript on the relevant polynomial. In this section, the transverse mode structure will be calculated using the paraxial approximation to the electromagnetic wave equations.

4.2.1 The Paraxial Approximation

The general electromagnetic wave equation for a laser material is given as

$$\nabla^2 \vec{\mathcal{E}} = \left(\mu_o \epsilon_h \frac{\partial^2 \vec{\mathcal{E}}}{\partial t^2} + \mu_o \frac{\partial^2 \vec{\mathcal{P}}}{\partial t^2} + \mu_o \sigma \frac{\partial \vec{\mathcal{E}}}{\partial t} \right) \tag{4.1}$$

where $\vec{\mathcal{E}}$ is the electric intensity vector and $\vec{\mathcal{P}}$ is the electric polarization vector given by[1]

$$\vec{\mathcal{P}} = \epsilon_o \chi \vec{\mathcal{E}} \tag{4.2}$$

where χ is the complex susceptibility, σ is the conductivity, and ϵ_h is the dielectric permittivity of the host crystal alone. (Note that in some references[2] the permittivity of free space may be included in the definition of χ, giving the slightly altered definition $\vec{\mathcal{P}} = \chi \vec{\mathcal{E}}$.)

The field $\vec{\mathcal{E}}$ is basically a plane wave of the form $\mathcal{E}_o e^{j(\omega t - k_c z)}$ and Equation (4.1) is often written in the phasor form

$$\left(\nabla^2 + \omega^2 \mu_o \epsilon_h \left(1 + \chi - j \frac{\sigma}{\omega \epsilon_h} \right) \right) \tilde{E} = 0 \tag{4.3}$$

or

$$\left(\nabla^2 + k_c^2 \right) \tilde{E} = 0 \tag{4.4}$$

---

[1] This definition of the polarizability is used by Amnon Yariv, *Optical Electronics*, 4th ed. (Philadelphia, PA: Saunders College Publishing, 1991) p. 6, and Anthony Siegman, *Lasers* (Mill Valley, CA: University Science Books, 1986), p. 103.

[2] This definition of the polarizability is used by Robert W. Boyd, *Non-Linear Optics* (Boston, MA: Academic Press, 1992), p. 2.

where $k_c = \omega\sqrt{\mu_o\epsilon_m}$, $\epsilon_m = \epsilon_o\epsilon_r$, $\epsilon_r = 1 + \chi$ and $\tilde{E}$ is the phasor representing the electric field. (The derivation of this equation is discussed in more detail in the Appendix.)

For deriving the transverse mode behavior, consider the transverse variation to be given by $U(x, y, z)$ so that the phasor expression for $\tilde{U}(x, y, z)$ has the form

$$E = \tilde{U}(x, y, z)e^{-jk_c z} \tag{4.5}$$

where $k_c = \omega\sqrt{\mu_o\epsilon_m}$. Now, expand the first term of the equation $\nabla^2 \tilde{U}(x, y, z)e^{-jk_c z}$ as

$$\begin{aligned}\nabla^2\tilde{U}(x, y, z)e^{-jk_c z} &= \frac{\partial^2\tilde{U}(x, y, z)}{\partial x^2}e^{-jk_c z} + \frac{\partial^2\tilde{U}(x, y, z)}{\partial y^2}e^{-jk_c z}\\ &\quad + \frac{\partial^2\tilde{U}(x, y, z)}{\partial z^2}e^{-jk_c z} + (-2jk_c)\frac{\partial\tilde{U}(x, y, z)}{\partial z}e^{-jk_c z}\\ &\quad + (-jk_c)^2\tilde{U}(x, y, z)e^{-jk_c z}.\end{aligned} \tag{4.6}$$

Rewriting this more conventionally,

$$\nabla^2\tilde{U}(x, y, z)e^{-jk_c z} = \left(\frac{\partial^2}{\partial x^2} + \frac{\partial^2}{\partial y^2} + \frac{\partial^2}{\partial z^2} + (-2jk_c)\frac{\partial}{\partial z} - k_c^2\right)\tilde{U}(x, y, z)e^{-jk_c z}. \tag{4.7}$$

In the *paraxial approximation*, it is assumed that the variation in the transverse structure with z is extremely small. Thus, the term $(\partial^2/\partial z^2)\tilde{U}(x, y, z)e^{-jk_c z}$ is quite small in comparison with the remaining terms, and so the previous expression is written as

$$\nabla^2\tilde{U}(x, y, z)e^{-jk_c z} = \left(\frac{\partial^2}{\partial x^2} + \frac{\partial^2}{\partial y^2} + (-2jk_c)\frac{\partial}{\partial z} - k_c^2\right)\tilde{U}(x, y, z)e^{-jk_c z}. \tag{4.8}$$

Substituting this back into the original phasor form for the wave equation and making the appropriate cancellations gives the full paraxial approximation to the wave equation as

$$\left(\frac{\partial^2}{\partial x^2} + \frac{\partial^2}{\partial y^2} + -2jk_c\frac{\partial}{\partial z}\right)\tilde{E} = 0 \tag{4.9}$$

and substituting in for k_c yields

$$\left(\frac{\partial^2}{\partial x^2} + \frac{\partial^2}{\partial y^2} + (-2j\omega\sqrt{\mu_o\epsilon_h})\frac{\partial}{\partial z} + \omega^2\mu_o\epsilon_h\left(\chi - j\frac{\sigma}{\omega\epsilon_h}\right)\right)\tilde{E} = 0 \tag{4.10}$$

which is often termed the transverse wave equation.

For propagation in free space, this is often written in a form that emphasizes its similarity to Schrodinger's equation.[3,4] In this formalism, $\psi = \tilde{E}$ and $\chi = \sigma = 0$ as

$$\left(\frac{\partial^2}{\partial x^2} + \frac{\partial^2}{\partial y^2} + (-2j\omega\sqrt{\mu_o\epsilon_h})\frac{\partial}{\partial z}\right)\psi = 0. \tag{4.11}$$

---

[3] Joseph T. Verdeyen, *Laser Electronics*, 2d ed. (Englewood Cliffs, NJ: Prentice Hall, 1989), p. 65.

[4] H. Kogelnik and T. Li, "Laser Beams and Resonators," *Proc. IEEE* 54:1312 (1966), eq. 11.

4.2.2 Mathematical Treatment of the Transverse Modes

There are two distinct sets of solutions to the higher-order modes. These solutions depend on whether the laser has a square transverse geometry (excimers, waveguide CO_2 lasers, etc.) or a circular transverse geometry (the majority of other lasers).

For a circular transverse geometry, the general transverse wave equation is expressed in cylindrical coordinates as[5]

$$\frac{1}{r}\frac{\partial}{\partial r}\left(r\frac{\partial \psi}{\partial r}\right) - j2k_c\frac{\partial \psi}{\partial z} = 0 \tag{4.12}$$

and has the general solution (where p is the radial integer, l is the angular integer, and r and ϕ are polar coordinates)[6]

$$\psi_{l,p} = \left(\frac{\sqrt{2}r}{w(z)}\right)^l L_p^l\left(\frac{2r^2}{w^2(z)}\right)\left[\exp\left(-j\left[k_c z - (1+l+2p)\tan^{-1}\left(\frac{\lambda z}{\pi w_o^2}\right)\right]\right)\right]$$
$$\cdot\left[\frac{w_o}{w(z)}\exp\left(\frac{-r^2}{w^2(z)}\right)\right]\left[\exp\left(-j\left(\frac{k_c r^2}{2R(z)}\right)\right)\right] \tag{4.13}$$

where $r^2 = x^2 + y^2$, $w(z)$ is the $1/e$ beam radius (waist) as a function of z, w_o is the minimum beam waist, and $R(z)$ is the beam radius of curvature. (For a more complete discussion of the beam waist $w(z)$ and beam radius of curvature $R(z)$; see Section 4.3.)

$L_p^l\left(2r^2/w^2(z)\right)$ is a *Laguerre polynomial function* in $2r^2/w^2(z)$ and is given by the Rodrigues formula[7]

$$L_p^l(v) = \frac{1}{p!}e^v\frac{1}{v^l}\frac{d^p}{dv^p}\left(e^{-v}v^l v^p\right). \tag{4.14}$$

The values of low-order Laguerre polynomials are[8]

$$L_0^l(x) = 1 \tag{4.15}$$

$$L_1^l(x) = l + 1 - x \tag{4.16}$$

$$L_2^l(x) = 0.5(l+1)(l+2) - (l+2)x + 0.5x^2. \tag{4.17}$$

The intensity of the cylindrical modes is given by[9]

$$I_p(r,\phi,z) = I_o \cdot \left(\frac{2r^2}{w^2(z)}\right)^l \cdot \left(L_p^l\frac{2r^2}{w^2(z)}\right)^2 \cdot \cos^2(l\phi)\cdot\exp\left(\frac{-2r^2}{w^2(z)}\right) \tag{4.18}$$

and the resulting modes are illustrated in Figure 4.1.

[5] Joseph T. Verdeyen, *Laser Electronics*, 2d ed. (Englewood Cliffs, NJ: Prentice Hall, 1989), p. 65.

[6] H. Kogelnik and T. Li, "Laser Beams and Resonators," *Proc. IEEE* 54:1312 (1966), using eq. 39 and 36 in eq. 28.

[7] Milton Abramowitz and Irene Stegen, *Handbook of Mathematical Functions* (New York: Dover Publications Inc., 1972), Section 22.11.

[8] H. Kogelnik and T. Li, "Laser Beams and Resonators," *Proc. IEEE* 54:1312 (1966), eq. 38.

[9] Walter Koechner, *Solid State Laser Engineering*, 3d ed. (Berlin, Germany: Springer-Verlag, 1992), p. 190.

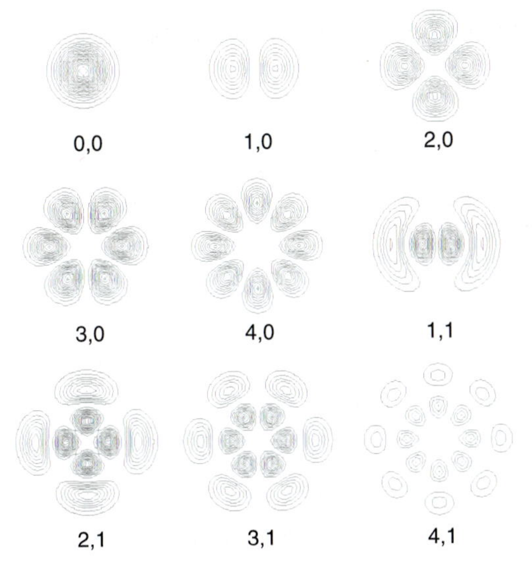

Figure 4.1 Transverse modes for a circular geometry laser.

For a rectangular transverse geometry, the general transverse wave equation is expressed in Cartesian coordinates as

$$\left(\frac{\partial^2}{\partial x^2} + \frac{\partial^2}{\partial y^2} + (-2j\omega\sqrt{\mu_o \epsilon_m})\frac{\partial}{\partial z}\right)\psi = 0 \tag{4.19}$$

and has the general solution (where m and n are the mode numbers for the x and y directions, respectively)[10]

$$\begin{aligned}\psi_{m,p} = {} & H_m\left(\frac{\sqrt{2}x}{w(z)}\right) H_p\left(\frac{\sqrt{2}y}{w(z)}\right) \\ & \cdot\left[\exp\left(-j\left[k_c z - (1+m+p)\tan^{-1}\left(\frac{\lambda z}{\pi w_o^2}\right)\right]\right)\right] \\ & \cdot\left[\frac{w_o}{w(z)}\exp\left(\frac{-(x^2+y^2)}{w^2(z)}\right)\right] \\ & \cdot\left[\exp\left(-j\left(\frac{k_c r^2}{2R(z)}\right)\right)\right]\end{aligned} \tag{4.20}$$

where $r^2 = x^2 + y^2$, $w(z)$ is the $1/e$ beam waist (radius) as a function of z, w_o is the minimum beam waist at the location z_o, and $R(z)$ is the beam, radius of curvature.

$H_m(\sqrt{2}x/w(z))$ and $H_p(\sqrt{2}y/w(z))$ are *Hermite polynomial functions* of $\sqrt{2}x/w(z)$ and $\sqrt{2}y/w(z)$ given by the Rodrigues formula[11]

$$H_m(v) = (-1)^m e^{v^2}\frac{d^m e^{-v^2}}{dv^m}. \tag{4.21}$$

[10] H. Kogelnik and T. Li, "Laser Beams and Resonators," *Proc. IEEE* 54:1312 (1966), eqs. 30 and 32.

[11] Milton Abramowitz and Irene Stegen, *Handbook of Mathematical Functions* (New York: Dover Publications Inc., 1972), Section 22.11.

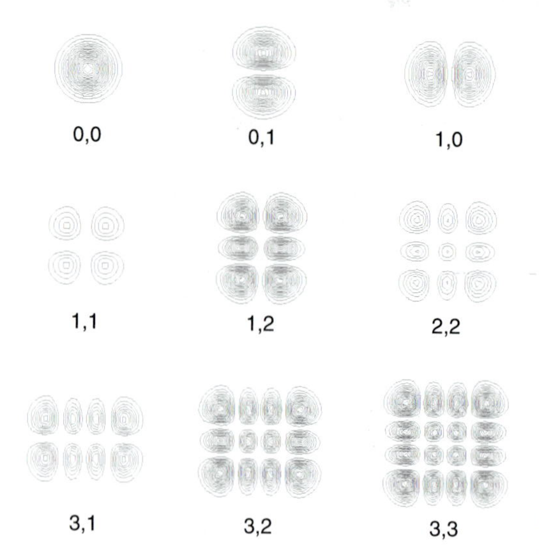

Figure 4.2 Transverse modes for a rectangular geometry laser.

The values of low-order Hermite polynomials are[12]

$$H_0(x) = 1 \tag{4.22}$$

$$H_1(x) = 2x \tag{4.23}$$

$$H_2(x) = 4x^2 - 2 \tag{4.24}$$

$$H_3(x) = 8x^3 - 12x. \tag{4.25}$$

The intensity of the rectangular modes is given by[13]

$$\boxed{\begin{aligned} I_{m,p}(x, y, z) = I_o \cdot &\left[H_m\left(\frac{\sqrt{2}x}{w(z)}\right) \cdot \exp\left(\frac{-x^2}{w^2(z)}\right)\right]^2 \\ \times &\left[H_p\left(\frac{\sqrt{2}y}{w(z)}\right) \cdot \exp\left(\frac{-y^2}{w^2(z)}\right)\right]^2 \end{aligned}} \tag{4.26}$$

and the resulting modes are illustrated in Figure 4.2.

4.3 TEM$_{0,0}$ GAUSSIAN BEAM PROPAGATION

4.3.1 The TEM$_{0,0}$ or Gaussian Transverse Mode

The lowest-order transverse mode is called the TEM$_{0,0}$ mode, the 00-mode, the lowest-order mode, or the *Gaussian mode*. This is the mode that is circular in transverse dimensions and has a Gaussian intensity profile. It is the mode that is most widely used in laser systems.

[12]H. Kogelnik and T. Li, "Laser Beams and Resonators," *Proc. IEEE* 54:1312 (1966), eq. 33.

[13]Walter Koechner, *Solid State Laser Engineering*, 3d ed. (Berlin, Germany: Springer-Verlag, 1992), p. 191.

Some important characteristics of this mode are its $1/e$ radius (called the beam waist $w(z)$) and its radius of curvature $R(z)$. Additional and more in-depth treatments of the Gaussian mode can be found in Yariv,[14] Verdeyen,[15] and in the classic paper by Kogelnik and Li.[16]

Mathematical treatment of the TEM$_{0,0}$ mode. The paraxial approximation to the transverse wave equation in free space is given by Equation (4.11) as

$$\left(\frac{\partial^2}{\partial x^2} + \frac{\partial^2}{\partial y^2} + (-2j\omega\sqrt{\mu_o \epsilon_m})\frac{\partial}{\partial z}\right)\psi = 0 \tag{4.27}$$

and is generally solved by converting the equation into cylindrical coordinates as

$$\frac{1}{r}\frac{\partial}{\partial r}\left(r\frac{\partial \psi}{\partial r}\right) - j2k\frac{\partial \psi}{\partial z} = 0. \tag{4.28}$$

The lowest-order solution to this equation has the general form

$$\psi = \exp\left(-j\left(P(z) + \frac{kr^2}{2q(z)}\right)\right) \tag{4.29}$$

and is the TEM$_{0,0}$ mode. (The transverse wave equation is discussed in the Appendix.)

Substituting Equation (4.29) back into the cylindrical form of the paraxial wave equation, Equation (4.28), permits the equation to be solved for the unknown beam parameter functions $P(z)$ and $q(z)$. After some mathematical manipulation it can be shown that

$$\frac{1}{q(z)} = \frac{1}{R(z)} - \frac{j\lambda_o}{\pi n w^2(z)} \tag{4.30}$$

and

$$jP(z) = \ln\left(1 - \frac{jz}{z_o}\right) \tag{4.31}$$

where λ_o is the free space wavelength, n is the index of refraction and where the physical parameters of the $1/e$ beam radius $w(z)$ and the beam curvature $R(z)$ have been introduced. For most calculations, this solution is written in a form that contains more physical meaning, namely

$$\psi = \left[\exp\left(-j\left[kz - \tan^{-1}\left(\frac{\lambda_o z}{\pi n w_o^2}\right)\right]\right)\right]\left[\frac{w_o}{w(z)}\exp\left(\frac{-r^2}{w^2(z)}\right)\right]\left[\exp\left(-j\left(\frac{kr^2}{2R(z)}\right)\right)\right]. \tag{4.32}$$

Examining the above equation more closely, notice that the first term

$$T_{\text{phase}} = \left[\exp\left(-j\left[kz - \tan^{-1}\left(\frac{\lambda_o z}{\pi n w_o^2}\right)\right]\right)\right] \tag{4.33}$$

---

[14] Amnon Yariv, *Optical Electronics*, 4th ed. (Philadelphia, PA: Saunders College Publishing, 1991), Chapter 2.

[15] Joseph T. Verdeyen, *Laser Electronics*, 2d ed. (Englewood Cliffs, NJ: Prentice Hall, 1989), Chapter 3.

[16] H. Kogelnik and T. Li, "Laser Beams and Resonators," *Proc. IEEE* 54:1312 (1966).

describes the longitudinal phase factor or the change in phase along the direction of propagation z. The second term

$$T_{\text{waist}} = \left[\frac{w_o}{w(z)} \exp\left(\frac{-r^2}{w^2(z)}\right)\right] \tag{4.34}$$

represents the cross-sectional $1/e$ radius of the beam (the beam waist $w(z)$) and describes how the waist changes as the beam propagates. The third term

$$T_{\text{curvature}} = \left[\exp\left(-j\left(\frac{kr^2}{2R(z)}\right)\right)\right] \tag{4.35}$$

describes the curvature of the phase front of the wave and indicates that the equiphase front is spherically curved with the radius $R(z)$.

The beam waist. The origin of the propagation axis z can be taken to be the place where the beam is the narrowest. At this point, the beam waist is termed the *minimum beam waist* and is identified as w_o. With this normalization, the $1/e$ beam radius $w(z)$ is given by

$$\boxed{w^2(z) = w_o^2\left(1 + \left(\frac{\lambda_o z}{\pi n w_o^2}\right)^2\right) = w_o^2\left(1 + \left(\frac{z}{z_R}\right)^2\right)} \tag{4.36}$$

which takes the form of a hyperbola with asymptotes at an angle

$$\boxed{\theta_a = \frac{\lambda_o}{\pi n w_o}} \tag{4.37}$$

an angle often termed the *far field diffraction angle* (see Figure 4.3).

Note that there is a potential confusion here, as θ_a is one-half the total radiated angle. With this formalism, the total diffracted angle is $2\theta_a$. Notice that it is not uncommon for

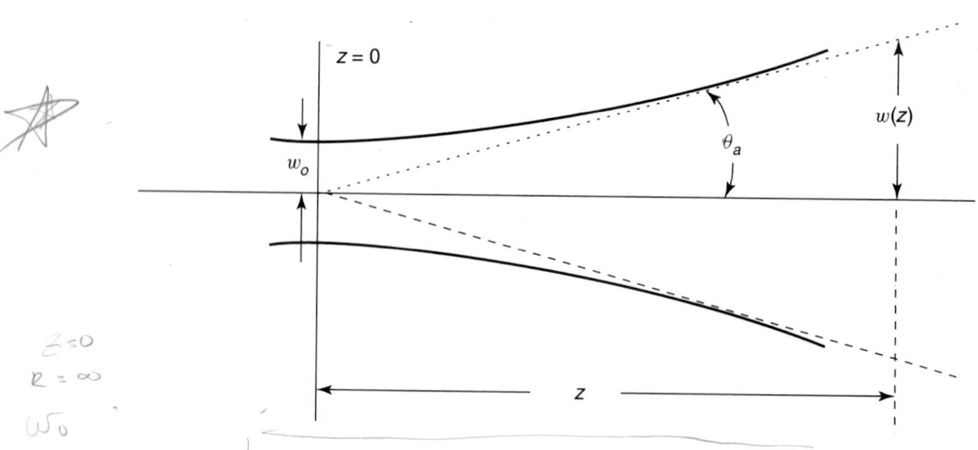

Figure 4.3 Major characteristics of the Gaussian beam waist $w(z)$.

this definition to be altered by a factor of 2 so that θ_a represents the *total* diffracted angle.[17] This is a common source of error in Gaussian beam calculations!

The $1/e$ beam radius $w(z)$ is called the *beam waist* (waist $\approx$ radius, $2 \cdot$ waist $\approx$ diameter). This can be quite confusing as it conflicts with the common use of the word "waist" as describing something approximately equal to a diameter. (For example, a wasp is not usually described as having a narrow "double waist.") Some care must be taken in making beam waist calculations to avoid factor-of-two errors due to this nomenclature!

The beam radius. The radius of curvature of the wavefront is given by

$$\boxed{R(z) = z\left(1 + \left(\frac{\pi n w_o^2}{\lambda_o z}\right)^2\right) = z\left(1 + \left(\frac{z_R}{z}\right)^2\right).} \tag{4.38}$$

Notice that the wavefront profile is planar at $z = 0$ ($R(z) \to \infty$) or where the beam is the narrowest (see Figure 4.4).

The distance

$$z_R = \frac{\pi n w_o^2}{\lambda_o} \tag{4.39}$$

is called the *Rayleigh range*. The Rayleigh range marks the transition from *near-field* to *far-field* behavior. This rather fuzzy definition can be clarified by examining a plot of $R(z)$ as a function of z for an argon-ion laser ($\lambda = 514.5$ nm) with beam waists of 3, 5, and 7 mm as shown in Figure 4.5. Notice that for distances smaller than the Rayleigh range, $R(z)$ is rapidly decreasing from infinity (a plane wave has $R(z) = \infty$). For distances larger than the Rayleigh range, $R(z)$ is slowly increasing and is approximately equal to z (a spherical wave from a point source at the origin).

Example 4.1

Consider an argon-ion laser ($\lambda_o = 514.5$ nm) with an initial beam waist of $w_o = 0.70$ mm at the end of the laser. What is the size of the beam waist $w(z)$ at a distance of $z = 50.0$ m from the front of the laser? What is the curvature of the beam waist $R(z)$ at that location? Assume the beam is traveling through air with an index of refraction $n = 1.0$.

Solution. First, calculate the Rayleigh range from Equation (4.39) as

$$z_R = \frac{\pi n w_o^2}{\lambda_o} = \frac{\pi \cdot 1.0 \cdot (0.70 \cdot 10^{-3}\ \text{m})^2}{514.5 \cdot 10^{-9}\ \text{m}} = 2.992\ \text{m}. \tag{4.40}$$

Then use Equation (4.36) for the waist in the form

$$w(z) = w_o\sqrt{1 + \left(\frac{z}{z_R}\right)^2} = (0.70 \cdot 10^{-3}\ \text{m})\sqrt{1 + \left(\frac{50.0\ \text{m}}{2.992\ \text{m}}\right)^2} = 11.719\ \text{mm} \tag{4.41}$$

and use Equation (4.38) for the radius as

$$R(z) = z\left(1 + \left(\frac{z_R}{z}\right)^2\right) = 50.0\ \text{m} \cdot \left(1 + \left(\frac{2.992\ \text{m}}{50.0\ \text{m}}\right)^2\right) = 50.179\ \text{m}. \tag{4.42}$$

[17] Joseph T. Verdeyen, *Laser Electronics*, 2d ed. (Englewood Cliffs, NJ: Prentice Hall, 1989), uses θ as the *full* angle, not the half-angle!

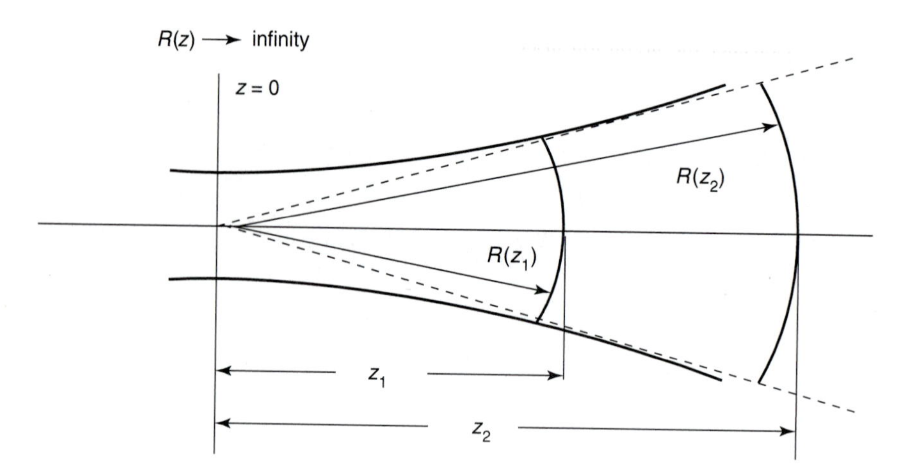

Figure 4.4 Major characteristics of the Gaussian beam radius of curvature $R(z)$.

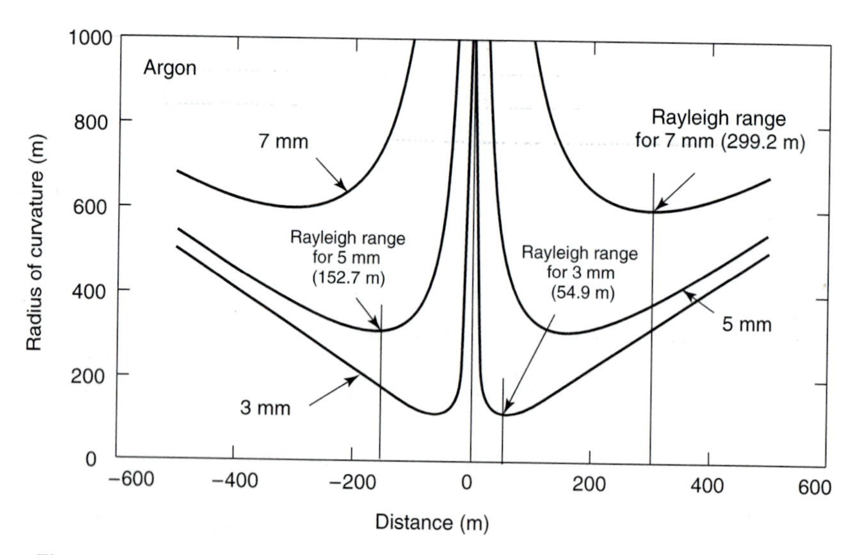

Figure 4.5 The Gaussian beam radius of curvature $R(z)$ plotted for an argon-ion laser running at λ =514.5 nm with three starting beam waists. The beam with a starting waist of $w_o = 3$ mm has a Rayleigh range of 54.955 m, the beam with a starting waist of $w_o = 5$ mm has a Rayleigh range of 152.653 m and the beam with a starting waist of $w_o = 7$ mm has a Rayleigh range of 299.199 m. Notice that the inflection point of the beam radius of curvature $R(z)$ corresponds to the Rayleigh range.

Example 4.2

Consider the argon-ion laser of Example 4.1. What is the size of the beam waist $w(z)$ at a distance of $z = 1.0$ m from the laser? What is the curvature of the beam waist $R(z)$ at that location? Assume the beam is traveling through air with an index of refraction $n = 1.0$.

Solution. Note that the Rayleigh range has not changed from $z_R = 2.992$ m of Example 4.1. Now use Equation (4.36) for the waist in the form

$$w(z) = w_o\sqrt{1 + \left(\frac{z}{z_R}\right)^2} = (0.70 \cdot 10^{-3}\ \text{m})\sqrt{1 + \left(\frac{1.0\ \text{m}}{2.992\ \text{m}}\right)^2} = 0.738\ \text{mm} \tag{4.43}$$

and use Equation (4.38) for the radius as

$$R(z) = z\left(1 + \left(\frac{z_R}{z}\right)^2\right) = 50.0\ \text{m} \cdot \left(1 + \left(\frac{2.992\ \text{m}}{50.0\ \text{m}}\right)^2\right) = 9.952\ \text{m}.$$

Example 4.3

Consider the argon-ion laser of Example 4.4. Assume that the beam waist is measured to be $w(z) = 0.80$ mm at a location z. How far is that location from the front of the laser? What is the curvature of the beam waist $R(z)$ at that location? Assume the beam is traveling through air with an index of refraction $n = 1.0$.

Solution. Note that the Rayleigh range has not changed from $z_R = 2.992$ m of Example (4.1). Now use Equation (4.36) in the slightly reorganized form

$$z = z_R\sqrt{\left(\frac{w(z)}{w_o}\right)^2 - 1} = (2.992\ \text{m})\sqrt{\left(\frac{0.80\ \text{mm}}{0.70\ \text{mm}}\right)^2 - 1} = 1.655\ \text{m} \tag{4.44}$$

and use Equation (4.38) for the radius as

$$R(z) = z\left(1 + \left(\frac{z_R}{z}\right)^2\right) = 1.655\ \text{m} \cdot \left(1 + \left(\frac{2.992\ \text{m}}{1.655\ \text{m}}\right)^2\right) = 7.063\ \text{m}. \tag{4.45}$$

Example 4.4

How accurate is the angular approximation for the beam waist given in Equation (4.37) versus the more accurate expression in Equation (4.36)? Plot both expressions for an argon-ion laser ($\lambda_o = 514.5$ nm) with an initial beam waist of $w_o = 1.0$ mm. Assume the beam is traveling through air with an index of refraction $n = 1.0$.

Solution. First, calculate the Rayleigh range from Equation (4.39) as

$$z_R = \frac{\pi n w_o^2}{\lambda_o} = \frac{\pi \cdot 1.0 \cdot (1.0 \cdot 10^{-3}\ \text{m})^2}{514.5 \cdot 10^{-9}\ \text{m}} = 6.106\ \text{m}. \tag{4.46}$$

The $1/e$ beam radius (waist) $w(z)$ is plotted using both Equation (4.36) and Equation (4.37) and displayed in Figure 4.6 for a range of $z = 1$ cm to 10 meters (up to approximately twice the Rayleigh range) and in Figure 4.7 for a range of $z = 1$ to 50 meters (up to approximately ten times the Rayleigh range).

Note that at $z = 5$ meters, the approximation gives $w(z) = 0.819$ mm while the more accurate expression gives $w(z) = 1.292$ mm for a factor of 1.578 difference. However, at $z =$ 25 meters, the approximation gives $w(z) = 4.094$ mm while the more accurate expression gives $w(z) = 4.215$ mm for a factor of 1.03 difference.

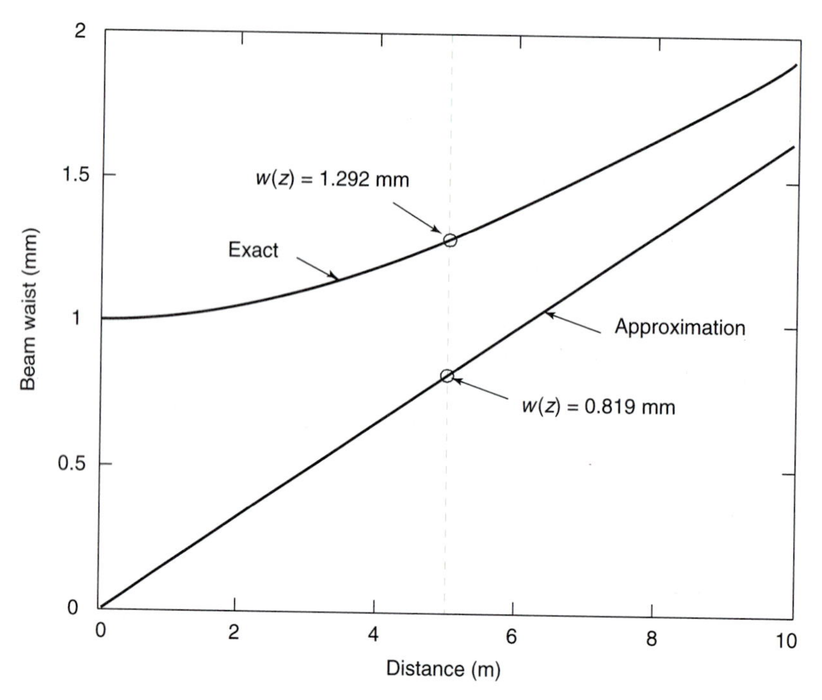

Figure 4.6 The $1/e$ beam radius (waist) $w(z)$ is plotted for an argon-ion laser (λ_o = 514.5 nm) using both the angular approximation and the more accurate expression for a range of z= 1 cm to 10 meters.

4.3.2 Properties of the TEM$_{0,0}$ Mode of the Laser

Beam waist and radius of curvature as a function of wavelength. Gaussian beams have a number of interesting properties. Consider first Equation (4.36) for the propagation of the beam waist $w(z)$

$$w(z) = w_o\sqrt{1+\left(\frac{\lambda_o z}{\pi n w_o^2}\right)^2} = w_o\sqrt{1+\left(\frac{z}{z_R}\right)^2}. \tag{4.47}$$

Notice if z/z_R is much greater than unity (i.e., the measurement is made in the far field region) then

$$w(z) = w_o\sqrt{1+\left(\frac{\lambda_o z}{\pi n w_o^2}\right)^2} \approx w_0\left(\frac{\lambda_o z}{\pi n w_o^2}\right). \tag{4.48}$$

This results in a situation where the beam waist is roughly proportional to the free space wavelength λ_o of the laser beam. This situation is illustrated graphically in Figure 4.8, which compares beam waist size as a function of z for an argon-ion beam at (λ_o = 514.5 nm), a HeNe beam at (λ_o = 632.8 nm), and a Nd:YAG beam at (λ_o = 1.064 μm). The starting beam waist is assumed to be 1 mm in each case. Notice that the Nd:YAG beam

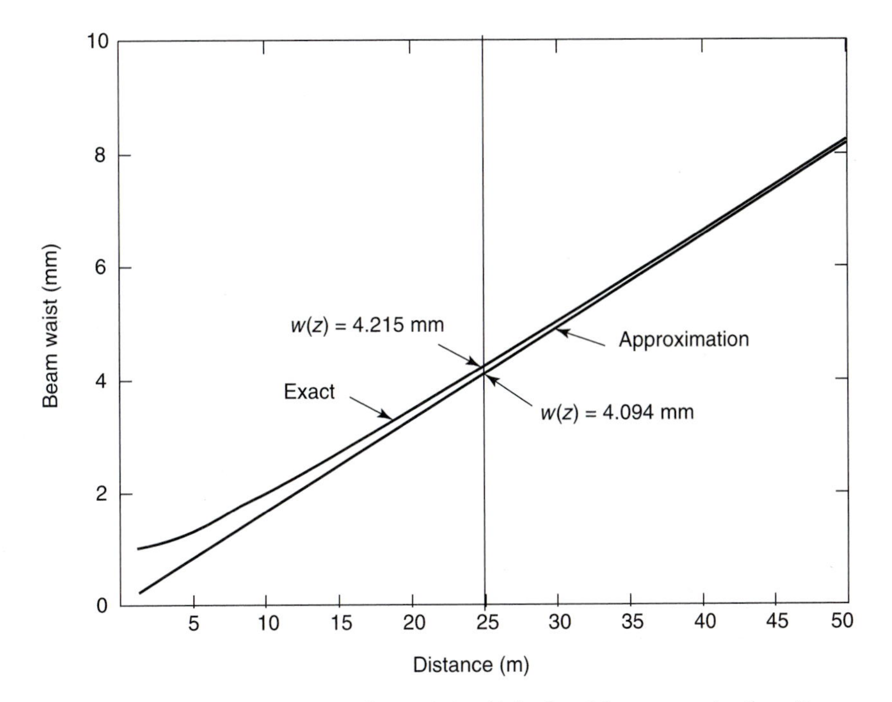

Figure 4.7 The $1/e$ beam radius (waist) $w(z)$ is plotted for an argon-ion laser (λ_o = 514.5 nm) using both the angular approximation and the more accurate expression for a range of z = 1 to 50 meters.

has expanded a factor of 3.5 over the 10 m range, while the argon-ion beam has roughly doubled in size.

A related situation exists with the radius of curvature $R(z)$. In this case, however, in the far field z_R/z is much less than 1, and the radius then becomes roughly equal to the distance from the beam waist as

$$R(z) = z\left(1 + \left(\frac{\pi n w_o^2}{\lambda_o z}\right)^2\right) = z\left(1 + \left(\frac{z_R}{z}\right)^2\right) \approx z. \tag{4.49}$$

Putting this more succinctly, the wavefront is acting like a *spherical wave* with its origin at the minimum beam waist.

This situation is illustrated in Figure 4.9, which compares beam radii of curvature as a function of z for an argon-ion beam at ($\lambda_o = 514.5$ nm), a HeNe beam at ($\lambda_o = 632.8$ nm), and a Nd:YAG beam at ($\lambda_o = 1.064$ μm). Again, the starting beam waist is assumed to be 1 mm in each case. Notice that all the radii begin to converge on a value where $R(z) \approx z$.

Example 4.5

Consider an argon-ion laser ($\lambda_o = 514.5$ nm) with three different beam waists, $w_{o1} = 0.50$ mm, $w_{o2} = 1.0$ mm, and $w_{o3} = 2.0$ mm, at the end of the laser. Plot the beam waist $w(z)$ as a function of z for each of these three starting waists. Assume the beam is traveling through air with an index of refraction $n = 1.0$.

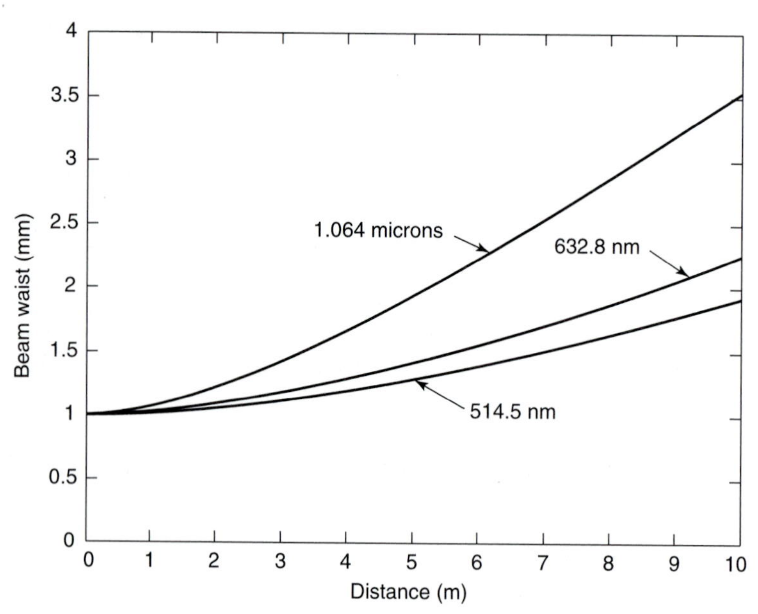

Figure 4.8 A comparison of beam waist size as a function of z for an argon-ion beam at ($\lambda = 514.5$ nm), a HeNe beam at ($\lambda = 632.8$ nm), and a Nd:YAG beam at ($\lambda = 1.064$ μm). The starting beam waist is assumed to be 1 mm in each case.

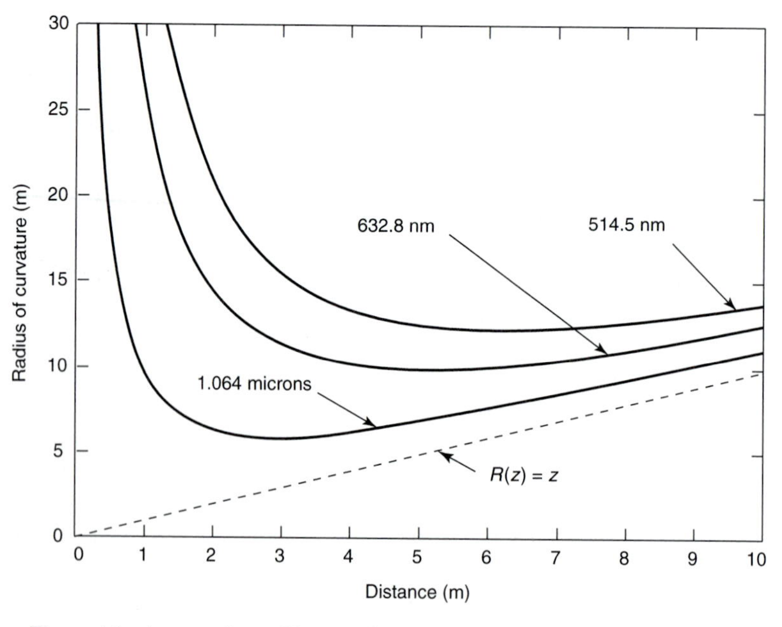

Figure 4.9 A comparison of beam radii of curvature $R(z)$ as a function of z for an argon-ion laser beam ($\lambda = 514.5$ nm), a HeNe beam ($\lambda = 632.8$ nm), and a Nd:YAG beam ($\lambda = 1.064$ μm); the starting beam waist is assumed to be 1 mm in each case.

Solution. The Rayleigh ranges can be calculated from Equation (4.39) as

$$z_{R1} = \frac{\pi n w_o^2}{\lambda_o} = \frac{\pi \cdot 1.0 \cdot (0.5 \cdot 10^{-3}\ \text{m})^2}{514.5 \cdot 10^{-9}\ \text{m}} = 1.527\ \text{m} \tag{4.50}$$

$$z_{R2} = \frac{\pi n w_o^2}{\lambda_o} = \frac{\pi \cdot 1.0 \cdot (1.0 \cdot 10^{-3}\ \text{m})^2}{514.5 \cdot 10^{-9}\ \text{m}} = 6.106\ \text{m} \tag{4.51}$$

$$z_{R3} = \frac{\pi n w_o^2}{\lambda_o} = \frac{\pi \cdot 1.0 \cdot (2.0 \cdot 10^{-3}\ \text{m})^2}{514.5 \cdot 10^{-9}\ \text{m}} = 24.424\ \text{m}. \tag{4.52}$$

The *1/e* beam radius (waist) $w(z)$ is plotted using Equation (4.36) and the beam curvature is plotted using Equation (4.38). The two plots are displayed in Figures 4.10 and 4.11 for a range of $z = 1$ cm to 10 meters. Notice that the wider the initial starting waist, the narrower the final beam waist.

Beam waist expansion during beam propagation. Consider an argon-ion laser ($\lambda_o = 514.5$ nm) where w_o is located at the output mirror of the laser and has the value $w_o = 0.4$ mm. The Rayleigh range is given by Equation (4.39) as

$$z_R = \frac{\pi n w_o^2}{\lambda_o} = \frac{\pi \cdot 1.0 \cdot (0.40 \cdot 10^{-3}\ \text{m})^2}{514.5 \cdot 10^{-9}\ \text{m}} = 0.977\ \text{m} \tag{4.53}$$

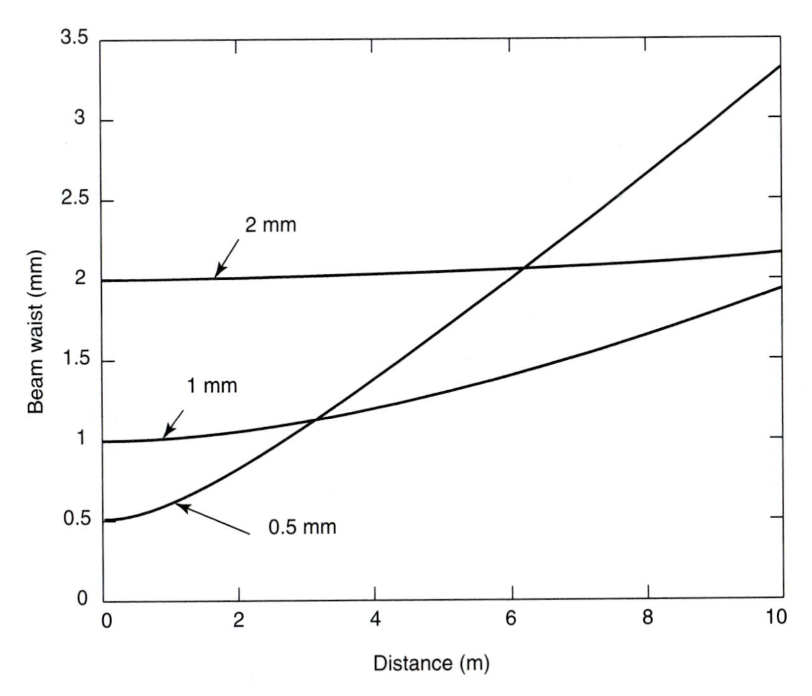

Figure 4.10 The beam waist $w(z)$ as a function of z is plotted for an argon-ion laser ($\lambda_o = 514.5$ nm) with three different starting minimum beam waists: $w_{o1} = 0.50$ mm, $w_{o2} = 1.0$ mm, and $w_{o3} = 2.0$ mm.

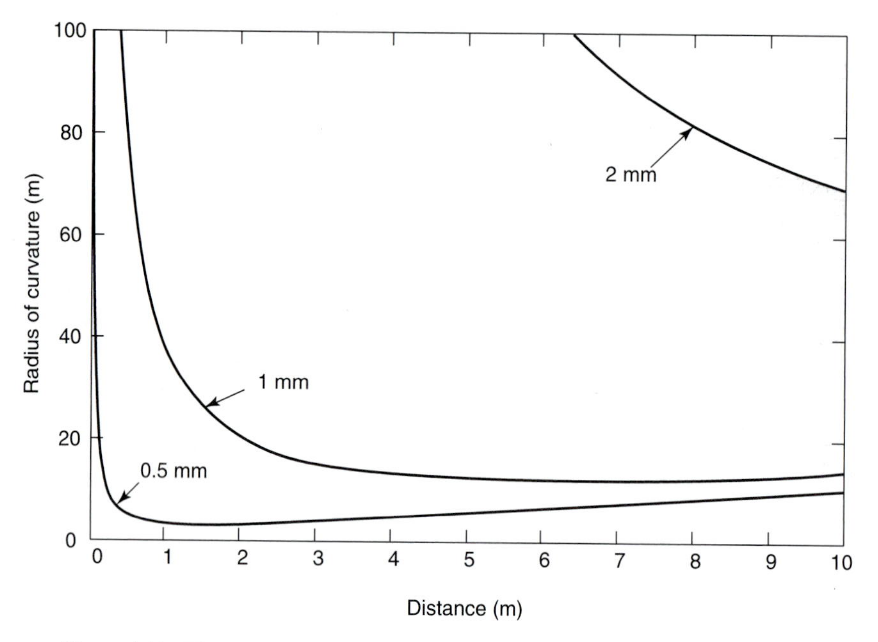

Figure 4.11 The beam curvature $R(z)$ as a function of z is plotted for an argon-ion laser ($\lambda_o = 514.5$ nm) with three different starting minimum beam waists, $w_{o1} = 0.50$ mm, $w_{o2} = 1.0$ mm, and $w_{o1} = 2.0$ mm.

and the beam waist at 100 m is given by Equation (4.36) as

$$w(z) = w_o\sqrt{1+\left(\frac{z}{z_R}\right)^2} = (0.40\cdot 10^{-3}\ \text{m})\sqrt{1+\left(\frac{100.0\ \text{m}}{0.977\ \text{m}}\right)^2} = 40.945\ \text{mm}. \tag{4.54}$$

In this case, the beam has expanded by a factor of

$$M = \frac{40.945\ \text{mm}}{0.4\ \text{mm}} = 102.361. \tag{4.55}$$

However, if the beam were expanded optically by a factor of ten by using a *beam expander* in front of the laser such that $w_o = 4$ mm (assuming the waist occurs at the end of the beam expander), then the Rayleigh range is given by Equation (4.39) as

$$z_R = \frac{\pi n w_o^2}{\lambda_o} = \frac{\pi\cdot 1.0\cdot(4.0\cdot 10^{-3}\ \text{m})^2}{514.50\cdot 10^{-9}\ \text{m}} = 97.698\ \text{m} \tag{4.56}$$

and the beam waist at 100 m is given by Equation (4.36) as

$$w(z) = w_o\sqrt{1+\left(\frac{z}{z_R}\right)^2} = (4.0\cdot 10^{-3}\ \text{m})\sqrt{1+\left(\frac{100.0\ \text{m}}{97.698\ \text{m}}\right)^2} = 5.724\ \text{mm}. \tag{4.57}$$

In this case, the beam has only expanded by a factor of

$$M = \frac{5.724\ \text{mm}}{0.4\ \text{mm}} = 1.431. \tag{4.58}$$

The $2z_R$ length region around the minimum beam waist is often assumed to be the most collimated region of a Gaussian laser beam. Notice that the beam will expand only by a factor of $\sqrt{2}$ over this region (see Figure 4.12).

Example 4.6

Consider an argon-ion laser (λ_o = 514.5 nm) that must propagate over 200 m. What is the critical beam waist size assuming that the total propagation distance is equal to two Rayleigh ranges and that the minimum waist occurs in the middle of this range? Plot the beam waist $w(z)$ as a function of z for this critical waist and for two additional beam waists, one a bit larger and one a bit smaller than the critical size. In all three cases, what is the total magnification of the beam? Assume the beam is traveling through air with an index of refraction $n = 1.0$.

Solution. The Rayleigh range is chosen so that the beam waist occurs at $z_R = 100$ m. From this, the critical waist size can be calculated from Equation (4.36) as

$$w_{\text{crit}} = \sqrt{\frac{\lambda_o z_R}{\pi \cdot n}} = \sqrt{\frac{514.50 \cdot 10^{-9}\ \text{m} \cdot 100\ \text{m}}{\pi \cdot 1.0}} = 4.047\ \text{mm}. \tag{4.59}$$

The $1/e$ beam radius (waist) $w(z)$ is plotted for $w_{\text{crit}} = 4.047$ mm, as well as $w_o = 3.0$ mm and $w_o = 5.0$ mm, using Equation (4.36). The plot is displayed in Figure 4.12 for a range of z = −100 to +100 meters.

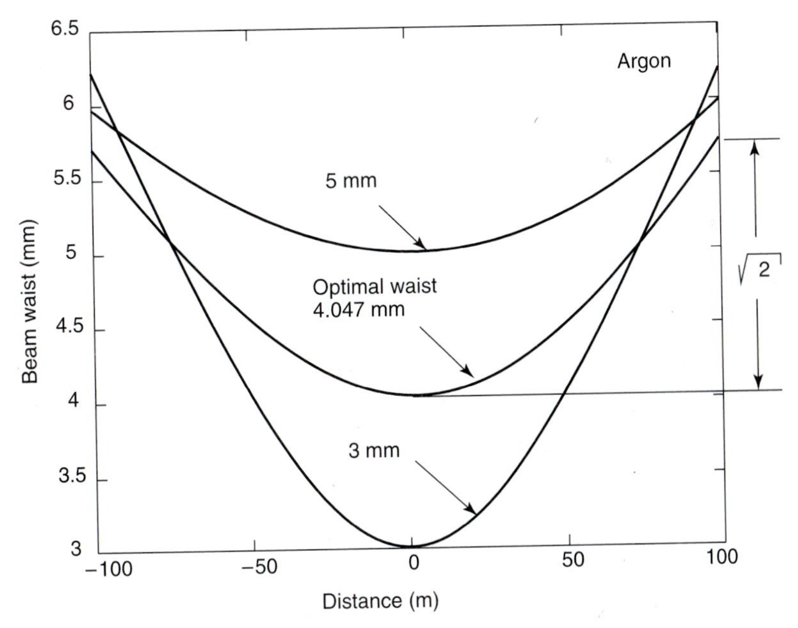

Figure 4.12 The beam waist $w(z)$ as a function of z is plotted for several starting minimum beam waists for an argon-ion laser (λ_o = 514.5 nm) propagating over a distance of 200 meters. The optimal waist $w_{\text{opt}} = 4.047$ mm, as well as the additional waists $w_o = 3.0$ mm and $w_o = 5.0$ mm, are illustrated.

The magnification of the three beams is given by

$$M_{\text{crit}} = \frac{w(100\text{ m})}{w_{Crit}} = \frac{5.723\text{ mm}}{4.047\text{ mm}} = 1.414 \tag{4.60}$$

$$M_{\text{smaller}} = \frac{w(100\text{ m})}{w_{\text{smaller}}} = \frac{6.229\text{ mm}}{3.0\text{ mm}} = 2.076 \tag{4.61}$$

$$M_{\text{bigger}} = \frac{w(100\text{ m})}{w_{\text{bigger}}} = \frac{5.977\text{ mm}}{5.0\text{ mm}} = 1.195. \tag{4.62}$$

Focusing the Gaussian beam. Consider the common requirement that a Gaussian laser beam be focused down to a small spot. Recall from Section 4.3 that the beam waist in the far field region can be expressed as

$$w(z) = w_o\sqrt{1 + \left(\frac{\lambda_o z}{\pi n w_o^2}\right)^2} \approx w_0\left(\frac{\lambda_o z}{\pi n w_o^2}\right). \tag{4.63}$$

In designing an optical system to focus a Gaussian laser beam down to a small spot, the beam waist $w(z)$ is used to represent the beam waist at the lens and z is taken to be the focal length f of the lens. Substituting these into Equation 4.63 gives

$$w(f) \approx w_o\left(\frac{\lambda_o f}{\pi n w_o^2}\right) \tag{4.64}$$

where w_o is now the spot size at the focus. Rewriting in terms of the minimum spot size w_o at the focus of the lens yields

$$w_o \approx \frac{\lambda_o f}{\pi n w(f)} \tag{4.65}$$

where f is the focal length of the lens and $w(z)$ is the radius at the lens. (Lenses are often characterized by their *f number*, defined as $f/$*aperture size*. Thus to achieve the smallest possible spot, the f number should be $< f/\pi\, w(z)$.)

Another interesting practical question is the *depth of field* of a Gaussian beam. Typically, the depth of field is assumed to be the Rayleigh range as[18]

$$z_R = \frac{\pi n w_o^2}{\lambda_o}. \tag{4.66}$$

Thus, if a beam is focused to a spot size of dimension $N\lambda_o$ then the depth of field is roughly $\pi n N^2 \lambda_o$. If the beam is focused down to the minimum spot size, then the depth of field is

$$z_R = \frac{\lambda_o f^2}{\pi n (w(f))^2}. \tag{4.67}$$

Example 4.7

Consider an argon-ion laser ($\lambda_o = 514.5$ nm) focused through an $f = 10$ cm focal length lens and possessing a beam waist of $w(z) = 5$ mm at the lens. What is the smallest possible spot permitted by the properties of the Gaussian beam? What is the smallest possible spot permitted by fundamental diffraction theory? How do these two compare?

---

[18] Anthony Siegman, *Lasers* (Mill Valley, CA: University Science Books, 1986), p. 677.

Solution. The smallest possible spot predicted using Gaussian beam theory is given by Equation (4.65) as

$$w_o \approx \frac{\lambda_o f}{\pi n w(z)} = \frac{514.50 \cdot 10^{-9}\ \text{m} \cdot 10\ \text{cm}}{\pi \cdot 1.0 \cdot 5.0\ \text{mm}} = 3.275\ \mu\text{m} \tag{4.68}$$

while the smallest spot permitted by the diffraction limit can be obtained from classical optics as[19]

$$w_o \approx 1.22 \cdot \frac{\lambda_o f}{2 n w(z)} = \frac{1.22 \cdot 514.50 \cdot 10^{-9}\ \text{m} \cdot 10\ \text{cm}}{2 \cdot 1.0 \cdot 5.0\ \text{mm}} = 6.277\ \mu\text{m}. \tag{4.69}$$

The diffraction-limited spot size is the smallest waist possible, so in this case the beam waist of the spot will be limited to 6.277 μm.

4.4 RAY MATRICES TO ANALYZE PARAXIAL LENS SYSTEMS

In laser engineering, it is frequently necessary to analyze beam propagation through a system of lenses and mirrors. If there are more than one or two elements in the system, a classical analysis can be complex. Luckily, a matrix method exists for analyzing complex optical systems.[20] This method (termed the *ray matrix, lens matrix*, or *ABCD matrix* method) approximates the electromagnetic wave by groups of rays and uses matrices to keep track of the location and slope of the rays.

The ray matrix method is directly applicable to the analysis of systems with Gaussian beams because the simple matrices developed for the ray matrix approach can be used in the Gaussian *ABCD* law (see Section 4.4.3). Thus, the simple ray tracing method has utility far beyond the conditions under which it is derived.

The information required to specify the path of a ray traveling from transverse plane 1 in free space to transverse plane 2 is the radial location (r_1 and r_2) and its slope (r_1' and r_2') (see Figure 4.13). With this information, a ray matrix can be written expressing the relationship between r_1, r_1' and r_2, r_2' as

$$\begin{bmatrix} r_2 \\ r_2' \end{bmatrix} = \begin{bmatrix} A & B \\ C & D \end{bmatrix} \begin{bmatrix} r_1 \\ r_1' \end{bmatrix} \tag{4.70}$$

where A can be interpreted as the lateral magnification, B as the effective thickness, C as the reciprocal length, and D as the angular magnification.

So, ray tracing through a large number of components (see Figure 4.14) consists simply of determining the *ABCD* matrix for each component and multiplying the matrices as

$$\begin{bmatrix} r_6 \\ r_6' \end{bmatrix} = \begin{bmatrix} A & B \\ C & D \end{bmatrix}_{5,6} \begin{bmatrix} A & B \\ C & D \end{bmatrix}_{4,5} \begin{bmatrix} A & B \\ C & D \end{bmatrix}_{3,4} \begin{bmatrix} A & B \\ C & D \end{bmatrix}_{2,3} \begin{bmatrix} A & B \\ C & D \end{bmatrix}_{1,2} \begin{bmatrix} r_1 \\ r_1' \end{bmatrix}. \tag{4.71}$$

Notice that the ray matrices are multiplied in *reverse* order.[21] The *rightmost* matrix in the expression is the *first* optical element in encountered in the system; the leftmost matrix is

[19] B. E. A. Salehy and M. C. Teich, *Fundamentals of Photonics* (New York: John Wiley and Sons, 1991), p. 131.

[20] H. Kogelnik and T. Li, "Laser Beams and Resonators," *Proc. IEEE* 54:1312 (1966).

[21] Joseph T. Verdeyen, *Laser Electronics*, 2d ed. (Englewood Cliffs, NJ: Prentice Hall, 1989), p. 38.

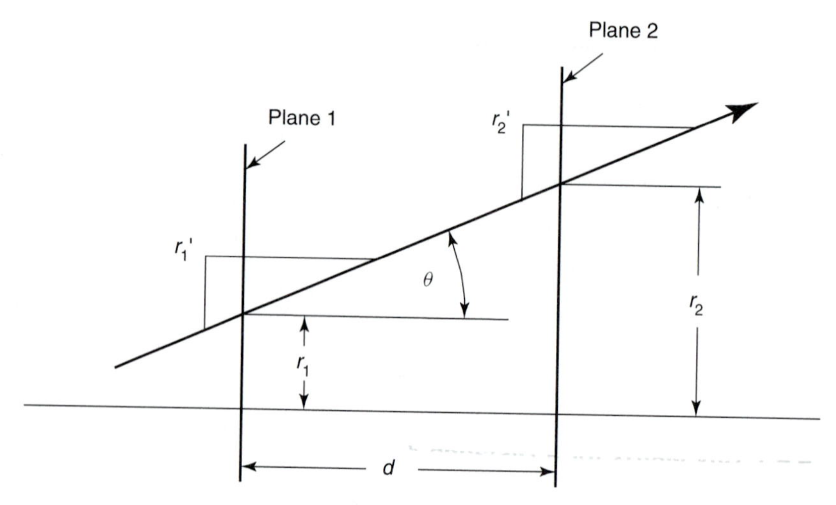

Figure 4.13 The minimum information required to specify the path of a ray traveling from plane 1 in free space to plane 2 is the radial location, r_1 and r_2, and its slope, r_1' and r_2'.

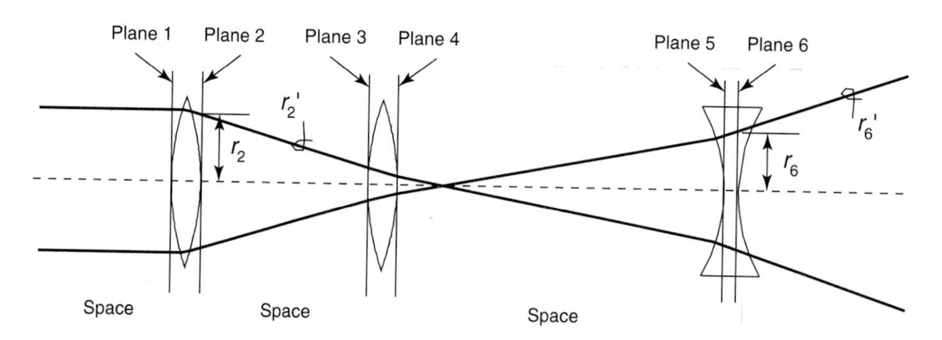

Figure 4.14 Ray tracing through a large number of components can be accomplished by multiplying the *ABCD* matrices for the various components (including spaces!).

the *last*. Also notice that the ray matrices are always written from plane to plane, not from element to element. Thus, ray matrices must be included for the *space* between elements as well as the elements themselves.

Example 4.8

Consider a simple paraxial optical system consisting of two lenses separated by a space d. Assume that the ray matrices for the lenses are given by

$$L_1 = \begin{bmatrix} 1 & 0 \\ -0.5 & 1 \end{bmatrix} \tag{4.72}$$

$$L_2 = \begin{bmatrix} 1 & 0 \\ -0.2 & 1 \end{bmatrix} \tag{4.73}$$

and assume that the ray matrix for the space is given by

$$d = \begin{bmatrix} 1 & 10 \\ 0 & 1 \end{bmatrix}. \tag{4.74}$$

Construct the ray matrix for the system from the front of lens 1 to the back of lens 2.

Solution. The light path starts at the front of lens 1 travels through the space between the lenses, travels through lens 2, and ends up at the back of lens 2. Thus, the ray matrix for the system looks like

$$\mathbf{S} = \text{lens 2} \cdot \text{space} \cdot \text{lens 1} = L_2 \cdot d \cdot L_1 \tag{4.75}$$

$$\mathbf{S} = \begin{bmatrix} 1 & 0 \\ -0.2 & 1 \end{bmatrix} \begin{bmatrix} 1 & 10 \\ 0 & 1 \end{bmatrix} \begin{bmatrix} 1 & 0 \\ -0.5 & 1 \end{bmatrix} = \begin{bmatrix} -4 & 10 \\ 0.3 & -1 \end{bmatrix}. \tag{4.76}$$

4.4.1 Ray Matrix for a Distance *d*

Consider the question of deriving a ray matrix for the space of length d (see Figure 4.15). Imagine a ray traveling through free space between two planes 1 and 2. The ray is a distance r_1 from the zero axis at plane 1 and a distance r_2 from the zero axis of plane 2. The distance between planes 1 and 2 is d. The slope is considered to be positive when r_2 (at plane 2) is greater than r_1 (at plane 1).

From Figure 4.15 the distance r_2 can be seen to be

$$r_2 = r_1 + d \tan\theta. \tag{4.77}$$

However, notice that the slope of the ray r_1' is given by

$$r_1' = \frac{d \tan\theta}{d} = \tan\theta. \tag{4.78}$$

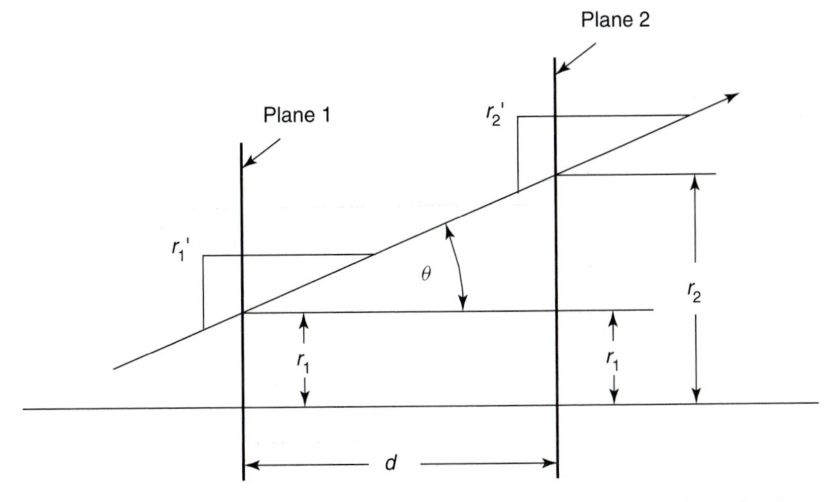

Figure 4.15 A ray traveling through free space between two transverse planes 1 and 2. The ray is a distance r_1 from the zero axis at plane 1 and a distance r_2 from the zero axis of plane 2. The distance between planes 1 and 2 is d.

Thus, an alternative expression for r_2 (notice without approximation) is

$$r_2 = r_1 + dr_1'. \tag{4.79}$$

Now, for this example, the slope of the ray 2 r_2' is equal to the slope of ray 1 so

$$r_2' = r_1' \tag{4.80}$$

or rewriting as simultaneous equations

$$r_2 = 1 \cdot r_1 + d \cdot r_1' \tag{4.81}$$

and the slope r_2' by

$$r_2' = 0 \cdot r_1 + 1 \cdot r_1'. \tag{4.82}$$

So, the ray matrix for a distance d of free space is given as

$$\begin{bmatrix} r_2 \\ r_2' \end{bmatrix} = \begin{bmatrix} 1 & d \\ 0 & 1 \end{bmatrix} \begin{bmatrix} r_1 \\ r_1' \end{bmatrix}. \tag{4.83}$$

4.4.2 Ray Matrix for a Lens

For a more complex example, consider a lens (see Figure 4.16). The ray matrix will be determined between the two evaluation planes on either side of the thin lens. For a thin lens, the assumption is that the ray location does not change as the ray crosses the lens. Thus,

$$r_2 = r_1. \tag{4.84}$$

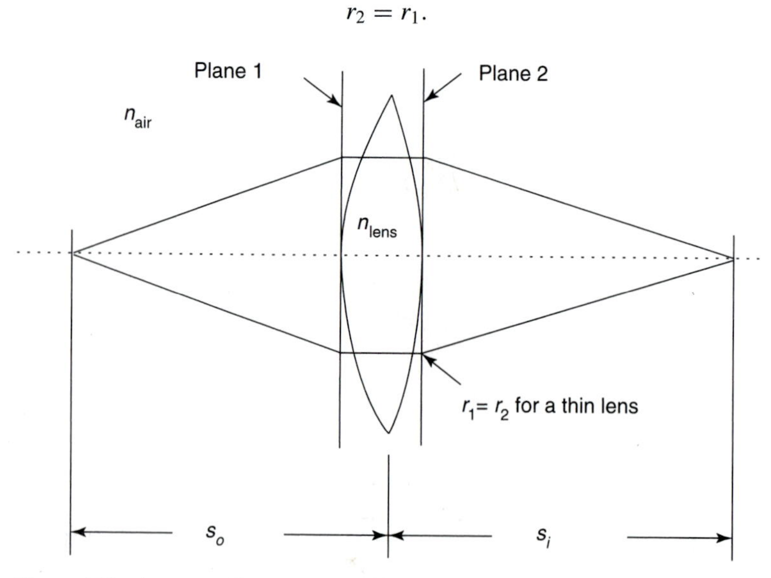

Figure 4.16 A ray traveling through a thin lens (drawn schematically as a thick lens!) between two planes 1 and 2, where s_o is the distance to the object and s_i is the distance to the image.

The major changes are in the slope of the ray. The *lensmaker's* equation (Equation (A.50) (see the Appendix) takes the form

$$\frac{1}{s_o} + \frac{1}{s_i} = (n_{\text{lens}} - 1)\left(\frac{1}{\mathbf{R}_1} + \frac{1}{\mathbf{R}_2}\right) \tag{4.85}$$

where s_o is the distance to the object, s_i is the distance to the image, and $\mathbf{R}_1$ and $\mathbf{R}_2$ are the curvatures of the two spherical lens surfaces. The focal length f is the image distance s_i when the object distance $s_o \rightarrow \infty$, or

$$\frac{1}{f} = (n_{\text{lens}} - 1)\left(\frac{1}{\mathbf{R}_1} + \frac{1}{\mathbf{R}_2}\right). \tag{4.86}$$

In other words, the focal length of a thin lens is the distance the focus is from the lens given a collimated input beam. Another way of saying this is that if the input ray has zero slope, the focal length is the distance from the back plane to where the output ray crosses the zero axis at the center of the lens (see Figure 4.17). This is also true in reverse. In other words, if the ray crosses the front focus of the lens, then it will emerge from the lens as an output ray of zero slope (see Figure 4.18). These factors make possible a very simple solution to the lens matrix of a thin lens.

Assume that the slope r_2' of the ray is given by the equation

$$r_2' = Cr_1 + Dr_1'. \tag{4.87}$$

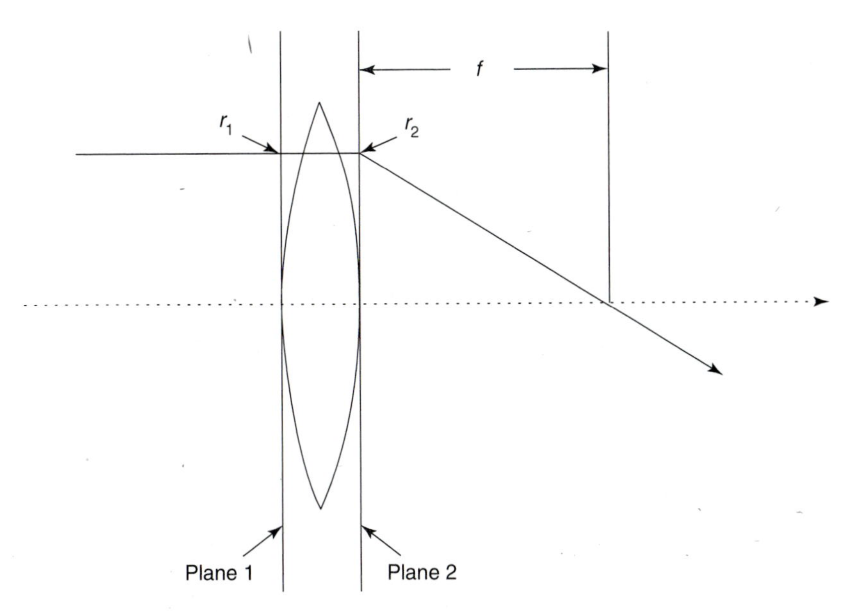

Figure 4.17 Given a collimated input beam, the focal length of a thin lens is the distance from the back plane to the focus.

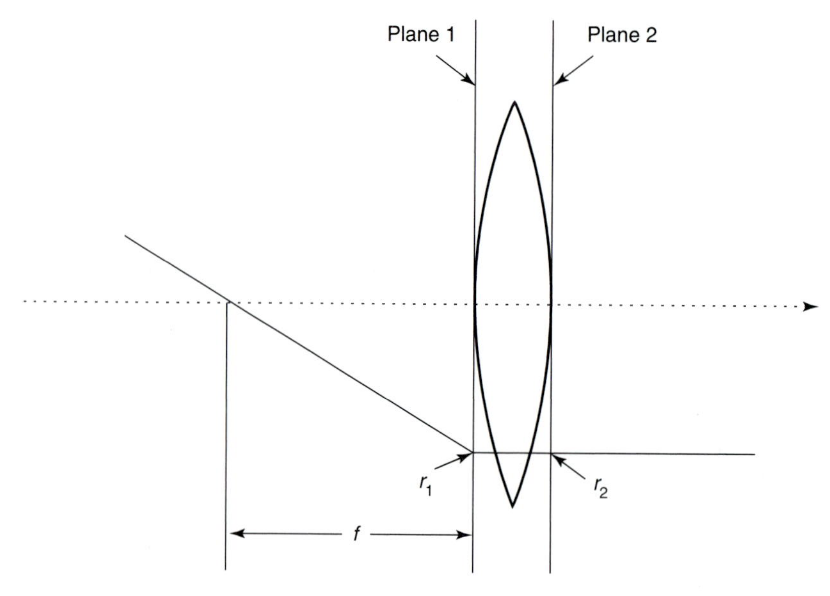

Figure 4.18 Given a collimated output beam, the focal length of a thin lens is the distance to the source.

Consider a collimated ray (which has a slope $r_1' = 0$) entering the lens at plane 1 (see Figure 4.17). The output ray will focus at a distance f, and so the slope of the output ray will be

$$r_2' = -\frac{r_1}{f}. \tag{4.88}$$

Substituting these parameters into the general equation for r_2', Equation (4.87), facilitates evaluation of the coefficient C as

$$r_2' = Cr_1 + Dr_1' = Cr_1 + D(0) = -\frac{r_1}{f} \tag{4.89}$$

or

$$C = -\frac{1}{f}. \tag{4.90}$$

To obtain the other coefficient D, consider a ray that is collimated (has a slope $r_2' = 0$) when leaving the lens at plane 2 (see Figure 4.18). The input ray must start at the distance f, and so the slope of the input ray will be

$$r_1' = \frac{r_1}{f}. \tag{4.91}$$

Substituting these parameters into the general equation for r_2', Equation (4.87), facilitates evaluation of the coefficient D as

$$r_2' = -\frac{1}{f}r_1 + D\frac{r_1}{f} = 0 \tag{4.92}$$

or

$$D = 1 \tag{4.93}$$

yielding a general equation

$$r_2' = -\frac{1}{f}r_1 + r_1'. \tag{4.94}$$

This is consistent for both cases. If the input beam is collimated, then r_1' is zero, and $r_2' = -\ 1/f\,r_1$ (which checks). If the output beam is collimated, then $r_1' = r_1/f$ and $r_2' = 0$ (which checks).

Thus, the set of equations describing the thin lens with plane 1 at one edge and plane 2 at the other is given as

$$r_2 = r_1 \tag{4.95}$$

$$r_2' = -\frac{1}{f}r_1 + r_1'. \tag{4.96}$$

So, the ray matrix for a thin lens is

$$\begin{bmatrix} r_2 \\ r_2' \end{bmatrix} = \begin{bmatrix} 1 & 0 \\ \frac{-1}{f} & 1 \end{bmatrix} \begin{bmatrix} r_1 \\ r_1' \end{bmatrix}. \tag{4.97}$$

Example 4.9

Consider a lens with some space d preceding the lens (see Figure 4.19). Derive the ray matrix for this combination.

Solution. The ray matrix for the combination is given by

$$\begin{bmatrix} r_3 \\ r_3' \end{bmatrix} = \begin{bmatrix} A & B \\ C & D \end{bmatrix} \begin{bmatrix} r_1 \\ r_1' \end{bmatrix} = \begin{bmatrix} 1 & 0 \\ \frac{-1}{f} & 1 \end{bmatrix} \begin{bmatrix} 1 & d \\ 0 & 1 \end{bmatrix} \begin{bmatrix} r_1 \\ r_1' \end{bmatrix} \tag{4.98}$$

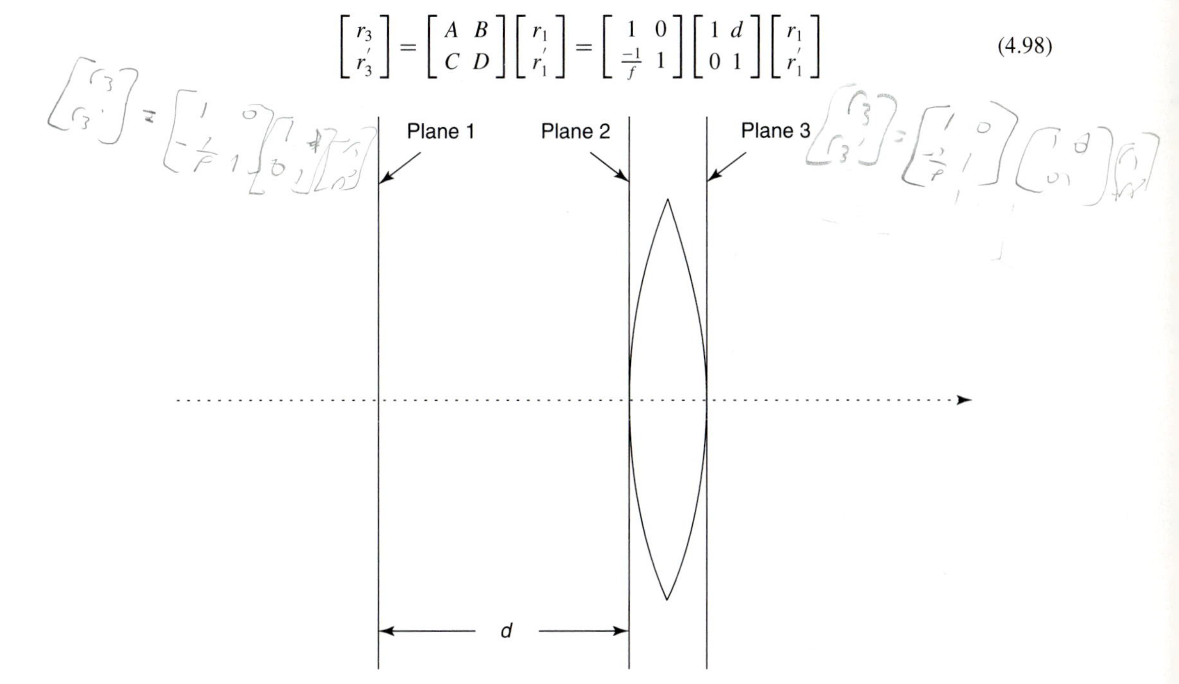

Figure 4.19 A simple lens system consists of a space d preceding a thin lens.

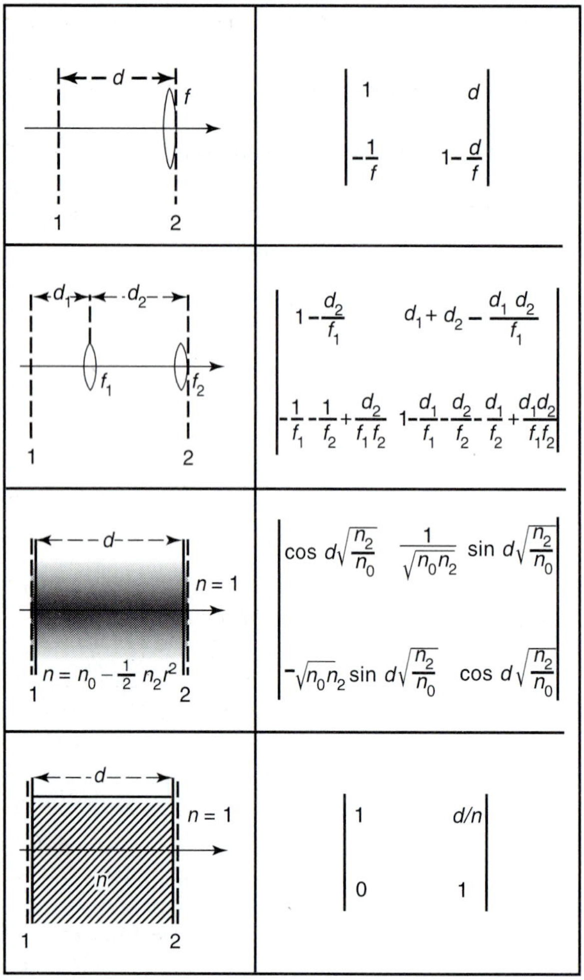

d, f, 1, 2	$\begin{vmatrix} 1 & d \\ -\frac{1}{f} & 1-\frac{d}{f} \end{vmatrix}$
d_1, d_2, f_1, f_2, 1, 2	$\begin{vmatrix} 1-\frac{d_2}{f_1} & d_1+d_2-\frac{d_1 d_2}{f_1} \\ -\frac{1}{f_1}-\frac{1}{f_2}+\frac{d_2}{f_1 f_2} & 1-\frac{d_1}{f_1}-\frac{d_2}{f_2}-\frac{d_1}{f_2}+\frac{d_1 d_2}{f_1 f_2} \end{vmatrix}$
d, $n = 1$, $n = n_0 - \frac{1}{2} n_2 r^2$, 1, 2	$\begin{vmatrix} \cos d\sqrt{\frac{n_2}{n_0}} & \frac{1}{\sqrt{n_0 n_2}} \sin d\sqrt{\frac{n_2}{n_0}} \\ -\sqrt{n_0 n_2} \sin d\sqrt{\frac{n_2}{n_0}} & \cos d\sqrt{\frac{n_2}{n_0}} \end{vmatrix}$
d, $n = 1$, n, 1, 2	$\begin{vmatrix} 1 & d/n \\ 0 & 1 \end{vmatrix}$

Figure 4.20 The method of constructing more elaborate matrices from simpler ones is key to developing a list of general lens matrices. From H. Kogelnik and T. Li, "Laser Beams and Resonators," *Proc. IEEE* 54:1312 (1966), Table I. ©1966 IEEE.

or

$$\begin{bmatrix} r_3 \\ r_3' \end{bmatrix} = \begin{bmatrix} 1 & d \\ \frac{-1}{f} & 1-\frac{d}{f} \end{bmatrix} \begin{bmatrix} r_1 \\ r_1' \end{bmatrix}. \tag{4.99}$$

This method of constructing more elaborate matrices from simpler ones is key to developing a list of general lens matrices. An example of such a list is given in Figure 4.20.

4.4.3 *ABCD* Law Applied to Simple Lens Systems

A lens or mirror can be used to focus or diverge a Gaussian beam. When this is done, a $TEM_{0,0}$ or $TEM_{m,n}$ beam still remains a $TEM_{0,0}$ or $TEM_{m,n}$ beam, but with the beam parameters $R(z)$ and $w(z)$ altered. Although the classical optics equations in the Appendix may be used to model the beam propagation under these circumstances, more accurate results can be obtained by using the properties of the Gaussian beam in conjunction with the *ABCD* matrices.

The process of using *ABCD* matrices for the analysis of Gaussian beam parameters is a fine art. The interested reader may wish to examine the treatments given by Yariv,[22] Siegman,[23] Verdeyen,[24] and Koechner,[25] as well as the classic treatment of Kogelnik and Li.[26]

In general, the transformation of a Gaussian beam by an optical component is described by the *ABCD* law that states

$$\boxed{q_2(z) = \frac{Aq_1(z) + B}{Cq_1(z) + D}} \tag{4.100}$$

where $q(z)$ is the *complex beam parameter* given by

$$\boxed{\frac{1}{q(z)} = \frac{1}{R(z)} - j\frac{\lambda}{\pi w^2(z)}} \tag{4.101}$$

and where $q_1(z)$ represents the complex beam parameter prior to the optical component and $q_2(z)$ represents the complex beam parameter after the optical component. Recall that the *A, B, C,* and *D* terms are the matrix elements of the *ABCD* lens matrix for the optical component. Notice that the *ABCD* matrices used in the Gaussian *ABCD* law are exactly the same as those used in the simple ray tracing approach! Thus, the simple ray tracing method has utility far beyond the conditions under which it was derived.

As an example, consider a simple lens system composed of a single thin lens such as illustrated in Figure 4.21. Assume a Gaussian laser beam with a beam waist of w_{o1} is incident on the lens. Further assume that the incident beam has a perfectly planar wavefront upon incidence on the lens (in other words, $R \to \infty$ at the input surface of the lens).

Under these conditions, the complex beam parameter q_1 at the front of the lens is given by Equation (4.101) with the appropriate substitutions for $R(z)$ and $w^2(z)$ as

$$\frac{1}{q_1(z)} = \frac{1}{R(z)} - j\frac{\lambda}{\pi w^2(z)} = \frac{1}{\infty} - j\frac{\lambda_o}{\pi w_{o1}^2 n} = -j\frac{\lambda_o}{\pi w_{o1}^2 n}. \tag{4.102}$$

The equation for the complex beam parameter a distance z after the lens is given by the Equation (4.100) with the appropriate substitutions for the *ABCD* matrix of a thin lens

[22] Amnon Yariv, *Optical Electronics*, 4th ed. (Philadelphia, PA: Saunders College Publishing, 1991), p. 49–72.

[23] Anthony Siegman, *Lasers* (Mill Valley, CA: University Science Books, 1986), Chapter 21.

[24] Joseph T. Verdeyen, *Laser Electronics*, 2d ed. (Englewood Cliffs, NJ: Prentice Hall, 1989), p. 91–101.

[25] Walter Koechner, *Solid State Laser Engineering*, 3d ed. (Berlin, Germany: Springer-Verlag, 1992), Chapter 5.

[26] H. Kogelnik and T. Li, "Laser Beams and Resonators," *Proc. IEEE* 54:1312 (1966).

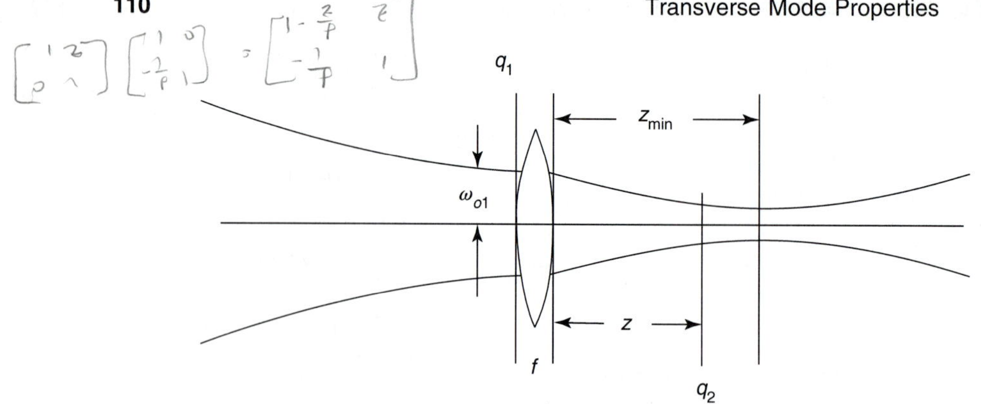

Figure 4.21 A simple lens system composed of a single thin lens. A Gaussian laser beam with a beam waist of w_{o1} is incident on the lens. The incident beam has a perfectly planar wavefront upon incidence on the lens (in other words, $R \to \infty$ at the input surface of the lens).

followed by a space

$$q_2(z) = \frac{Aq_1(z) + B}{Cq_1(z) + D} = \frac{\left(1 - \frac{z}{f}\right) \cdot q_1(z) + z}{\frac{-1}{f} \cdot q_1(z) + 1}. \tag{4.103}$$

Substituting Equation (4.102) into Equation (4.103), solving for the real and imaginary parts, and then equating the real part to $1/R(z)$ and the imaginary part to $-\lambda_o/\pi n w^2(z)$ gives

$$\frac{1}{R(z)} = \frac{\frac{-1}{f} + z\left(\frac{1}{f^2} + \frac{1}{z_{01}^2}\right)}{\left(1 - \frac{z}{f}\right)^2 + \left(\frac{z}{z_{01}}\right)^2} \tag{4.104}$$

and

$$\frac{\lambda_o}{\pi w^2(z) n} = \frac{\frac{1}{z_{01}}}{\left(1 - \frac{z}{f}\right)^2 + \left(\frac{z}{z_{01}}\right)^2} \tag{4.105}$$

where z_{01} is given by

$$z_{01} = \frac{\pi w_{01}^2 n}{\lambda_o}. \tag{4.106}$$

As an aside, notice that the minimum spot size following the lens does not occur at f but rather at

$$z_{\min} = \frac{f}{1 + \left(\frac{f}{z_{01}}\right)^2}. \tag{4.107}$$

4.5 GAUSSIAN BEAMS IN RESONANT CAVITIES

The optical resonant cavity (or resonator) provides a regenerative (closed loop) path for the photons. The resonator both physically shortens the laser and permits tailoring of the optical

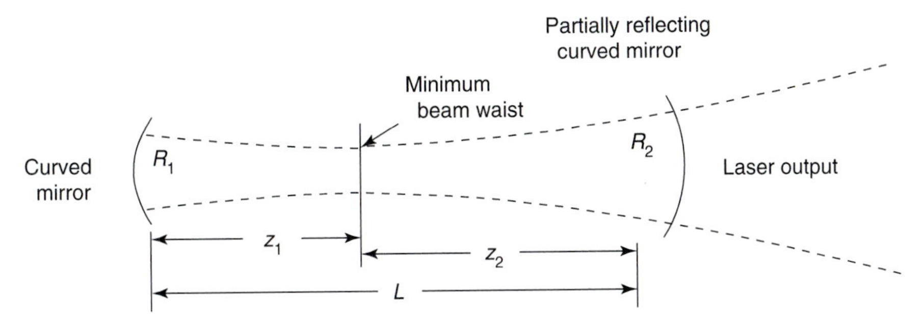

Figure 4.22 Energy can be extracted from a stable resonator by fabricating one of the end mirrors as a partial reflector.

mode. Although there are lasers without optical resonators (x-ray lasers, for example), most commercial lasers incorporate a resonator.

Historically, Dicke,[27] Prokhorov,[28] and Schawlow and Townes[29] independently proposed using a Fabry-Perot interferometer (see Section 3.2) as a laser resonator. In 1960, Fox and Li[30] analyzed the laser modes of such a structure. In 1961, Boyd and Gordon[31] extended this analysis into resonators with spherical mirrors. This treatment was expanded in 1962 by Boyd and Kogelnik.[32] Also in 1961, Goubau and Schwering[33] proposed the idea of using sequences of lenses to model the resonator structure. In 1966, Kogelnik and Li teamed up to write the classical reference on laser beams and resonators.[34]

One of the critical questions for a laser resonator is whether the beam is permanently trapped inside the resonator, or if the beam can escape after a certain number of reflections off the resonator mirrors. A resonator where the beam is permanently trapped is called a *stable* resonator. A resonator where the beam can escape after a certain number of round trips is termed an *unstable* resonator.[35] A resonator exactly between the two conditions is termed *marginally stable.* (In practice, marginally stable resonators are actually unstable because they are extremely sensitive to minor variations in alignment and secondary optical effects.)

Typically, optical energy is extracted from a stable resonator by fabricating one of the end mirrors as a partial reflector (see Figure 4.22). For example, a resonator can be formed by two mirrors, one flat with a reflectance of 95% and the other curved with a reflectance of 100%. The 5% transmitted energy through the flat reflector (output coupler) is the laser output.

---

[27] R. H. Dicke, "Molecular Amplification and Generation Systems and Methods," U.S. Patent 2,851,652, September 9, 1958.

[28] A. M. Prokhorov, *JETP* 34:1658 (1958); and *Soviet Physics JETP* 7:1140 (1958).

[29] A. L. Schawlow and C. H. Townes, *Phys. Rev.* 29:1940 (1958).

[30] A. G. Fox and T. Li, *Bell Sys. Tech. J.* 40:453 (1961).

[31] G. D. Boyd and J. P. Gordon, *Bell Sys. Tech. J.* 40:489 (1961).

[32] G. D. Boyd and H. Kogelnik, *Bell Sys. Tech. J.* 41:1347 (1962).

[33] G. Goubau and F. Schwering, *IRE Trans. on Antennas and Propagation* AP-9:248 (1961).

[34] H. Kogelnik and T. Li, "Laser Beams and Resonators," *Proc. IEEE* 54:1312 (1966).

[35] A. E. Siegman, *Appl. Opt.* 13:353 (1974).

Energy is generally extracted from an unstable resonator by using the energy that escapes after a certain number of round trips. For example, a resonator can be formed by a large mirror on one end and a small mirror on the other. The energy that escapes around the smaller mirror can be used as a donut shaped output beam (see Figure 4.23).

Most laser resonators are constructed with some combination of spherical and flat mirrors. Figure 4.24 illustrates a few of the possibilities. Notice that it is not intuitively obvious which resonators are stable, marginally stable, or unstable! The analysis of the next section addresses the issues of determining the stability of a resonator and calculating the beam waist for Gaussian beams in the resonator.

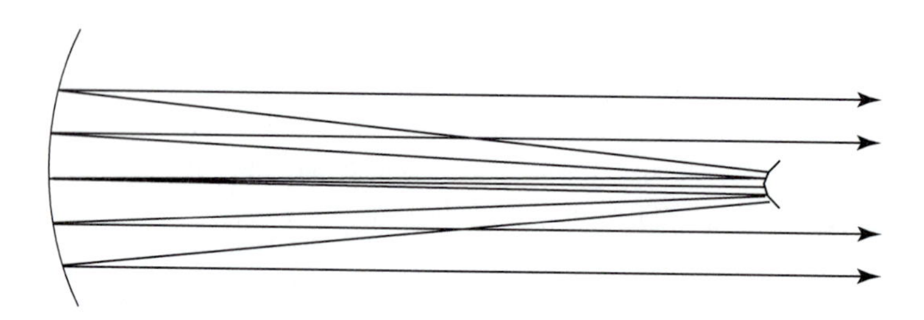

Figure 4.23 An unstable resonator can be formed by a large mirror on one end and a small mirror (or polka-dot coating) on the other. The energy that escapes around the smaller mirror can be used as a donut-shaped output beam.

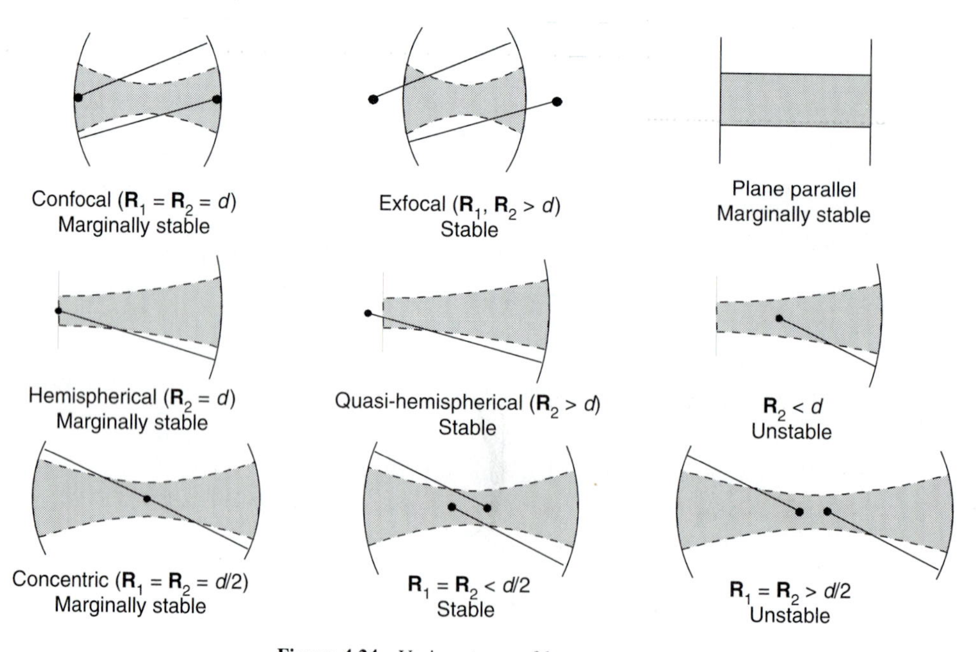

Figure 4.24 Various types of laser resonators.

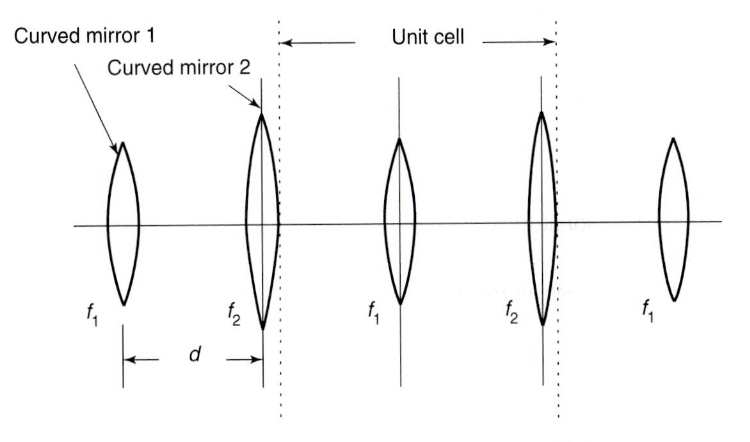

Figure 4.25 A periodic system of lenses can be used to illustrate multiple ray bounces in a laser resonator.

4.5.1 Modeling the Stability of the Laser Resonator

The laser resonator can be modeled as a periodic system of lenses such as that illustrated in Figure 4.25. The periodic lens system is composed of one lens with focal length f_1 (corresponds to a spherical mirror with radius $\mathbf{R}_1/2$), a second lens with a focal length f_2 (corresponds to a spherical mirror with radius $\mathbf{R}_2/2$), and a distance of length d separating the lenses.

The ray matrix for a single cell of this system takes the form

$$\begin{bmatrix} r_2 \\ r_2' \end{bmatrix} = \begin{bmatrix} 1 & d \\ -\frac{1}{f_1} & 1-\frac{d}{f_1} \end{bmatrix} \begin{bmatrix} 1 & d \\ -\frac{1}{f_2} & 1-\frac{d}{f_2} \end{bmatrix} \begin{bmatrix} r_1 \\ r_1' \end{bmatrix}, \tag{4.108}$$

or, multiplying this out

$$\begin{bmatrix} r_2 \\ r_2' \end{bmatrix} = \begin{bmatrix} 1-\frac{d}{f_2} & 2d-\frac{d^2}{f_2} \\ -\frac{1}{f_1}-\frac{1}{f_2}+\frac{d}{f_1 f_2} & 1-\frac{2d}{f_1}-\frac{d}{f_2}+\frac{d^2}{f_1 f_2} \end{bmatrix} \begin{bmatrix} r_1 \\ r_1' \end{bmatrix} = \begin{bmatrix} A & B \\ C & D \end{bmatrix} \begin{bmatrix} r_1 \\ r_1' \end{bmatrix}. \tag{4.109}$$

For m periodic sequences of this unit cell, the ray matrix expression takes the form

$$\begin{bmatrix} r_m \\ r_m' \end{bmatrix} = \begin{bmatrix} A & B \\ C & D \end{bmatrix}^m \begin{bmatrix} r_1 \\ r_1' \end{bmatrix}. \tag{4.110}$$

Notice that the determinant of the ray matrix $(AD - BC)$ is equal to one. Sylvester's theorem for calculating the mth power of a unimodular matrix can be applied to this problem as[36]

$$\begin{bmatrix} A & B \\ C & D \end{bmatrix}^m = \frac{1}{\sin\theta} \cdot \begin{bmatrix} A\sin(m\theta) - \sin((m-1)\theta) & B\sin(m\theta) \\ C\sin(m\theta) & D\sin(m\theta) - \sin((m-1)\theta) \end{bmatrix} \tag{4.111}$$

where

$$\theta = \cos^{-1}\left(\frac{A+D}{2}\right). \tag{4.112}$$

[36]H. Kogelnik and T. Li, "Laser Beams and Resonators," *Proc. IEEE* 54:1312 (1966), eq. 5.

This system only has solutions for the condition

$$-1 < \cos\theta < 1 \tag{4.113}$$

or

$$-1 < \frac{(A+D)}{2} < 1. \tag{4.114}$$

The physical meaning of the existence of a solution is that the ray is still contained in the periodic sequence. If the sequence does not have a solution, then the ray has escaped the periodic system of lenses.

Thus, the ray only remains in the periodic system of lenses if

$$-1 < \frac{(A+D)}{2} < 1. \tag{4.115}$$

Adding 1 to each side and dividing by 2 gives

$$0 < \frac{(A+D+2)}{4} < 1 \tag{4.116}$$

which is evaluated as

$$0 < 1 - \frac{d}{2f_1} - \frac{d}{2f_2} + \frac{d^2}{4f_1 f_2} < 1 \tag{4.117}$$

$$0 < \left(1 - \frac{d}{2f_1}\right)\left(1 - \frac{d}{2f_2}\right) < 1 \tag{4.118}$$

where Equation (4.118) represents the stable solutions for the periodic lens system.

Since a laser resonator can be considered to be a periodic lens system composed of one lens with focal length f_1 corresponding to a spherical mirror with radius $\mathbf{R}_1/2$ and a second lens with a focal length f_2 corresponding to a spherical mirror with radius $\mathbf{R}_2/2$, then the stability criterion for a laser resonator can be written directly from Equation (4.118) as

$$\boxed{0 < \left(1 - \frac{d}{\mathbf{R}_1}\right)\left(1 - \frac{d}{\mathbf{R}_2}\right) < 1.} \tag{4.119}$$

Equation (4.119) is important, as it represents the various mirror and length combinations that lead to stable resonators. In general, this equation is presented in the form of a graph, where the stable regions are white and the unstable regions are shaded, as illustrated in Figure 4.26. Notice the marginally stable cases of a *confocal cavity* (where $\mathbf{R}_1 = \mathbf{R}_2 = d$), a cavity with flat-flat reflectors, and a *concentric cavity* (where $\mathbf{R}_1 = \mathbf{R}_2 = d/2$).

Example 4.10

Consider an unstable laser resonator with a front mirror with a radius of curvature $\mathbf{R}_1 = 30$ cm, a back mirror with a radius of curvature $\mathbf{R}_2 = 60$ cm, and a length of $L = 100$ cm. Assume that both mirrors are 1.75 cm in radius. Assume that the beam starts at the front mirror at a distance of 0.2 cm from the central axis and possesses no initial slope. How many single passes of the cavity does the beam make before it leaves the resonator? At which mirror does it leave the resonator? Does the beam exit the resonator above or below this mirror?

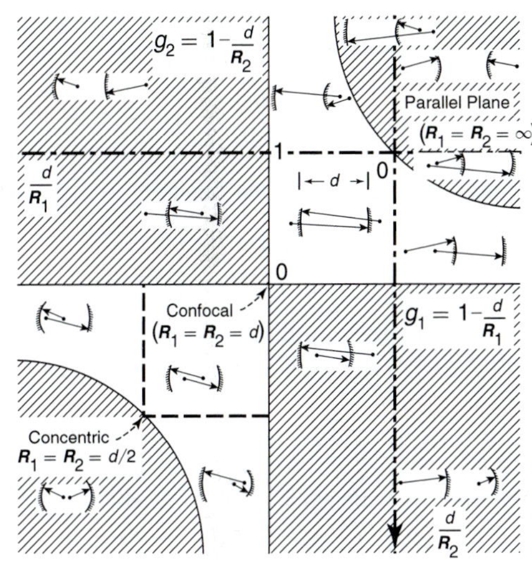

Figure 4.26 The traditional stability graph for laser resonators. From H. Kogelnik and T. Li, "Laser Beams and Resonators," *Proc. IEEE* 54:1312 (1966), Fig. 4. ©1966 IEEE.

Solution. Although this example can be solved in a number of ways, one of the most illuminating is to multiply out the various lens matrices and observe the ray's progress through the system. The initial ray matrix is given as

$$r_{\text{start}} = \begin{bmatrix} 0.2 \text{ cm} \\ 0 \end{bmatrix} \tag{4.120}$$

while the matrix representing the length of the resonator is

$$\mathbf{S}_{\text{space}} = \begin{bmatrix} 1 & 100 \text{ cm} \\ 0 & 1 \end{bmatrix}. \tag{4.121}$$

The front mirror matrix is given by

$$\mathbf{S}_{\text{front}} = \begin{bmatrix} 1 & 0 \\ \frac{-1}{15\text{cm}} & 1 \end{bmatrix} = \begin{bmatrix} 1 & 0 \\ \frac{-0.067}{\text{cm}} & 1 \end{bmatrix} \tag{4.122}$$

and, finally, the back mirror matrix is

$$\mathbf{S}_{\text{back}} = \begin{bmatrix} 1 & 0 \\ \frac{-1}{30\text{cm}} & 1 \end{bmatrix} = \begin{bmatrix} 1 & 0 \\ \frac{-0.033}{\text{cm}} & 1 \end{bmatrix}. \tag{4.123}$$

The first pass (front mirror to back mirror) is given by

$$\mathbf{S} = \mathbf{S}_{\text{space}} \cdot r_{\text{start}} = \begin{bmatrix} 0.2 \text{ cm} \\ 0 \end{bmatrix}. \tag{4.124}$$

After the ray reflects off the back mirror, the result is

$$\mathbf{S} = \mathbf{S}_{\text{back}} \cdot \mathbf{S} = \begin{bmatrix} 0.2 \text{ cm} \\ \text{-0.007} \end{bmatrix}. \tag{4.125}$$

After traveling from back mirror to front mirror, the result is

$$\mathbf{S} = \mathbf{S}_{\text{space}} \cdot \mathbf{S} = \begin{bmatrix} \text{-0.467 cm} \\ \text{-0.007} \end{bmatrix}. \tag{4.126}$$

After reflecting of the front mirror, the result is

$$\mathbf{S} = \mathbf{S}_{\text{front}} \cdot \mathbf{S} = \begin{bmatrix} \text{-0.467 cm} \\ 0.024 \end{bmatrix}. \tag{4.127}$$

After traveling from front mirror to back mirror, the result is

$$\mathbf{S} = \mathbf{S}_{\text{space}} \cdot \mathbf{S} = \begin{bmatrix} \text{1.978 cm} \\ 0.024 \end{bmatrix}. \tag{4.128}$$

So, the ray exits the resonator at the back mirror (1.978 cm >1.75 cm) with a slope of 0.024 after three single passes of the cavity, and exits over the top of the mirror.

Example 4.11

Consider the same laser resonator as in Example 4.10. As an alternative way of visualizing the behavior of the ray in the resonator, plot the position of the ray in the cavity after each single pass.

Solution. This is solved in the same way as Example 4.10. However, in this case, the individual distances are retained and plotted. The result is illustrated in Figure 4.27.

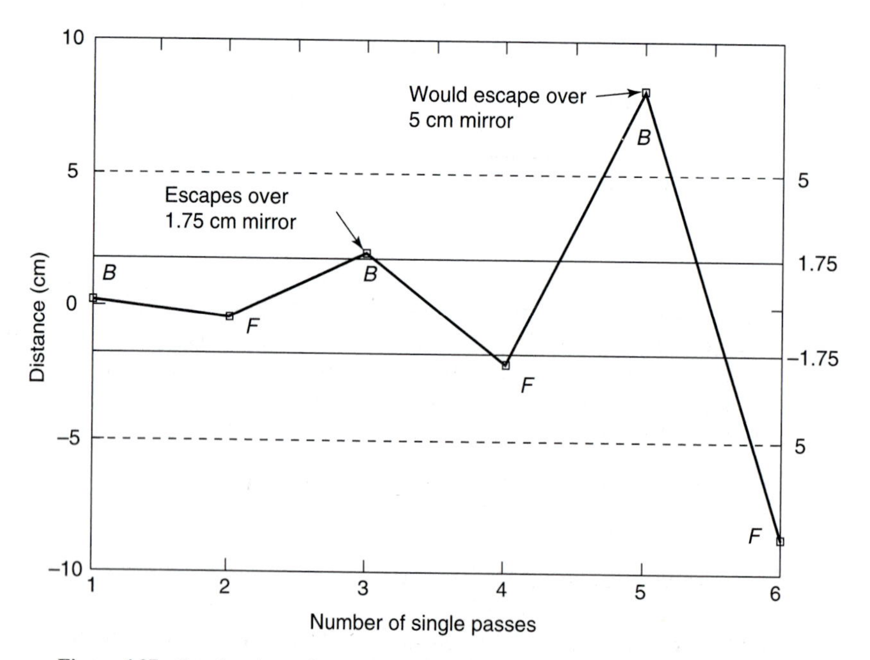

Figure 4.27 Ray locations after each single pass of an unstable resonator with a front mirror with radius of curvature $\mathbf{R}_1 = 30$ cm, a back mirror with radius of curvature $\mathbf{R}_2 = 60$ cm, and a length of $L = 100$ cm. The beam is assumed to start at the front mirror at a distance of 0.2 cm from the central axis and possess no initial slope. Positive numbers are distances above the center axis and negative numbers are distances below the center axis.

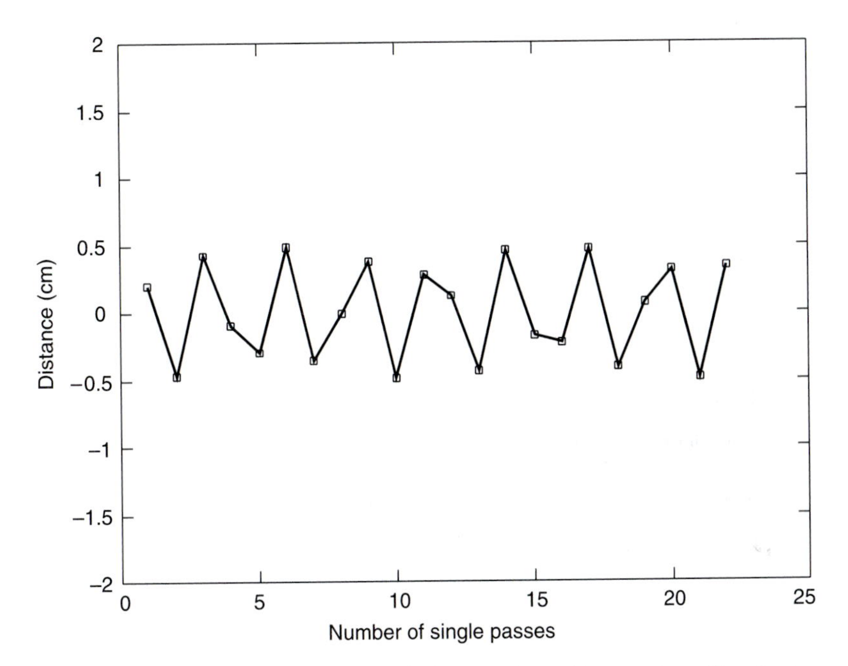

Figure 4.28 Ray locations after each single pass of a stable resonator with a front mirror with a radius of curvature $\mathbf{R}_1 = 60$ cm, a back mirror with a radius of curvature $\mathbf{R}_2 = 60$ cm, and a length of $L = 100$ cm. The beam is assumed to start at the front mirror at a distance of 0.2 cm from the central axis, and possess no initial slope. Positive numbers are distances above the center axis and negative numbers are distances below the center axis.

Example 4.12

Consider a stable laser resonator with a front mirror with a radius of curvature $\mathbf{R}_1 = 60$ cm, a back mirror with a radius of curvature $\mathbf{R}_2 = 60$ cm, and a length of $L = 100$ cm. Assume that both mirrors are 1.75 cm in radius. Assume that the beam starts at the front mirror at a distance of 0.2 cm from the central axis and possesses no initial slope. As an alternative way of visualizing the behavior of the ray in the resonator, plot the position of the ray in the cavity after each single pass.

Solution. This is solved in the same way as Example 4.11. However, in this case, the ray never escapes. The result is illustrated in Figure 4.28.

4.5.2 *ABCD* Law Applied to Resonators

In a stable laser resonator, the Gaussian beam will adapt itself to the resonator mirror configuration. In other words, the equiphase contours $R(z)$ will match the mirror radii of curvature. For example, if the resonator has a flat front mirror and a curved rear mirror, then the beam waist of the Gaussian beam (where $R(z) \to \infty$) will be at the flat mirror and the radius of beam curvature $R_2(L)$ at the rear mirror will equal the curvature of the rear mirror. For a cavity with two curved mirrors, the Gaussian beam will adapt itself so that both mirror curvatures match the curvature of the Gaussian beam at the mirrors. This property makes it relatively straightforward to calculate the beam waists and curvatures inside the laser resonator.

Consider an arbitrary resonator containing a stable mode of the resonator. The radius $R(z)$ and the waist $w(z)$ can be found at any point z in the cavity (referenced from the edge of the unit cell in the given *ABCD* matrix) by requiring that the complex beam parameter $q(z)$ be the same after one pass of the cavity. Thus, the *ABCD* rule can be applied giving the general form

$$q_2(z) = \frac{Aq_1(z) + B}{Cq_1(z) + D}. \tag{4.129}$$

Since $q_2(z) = q_1(z) = q(z)$,

$$q(z) = \frac{Aq(z) + B}{Cq(z) + D}. \tag{4.130}$$

Equation (4.130) can be solved as a quadratic in $q(z)$ yielding

$$\frac{1}{q(z)} = \frac{D - A}{2B} \mp j\frac{\sqrt{4 - (A + D)^2}}{2B} \tag{4.131}$$

which gives the relations for $R(z)$ and $w(z)$ as

$$R(z) = \frac{2B}{D - A} \tag{4.132}$$

and

$$w(z)^2 = \frac{2\lambda_o B}{n\pi\sqrt{4 - (A + D)^2}}. \tag{4.133}$$

For this to work properly, some care must be taken with setting up the *ABCD* matrices. In particular, it is important to artificially introduce the distance variable z so that the final expressions are a function of the distance.

As an example, consider a laser with a flat output coupler and a curved rear mirror, as shown in Figure 4.29. This is a common configuration because the high reflecting rear mirror sets the structure of the cavity mode and the partially reflecting output coupler sets the coupling for the output laser power.

The lens equivalent to the resonator in Figure 4.29 is illustrated in Figure 4.30. Notice that the distance z has been incorporated by redefinition of the repeating unit cell of the periodic sequence.

Remember that the ray matrices stack in the reverse order to the propagating beam; thus, the ray matrices for Figure 4.30 referenced to z are

$$\mathbf{S} = \begin{bmatrix} 1 & L + z \\ 0 & 1 \end{bmatrix} \begin{bmatrix} 1 & L - z \\ -\frac{1}{f_1} & 1 - \frac{L-z}{f_1} \end{bmatrix}. \tag{4.134}$$

Recall from Equation (4.132)

$$R(z) = \frac{2B}{D - A} \tag{4.135}$$

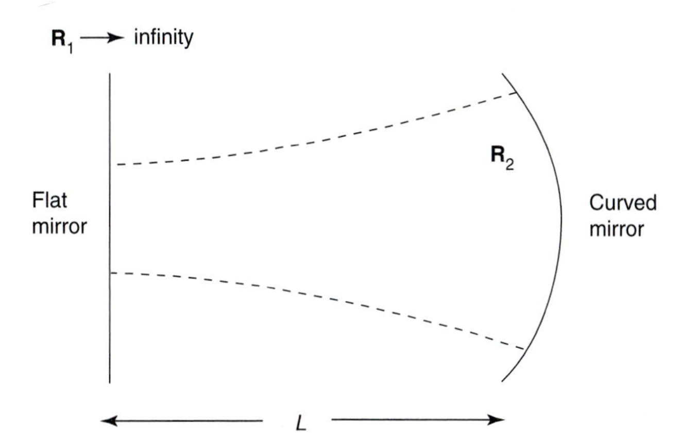

Figure 4.29 A typical resonator with a flat and a curved mirror. The minimum beam waist will fall on the flat mirror.

and Equation (4.133)

$$w(z)^2 = \frac{2\lambda_o B}{n\pi\sqrt{4-(A+D)^2}}. \quad (4.136)$$

Thus, the radius is given by

$$R(z) = z\left(1+\left(\frac{z_o}{z}\right)^2\right) \quad (4.137)$$

where the distance z_o is given by

$$z_o = \sqrt{L\mathbf{R}_2}\left(1-\frac{L}{\mathbf{R}_2}\right)^{1/2} \quad (4.138)$$

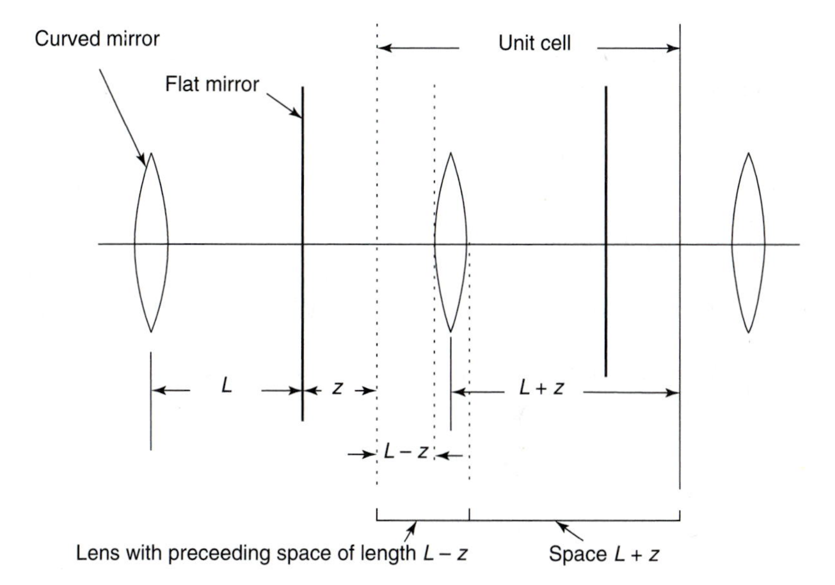

Figure 4.30 A lens equivalent to the resonator with a flat and a curved mirror.

and the beam waist $w(z)$ by

$$w(z)^2 = w_o^2\left(1+\left(\frac{z}{z_o}\right)^2\right) \tag{4.139}$$

and where the minimum beam waist w_o is expressed as

$$w_o^2 = \frac{\lambda_o\sqrt{L\mathbf{R}_2}}{n\pi}\left(1-\frac{L}{\mathbf{R}_2}\right)^{1/2} \tag{4.140}$$

and is located at the flat output coupler mirror.

As another example, consider a laser with two curved mirrors as shown in Figure 4.31. The lens equivalent to the resonator in Figure 4.31 is illustrated in Figure 4.32.

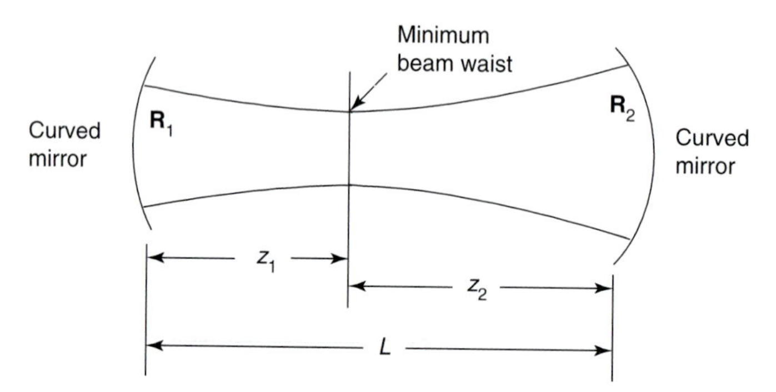

Figure 4.31 A typical resonator with two curved mirrors. The minimum beam waist will fall somewhere in the middle of the cavity.

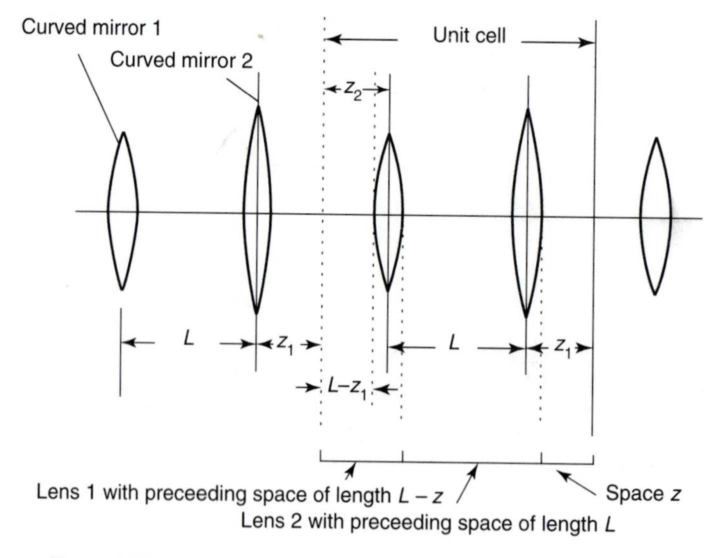

Figure 4.32 A lens equivalent to the resonator with two curved mirrors.

Define the location where the beam waist is minimum to be $z = 0$, with a distance of z_1 to mirror $\mathbf{R}_1$ and z_2 to mirror $\mathbf{R}_2$. Assume that both mirror $\mathbf{R}_1$ and $\mathbf{R}_2$ have positive values for the curvatures shown in Figure 4.31. Assume that both z_1 and z_2 are positive such that $z_1 + z_2 = L$. Under these conditions, then the beam radii on the two mirrors is given by

$$R(z_1) = z_1\left(1 + \left(\frac{z_o}{z_1}\right)^2\right) \tag{4.141}$$

$$R(z_2) = z_2\left(1 + \left(\frac{z_o}{z_2}\right)^2\right) \tag{4.142}$$

where z_o is[37]

$$z_o^2 = \frac{L(\mathbf{R}_1 - L)(\mathbf{R}_2 - L)(\mathbf{R}_1 + \mathbf{R}_2 - L)}{(\mathbf{R}_1 + \mathbf{R}_2 - 2L)^2} \tag{4.143}$$

and the locations $z_1$and z_2 are given by

$$z_1 = \frac{L(\mathbf{R}_2 - L)}{(\mathbf{R}_1 + \mathbf{R}_2 - 2L)} \tag{4.144}$$

$$z_2 = \frac{L(\mathbf{R}_1 - L)}{(\mathbf{R}_1 + \mathbf{R}_2 - 2L)} \tag{4.145}$$

and where the minimum beam waist w_o is given by

$$w_o^4 = \left(\frac{\lambda_o}{n\pi}\right)^2 \frac{L(\mathbf{R}_1 - L)(\mathbf{R}_2 - L)(\mathbf{R}_1 + \mathbf{R}_2 - L)}{(\mathbf{R}_1 + \mathbf{R}_2 - 2L)^2} \tag{4.146}$$

and, finally, where the beam waists w_1 and w_2 on mirrors $\mathbf{R}_1$ and $\mathbf{R}_2$ are given by

$$w_1^4 = \left(\frac{\lambda_o \mathbf{R}_1}{n\pi}\right)^2 \left(\frac{\mathbf{R}_2 - L}{\mathbf{R}_1 - L}\right)\left(\frac{L}{(\mathbf{R}_1 + \mathbf{R}_2 - L)}\right) \tag{4.147}$$

$$w_2^4 = \left(\frac{\lambda_o \mathbf{R}_2}{n\pi}\right)^2 \left(\frac{\mathbf{R}_1 - L}{\mathbf{R}_2 - L}\right)\left(\frac{L}{(\mathbf{R}_1 + \mathbf{R}_2 - L)}\right). \tag{4.148}$$

Example 4.13

Consider an argon-ion laser ($\lambda_o = 514.5$ nm) resonator that is 40 cm long with a front mirror of radius of curvature of $\mathbf{R}_1 = 55$ cm and a back mirror of radius of curvature of $\mathbf{R}_2 = 45$ cm. Assume that the index of refraction in the resonator is $n = 1.0$. Determine the location of the beam waist in the cavity, the minimum beam waist, and the size of the beam waist at each mirror.

---

[37] Amnon Yariv, *Optical Electronics*, 4th ed. (Philadelphia, PA: Saunders College Publishing, 1991), p. 121. Note that Yariv uses the convention that R_1 is less than zero.

Solution. The location of the beam waist is determined from Equations (4.144) and (4.145) as

$$z_1 = \frac{L(\mathbf{R}_2 - L)}{(\mathbf{R}_1 + \mathbf{R}_2 - 2L)} = \frac{40.0\text{ cm} \cdot (45\text{ cm} - 40.0\text{ cm})}{(55.0\text{ cm} + 45\text{ cm} - 2 \cdot 40\text{ cm})} = 10.0\text{ cm} \tag{4.149}$$

$$z_2 = \frac{L(\mathbf{R}_1 - L)}{(\mathbf{R}_1 + \mathbf{R}_2 - 2L)} = \frac{40.0\text{ cm} \cdot (55\text{ cm} - 40.0\text{ cm})}{(55.0\text{ cm} + 45\text{ cm} - 2 \cdot 40\text{ cm})} = 30.0\text{ cm}. \tag{4.150}$$

The minimum beam waist is determined from Equation (4.146) as

$$w_o = \left[\left(\frac{\lambda_o}{n\pi}\right)^2 \frac{L(\mathbf{R}_1 - L)(\mathbf{R}_2 - L)(\mathbf{R}_1 + \mathbf{R}_2 - L)}{(\mathbf{R}_1 + \mathbf{R}_2 - 2L)^2}\right]^{1/4}$$

$$w_o = \left[\left(\frac{\lambda_o}{n\pi}\right)^2 \cdot \frac{40.0\text{ cm} \cdot (55\text{ cm} - 40.0\text{ cm})(45\text{ cm} - 40.0\text{ cm})(55.0\text{ cm} + 45\text{ cm} - 40\text{ cm})}{(55.0\text{ cm} + 45\text{ cm} - 2 \cdot 40\text{ cm})^2}\right]^{1/4} \tag{4.151}$$

$$w_o = 0.18639\text{ mm}.$$

The beam waists at the two mirrors are determined by Equations (4.147) and (4.148) as

$$w_1 = \left[\left(\frac{\lambda_o \mathbf{R}_1}{n\pi}\right)^2 \left(\frac{\mathbf{R}_2 - L}{\mathbf{R}_1 - L}\right)\left(\frac{L}{(\mathbf{R}_1 + \mathbf{R}_2 - L)}\right)\right]^{1/4} \tag{4.152}$$

$$w_1 = \left[\left(\frac{\lambda_o}{n\pi} \cdot 55\text{ cm}\right)^2 \left(\frac{(45\text{ cm} - 40.0\text{ cm})}{(55\text{ cm} - 40.0\text{ cm})}\right)\left(\frac{40.0\text{ cm}}{(55.0\text{ cm} + 45\text{ cm} - 40\text{ cm})}\right)\right]^{1/4} \tag{4.153}$$

$$w_1 = 0.20606\text{ mm}$$

$$w_2 = \left[\left(\frac{\lambda_o \mathbf{R}_2}{n\pi}\right)^2 \left(\frac{\mathbf{R}_1 - L}{\mathbf{R}_2 - L}\right)\left(\frac{L}{(\mathbf{R}_1 + \mathbf{R}_2 - L)}\right)\right]^{1/4}$$

$$w_2 = \left[\left(\frac{\lambda_o}{n\pi} \cdot 45\text{ cm}\right)^2 \left(\frac{(55\text{ cm} - 40.0\text{ cm})}{(45\text{ cm} - 40.0\text{ cm})}\right)\left(\frac{40.0\text{ cm}}{(55.0\text{ cm} + 45\text{ cm} - 40\text{ cm})}\right)\right]^{1/4} \tag{4.154}$$

$$w_2 = 0.32284\text{ mm}.$$

SYMBOLS USED IN THE CHAPTER

$\vec{\mathcal{E}}$: Electric field (V/cm)

$\vec{\mathcal{P}}$: Electric polarization vector (C/m^2)

The vector components of $\vec{\mathcal{D}}$, $\vec{\mathcal{B}}$, $\vec{\mathcal{E}}$, $\vec{\mathcal{H}}$, $\vec{\mathcal{J}}$, $\vec{\mathcal{P}}$, and $\vec{\mathcal{M}}$ are indicated by subscripts without the vector symbol as $\vec{\mathcal{D}} \to \mathcal{D}_x$, $\mathcal{D}_y$, and $\mathcal{D}_z$

$\tilde{E}$, ψ: Electric field phasor (V/cm)

ϵ_r: Relative permittivity of a material: used as $\epsilon_m = \epsilon_r \epsilon_o$ (no units)

ϵ_m: Permittivity of a material: used as $\epsilon_m = \epsilon_r \epsilon_o$ (farad/meter)

ϵ_h: Non-resonant permittivity of a host crystal (farad/meter)

σ: Conductivity (Ω-m)

$\chi = \chi' \pm j\chi''$: Complex susceptibility (unitless)

$\varepsilon = \varepsilon' + j\varepsilon'' = \varepsilon_o(1 + \chi)$: Complex relative dielectric constant (farad/meter)

$n_c = n + j\kappa$: Complex index of refraction (unitless)

$k_c = \sqrt{\mu_o \varepsilon_m}$: Complex wavenumber ($m^{-1}$ or cm^{-1})

z: Spatial variable representing the direction of wave propagation (cm or m)

$U(x, y, z)$: Transverse variation in the electric field of the waveform (V/cm)

r and ϕ: Polar coordinates for the transverse waveform (cm or m and rad or deg)

l, p: Radial (p) and angular (l) coordinate numbers for a circular waveform

$\psi_{l,p}$: General wavefunction for the circular transverse waveform (V/cm)

$w(z)$: $1/e$ laser beam radius, called the beam waist (cm or m)

w_o: Minimum beam waist, the minimum 1/e radius of the beam (cm or m)

$L_p^l(\nu)$: Laguerre polynomials

$I_p(r, \phi, z)$: Intensity component for cylindrical laser modes (watts)

m, p: x (m) and y (p) coordinate numbers for a rectangular waveform

$\psi_{m,p}$: General wavefunction for the rectangular transverse waveform (V/cm)

$H_m(\nu)$: Hermite polynomials

$R(z)$: Laser beam curvature (cm or m)

$I_{m,p}(x, y, z)$: Intensity component for rectangular laser modes (watts)

ω: Frequency (rad/sec)

k: Analytic (real) wavenumber ($2\pi/\lambda$) (m^{-1} or cm^{-1})

$\tilde{E}$, ψ: Electric field phasor (V/cm)

ϵ_r: Relative permittivity of a material: used as $\epsilon_m = \epsilon_r \epsilon_o$ (no units)

ϵ_m: Permittivity of a material: used as $\epsilon_m = \epsilon_r \epsilon_o$ (farad/meter)

ϵ_h: Non-resonant permittivity of a host crystal (farad/meter)

z: Spatial variable representing the direction of wave propagation (cm or m)

r and ϕ: Polar coordinates for the transverse waveform (cm or m and rad or deg)

$R(z)$: Laser beam curvature (cm or m)

$q(z)$: Beam parameter function q

$P(z)$: Beam parameter function P

λ_o: Free space wavelength (cm or m)

λ: Material wavelength (cm or m)

k_o: Analytic real wavenumber in free space ($2\pi/\lambda_o$) (m^{-1} or cm^{-1})

$w(z)$: Beam waist as a function of distance (cm or m)

w_o: Minimum beam waist as a function of distance (cm or m)

n: General real refractive index

T_{phase}, T_{waist}, $T_{\text{curvature}}$: Terms in the transverse wave equation

θ_a: Angle of asymptotes to hyperbolic beam waist equation (degrees or radians)

z_R: Rayleigh range (cm or m)

r_1 and r_2: Radial location of a ray (cm or m)

r_1' and r_2': Slope of a ray (unitless)

A, B, C, D: Matrix elements of an *ABCD* or lens matrix

S: General *ABCD* matrix

θ: Angle used in ray matrix analysis, $r_1' = \tan\theta$ (deg or rad)

$\mathbf{R}_1$, $\mathbf{R}_2$: Radii of curvature of lenses, both positive for a biconvex lens (cm or m)

$\mathbf{R}_1$, $\mathbf{R}_2$: Radii of curvature of mirrors, positive for a concave mirror (cm or m)

n_{air}: Real index of refraction for air, generally taken as 1.0 (unitless)

n_{lens}, n_{glass}, n_{film}: Real indices of refraction for various materials (unitless)

s_o: Object distance for a lens, positive to the left if light is traveling from left to right (cm or m)

s_i: Image distance for a lens, positive to the right if light is traveling from left to right (cm or m)

f, f_1, f_2: Focal length, positive for converging and negative for diverging lenses (cm or m)

d: Separation between two lenses in a lens system (cm or m)

z_{o1}: Rayleigh range for Gaussian focusing (cm or m)

z_{min}: Focusing distance for Gaussian focusing (cm or m)

m: Integer describing the number of passes of a cavity (unitless)

L: General length (cm or m)

z_1, z_2: Distance from the minimum beam waist to mirrors $\mathbf{R}_1$, $\mathbf{R}_2$ (cm or m)

w_1, w_2: Beam waist size on mirrors $\mathbf{R}_1$, $\mathbf{R}_2$ (cm or m)

EXERCISES

TEM$_{x,y}$ transverse modes

4.1 Consider a cylindrical laser operating on the TEM$_{x,y}$ mode. Using the exact mathematical expression and your choice of computer tool, create a contour or 3D plot of the normalized intensity I/I_o as a function of the polar coordinates r and ϕ. Assume $z = 0$ and that the minimum beam waist is 5 mm. Plot and label three different transverse mode patterns that are not included in the figures in Chapter 2.

4.2 Consider a rectangular laser operating on the TEM$_{x,y}$ mode. Using the exact mathematical expression and your choice of computer tool, create a a contour or 3D plot of the normalized intensity I/I_o as a function of the Cartesian coordinates x and y. Assume $z = 0$ and that the minimum beam waist is 5 mm. Plot and label three different transverse mode patterns that are not included in the figures in Chapter 2.

$TEM_{0,0}$ Gaussian beam propagation

4.3 Consider an argon-ion laser with $\lambda_o = 488$ nm. Assume the minimum beam waist is 1 mm and $n = 1$.

(a) Compare the beam waist calculated using the exact expression in Equation (4.36) with that calculated using the approximate angular expression in Equation (4.37). Assume you are measuring the beam at a distance of 2 meters from the minimum waist.

(b) Repeat the calculation in part (a), except assume that you are measuring at a distance of 50 meters.

(c) From the answers to parts (a,b) formulate a rule of thumb for when it is appropriate to use the approximate angular expression.

4.4 Consider an argon-ion laser with $\lambda_o = 488$ nm.

(a) Plot $R(z)$ as a function of z for three different minimum beam waists. (Similar to Figure 4.5). Use 2 mm, 3 mm, and 5 mm as the values of the minimum beam waist for the three plots. Your graph should have a range of -200 meters to + 200 meters.

(b) Calculate the Rayleigh range for each of the three beam waists and locate the Rayleigh range on the graphs.

(c) What is the relationship between the Rayleigh range and the character of $R(z)$?

4.5 Assume that you want to bounce a laser beam off the Moon. Assume the laser beam is $TEM_{0,0}$ and that the Moon is a distance of $3.8 \cdot 10^8$ meters from Earth. Calculate the following.

(a) Assume that the initial beam waist of a CO_2 laser running on the 10.6 μm line is 15 cm. How large is the beam waist when the beam hits the Moon? What is the radius of curvature?

(b) Assume that the initial beam waist of a CO_2 laser running on the 10.6 μm line is 1.5 meters. How large is the beam waist when the beam hits the Moon? What is the radius of curvature?

(c) Repeat part (a) for an excimer laser operating at 254.0 nm; use the same beam waist.

(d) Repeat part (b) for an excimer laser operating at 254.0 nm; use the same beam waist.

(e) From your answers to parts (a–d), speculate on what kind of laser and what type of associated optical system you would want to use in order to optimally bounce a laser beam off the Moon.

4.6 Consider an argon-ion laser running at 514.5 nm with a minimum beam waist (radius) of $w_o = 0.6$ mm, where the minimum waist is located at the flat front mirror of the laser.

(a) You measure the beam waist some distance away from the front of the laser. The beam waist measures 0.8 mm. How far are you away from the laser? What is the radius of curvature at this location?

(b) Assume the minimum waist is reduced to 0.15 mm. You then find the distance from the laser where the beam waist measures 4.5 cm. How far are you away from the laser? What is the radius of curvature at this location?

(c) Assume the minimum waist is increased to 3 cm. You then find the distance from the laser where the beam waist measures 4.5 cm. How far are you away from the laser? What is the radius of curvature at this location?

4.7 It is very difficult to determine the location of the minimum beam waist by simply measuring the waist at various points inside the Rayleigh range and looking for the smallest one. Instead, it is much easier to go outside the Rayleigh range (where the beam is expanding) and measure the waist at several locations (see Figure 4.33). The location of the beam waist can then be calculated relative to the measured locations.

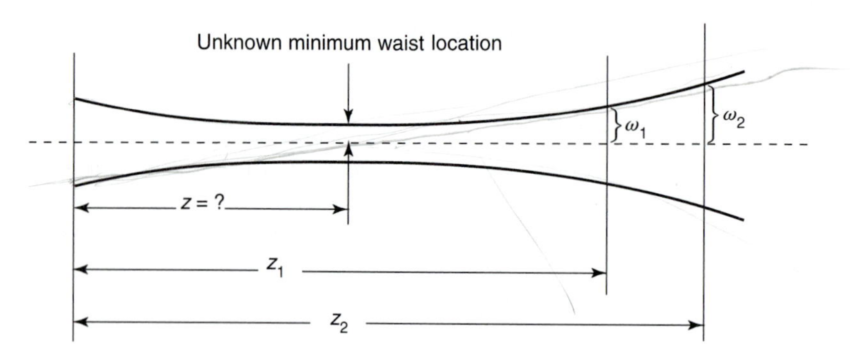

Figure 4.33 Calculating the location of the beam waist z relative to the measured locations (z_1, z_2) and the waists (w_1, w_2).

(a) Use approximations to derive an equation that can be used to calculate the location and size of the minimum beam waist given the measured waist size at two locations (as referenced from an arbitrary point that is not the location of the minimum beam waist).

(b) Pick a set of values and check your equation against the exact Gaussian expression of Equation (4.36).

(c) Is there an exact analytic way to solve this problem? What about an exact numerical way?

4.8 In many applications, the laser beam interacts with a volume of material in a cell (see Figure 4.34). Assume that the laser is an argon-ion with a wavelength of $\lambda_o = 496.5$ nm. Assume that the cell is 5 cm long. Assume that the diameter of the ends of the cell has been chosen so that the beam clips at the $1/e$ radius.

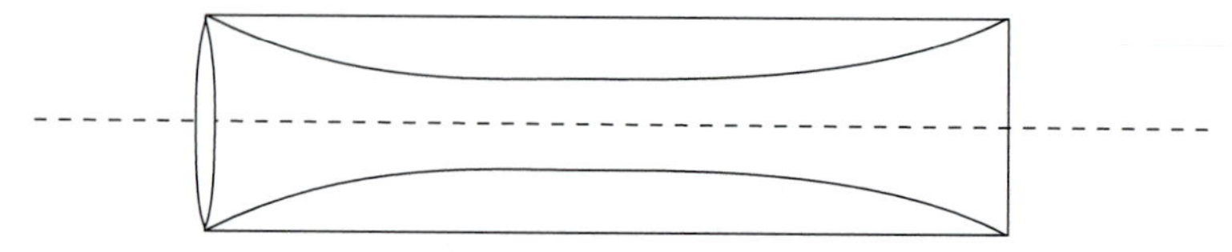

Figure 4.34 In many applications, the laser beam interacts with a volume of material in a cell.

(a) Numerically or analytically calculate the interaction volume assuming that the beam waist in the center of the cell is 0.2 mm.

(b) Repeat part (a), assuming the cell is 1 meter long.

(c) Develop a qualitative statement about the interaction volume when interacting laser beams with short and long cells.

4.9 (design) A common problem in the laser laboratory is to expand a collimated laser beam to a collimated larger size beam.

(a) Using a Melles Griot, Edmund Scientific (or equivalent!) catalog, design a simple lens system that will expand a collimated 0.5 mm 50 mW argon-ion beam up to a collimated 1 cm beam. Your answer should include:

i. a sketch of the system,
ii. a list of the parts you need,
iii. the cost, and
iv. supporting calculations that demonstrate that the system will perform to specification.

(b) Repeat the design in part (a), except expand an 0.5 mm 50 mW beam up to 4 inches in diameter. Your answer should include:

i. a sketch of the system,
ii. a list of the parts you need,
iii. the cost, and
iv. supporting calculations that demonstrate that the system will perform to the design specifications. In addition, your answer should include a calculation of the power density (watts/cm^2) for the beam and a comparison of this power density with the 0.5 mm to 1 cm case in part (a).

$TEM_{0,0}$ Gaussian beam propagation—focusing Gaussian beams

4.10 Consider a laser diode running at 870 nm. Given a lens of f number = 2.2, compare the minimum size calculated using Gaussian beam theory with that calculated using diffraction theory. (Remember that the f number is the focal length divided by the aperture size.)

(a) What will the minimum spot size be?
(b) Repeat using a lens of f number = 16.
(c) Repeat the two previous calculations (one for f number = 2.2 and one for f number = 16) assuming the laser is an argon-ion laser operating at 457.9 nm.
(d) Summarize what the calculation has taught you about focusing laser beams.

4.11 Recall that basic ray tracing theory says that the focal spot of a collimated beam will be located at the focal length of the lens. However, for Gaussian beams, the minimum spot size may not be located at the focal length of the lens. Assume that you are using an argon-ion laser with a wavelength of $\lambda_o = 488$ nm and that you are propagating in free space with an index of $n = 1$.

(a) Plot the location of the minimum spot size as a function of the minimum beam waist (w_o) for a given focal length lens. Your plot should have $z_{\min}$ on the y-axis and minimum beam waist on the x-axis. Assume that the lens has a focal length of 30 cm.
(b) From the graph in (a), develop a qualitative statement about the relationship between minimum beam waist and the location of the focus spot for Gaussian beams.

4.12 (design) A critical question in modern lasers is the minimum size of spot that can be focused upon a surface. For example, in excimer lasers used for hole drilling, the minimum spot size essentially determines the minimum hole size.

(a) Assume that you have an argon-ion laser running at 488 nm with a 1 mm beam waist located at the end mirror of the laser. Using a Melles Griot, Edmund Scientific (or equivalent!) catalog, design an optical system to focus a Gaussian beam from the argon-ion laser down to the minimum spot size. How much does the system cost?
(b) Repeat the design in part (a), this time designing for minimal cost. Your answer should include:

i. your first design, created under the criterion of minimal spot size, and
ii. your second design, created under the criterion of minimal cost.

Ray matrices to analyze lens systems

4.13 Consider a lens with a space d following the lens (see Figure 4.35). Derive the ray matrix for this combination. Your answer should be a 2 × 2 matrix with variables representing the focal length of the lens f_1 and the distance d.

4.14 Consider the lens system illustrated in Figure 4.36. Derive the lens matrix (the *ABCD* matrix) for the system with a ray traveling from transverse plane 1 to plane 2. Your answer should be the numerical values in a 2 × 2 matrix.

4.15 Consider the infinite lens series shown in Figure 4.37. The infinite lens series is composed of *unit cells* that are repeated infinitely. A single unit cell of this system consists of two lenses of

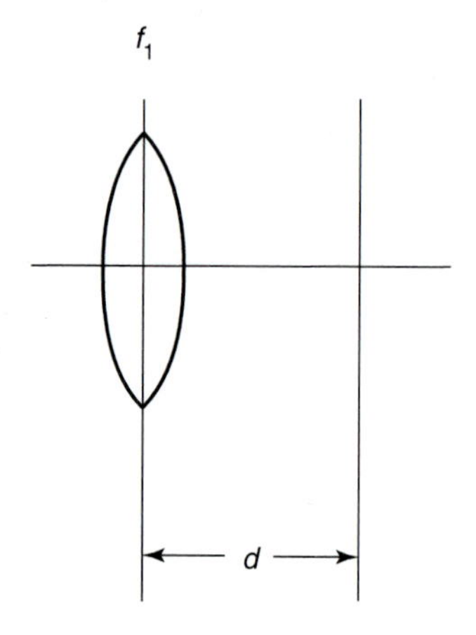

Figure 4.35 A lens with a space d following the lens.

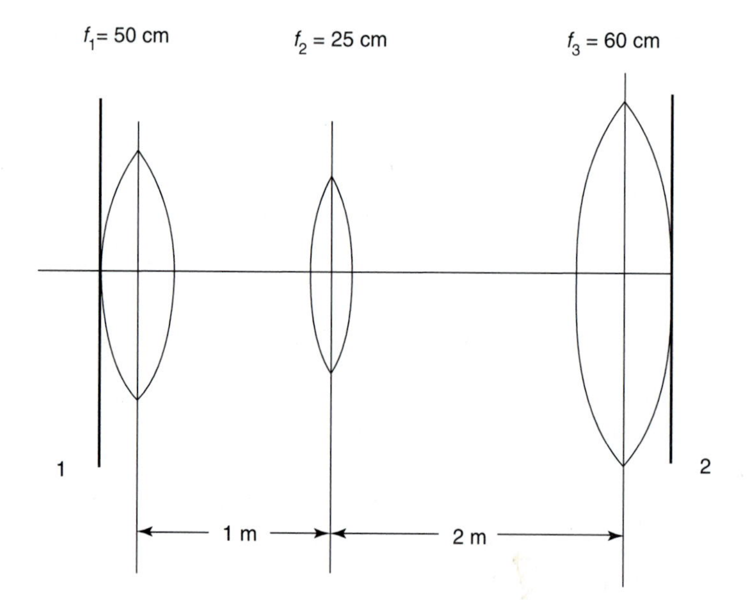

Figure 4.36 A simple lens system.

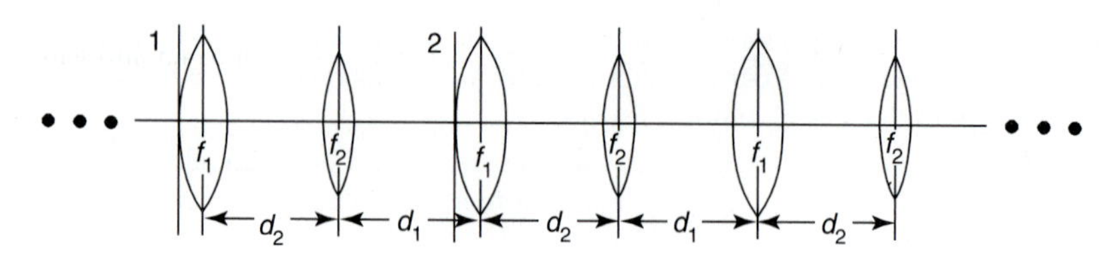

Figure 4.37 An infinite lens series.

focal lengths f_1 and f_2 and two spaces, d_1 and d_2. The unit cell is not unique. One possible unit cell is indicated in the figure. There are three other different possible unit cells that can be used to describe the infinite lens series. Sketch the four possible unit cells and derive the lens matrix (the *ABCD* matrix) expression for each.

4.16 Given an *ABCD* matrix for a system, determine the ray matrix for the system assuming that the light is propagating in the reverse direction.

4.17 Determine a ray matrix (an *ABCD* matrix) for a thick lens (see Figure 4.38). A thick lens is a lens that is thick enough that the ray exiting the lens is not at the same location as the ray entering the lens. Assume the lens is a thickness t. Find the *ABCD* matrix in terms of R_1 and R_2.

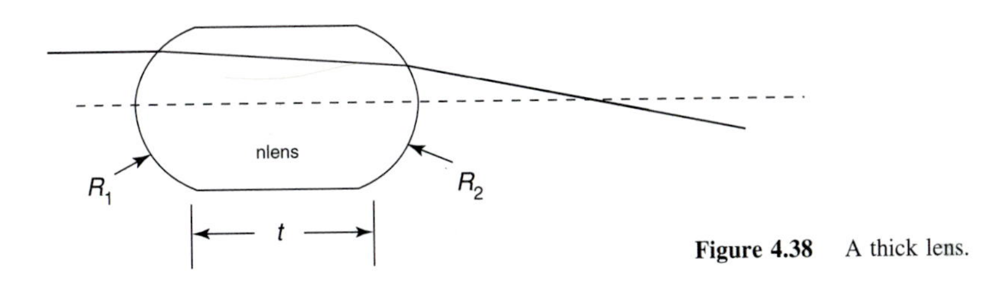

Figure 4.38 A thick lens.

Gaussian Beams in Resonators

4.18 Determine whether the following resonators are stable, marginally stable, or unstable.

Radius, mirror 1	Radius, mirror 2	Length	Stability
Flat	65 cm	45 cm	
45 cm	Flat	65 cm	
40 cm	40 cm	45 cm	
40 cm	50 cm	45 cm	
50 cm	50 cm	45 cm	

4.19 Consider an unstable cavity of the configuration given in Figure 4.39. Assume that a ray starts out at the front mirror with a zero slope at a position 25 mm (0.25 cm) from the center of the front mirror.

(a) Where does the ray leave the cavity, at the front mirror or at the back mirror?

(b) How many single passes does the ray make before it leaves the cavity?

(c) Does the ray leave the cavity over or under the mirror?

4.20 For Exercise 4.19, create a plot of the distance the ray is from the axis after each bounce; in other words, a plot similar to Figure 4.27.

4.21 **(a)** Consider a 75 cm long argon-ion laser cavity (operating at 514.5 nm) with one curved back mirror and one flat front mirror. The radius of curvature of the curved mirror is 85 cm.

i. Compute the size of the minimum beam waist on the flat mirror.

ii. Compute the size of the beam waist on the curved mirror.

(b) Now consider the same system as in part (a), but replace the back mirror with a 10 meter radius of curvature back mirror.

i. Compute the size of the minimum beam waist on the flat mirror.

ii. Compute the size of the beam waist on the curved mirror.

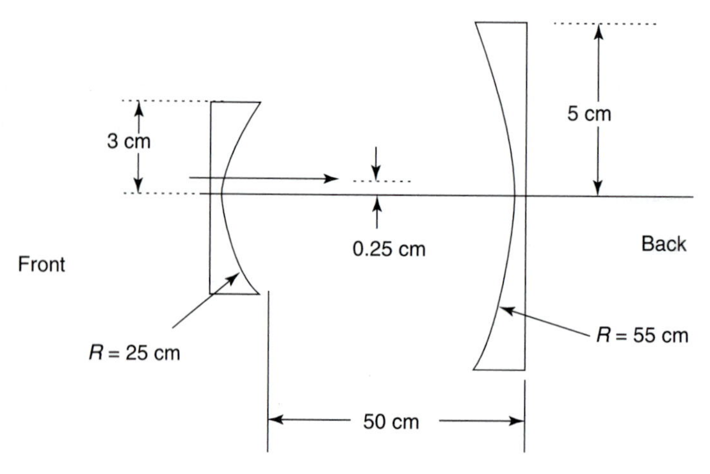

Figure 4.39 An unstable cavity.

4.22 Consider a 75 cm long argon-ion laser cavity (operating at 514.5 nm) with two curved mirrors, the front mirror of radius 55 cm and the rear mirror of radius 45 cm.

(a) Compute the location of the minimum beam waist inside the cavity.

(b) Compute the size of the minimum beam waist.

(c) Compute the size of the beam waist on the front mirror.

(d) Compute the size of the beam waist on the rear mirror.

5 Gain Saturation

Objectives

- To set up rate equations for a laser system with a reasonable number of states.
- To simplify rate equations by neglecting inconsequential terms, and to justify why these terms can be neglected.
- To compute the optimal mirror transmittance for a laser system and the optimal output power associated with that transmittance.
- To summarize the physical origin of the various terms comprising the overall laser efficiency η_{laser}.
- To calculate and plot the output power P_{out} versus the input power P_{in}, using either an engineering approach or the Rigrod approach.
- To determine η_{laser}, the saturation intensity I_{sat}, and the loss α, given P_{out} versus P_{in} for several mirror transmittances.

5.1 SATURATION OF THE EXPONENTIAL GAIN PROCESS

Consider a laser resonator of length L with an electromagnetic wave of an intensity I_{start} beginning at one end of the resonator (see Figure 5.1). Recall that the intensity I_{end} just before the ray is incident on the front mirror is expressed as

$$I_{\text{end}} = I_{\text{start}} e^{\gamma(\nu)L} \tag{5.1}$$

where γ (the gain coefficient) is a function of the frequency ν and is given by

$$\gamma(\nu) = g(\nu)\left(\frac{A_{21}\lambda_o^2}{8\pi n^2}\right)\left(N_2 - N_1\left(\frac{g_2}{g_1}\right)\right) \tag{5.2}$$

and where $g(\nu)$ is the normalized gain curve, A_{21} is the Einstein coefficient for the transition (which is a property of the gain material), λ_o is the free space wavelength of the laser

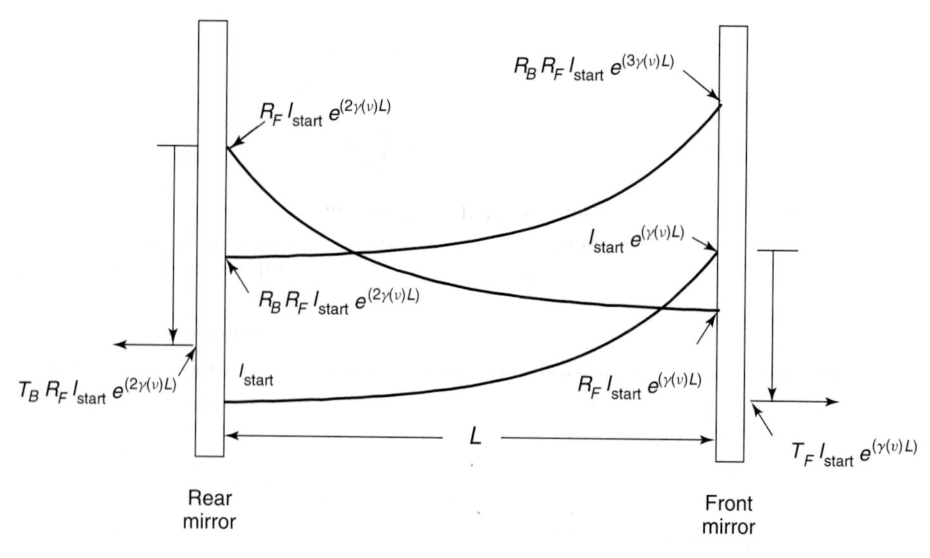

Figure 5.1 Schematic diagram of a laser beam after multiple reflections in the resonator.

transition, n is the index of refraction, N_2 and N_1 are the electron population densities for states 2 and 1, respectively (which are determined by the pumping of the laser), and g_2 and g_1 are the degeneracies (the number of states at the same energy) for states 2 and 1 (again a property of the gain material).

Once the beam is incident on the front mirror of the resonator, a certain fraction is reflected back into the resonator and a certain fraction is transmitted. The transmitted light,

$$I_t = T_F \cdot I_{\text{start}} e^{\gamma(\nu)L} = (1 - R_F) \cdot I_{\text{start}} e^{\gamma(\nu)L} \tag{5.3}$$

is the useful external light from the laser. The reflected light,

$$I_r = R_F \cdot I_{\text{start}} e^{\gamma(\nu)L} \tag{5.4}$$

remains inside the resonator to stimulate more transitions.

Thus, a round-trip pass of the laser resonator results in a final intensity (inside the resonator) of

$$I_{rt} = I_{\text{start}} \cdot R_F R_B \cdot e^{2\gamma(\nu)L} \tag{5.5}$$

where R_F and R_B are the reflectances of the front and back mirrors and the round-trip gain is

$$G_{rt} = \frac{I_{\text{nt}}}{I_{\text{start}}} = R_F R_B e^{2\gamma(\nu)L}. \tag{5.6}$$

For oscillation to exist, the round-trip gain must be greater than or equal to unity as

$$R_F R_B e^{2\gamma(\nu)L} \geq 1 \tag{5.7}$$

or

$$\gamma(\nu) \geq \frac{1}{2L} \ln\left(\frac{1}{R_F R_B}\right). \tag{5.8}$$

Now, consider a laser cavity being pumped by an arbitrary pump source. The function of the pump source is to make

$$\left(N_2 - N_1\left(\frac{g_2}{g_1}\right)\right) > 0 \tag{5.9}$$

or (in other words) to create a population inversion.

As the laser gain material is pumped, spontaneous emission of radiation causes photons to be emitted at many frequencies in all directions. All photons with energies such that $g(\nu)\left(A_{21}\lambda_o^2/8\pi n^2\right)(N_2 - N_1(g_2/g_1)) > 0$ will be subject to gain via stimulated emission. However, only some of these photons are traveling in a regenerative path. (That is, a path that eventually returns on itself.) The design intention is that the most optimal (lowest loss and maximum gain) of these regenerative paths is the path between the laser mirrors.

Recall that the regenerative path between the two mirrors has the characteristic that it can only support longitudinal modes spaced by $c_o/2nL$. Thus (if the laser has been designed properly), the most optimal path with gain consists of the longitudinal modes between the two mirrors. Now, from the above condition, the only longitudinal modes that can lase are those whose round-trip gain is

$$R_F R_B e^{2\gamma(\nu)L} \geq 1. \tag{5.10}$$

Now, consider the following situation. Assume that five modes were initially greater than threshold, and that the $\gamma_o \cdot L$ for each of these modes is: mode 1 = 1.2, mode 2 = 1.3, mode 3 = 1.4, mode 4 = 1.3, and mode 5 = 1.2. After ten round-trip passes of the cavity, the overall gain for each mode would be

$$\left(\frac{I_{\text{end}}}{I_{\text{start}}}\right)_1 = (R_F R_B)^{10} e^{20 \cdot 1.1} = (R_F R_B)^{10}\, 3.59 \cdot 10^9 \tag{5.11}$$

$$\left(\frac{I_{\text{end}}}{I_{\text{start}}}\right)_2 = (R_F R_B)^{10} e^{20 \cdot 1.3} = (R_F R_B)^{10}\, 1.96 \cdot 10^{11} \tag{5.12}$$

$$\left(\frac{I_{\text{end}}}{I_{\text{start}}}\right)_3 = (R_F R_B)^{10} e^{20 \cdot 1.5} = (R_F R_B)^{10}\, 1.7 \cdot 10^{13} \tag{5.13}$$

$$\left(\frac{I_{\text{end}}}{I_{\text{start}}}\right)_4 = (R_F R_B)^{10} e^{20 \cdot 1.3} = (R_F R_B)^{10}\, 1.96 \cdot 10^{11} \tag{5.14}$$

$$\left(\frac{I_{\text{end}}}{I_{\text{start}}}\right)_5 = (R_F R_B)^{10} e^{20 \cdot 1.2} = (R_F R_B)^{10}\, 3.59 \cdot 10^9. \tag{5.15}$$

Even after just ten round trips, it is clear that the final intensity is unreasonably large! The intensity *cannot* keep increasing exponentially, because eventually enough stimulated transitions from E_2 to E_1 will occur to drop the population inversion below the threshold for lasing.

What happens instead is that the entire gain curve $\gamma(\nu)$ saturates until the round-trip gain (at the laser mode frequency) *exactly* equals 1. Essentially, a feedback mechanism is operating where either the gain is just slightly greater than 1 and the intensity increases

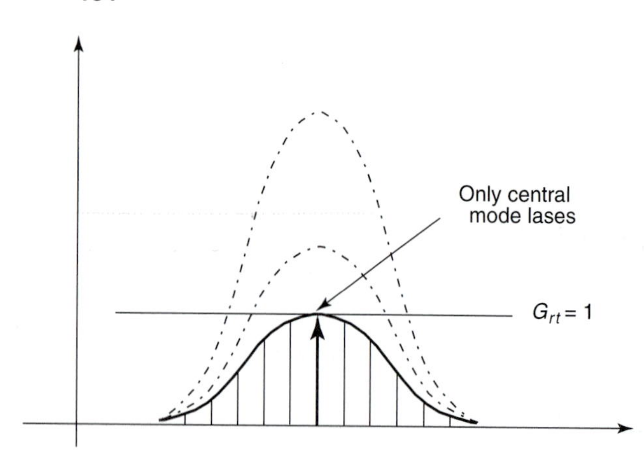

Figure 5.2 In a homogeneous laser, the entire gain curve saturates proportionally. As a consequence, only the central longitudinal mode can sustain a round trip gain of 1.

rapidly (thus saturating the transition and dropping the gain back to 1) or the gain oversaturates and drops below 1 and (so long as the pumping continues) the population inversion is rapidly re-established until the gain equals 1. This feedback loop then operates to keep each longitudinal mode precisely at a round-trip gain of unity.

5.1.1 Gain Saturation for the Homogeneous Line

In a homogeneous laser, the entire gain curve saturates proportionally (see Figure 5.2). As a consequence, only the central longitudinal mode can sustain a round-trip gain of 1. All other modes have round trip gains of less than unity. Thus the output of a homogeneous laser is a single large central mode with very tiny side modes. This output is independent of the actual number of longitudinal modes that are permitted by the $c_o/2nL$ condition to run under the gain curve.

5.1.2 Gain Saturation for the Inhomogeneous Line

In an inhomogeneous laser, saturation at one particular frequency causes a reduction in the gain profile only *near* that frequency (see Figure 5.3). So, the gain γ in the cavity drops only for those frequencies corresponding to the longitudinal modes of the cavity. Effectively, holes are burned in the gain profile such that the frequencies corresponding to the longitudinal modes are precisely at threshold. The gain profile at other frequencies retains a high value.

5.1.3 The Importance of Rate Equations

The saturation of the exponential small signal gain coefficient $\gamma(\nu)$ can be calculated using *rate equations*. Rate equations are a system of coupled equations that describe the dynamics of the various state populations. The rate equations can be solved for the steady state (this chapter) or for the transient state (Chapter 6). The *steady-state* solutions permit calculation of output power versus input power as well as optimal output coupler transmittance. The *transient solutions* permit calculation of the magnitude and time-dependence of relaxation oscillations as well as the magnitude, time dependence, and temporal width of pulses.

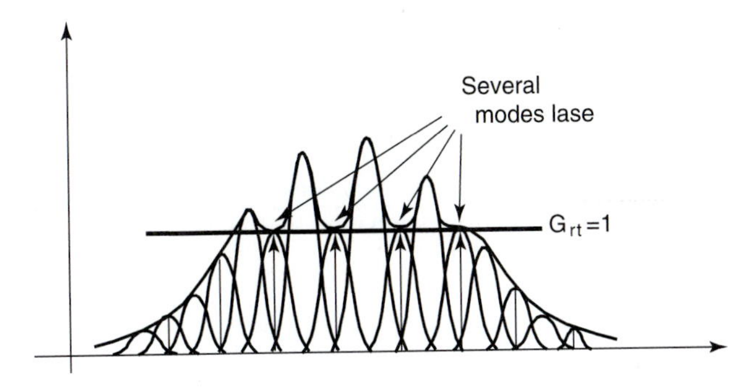

Figure 5.3 In an inhomogeneous laser, saturation at one particular frequency causes a reduction in the gain profile only *near* that frequency. Effectively, holes are burned in the gain profile such that the frequencies corresponding to the longitudinal modes are precisely at threshold.

5.2 SETTING UP RATE EQUATIONS

Although rate equations look complex, they are actually simple to set up. Writing them requires no more than careful accounting of the various state populations.

As an example, consider the arbitrary three state system illustrated in Figure 5.4. Assume that this system has a pump running from state $|1\rangle$ to $|3\rangle$ and lases from state $|2\rangle$ to $|1\rangle$.

There are two categories of processes occurring in this system. The first category contains spontaneous processes (such as spontaneous emission from state $|2\rangle$ to $|1\rangle$ and thermal transitions from state $|1\rangle$ to $|2\rangle$) . The second category contains stimulated processes. Notice that stimulated processes include both pumping processes and stimulated emission processes. Thus, both pumping from state $|1\rangle$ to $|3\rangle$ and lasing from state $|2\rangle$ to $|1\rangle$ are stimulated processes.

Recognize that there is a duality between processes. For every stimulated upward transition from state $|a\rangle$ to $|b\rangle$, there is a corresponding stimulated downward transition from state $|b\rangle$ to $|a\rangle$.

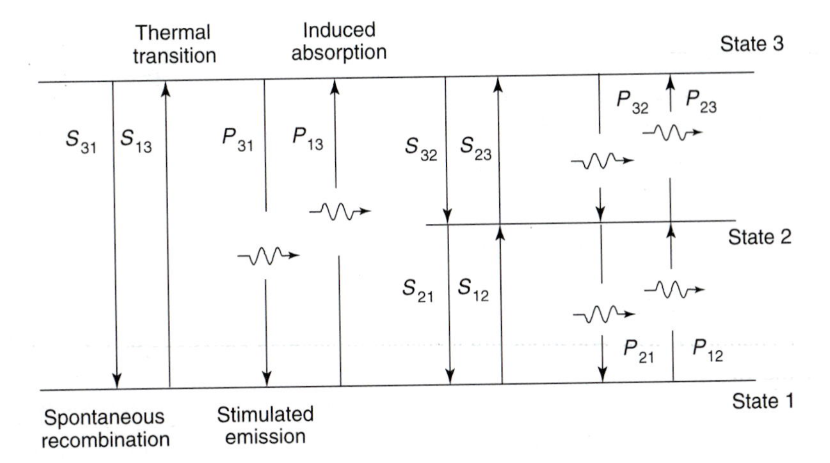

Figure 5.4 An arbitrary nondegenerate homogeneously broadened three-state system with an excitation pump running from state $|1\rangle$ to $|3\rangle$ and lasing from state $|2\rangle$ to $|1\rangle$.

Consider the most general rate equation for state $|1\rangle$. The equation describes the change in the electron population in state $|1\rangle$ per second. The equation consists of a series of transition probabilities (probability of making a transition per second) multiplied times the state populations (numbers of electrons). Using P to represent the stimulated transition probabilities (think of "pumping") and S to represent the spontaneous transition probabilities (think of "spontaneous"), then the change in the electron population in state $|1\rangle$ can be written most generally as[1]

$$\frac{dN_1}{dt} = -(P_{12} + P_{13} + S_{12} + S_{13})\,N_1 + (P_{21} + S_{21})\,N_2 + (P_{31} + S_{31})\,N_3 \tag{5.16}$$

where the various N_x terms are the state populations (numbers of electrons).

The first term represents the various ways that electrons can leave state $|1\rangle$ (see Figure 5.4). They can be stimulated by the laser transition from $|1\rangle$ to $|2\rangle$ (P_{12}) or by the pump from $|1\rangle$ to $|3\rangle$ (P_{13}). They can thermally transition from $|1\rangle$ to $|2\rangle$ (S_{12}) or from $|1\rangle$ to $|3\rangle$ (S_{13}). (The probability of a thermal transition from $|1\rangle$ to $|2\rangle$ or from $|1\rangle$ to $|3\rangle$ is unlikely unless the states are spaced by a few $k_B T$ or less.)

The second term represents the electrons that enter state $|1\rangle$ from state $|2\rangle$. These include electrons stimulated by the laser transition from $|2\rangle$ to $|1\rangle$ (P_{21}), or spontaneously recombining from $|2\rangle$ to $|1\rangle$ (S_{21}).

The third term represents the electrons that enter state $|1\rangle$ from state $|3\rangle$. These include electrons stimulated from $|3\rangle$ to $|1\rangle$ (P_{31}), or spontaneously recombining from $|3\rangle$ to $|1\rangle$ (S_{31}).

Equations for the changes in the other state populations can be written in a similar way as

$$\frac{dN_2}{dt} = -(P_{21} + P_{23} + S_{21} + S_{23})\,N_2 + (P_{12} + S_{12})\,N_1 + (P_{32} + S_{32})\,N_3 \tag{5.17}$$

$$\frac{dN_3}{dt} = -(P_{31} + P_{32} + S_{31} + S_{32})\,N_3 + (P_{13} + S_{13})\,N_1 + (P_{23} + S_{23})\,N_2. \tag{5.18}$$

The continuity equation

$$N_1 + N_2 + N_3 = N \tag{5.19}$$

is also needed to reduce the system to standard form.

Now, thermal transition probabilities $(S_{ij(i<j)})$ between states i and j are related by the Boltzman equation[2]

$$\frac{S_{ij}}{S_{ji}} = \frac{g_j}{g_i} \cdot \exp\left(\frac{-(E_j - E_i)}{k_B T}\right) \tag{5.20}$$

where E_j and E_i are the state energies and g_j and g_i are the state degeneracies. (Recall that the degeneracy of a state is the number of possible states that exist at exactly the same energy. Typical degeneracies are integers on the order of 2 to 5.) Notice that Equation (5.20) shows that thermally-induced transition probabilities for upward transitions are vanishingly small in real laser systems!

---

[1] Anthony Siegman, *Lasers* (Mill Valley, CA: University Science Books, 1986), p. 214.

[2] Anthony Siegman, *Lasers* (Mill Valley, CA: University Science Books, 1986), p. 213.

The simulated transition probabilities P_{ij} and P_{ji} are related by the state degeneracies as[3]

$$\frac{P_{ij}}{P_{ji}} = \frac{g_j}{g_i}. \tag{5.21}$$

Assuming that the degeneracy of both states is 1 (in other words, no degenerate states exist), then $P_{ji} = P_{ij}$.

Typically, a time constant (often termed the *relaxation lifetime*) is used to describe the spontaneous transition probability. Thus, the S_{ij} terms often take the form of $S_{ij} = 1/\tau_{ij}$. Additionally, the stimulated transitions either originate in laser action (in this chapter taking the form $R_{\text{trans},ij}$) or by means of intentional pumping (taking the form P_{ij}). Rewriting the set of three rate equations in this more conventional form gives

$$\frac{dN_1}{dt} = -\left(R_{\text{trans},12} + P_{13} + \frac{1}{\tau_{12}} + \frac{1}{\tau_{13}}\right) N_1 + \left(R_{\text{trans},21} + \frac{1}{\tau_{21}}\right) N_2 + \left(P_{31} + \frac{1}{\tau_{31}}\right) N_3 \tag{5.22}$$

$$\frac{dN_2}{dt} = -\left(R_{\text{trans},21} + P_{23} + \frac{1}{\tau_{21}} + \frac{1}{\tau_{23}}\right) N_2 + \left(R_{\text{trans},12} + \frac{1}{\tau_{12}}\right) N_1 + \left(P_{32} + \frac{1}{\tau_{32}}\right) N_3 \tag{5.23}$$

$$\frac{dN_3}{dt} = -\left(P_{31} + P_{32} + \frac{1}{\tau_{31}} + \frac{1}{\tau_{32}}\right) N_3 + \left(P_{13} + \frac{1}{\tau_{13}}\right) N_1 + \left(P_{23} + \frac{1}{\tau_{23}}\right) N_2. \tag{5.24}$$

5.2.1 Rate Equations for Four-State Lasers

Consider a nondegenerate homogeneously broadened four-state laser system (illustrated in Figure 5.5), where τ_{21} is the total relaxation lifetime from state $|2\rangle$ to state $|1\rangle$ due to both spontaneous emission and other processes, τ_{20} is the total relaxation lifetime from state $|2\rangle$ to any state other than $|1\rangle$, $1/\tau_2 = 1/\tau_{20} + 1/\tau_{21}$ is the total relaxation lifetime from state $|2\rangle$, τ_1 is the total relaxation lifetime from state $|1\rangle$ to state $|0\rangle$ due to both spontaneous emission and other processes, P_2 is the pumping rate to state $|2\rangle$, and P_1 is the pumping rate to state $|1\rangle$.

Unlike a three-state laser system, in a four-state laser system the populations of states $|0\rangle$ and $|4\rangle$ are not significantly changed when the laser is pumped to threshold. Thus, only the populations in the laser states need to be included in the rate equations. Writing the

[3]Anthony Siegman, *Lasers* (Mill Valley, CA: University Science Books, 1986), p. 212.

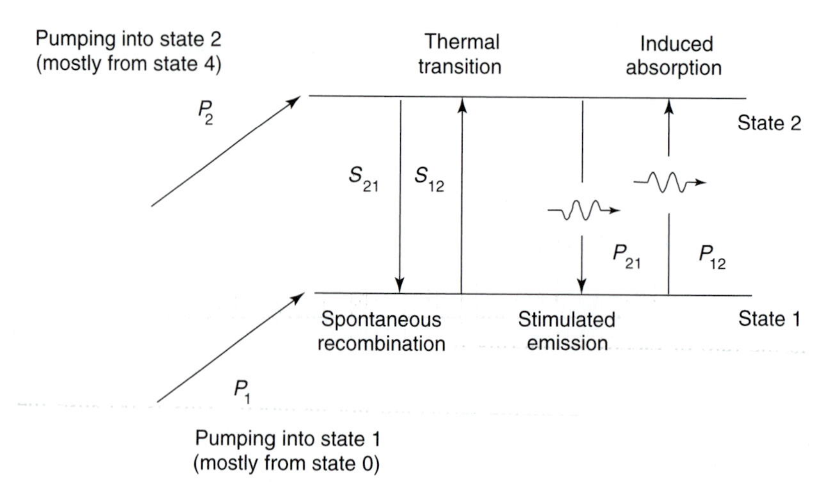

Figure 5.5 An arbitrary nondegenerate homogeneously broadened four-state laser system with a pump running from state $|1\rangle$ to $|4\rangle$ and lasing from state $|2\rangle$ to $|1\rangle$.

general rate equations for a two-state system gives

$$\frac{dN_1}{dt} = -\left(R_{\text{trans},12} + \frac{1}{\tau_{12}}\right)N_1 + \left(R_{\text{trans},21} + \frac{1}{\tau_{21}}\right)N_2 \tag{5.25}$$

$$\frac{dN_2}{dt} = -\left(R_{\text{trans},21} + \frac{1}{\tau_{21}}\right)N_2 + \left(R_{\text{trans},12} + \frac{1}{\tau_{12}}\right)N_1 \tag{5.26}$$

where[4]

$$R_{\text{trans},ij} = \frac{\sigma_{ij} I_\nu}{(h\nu)_{ij}} \tag{5.27}$$

represents the transition rate due to laser action. I_ν is the intensity of the laser beam in the cavity and σ_{ij} is the cross-section between states i and j.

In a typical laser system, states $|1\rangle$ and $|2\rangle$ are separated by many $k_B T$. Thus, the upward thermal transition probability $1/\tau_{12}$ will be negligible. However, since the system does have a $|0\rangle$ state, there is the possibility of spontaneous decay from $|1\rangle$ and $|0\rangle$. Therefore, the rate equation for the lower state must be modified to include this possibility by adding a term τ_1. Furthermore, the additional states permit spontaneous decay mechanisms from both $|2\rangle$ to $|1\rangle$ and from $|2\rangle$ to $|0\rangle$. Thus, $1/\tau_2 = 1/\tau_{20} + 1/\tau_{21}$ is used as the total relaxation lifetime from state $|2\rangle$. Finally, assume that the degeneracies of both states are 1 ($R_{\text{trans},12} = R_{\text{trans},21} = R_{\text{trans}}$). With these changes, the N_1 equation takes the form

$$\frac{dN_1}{dt} = -\left(R_{\text{trans}} + \frac{1}{\tau_1}\right)N_1 + \left(R_{\text{trans}} + \frac{1}{\tau_{21}}\right)N_2. \tag{5.28}$$

The $R_{\text{trans}}N_1$ term is the rate of electrons leaving the state by laser action, $\frac{1}{\tau_1}N_1$ is the rate of electrons leaving the state (to go either to state $|2\rangle$, not likely, or $|0\rangle$), the $R_{\text{trans}}N_2$

[4]Walter Koechner, *Solid State Laser Engineering*, 3d ed. (Berlin: Springer-Verlag, 1992), p. 15.

term is the rate of electrons entering the state by laser action (stimulated from $|2\rangle$ to $|1\rangle$ by the laser light) and the $\frac{1}{\tau_{21}}N_2$ term is the rate of electrons entering the state by spontaneous recombination from $|2\rangle$ to $|1\rangle$.

The N_2 equation takes the form

$$\frac{dN_2}{dt} = -\left(R_{\text{trans}} + \frac{1}{\tau_2}\right)N_2 + (R_{\text{trans}})\,N_1. \tag{5.29}$$

The $R_{\text{trans}}N_2$ term is the rate of electrons leaving the state by laser action, $(1/\tau_2)N_2$ is the rate of electrons leaving the state (to go either to state $|1\rangle$ or $|0\rangle$), and the $R_{\text{trans}}N_1$ term is the rate of electrons entering the state by laser action (stimulated from $|1\rangle$ to $|2\rangle$ by the laser light).

Notice that the pumping terms are not included. This is because the system fills by being pumped from states *different* from $|1\rangle$ and $|2\rangle$. This external pumping is most readily included by adding constant pumping rate terms P_2 and P_1 as[5]

$$\frac{dN_1}{dt} = P_1 - \left(R_{\text{trans}} + \frac{1}{\tau_1}\right)N_1 + \left(R_{\text{trans}} + \frac{1}{\tau_{21}}\right)N_2 \tag{5.30}$$

$$\frac{dN_2}{dt} = P_2 - \left(R_{\text{trans}} + \frac{1}{\tau_2}\right)N_2 + (R_{\text{trans}})\,N_1. \tag{5.31}$$

Now, consider the steady-state solution where all the d/dt terms go to zero as

$$\frac{dN_1}{dt} = P_1 - \left(R_{\text{trans}} + \frac{1}{\tau_1}\right)N_1 + \left(R_{\text{trans}} + \frac{1}{\tau_{21}}\right)N_2 = 0 \tag{5.32}$$

$$\frac{dN_2}{dt} = P_2 - \left(R_{\text{trans}} + \frac{1}{\tau_2}\right)N_2 + (R_{\text{trans}})\,N_1 = 0. \tag{5.33}$$

To solve this, reorganize into the standard simultaneous equation form

$$\left(\frac{1}{\tau_1} + R_{\text{trans}}\right)N_1 + \left(-\frac{1}{\tau_{21}} - R_{\text{trans}}\right)N_2 = P_1 \tag{5.34}$$

$$(-R_{\text{trans}})\,N_1 + \left(\frac{1}{\tau_2} + R_{\text{trans}}\right)N_2 = P_2 \tag{5.35}$$

and solve using Cramer's rule[6,7]

$$(N_2 - N_1) = \frac{P_2\tau_2\left(1 - \frac{\tau_1}{\tau_{21}}\right) - P_1\tau_1}{1 + \left(\tau_1 + \tau_2 - \frac{\tau_1\tau_2}{\tau_{21}}\right)R_{\text{trans}}}. \tag{5.36}$$

[5] Amnon Yariv, *Optical Electronics*, 4th ed. (Philadelphia, PA: Saunders College Publishing, 1991), p. 167.

[6] Amnon Yariv, *Optical Electronics*, 4th ed. (Philadelphia, PA: Saunders College Publishing, 1991), p. 168.

[7] Joseph T. Verdeyen, *Laser Electronics*, 2d ed. (Englewood Cliffs, NJ: Prentice Hall, 1989), p. 194.

Now, recall the form of the gain coefficient for a nondegenerate system

$$\gamma(\nu) = g(\nu)\left(\frac{A_{21}\lambda_o^2}{8\pi n^2}\right)(N_2 - N_1) \tag{5.37}$$

and substitute in for $(N_2 - N_1)$ giving

$$\gamma(\nu) = g(\nu)\left(\frac{A_{21}\lambda_o^2}{8\pi n^2}\right)\left(\frac{P_2\tau_2\left(1-\frac{\tau_1}{\tau_{21}}\right) - P_1\tau_1}{1+\left(\tau_1+\tau_2-\frac{\tau_1\tau_2}{\tau_{21}}\right)R_{\text{trans}}}\right). \tag{5.38}$$

Now, γ_o is defined where R_{trans} goes to zero, giving[8]

$$\gamma_o = \left(\frac{g(\nu)A_{21}\lambda_o^2}{8\pi n^2}\right)\left(P_2\tau_2\left(1-\frac{\tau_1}{\tau_{21}}\right) - P_1\tau_1\right) \tag{5.39}$$

and the equation for $\gamma(\nu)$ can be written in the form

$$\gamma(\nu) = \frac{\gamma_o}{1+\left(\tau_1+\tau_2-\frac{\tau_1\tau_2}{\tau_{21}}\right)R_{\text{trans}}} \tag{5.40}$$

or by using the *saturation intensity* I_{sat} as

$$\boxed{\gamma(\nu) = \frac{\gamma_o}{1+\frac{I}{I_{\text{sat}}}}} \tag{5.41}$$

where[9]

$$I_{\text{sat}} = \left(\frac{8\pi h n^2 \nu^3}{g(\nu)A_{21}c_o^2\tau_2}\right)\left(\frac{1}{1+\frac{\tau_1}{\tau_2}\left(1-\frac{\tau_2}{\tau_{21}}\right)}\right). \tag{5.42}$$

Notice what this shows. If the intensity in the laser is very low, then the gain $\gamma(\nu) = \gamma_o$. However, as the intensity in the cavity increases, then the gain begins to drop. At a cavity intensity of I_{sat} the gain has dropped by half to $\gamma(\nu) = \gamma_o/2$. The gain will continue to decrease until the saturated gain equals the cavity losses.

Example 5.1

Use a computational solver (such as *Mathematica* or *Maple*) to solve the steady-state rate-equation system given by Equations (5.32) and (5.33) for N_1 and N_2.

Solution. The results are

$$N_1 = \frac{\tau_1(P_2\tau_2 + P_1\tau_{21} + P_1R_{\text{trans}}\tau_2\tau_{21} + P_2R_{\text{trans}}\tau_2\tau_{21})}{-R_{\text{trans}}\tau_1\tau_2 + \tau_{21} + R_{\text{trans}}\tau_1\tau_{21} + R_{\text{trans}}\tau_2\tau_{21}} \tag{5.43}$$

$$N_2 = \frac{(P_2 + P_1R_{\text{trans}}\tau_1 + P_2R_{\text{trans}}\tau_1)\tau_2\tau_{21}}{-R_{\text{trans}}\tau_1\tau_2 + \tau_{21} + R_{\text{trans}}\tau_1\tau_{21} + R_{\text{trans}}\tau_2\tau_{21}} \tag{5.44}$$

[8] Joseph T. Verdeyen, *Laser Electronics*, 2d ed. (Englewood Cliffs, NJ: Prentice Hall, 1989), p. 195.

[9] Amnon Yariv, *Optical Electronics*, 4th ed. (Philadelphia, PA: Saunders College Publishing, 1991), p. 168.

Example 5.2

Calculate γ_o and I_{sat} using the four-state expressions given in Equations (5.39) and (5.42). Assume that the laser is a low-power Nd:YAG operating at $\lambda_o = 1.064\ \mu$m, with $\eta = 0.025$, the index of refraction $n = 1.5$, $\tau_{21} = \tau_2 = 230\ \mu$s, $\tau_1 = 120$ ns, the loss per unit length $\alpha = 0.00031$ cm^{-1}, and the gain curve is Lorentzian with a linewidth of 119.165 GHz. Assume that $P_2 = 3.555 \cdot 10^{18}$ cm^{-3}sec^{-1} and $P_1 = 0$.

Solution. For a Lorentzian laser, $g(\nu)$ at the peak of the gain curve is given as

$$g = \frac{2}{\pi \cdot \Delta\nu} = \frac{2}{\pi \cdot 119.165\ \text{GHz}} = 5.342\ \text{ps}$$

Use Equation (5.39) to calculate γ_o for the four-state laser as

$$\gamma_o = \left(\frac{g(\nu)A_{21}\lambda_o^2}{8\pi n^2}\right)\left(P_2\tau_2\left(1 - \frac{\tau_1}{\tau_{21}}\right) - P_1\tau_1\right) \tag{5.45}$$

$$\gamma_o = \left(\frac{(5.342\ \text{ps})(1/230\ \mu\text{s})(1.064\ \mu\text{m})^2}{8\pi(1.5)^2}\right) \tag{5.46}$$

$$\cdot\left((3.555 \cdot 10^{18}\ \text{cm}^{-3}\ \text{sec}^{-1})(230\ \mu\text{s})\left(1 - \frac{120\ \text{ns}}{230\ \mu\text{s}}\right) - 0\right)$$

$$\gamma_o = 0.0038\ \text{cm}^{-1} \tag{5.47}$$

and use Equation (5.42) to calculate I_{sat} as

$$I_{\text{sat}} = \left(\frac{8\pi h n^2 \nu^3}{g(\nu)A_{21}c_o^2\tau_2}\right)\left(\frac{1}{1 + \frac{\tau_1}{\tau_2}\left(1 - \frac{\tau_2}{\tau_{21}}\right)}\right) \tag{5.48}$$

$$I_{\text{sat}} = \left(\frac{8\pi h(1.5)^2(c_o/1.064\ \mu\text{m})^3}{(5.342\ \text{ps})(1/230\ \mu\text{s})c_o^2(230\ \mu\text{s})}\right)\left(\frac{1}{1 + \frac{120\ \text{ns}}{230\ \mu\text{S}}\left(1 - \frac{230\ \mu\text{S}}{230\ \mu\text{S}}\right)}\right) \tag{5.49}$$

$$I_{\text{sat}} = 174.57\ \frac{\text{watt}}{\text{cm}^2}.$$

Example 5.3

Use the results from Example 5.2 to plot the gain as a function of the intensity in the laser cavity. Assume the same conditions as in Example 5.2.

Solution. Equation (5.41) can be used to plot the gain as a function of the intensity in the cavity

$$\gamma = \frac{\gamma_o}{1 + \frac{I}{I_{\text{sat}}}}.$$

These results are plotted in Figure 5.6 for $\gamma_o = 0.0038$ cm^{-1} and $I_{\text{sat}} = 174.57$ watt/cm^2.

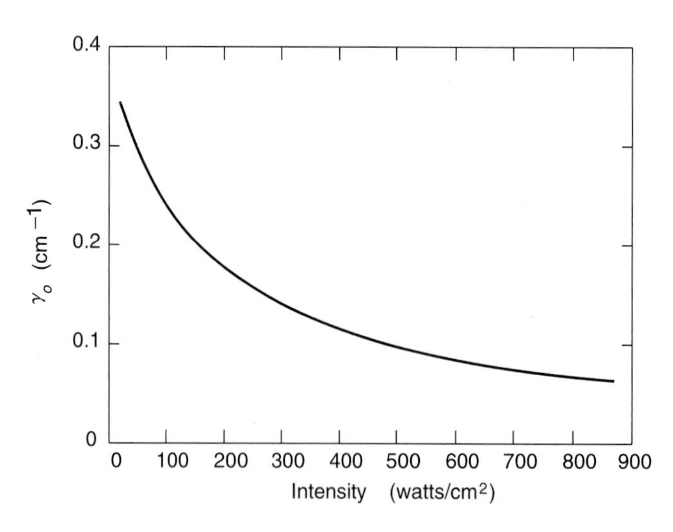

Figure 5.6 The gain for a low-power Nd:YAG laser as a function of the intensity in the laser cavity for $\gamma_o = 0.0038 \text{ cm}^{-1}$ and $I_{\text{sat}} = 174.57 \text{ watt/cm}^2$.

5.3 LASER OUTPUT POWER CHARACTERISTICS

5.3.1 Optimal Coupling, a Simple Approach

Consider again the simple resonator, L in length, consisting of two mirrors of reflectivity R_F and R_B (see Figure 5.7). Assume that the rear mirror has a reflectivity of 100%. If the front mirror (the output coupler) also had a reflectivity of 100%, then no light would be extracted from the laser. Alternatively, if the output coupler had a reflectivity of 0%, then round-trip gain would not be possible, and the laser would have to have an exceptionally high gain coefficient for there to be any useful output. Thus, there must exist some optimal output coupler reflectivity R_{opt} which yields the maximum possible output P_{out} from the laser system. The following procedure illustrates how to calculate R_{opt} by using the expressions for I_{sat} derived from the laser rate equations.

Recall the equation for the round-trip gain in a laser

$$G_{rt} = \frac{I_{\text{end}}}{I_{\text{start}}} = R_F R_B e^{2\gamma(\nu)L}. \tag{5.50}$$

If a loss per unit length α (due to polarization losses, dirt, dust, internal reflections, etc.) is included, then

$$G_{rt} = \frac{I_{\text{end}}}{I_{\text{start}}} = R_F R_B e^{2(\gamma(\nu)-\alpha)L}. \tag{5.51}$$

The threshold oscillation condition is then obtained by setting $G_{rt} = 1$ and solving for $\gamma = \gamma_t$,

$$R_F R_B e^{2(\gamma(\nu)-\alpha)L} = 1 \tag{5.52}$$

$$\gamma_t = \alpha + \frac{1}{2L}\ln\left(\frac{1}{R_F R_B}\right) = \alpha - \frac{1}{2L}\ln(R_F R_B). \tag{5.53}$$

The logarithmic term can be approximated (for small x) using the series expansion

$$\ln(1+x) = x - \frac{x^2}{2} + \frac{x^3}{3} + \cdots \tag{5.54}$$

For this case $1 + x = R_F R_B$, so $x = R_F R_B - 1$ and

$$\gamma_t = \alpha + \frac{1}{2L} \cdot (1 - R_F R_B). \tag{5.55}$$

Now, recall that the saturated gain γ_t for a homogeneously broadened system is given by Equation (5.41)

$$\gamma_t = \frac{\gamma_o}{1 + \frac{I}{I_{sat}}}. \tag{5.56}$$

This expression describes the gain saturation for a single wave propagating through the material. However, in a real laser cavity, there is one wave propagating forward and one wave propagating backward as illustrated in Figure 5.7.

Thus, the equation describing the gain in a laser cavity can more accurately be written as

$$\gamma_t = \frac{\gamma_o}{1 + \frac{I_+ + I_-}{I_{sat}}} \tag{5.57}$$

where I_+ is the right traveling wave and I_- is the left traveling wave. A reasonable assumption is to assume an average cavity intensity I_{circ} where $2I_{circ} = I_+ + I_-$ and to rewrite this equation as

$$\gamma_t = \frac{\gamma_o}{1 + \frac{2I_{circ}}{I_{sat}}}. \tag{5.58}$$

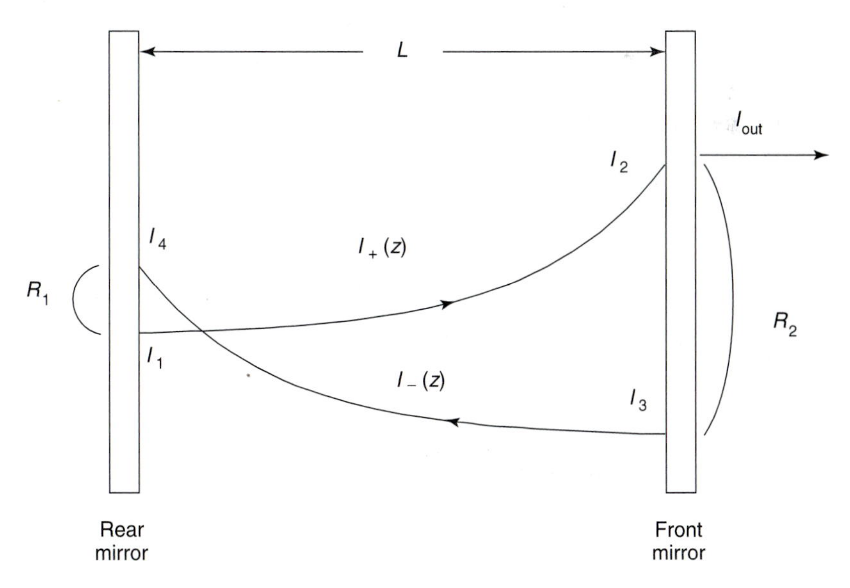

Figure 5.7 In a typical laser cavity, there is one wave propagating forward and one wave propagating backward.

Solving for I_{circ} yields

$$I_{\text{circ}} = \frac{I_{\text{sat}}}{2}\left(\frac{\gamma_o}{\gamma_t} - 1\right). \tag{5.59}$$

Substituting for γ_t gives

$$I_{\text{circ}} = \frac{I_{\text{sat}}}{2}\left(\frac{\gamma_o}{\alpha + \frac{1}{2L}\cdot(1 - R_F R_B)} - 1\right). \tag{5.60}$$

Now, the output power from the laser at the output mirror is simply the intensity in the cavity times the transmittance of the output coupler times the area or

$$P_{\text{out}} = A \cdot T_F \cdot I_{\text{circ}} = \frac{A I_{\text{sat}} T_F}{2}\left(\frac{\gamma_o}{\alpha + \frac{1}{2L}\cdot(1 - R_F R_B)} - 1\right). \tag{5.61}$$

Assuming that R_B is a high reflector, then $R_B = 1$ and

$$\boxed{P_{\text{out}} = \frac{A I_{\text{sat}} T_F}{2}\left(\frac{\gamma_o}{\alpha + \frac{1}{2L}\cdot T_F} - 1\right).} \tag{5.62}$$

The optimal output intensity condition as a function of mirror transmittance can be obtained by taking the partial derivative and setting it equal to zero as

$$\frac{\partial P_{\text{out}}}{\partial T_F} = 0 \tag{5.63}$$

giving the optimal transmittance as

$$\boxed{T_{\text{opt}} = 2L\sqrt{\gamma_o \alpha} - 2\alpha L.} \tag{5.64}$$

Substituting this back into Equation (5.62) for the output power gives the optimal output power as

$$P_{\text{out(opt)}} = \frac{A I_{\text{sat}}\left(2L\sqrt{\gamma_o\alpha} - 2\alpha L\right)}{2}\left(\frac{\gamma_o}{\alpha + \frac{1}{2L}\cdot\left(2L\sqrt{\gamma_o\alpha} - 2\alpha L\right)} - 1\right) \tag{5.65}$$

or

$$\boxed{P_{\text{out(opt)}} = A I_{\text{sat}} L\left(\sqrt{\gamma_o} - \sqrt{\alpha}\right)^2.} \tag{5.66}$$

Example 5.4

Calculate and plot the output power from a low-power Nd:YAG laser as a function of the transmittance of the output mirror. Assume that the Nd:YAG laser is a four-state homogeneous laser operating at $\lambda_o = 1.064\ \mu\text{m}$, with $\gamma_o = 0.0038\ \text{cm}^{-1}$, $I_{\text{sat}} = 174.57\ \text{watt/cm}^2$, the loss per unit length $\alpha = 0.00031\ \text{cm}^{-1}$, a length of 150 cm, and an output area of 0.5 cm^2.

Solution. Equation (5.62) can be used to calculate P_{out} as a function of T_F

$$P_{\text{out}} = \frac{A I_{\text{sat}} T_F}{2}\left(\frac{\gamma_o}{\alpha + \frac{1}{2L}\cdot T_F} - 1\right). \tag{5.67}$$

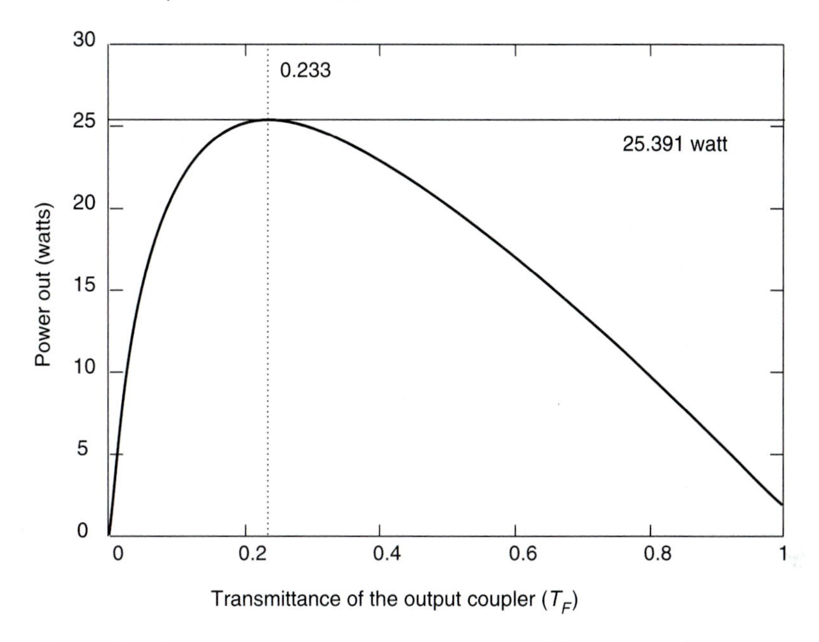

Figure 5.8 The output power from a low-power Nd:YAG laser as a function of the transmittance of the output mirror.

These results are plotted in Figure 5.8. The optimal output coupler and output power can be calculated from Equations (5.64) and (5.66) as

$$T_{\text{opt}} = 2L\sqrt{\gamma_o \alpha} - 2\alpha L \tag{5.68}$$

$$T_{\text{opt}} = 2(150 \text{ cm})\sqrt{(0.0038 \text{ cm}^{-1})(0.00031 \text{ cm}^{-1})} - 2(0.00031 \text{ cm}^{-1})(150 \text{ cm}) = 0.233 \tag{5.69}$$

$$P_{\text{out(opt)}} = A I_{\text{sat}} L \left(\sqrt{\gamma_o} - \sqrt{\alpha}\right)^2 \tag{5.70}$$

$$P_{\text{out(opt)}} = (0.5 \text{ cm}^{-2})(174.57 \text{ watt/cm}^2)(150 \text{ cm}) \left(\sqrt{(0.0038 \text{ cm}^{-1})} - \sqrt{(0.00031 \text{ cm}^{-1})}\right)^2 \tag{5.71}$$

$$P_{\text{out(opt)}} = 25.391 \text{ watt} \tag{5.72}$$

Example 5.5

It is interesting to observe the behavior of the optimal output coupler as a function of gain, loss, and length. For the same parameters as in the previous example, plot the optimal output transmittance as a function of γ_o with L as a parameter. Also plot the optimal output transmittance as a function of α with L as a parameter.

Solution. The optimal output coupler can be calculated from Equation (5.64). For $\alpha = 0.00031$ cm^{-1}, over a range of $\gamma_o = 0$ to 0.003, and using values of $L = 25$ cm, 50 cm, 100 cm and 200 cm gives the results plotted in Figure 5.9. For $\gamma^o = 0.0038$ cm^{-1}, over a range of $\alpha = 0$ to 0.003, and using values of $L = 25$ cm, 50 cm, 100 cm, and 200 cm gives the results plotted in Figure 5.10.

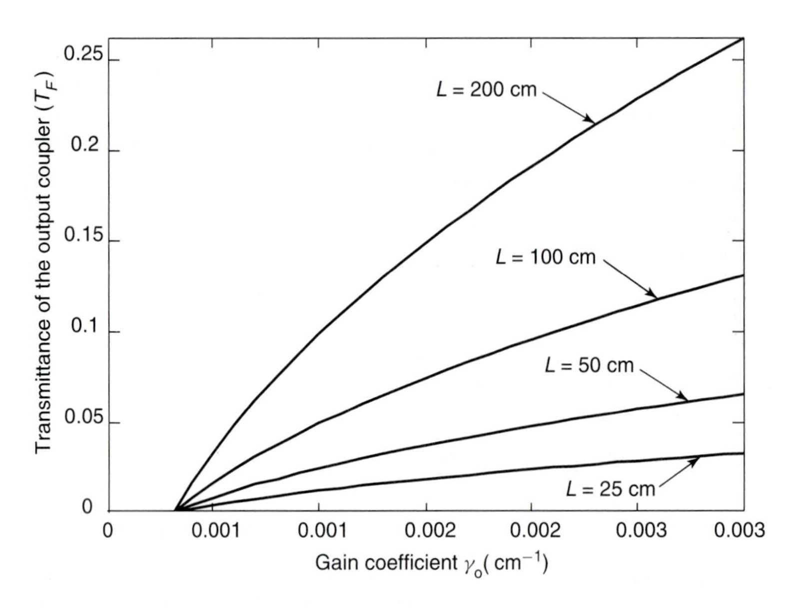

Figure 5.9 The value for the optimal output coupler for a low-power Nd:YAG laser as a function of gain, loss, and length. Calculated for $\alpha = 0.00031$ cm^{-1}, over a range of $\gamma_o = 0$ to 0.003, and using values of $L = 25$ cm, 50 cm, 100 cm, and 200 cm.

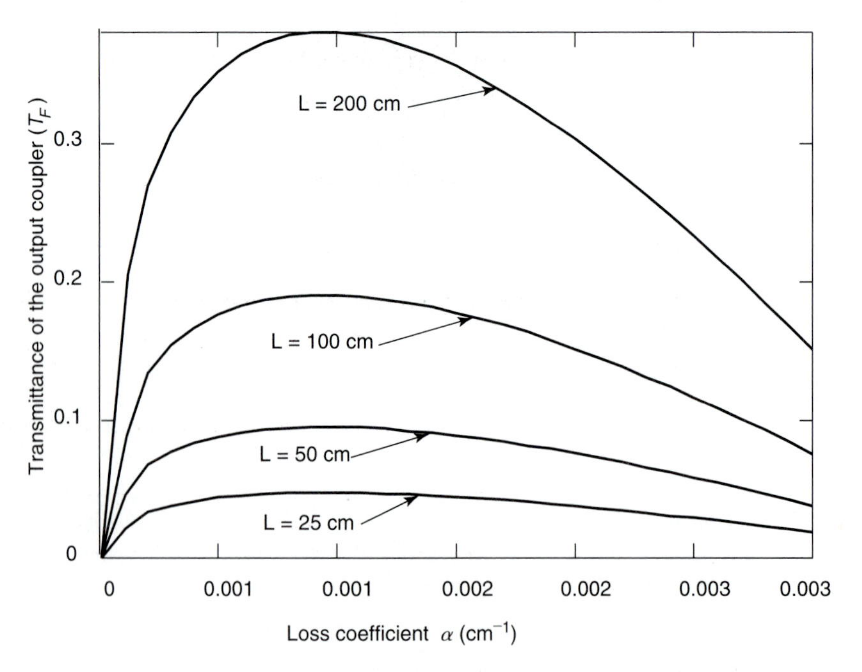

Figure 5.10 The optimal output transmittance for a low-power Nd:YAG laser as a function of α with L as a parameter. Calculated for $\gamma_o = 0.0038$ cm^{-1}, over a range of $\alpha = 0$ to 0.003, and using values of $L = 25$ cm, 50 cm, 100 cm, and 200 cm.

5.3.2 P_{out} versus P_{in}, An Engineering Approach

The output power for a four-state homogeneously broadened laser is a straight line described by

$$P_{out} = \sigma_s (P_{in} - P_{thres}) \tag{5.73}$$

which begins at the *threshold power* P_{thres} and has a *slope efficiency* of σ_s (the slope of the P_{in} versus P_{thres} line; see Example 5.6). The following procedure demonstrates this by defining some additional terms and incorporating them into the basic output power Equation (5.62). This discussion follows the approach of Koechner.[10]

Consider γ_o in cm^{-1} for a four-state homogeneously broadened, continuous wave (steady-state) laser system, as determined from the rate equations

$$\gamma_o = \left(\frac{g(\nu)A_{21}\lambda_o^2}{8\pi n^2}\right)\left(P_2\tau_2\left(1 - \frac{\tau_1}{\tau_{21}}\right) - P_1\tau_1\right). \tag{5.74}$$

Under most conditions, the spontaneous relaxation time from state $|1\rangle$ will be very short and the relaxation time between states $|2\rangle$ and $|1\rangle$ will be very long. Thus, this equation can be approximated rather well by

$$\gamma_o = \left(\frac{g(\nu)A_{21}\lambda_o^2}{8\pi n^2}\right) P_2\tau_2 \tag{5.75}$$

where P_2 is an energy pumping rate into state $|2\rangle$ with the units of atoms/second-cm^3 and where P_2 is related to the input electrical power (in watts) P_{in} into the pumping source by

$$P_2 = \frac{\eta_1\eta_2\eta_3\eta_4\eta_5 P_{in}}{h\nu V} \tag{5.76}$$

where V is the volume of the laser gain material, P_{in} is the electrical power in watts, ν_o is the center laser frequency, η_1 is the quantum efficiency (the percentage of electrons pumped to the pump band which make it to the upper laser state), η_2 is the absorption efficiency by the laser rod of the light that is within the pumpbands, η_3 is the transfer efficiency of light from the pump source to the laser rod, η_4 is the percentage of pump source light that falls in the pump bands, and η_5 is the conversion efficiency of the electrical input into the pump source into light output from the pump source.

Now, it is conventional to lump all these efficiency and gain factors together and consider the single pass gain $\gamma_o L$ to be given by

$$\gamma_o L = \left(\frac{g(\nu)A_{21}\lambda_o^2}{8\pi n^2}\right) \frac{\tau_2\eta_1\eta_2\eta_3\eta_4\eta_5 P_{input} L}{h\nu V} = K P_{in} \tag{5.77}$$

or more simply by

$$\gamma_o L = \sigma_{21} \frac{\tau_2 \eta_{laser} P_{input} L}{h\nu V} = K P_{in} \tag{5.78}$$

where K can be written as

$$K = \sigma_{21} \frac{\tau_2 \eta_{laser} L}{h\nu V} \tag{5.79}$$

[10] Walter Koechner, *Solid State Laser Engineering*, 3d ed. (Berlin: Springer-Verlag, 1992), pp. 96–8.

where σ_{21} is the cross-section in cm^2, $\eta_{\text{laser}} = \eta_1\eta_2\eta_3\eta_4\eta_5$ is the total system efficiency, and K is given in watt^{-1}.

Notice that I_{sat} for a four-state laser system is in watts/cm^2 and is given by

$$I_{\text{sat}} = \left(\frac{8\pi h n^2 \nu^3}{g(\nu) A_{21} c_o^2 \tau_2}\right)\left(\frac{1}{1 + \frac{\tau_1}{\tau_2}\left(1 - \frac{\tau_2}{\tau_{21}}\right)}\right) \tag{5.80}$$

If τ_1 is very small, then this can be approximated as

$$I_{\text{sat}} = \left(\frac{8\pi h n^2 \nu^3}{g(\nu) A_{21} c_o^2 \tau_2}\right) \tag{5.81}$$

or, recalling the definition of σ_{21},

$$I_{\text{sat}} = \frac{h\nu}{\sigma_{21}\tau_2} \tag{5.82}$$

and substituting Equation (5.82) into Equation (5.79), permits K to be written as

$$K = \frac{\eta_{\text{laser}}}{I_{\text{sat}} A} \tag{5.83}$$

where K is given in watts^{-1} and A is the area of the rod in cm^2.

Recall that the threshold oscillation condition is that the round-trip gain equals 1. If losses in the cavity are included

$$R_F R_B e^{2(\gamma(\nu)_t - \alpha) L} = 1 \tag{5.84}$$

or

$$2\gamma(\nu)_t L = \ln\left(\frac{1}{R_F R_B}\right) + 2\alpha L \tag{5.85}$$

where α is the net loss due to everything (ohmic loss, reflections, impurities, etc.). Substituting the expression for the single pass gain (at threshold) into the above equation gives

$$2K P_{\text{thres}} = \ln\left(\frac{1}{R_F R_B}\right) + 2\alpha L \tag{5.86}$$

or at threshold

$$\ln\left(\frac{1}{R_F R_B}\right) = -\ln(R_F R_B) = 2K P_{\text{thres}} - 2\alpha L \tag{5.87}$$

where P_{thres} is in units of watts.

Now, notice that it is possible to run an experiment where the threshold power of the laser is measured for several different mirrors.[11] If the results are plotted in a graph of P_{thres} versus $-\ln R_F R_B$, the slope of the graph is $2K$ and the extrapolation of the curve to $P_{\text{in}} = 0$ gives $2\alpha L$ (see Figure 5.19).

Now, the basic output power Equation (5.62) is

$$P_{\text{out}} = \frac{A T_F I_{\text{sat}}}{2}\left(\frac{\gamma_o}{\alpha + \frac{1}{2L} \cdot T_F} - 1\right) \tag{5.88}$$

[11] The original version of this experiment was first proposed by D. Findlay and R. A. Clay, *Phys. Lett.* 20:277 (1966).

and using $\gamma_o L = K P_{\text{in}}$ gives

$$P_{\text{out}} = \frac{A T_F I_{\text{sat}}}{2} \left(\frac{K P_{\text{in}}}{\left(\alpha + \frac{1}{2L} \cdot T_F\right) L} - 1 \right) \tag{5.89}$$

or rewriting

$$P_{\text{out}} = A T_F I_{\text{sat}} \frac{K P_{\text{in}}}{(2\alpha L + T_F)} - \frac{A T_F I_{\text{sat}}}{2}. \tag{5.90}$$

The threshold power for the laser to lase (P_{thres}) can be obtained by solving Equation (5.90) for $P_{\text{in}} = P_{\text{thres}}$ at $P_{\text{out}} = 0$ as

$$A T_F I_{\text{sat}} \frac{K P_{\text{thres}}}{(2\alpha L + T_F)} - \frac{A T_F I_{\text{sat}}}{2} = 0. \tag{5.91}$$

This can be simplified to give P_{thres} in terms of K, or, by using Equation (5.83) in terms of I_{sat}

$$P_{\text{thres}} = \frac{(2\alpha L + T_F)}{2K} = \frac{I_{\text{sat}} A}{2\eta_{\text{laser}}} (2\alpha L + T_F). \tag{5.92}$$

Now, substitute P_{thres} back into Equation (5.90), giving

$$P_{\text{out}} = A T_F I_{\text{sat}} \frac{K P_{\text{in}}}{(2\alpha L + T_F)} - A T_F I_{\text{sat}} \frac{K P_{\text{thres}}}{(2\alpha L + T_F)}. \tag{5.93}$$

Rewriting yields

$$P_{\text{out}} = \left(A T_F I_{\text{sat}} \frac{K}{(2\alpha L + T_F)} \right) (P_{\text{in}} - P_{\text{thres}}) = \sigma_s (P_{\text{in}} - P_{\text{thres}}). \tag{5.94}$$

The term σ_s is the slope efficiency and can be simplified using Equation (5.83) as

$$\sigma_s = \left(A T_F I_{\text{sat}} \frac{K}{(2\alpha L + T_F)} \right) = \left(T_F \frac{\eta_{\text{laser}}}{(2\alpha L + T_F)} \right). \tag{5.95}$$

So, for a four-state homogeneously broadened, continuous wave laser using the $2I_{\text{circ}} = I_+ + I_-$ and the ln approximation indicated in Equation (5.54), the laser output power is a straight line described by

$$\boxed{P_{\text{out}} = \sigma_s (P_{\text{in}} - P_{\text{thres}})} \tag{5.96}$$

which begins at P_{thres} and has a slope of σ_s. The threshold power (P_{thres}) can be evaluated using the saturation intensity I_{sat}, the losses per unit length α, the transmittance of the output coupler T_F, and the laser system efficiency η_{laser} as

$$\boxed{P_{\text{thres}} = \frac{I_{\text{sat}} A}{2\eta_{\text{laser}}} (2\alpha L + T_F).} \tag{5.97}$$

The slope efficiency σ_s can be expressed using the same parameters

$$\boxed{\sigma_s = \left(T_F \frac{\eta_{\text{laser}}}{(2\alpha L + T_F)} \right).} \tag{5.98}$$

Example 5.6

Plot P_{out} versus P_{in} for the same low-power Nd:YAG laser as in Example 5.2. Assume $I_{sat} =$ 174.57 watt/cm^2, $L = 150$ cm, the output area is 0.5 cm^2, $\eta_{laser} = 0.025$, $\alpha = 0.00031$ cm^{-1}, and $T_F = 0.233$.

Solution. The threshold power P_{thres} can be calculated from Equation (5.97),

$$P_{thres} = \frac{I_{sat} A}{2\eta_{laser}} (2\alpha L + T_F) \tag{5.99}$$

$$P_{thres} = \frac{(174.57 \text{ watt/cm}^2)(0.5 \text{ cm}^2)}{2(0.025)} \left(2(0.00031 \text{ cm}^{-1})(150 \text{ cm}) + (0.233)\right) \tag{5.100}$$

$$P_{thres} = 569.098 \text{ watt} \tag{5.101}$$

and the slope efficiency σ_s from Equation (5.98) as

$$\sigma_s = \left(T_F \frac{\eta_{laser}}{(2\alpha L + T_F)} \right) \tag{5.102}$$

$$\sigma_s = \left((0.233) \frac{(0.025)}{\left(2(0.00031 \text{ cm}^{-1})(150 \text{ cm}) + (0.233)\right)} \right) \tag{5.103}$$

$$\sigma_s = 0.018. \tag{5.104}$$

Using

$$P_{out} = \sigma_s (P_{in} - P_{thres}) \tag{5.105}$$

over a range of $P_{in} = 0$ to 3 kW gives the results plotted in Figure 5.11.

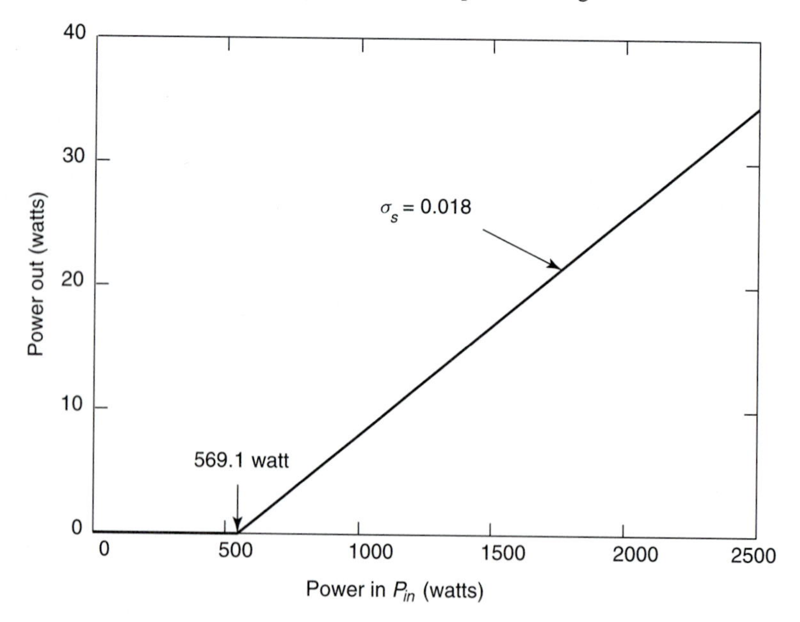

Figure 5.11 P_{out} versus P_{in} for a low-power Nd:YAG laser with $I_{sat} = 174.57$ watt/cm^2, $L = 150$ cm, an output area of 0.5 cm^2, $\eta_{laser} = 0.025$, $\alpha = 0.00031$ cm^{-1}, and $T_F = 0.233$.

Example 5.7

Plot P_{thres} and σ_s versus mirror transmittance for the same Nd:YAG laser as Example 5.2. Assume $I_{\text{sat}} = 174.57$ watt/cm^2, $L = 150$ cm, the output area is 0.5 cm^2, $\eta_{\text{laser}} = 0.025$, and $\alpha = 0.00031$ cm^{-1}.

Solution. The threshold power P_{thres} can be calculated from Equation (5.97) as

$$P_{\text{thres}} = \frac{I_{\text{sat}} A}{2\eta_{\text{laser}}} (2\alpha L + T_F) \tag{5.106}$$

and the slope efficiency σ_s from Equation (5.98) as

$$\sigma_s = \left(T_F \frac{\eta_{\text{laser}}}{(2\alpha L + T_F)} \right). \tag{5.107}$$

Varying the output transmittance from $T_F = 0$ to 1 gives the results plotted in Figures 5.12 and 5.13.

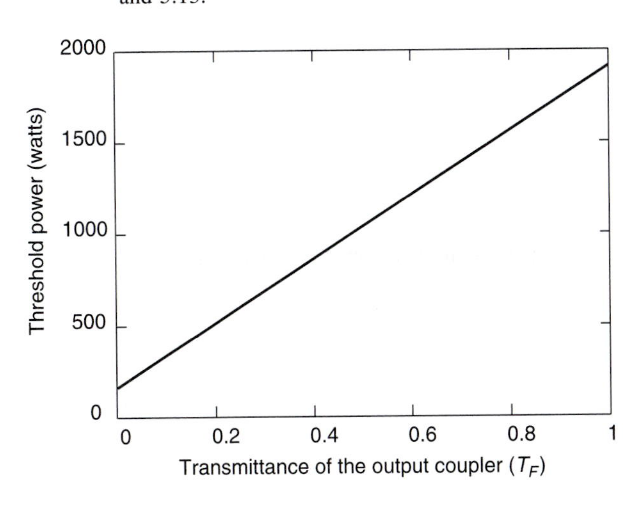

Figure 5.12 P_{thres} versus mirror transmittance for a low-power Nd:YAG laser with $I_{\text{sat}} = 174.57$ watt/cm^2, $L = 150$ cm, an output area of 0.5 cm^2, $\eta_{\text{laser}} = 0.025$, and $\alpha = 0.00031$ cm^{-1}.

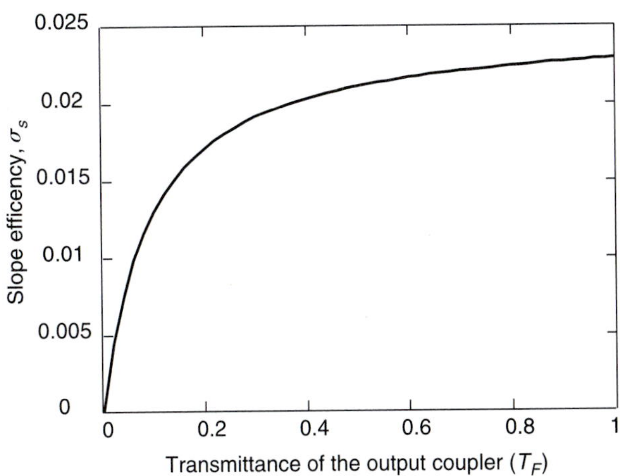

Figure 5.13 σ_s versus mirror transmittance for a low-power Nd:YAG laser with $I_{\text{sat}} = 174.57$ watt/cm^2, $L = 150$ cm, an output area of 0.5 cm^2, $\eta_{\text{laser}} = 0.025$, and $\alpha = 0.00031$ cm^{-1}.

5.3.3 P_{out} versus P_{in}, the Rigrod Approach

Integral to the previous analyses are two critical approximations. First, the left and right circulating waves in the laser cavity are approximated by $2I_{circ} = I_+ + I_-$. Second, the logarithmic term has been approximated by Equation (5.54) where $\ln(1+x) \approx x$. There is a slightly more accurate way to work this problem that does not require these approximations. This method was originally proposed by Rigrod.[12] The following derivation follows the modified Rigrod approach as developed by Siegman.[13]

Consider a real laser cavity with one wave propagating forward and one wave propagating backward, as illustrated in Figure 5.14. The intensities in the cavity can be modeled with a pair of differential equations as

$$\frac{dI_+(z)}{dz} = \gamma(z)I_+(z) \tag{5.108}$$

$$\frac{dI_-(z)}{dz} = -\gamma(z)I_-(z). \tag{5.109}$$

Notice that

$$\frac{d(I_+(z)I_-(z))}{dz} = -\gamma(z)I_+(z)I_-(z) + \gamma(z)I_+(z)I_-(z) = 0 \tag{5.110}$$

and thus $I_+(z)I_-(z)$ can be set equal to a constant

$$I_+(z)I_-(z) = \text{constant} = C. \tag{5.111}$$

[12] W. W. Rigrod, *J. Appl. Phys.* 36:2487 (1965).

[13] Anthony Siegman, *Lasers* (Mill Valley, CA: University Science Books, 1986), p. 487.

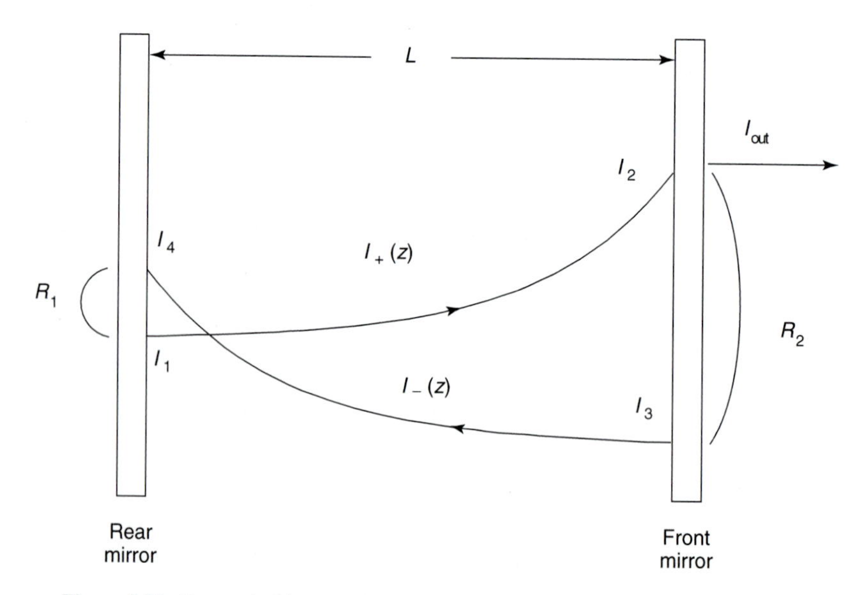

Figure 5.14 In a typical laser cavity, there is one wave propagating forward and one wave propagating backward.

Equation (5.57) can be used to describe the saturated gain in a laser cavity

$$\gamma(z) = \frac{\gamma_o}{1 + \frac{I_+(z)+I_-(z)}{I_{sat}}} \tag{5.112}$$

where I_+ is the right traveling wave and I_- is the left traveling wave.

So, the differential equation for the I_+ wave can be written as

$$\frac{dI_+(z)}{dz} = \frac{\gamma_o}{1 + \frac{I_+(z)+I_-(z)}{I_{sat}}} I_+(z) = \frac{\gamma_o}{1 + \frac{I_+(z)+C/I_+(z)}{I_{sat}}} I_+(z) \tag{5.113}$$

or

$$\left(\frac{1}{I_+(z)} + \frac{I_+(z) + C/I_+(z)}{I_{sat} I_+(z)}\right) dI_+(z) = \gamma_o dz. \tag{5.114}$$

Integrating over the length of the laser gives

$$\int_{I_1}^{I_2} \left(\frac{1}{I_+(z)} + \frac{I_+(z) + C/I_+(z)}{I_{sat} I_+(z)}\right) dI_+(z) = \int_0^L \gamma_o dz \tag{5.115}$$

and a similar equation for I_-.

Evaluating the integrals from the boundary conditions in Figure 5.14 yields a pair of coupled equations

$$\ln\left(\frac{I_2}{I_1}\right) + \frac{I_2}{I_{sat}} - \frac{I_1}{I_{sat}} - \frac{C}{I_{sat}}\left(\frac{1}{I_2} - \frac{1}{I_1}\right) = \gamma_o L \tag{5.116}$$

$$\ln\left(\frac{I_4}{I_3}\right) + \frac{I_4}{I_{sat}} - \frac{I_3}{I_{sat}} - \frac{C}{I_{sat}}\left(\frac{1}{I_4} - \frac{1}{I_3}\right) = \gamma_o L \tag{5.117}$$

where $I_1 = R_1 I_4$, $I_3 = R_2 I_2$, and $I_1 I_4 = I_2 I_3 = C$. Solving these for I_2 gives

$$I_2 = \frac{I_{sat}}{\left(1 + \sqrt{R_2}/\sqrt{R_1}\right)\left(1 - \sqrt{R_1} \cdot \sqrt{R_2}\right)} \cdot \left(\gamma_o L - \ln\left(\frac{1}{\sqrt{R_1} \cdot \sqrt{R_2}}\right)\right). \tag{5.118}$$

Now R_2 is the output mirror reflectivity and R_1 can be taken to be the internal loss. The output power is the output intensity times the area times the transmittance of the output coupler. With these changes, the output power can be written as

$$P_{out} = \frac{T_F I_{sat} A}{\left(1 + \sqrt{(1 - T_F)/(1 - 2\alpha L)}\right)\left(1 - \sqrt{(1 - 2\alpha L)(1 - T_F)}\right)} \times \left(\gamma_o L - \ln\left(\frac{1}{\sqrt{(1 - 2\alpha L)(1 - T_F)}}\right)\right) \tag{5.119}$$

and, using $\gamma_o L = K P_{in}$ gives and $K = (\eta_{laser}/I_{sat} A)$ gives

$$P_{out} = \frac{T_F \eta_{laser}}{\left(1 + \sqrt{(1 - T_F)/(1 - 2\alpha L)}\right)\left(1 - \sqrt{(1 - 2\alpha L)(1 - T_F)}\right)} \times \left(P_{in} - \frac{I_{sat} A}{\eta_{laser}} \ln\left(\frac{1}{\sqrt{(1 - 2\alpha L)(1 - T_F)}}\right)\right). \tag{5.120}$$

So, in the case of a four-state homogeneously broadened, continuous wave laser (using the Rigrod approach) the laser output power is a straight line described by

$$\boxed{P_{\text{out}} = \sigma_s \left(P_{\text{in}} - P_{\text{thres}}\right)} \tag{5.121}$$

that begins at P_{thres} and has a slope of σ_s. The threshold power P_{thres} can be evaluated using the saturation intensity I_{sat}, the losses per unit length α, the transmission of the output coupler T_F, and the laser system efficiency η_{laser} as

$$\boxed{P_{\text{thres}} = \frac{-A I_{\text{sat}} \ln \sqrt{(1 - T_F)(1 - 2\alpha L)}}{\eta_{\text{laser}}}.} \tag{5.122}$$

and the slope efficiency (σ_s) can be expressed using the same parameters,[14]

$$\boxed{\sigma_s = \left(\frac{T_F \eta_{\text{laser}}}{\left(1 + \sqrt{(1 - T_F)/(1 - 2\alpha L)}\right)\left(1 - \sqrt{(1 - T_F)(1 - 2\alpha L)}\right)}\right).} \tag{5.123}$$

Example 5.8

Plot P_{out} versus P_{in}, using the Rigrod approach, versus P_{out} versus P_{in}, using the approximate approach described in Section 5.3.2. Assume the same Nd:YAG laser as used before, with $I_{\text{sat}} = 174.57$ watt/cm^2, $L = 150$ cm, the output area is 0.5 cm^2, $\eta_{\text{laser}} = 0.025$, $\alpha = 0.00031$ cm^{-1}, and $T_F = 0.233$.

Solution. The threshold power P_{thres} and the slope efficiency σ_s were calculated using the approximate approach in Example 5.7 and shown to be $P_{\text{thres}} = 569.098$ watt and $\sigma_s = 0.018$. Using the Rigrod approach, P_{thres} can be calculated from Equation (5.122)

$$P_{\text{thres}} = \frac{-A I_{\text{sat}} \ln \sqrt{(1 - T_F)(1 - 2\alpha L)}}{\eta_{\text{laser}}} \tag{5.124}$$

$$P_{\text{thres}} = \frac{-(0.5\ \text{cm}^2)(174.57\ \text{watt/cm}^2) \ln \sqrt{(1 - (0.233))\left(1 - 2(0.00031\ \text{cm}^{-1})(150\ \text{cm})\right)}}{(0.025)} \tag{5.125}$$

$$P_{\text{thres}} = 633.482\ \text{watt} \tag{5.126}$$

and the slope efficiency σ_s from Equation (5.123) as

$$\begin{aligned}\sigma_s &= \left(\frac{T_F \eta_{\text{laser}}}{\left(1 + \sqrt{(1 - T_F)/(1 - 2\alpha L)}\right)\left(1 - \sqrt{(1 - T_F)(1 - 2\alpha L)}\right)}\right) \\ &= \left(\frac{(0.233)(0.025)}{\left(1 + \sqrt{(1 - (0.233))/\left(1 - 2(0.00031\ \text{cm}^{-1})(150\ \text{cm})\right)}\right)}\right) \\ &\cdot \left(\frac{1}{\left(1 - \sqrt{(1 - (0.233))\left(1 - 2(0.00031\ \text{cm}^{-1})(150\ \text{cm})\right)}\right)}\right) = 0.018.\end{aligned} \tag{5.127}$$

[14] Walter Koechner, *Solid State Laser Engineering*, 3d ed. (Berlin: Springer-Verlag, 1992), p. 100.

Using

$$P_{\text{out}} = \sigma_s \left(P_{in} - P_{\text{thres}}\right) \tag{5.128}$$

over a range of $P_{in} = 0$ to 3 kW for both values of σ_s and P_{thres} gives the results plotted in Figure 5.15. Notice that both the approximation and the Rigrod approach have the same slope efficiency. However, the Rigrod approach predicts a threshold power P_{thres}, which is about 11% greater than the approximation.

Example 5.9

Plot P_{thres} and σ_s versus mirror transmittance, using the Rigrod approach, versus P_{thres} and σ_s using the approximate engineering approach described in Section 5.3.2. Assume the same Nd:YAG laser as before with $I_{\text{sat}} = 174.57$ watt/cm^2, $L = 150$ cm, the output area is 0.5 cm^2, $\eta_{\text{laser}} = 0.025$, and $\alpha = 0.00031$ cm^{-1}.

Solution. The threshold power P_{thres} for the approximate approach can be calculated from Equation (5.97)

$$P_{\text{thres}} = \frac{I_{\text{sat}} A}{2\eta_{\text{laser}}} \left(2\alpha L + T_F\right). \tag{5.129}$$

The slope efficiency σ_s for the approximate approach can be calculated from Equation (5.98)

$$\sigma_s = \left(T_F \frac{\eta_{\text{laser}}}{(2\alpha L + T_F)}\right). \tag{5.130}$$

The threshold power P_{thres} for the Rigrod approach can be calculated from Equation (5.122)

$$P_{\text{thres}} = \frac{-A I_{\text{sat}} \ln \sqrt{(1 - T_F)(1 - 2\alpha L)}}{\eta_{\text{laser}}}. \tag{5.131}$$

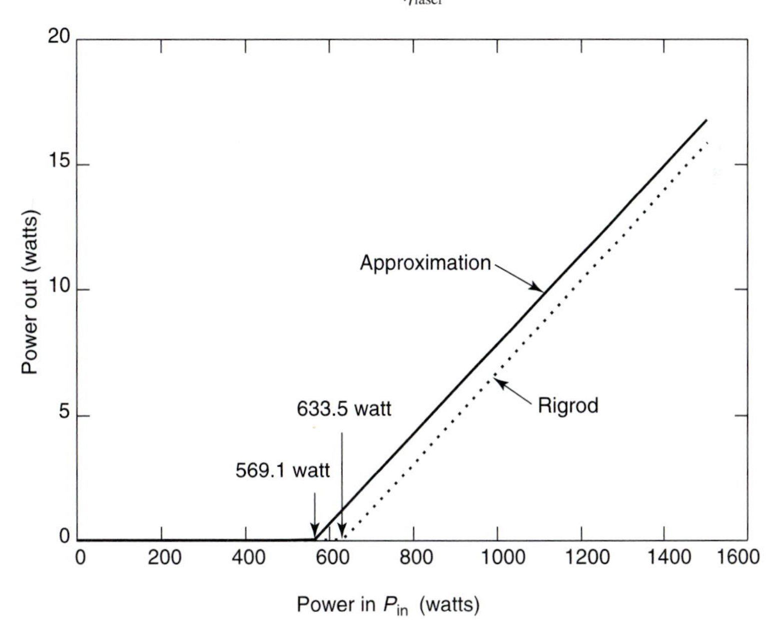

Figure 5.15 P_{out} versus P_{in} for a low-power Nd:YAG laser using the Rigrod approach, versus P_{out} versus P_{in} using the approximate engineering approach.

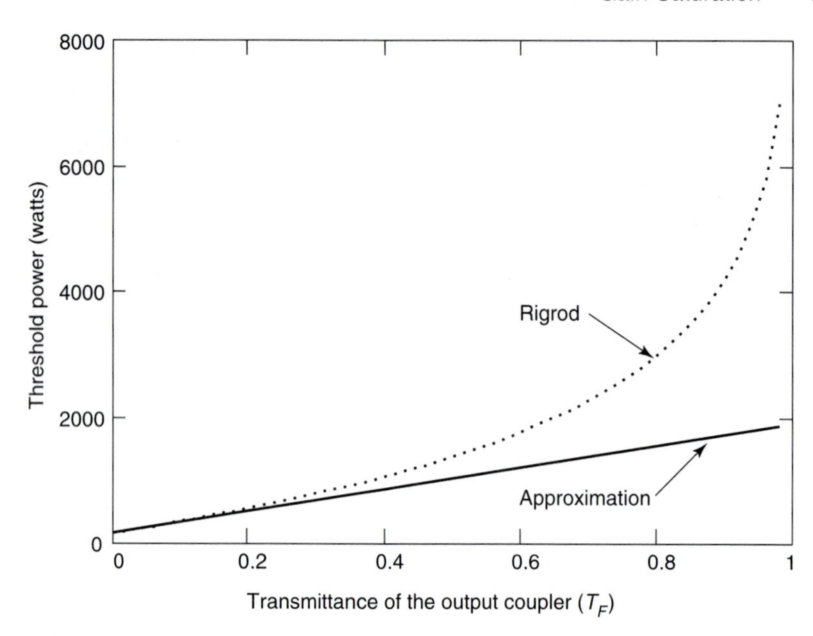

Figure 5.16 P_{thres} versus mirror transmittance using the Rigrod approach, versus P_{thres} versus mirror transmittance using the approximate engineering approach, for a low-power Nd:YAG laser.

The slope efficiency σ_s for the Rigrod approach can be calculated from Equation (5.123) as

$$\sigma_s = \left(\frac{T_F \eta_{\text{laser}}}{\left(1 + \sqrt{(1 - T_F) / (1 - 2\alpha L)}\right)\left(1 - \sqrt{(1 - T_F)(1 - 2\alpha L)}\right)} \right). \tag{5.132}$$

Varying the output transmittance from $T_F = 0$ to 1 gives the results plotted in Figures 5.16 and 5.17.

Notice that both methods yield similar answers for low values of mirror transmittance. However, as the mirror transmittance increases, the threshold power predicted by the Rigrod method deviates significantly from that predicted by the engineering approximation. A similar pattern can be observed with the slope efficiency, but the differences are much smaller. This suggests that the engineering approximation can be used for low-gain lasers (such as HeNe lasers) but should be replaced by the Rigrod analysis for higher-gain lasers (such as Nd:YAG lasers).

Example 5.10

It is possible to run an experiment where the threshold power of a laser is measured for several different mirrors. If these results are plotted in a graph of P_{thres} versus $-\ln R_F R_B$, the slope of the graph is $2K$ and the extrapolation of the curve to $P_{\text{input}} = 0$ gives $2\alpha L$. Using these results, it is possible to calculate I_{sat}, α, and η_{laser}.

Solution. Consider the following data set for a high-power, continuous wave Nd:YAG laser (plotted in Figure 5.18).

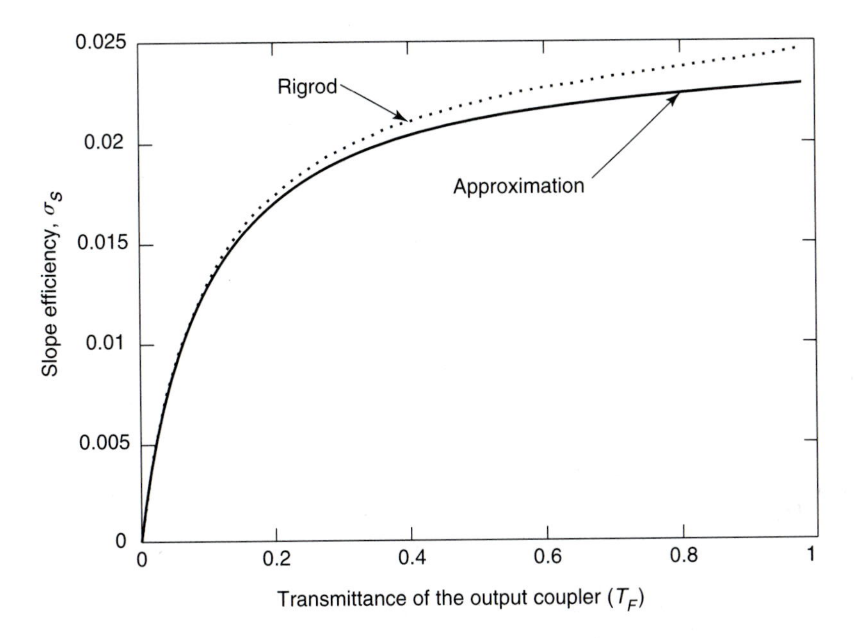

Figure 5.17 σ_s versus mirror transmittance using the Rigrod approach, versus σ_s versus mirror transmittance using the approximate engineering approach.

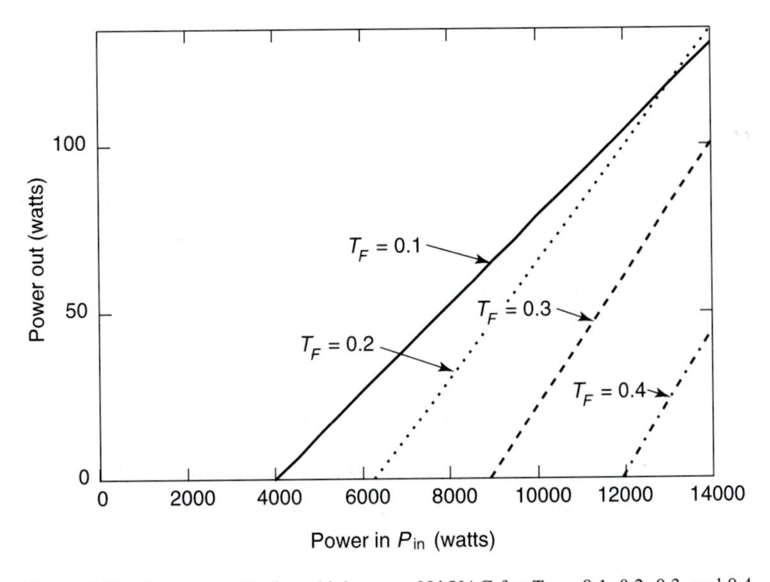

Figure 5.18 P_{out} versus P_{in} for a high-power Nd:YAG for $T_F = 0.1, 0.2, 0.3$, and 0.4.

Power out versus power in for a high-power Nd:YAG

P_{in} (kW)	P_{out} (watt) $T_F = 0.1$	$T_F = 0.2$	$T_F = 0.3$	$T_F = 0.4$
5	13.4			
6	26.4			
7	39.3	12.7		
8	52.3	30.1		
9	65.3	47.5	2.3	
10	78.3	64.9	21.9	
11	91.3	82.3	41.6	
12	104.2	99.7	61.3	2.2
13	117.2	117.1	80.9	23.2
14	130.2	134.5	100.6	44.2
15		151.9	120.2	65.3
16		169.3	139.8	86.3
17		186.7	159.5	107.3
18			179.1	128.3
19			198.8	149.4
20				170.4
21				191.4
22				212.4

Performing a linear regression on these values gives the threshold power P_{thres} and the slope efficiencies σ_s for each set of mirrors,

P_{thres} and σ_s versus T_F

T_F	P_{thres}	σ_s
0.1	3.968 kW	0.012978
0.2	6.271 kW	0.017401
0.3	8.882 kW	0.019649
0.4	11.895 kW	0.021024

The next step is to create a plot of P_{thres} versus $-\ln R_F R_B$, as illustrated in Figure 5.19. The slope of this line can be found from the plot or by a linear regression of the above values. A linear regression yields a slope of $2K = 5.114886 \cdot 10^{-5}$ watt^{-1} for $K = 2.55744 \cdot 10^{-5}$ watt^{-1}. The intercept (on the y-axis) is -0.098 for an $\alpha = 0.000325$ cm^{-1}.

Given α and the measured slope efficiencies σ_s for each set of mirrors, then the laser efficiency η_{laser} can be calculated from Equation (5.123).

$$\sigma_s = \left(\frac{T_F \eta_{laser}}{\left(1 + \sqrt{(1 - T_F/(1 - 2\alpha L)}\right)(1 - sqrt(1 - T_F)(1 - 2\alpha L))} \right). \tag{5.133}$$

For the four mirrors, the calculated efficiencies are listed in the Table.

T_F	σ_s	η_{laser}
0.1	0.012978	0.02563
0.2	0.017401	0.0254
0.3	0.019649	0.02528
0.4	0.021024	0.02521

The average $\eta_{laser} = 0.025$.

Given η_{laser} for each set of mirrors, I_{sat} can be calculated from Equation (5.83),

$$K = \frac{\eta_{laser}}{I_{sat} A}. \tag{5.134}$$

For the four mirrors, the calculated efficiencies are listed in the Table.

T_F	η_{laser}	I_{sat} (watt/cm^2)
0.1	0.02563	1993.645
0.2	0.0254	1975.656
0.3	0.02528	1966.514
0.4	0.02521	1960.921

The average $I_{sat} = 1974.184$ watt/cm^2.

SYMBOLS USED IN THE CHAPTER

I_{start}, I_{end}, I_t, I_{rt}, I_r, I_o, I: Various light intensities (watts/cm^2)

ν: Frequency (Hz)

$\gamma(\nu)$: Gain coefficient as a function of frequency (cm^{-1} or m^{-1})

L: Laser resonator length (cm or m)

$g(\nu)$: Function for the normalized gain curve (1/Hz = sec)

A_{12}, A_{21}: Spontaneous Einstein A coefficients (1/sec)

λ_o: Free space wavelength of light (m)

ν_o: Center frequency of a transition (Hz)

n: Real index of refraction (unitless)

N_3, N_2, N_1: State population densities (atoms/cm^3)

g_3, g_2, g_1: State degeneracies, the number of states at the same energy (unitless)

R_F, R_B: Front and back mirror reflectances, between 0 and 1

T_F, T_B: Front and back mirror transmittances, between 0 and 1

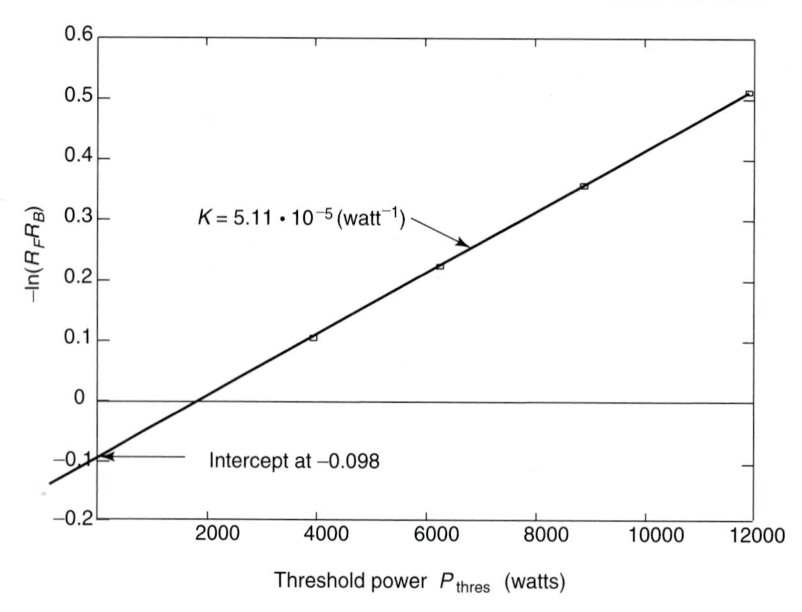

Figure 5.19 It is possible to run an experiment where the threshold power of the laser measured for several different mirrors. If these results are plotted in a graph of P_{thresh} versus $-ln R_F R_B$, the slope of the graph is $2K$ and the extrapolation of the curve to $P_{input} = 0$ gives $2\alpha L$. Using these results, it is possible to determine I_{sat}, α, and η_{laser}.

G_{rt}: Round-trip gain (unitless)

P_{12}, P_{21}, P_{13}, P_{31}, P_{23}, P_{32}: Stimulated transition probabilities (1/sec)

S_{12}, S_{21}, S_{13}, S_{31}, S_{23}, S_{32}: Spontaneous and thermal transition probabilities (1/sec)

N: Total population density (atoms/cm^3)

E_i, E_j: State energies (J or eV)

T: Temperature (K)

τ_{12}, τ_{13}, τ_{23}: Thermal transition times (sec)

τ_1, τ_{10}, τ_2, τ_{20}, τ_{21}, τ_{31}, τ_{32}: Spontaneous transition times (relaxation lifetimes) (sec)

R_{trans}, $R_{trans,12}$, $R_{trans,21}$: Stimulated transition probabilities due to laser action (1/sec)

P_1, P_2: External pumping density to states (four-state approach) (cm^{-3} sec^{-1})

σ_{ij}: Cross-section between states i and j (cm^2 or m^2)

I_ν: Intensity in the cavity due to laser action (watts/cm^2)

γ_o: Unsaturated gain coefficient (cm^{-1} or m^{-1})

I_{sat}: Saturation intensity (watts/cm^2)

α: Loss coefficient (cm^{-1} or m^{-1})

γ_t: Gain coefficient at threshold (cm^{-1} or m^{-1})

I_+, I_-, I_{circ}: Left, right, and circulating intensities (watts/cm^2)

A: Output coupler area (cm^2 or m^2)

P_{in}, P_{out}, P_{thres}: Input, output, and threshold power densities (watts/cm^2)

T_{opt}, $P_{out(opt)}$: Optimal output transmittance (unitless) and associated power (watts)

σ_s: Slope efficiency (unitless)

η_1, η_2, η_3, η_4, η_5, η_{laser}: Laser system efficiency terms (unitless)

V: Laser volume (cm^3 or m^3)

K: Power/reflectivity slope (watt^{-1})

R_1, R_2: Mirror reflectances used for Rigrod analysis, between 0 and 1

I_1, I_2, I_3, I_4: Various intensities used for the Rigrod analysis (watts/cm^2)

C: Constant used for Rigrod Analysis (watt^2 cm^{-6})

EXERCISES

Setting up rate equations

5.1 Write the system of rate equations for a homogeneously broadened four-level laser system without making the approximation to two levels discussed in the chapter. Name the states 1, 2, 3, and 4 and assume the laser action occurs between states 2 and 3. Your final answer should consist of four equations and be written in terms of R_{trans}, the various Ps, and the various relaxation times.

5.2 Consider two states separated by 1.3 eV. Assume that the degeneracy of the top state is 2 and the bottom state is 3. Assume that the spontaneous relaxation time of the top state is 1.5 microseconds. Assume the temperature is 300 K. Calculate the thermal upwards transition rate (S_{12}) from the bottom state.

Rate equations for four-state lasers

5.3 Consider a He-Cd laser with a transition wavelength of 441.6 nm and a cross-section of $8 \cdot 10^{-14}$ cm^2. Assume the laser is operating cw with an output intensity of 10 watt/cm^2. Calculate $R_{trans12}$.

5.4 Consider a Cr:LiSAF laser with an operating wavelength of 840 nm and a cross-section of $5 \cdot 10^{-20}$ cm^2. Assume the spontaneous upper state lifetime is $6.7 \cdot 10^{-5}$ sec and that t_1 is very small. Assume that the pumping rate to the upper state is 10^{18} cm^{-3} sec^{-1} and that the laser is four-level. Calculate the small signal gain and the saturation intensity. Plot the large signal gain versus cavity intensity. Discuss the similarities and differences to a Nd:YAG laser.

Optimal coupling and P_{out} versus P_{in}

5.5 Consider a copper vapor laser with an operating wavelength of 510.5 nm and a cross-section of $8 \cdot 10^{-14}$ cm^2. Assume a diameter of 1 cm and a length of 1 m. Assume the spontaneous upper state lifetime is $5 \cdot 10^{-7}$ sec and the pump rate is $1 \cdot 10^{18}$ cm^{-3} sec^{-1}. Assume the unintentional loss is $\alpha = 0.0005$ cm^{-1}. Derive Equation (5.61) without using the ln approximation given in Equation (5.54). Calculate and plot the output power versus transmittance using Equation (5.61). Calculate and plot (on the same graph) the output power versus transmittance using your equation. Compare these and comment on the quality of the approximation.

5.6 Consider a He Ne laser with an operating wavelength of 632.8 nm and a cross-section of $3 \cdot 10^{-13}$ cm^2. Assume a diameter of 2 mm and a length of 20 cm. Assume the spontaneous upper state lifetime is $1 \cdot 10^{-7}$ sec and the pump rate is $5 \cdot 10^{16}$ cm^{-3} sec^{-1}. Assume the unintentional loss is $\alpha = 0.0005$ cm^{-1}. Calculate and plot the output power versus mirror transmittance using the engineering approach of Section 5.3.2. In addition, calculate the optimum output coupler transmittance and the power at that transmittance. Discuss similarities and differences to a Nd:YAG laser.

5.7 Consider a Nd:YAG laser with an operating wavelength of 1.064 μm and a cross-section of $6.5 \cdot 10^{-19}$ cm^2. Assume a diameter of 5 mm and a length of 15 cm. Assume the spontaneous upper state lifetime is $2.3 \cdot 10^{-4}$ sec and the pump rate is $1 \cdot 10^{13}$ cm^{-3} sec^{-1}. Assume the unintentional loss is $\alpha = 0.001$ cm^{-1} and the laser efficiency is 3%. Plot the threshold power and slope efficiency as a function of the mirror transmittance using both the engineering approach of Section 5.3.2 and the Rigrod approach of Section 5.3.3.

5.8 Consider a Nd:glass laser with an operating wavelength of 1.055 μm and a cross-section of $4 \cdot 10^{-19}$ cm^2. Assume a diameter of 7 mm and a length of 25 cm. Assume the spontaneous upper state lifetime is $3 \cdot 10^{-4}$ sec and the pump rate is $1 \cdot 10^{15}$ cm^{-3} sec^{-1}. Assume the unintentional loss is α =0.001 cm^{-1}, the laser efficiency is 4%, and the output coupler is a 20% transmittance mirror. Plot the output power versus input power for the laser using both the engineering approach of Section 5.3.2 and the Rigrod approach of Section 5.3.3.

5.9 Consider a Nd:YAG laser with an operating wavelength of 1.064 μm. Assume a diameter of 5 mm and a length of 20 cm. Assume that the output power from the laser has been measured as a function of the input power for several different output couplers. The measured output power is given as in the Table.

Input power (watts)	Output power $T = 15\%$	$T = 25\%$	$T = 35\%$	$T = 45\%$
1000				
2000	20.391			
3000	48.319	23.606		
4000	76.246	52.402	22.736	
5000	104.174	81.198	51.925	16.587
6000	132.102	109.993	81.113	46.002
7000		138.789	110.301	75.417
8000			138.439	104.832
9000				134.246
10000				163.661

Use the Rigrod equations to calculate the loss per unit length α, the laser efficiency η_{laser}, and the saturation intensity I_{sat}.

6

Transient Processes

Objectives

- To qualitatively sketch the upper state population N_2 and the photon density ϕ for a laser system undergoing relaxation oscillations as a function of parameters such as the pumping, upper state lifetime, and initial conditions.
- To model relaxation oscillations in real systems using a numerical differential equation solver.
- To analytically compute the relaxation oscillation frequency and exponential decay constant.
- To analytically determine if a specific laser system is likely to display relaxation oscillations.
- To qualitatively sketch the upper state population N_2 and the photon density ϕ for a laser system undergoing Q-switching, as parameters such as the pumping, upper state lifetime, and initial conditions are varied.
- To model Q-switching in real systems using a numerical differential equation solver.
- To analytically compute the peak power and pulse width for a Q-switched pulse.
- To qualitatively summarize the main features of Q-switch design.
- To sketch and describe the operation of an electrooptic, acousto-optic, or dye Q-switch.
- To qualitatively describe the mode-locking process.
- To calculate the temporal width of a mode-locked pulse.
- To sketch and describe the operation of an electrooptic, acousto-optic, or dye mode-locker.

6.1 RELAXATION OSCILLATIONS

Short intense laser pulses are critical for applications as diverse as laser eye surgery and laser machining. Intense pulses are also essential for producing useful nonlinear effects, such as second harmonic generation and Raman shifting (see Chapter 7). The creation of short intense pulses from lasers derives from the ability to control the transient properties of the energy states.

Some lasers will spontaneously produce pulse trains in a process called *relaxation oscillation* (discussed in this section). This relaxation oscillation process can be harnessed to produce short energetic pulses by selectively blocking the cavity with a device called a Q-switch (see Section 6.2). Even shorter pulses can be obtained by forcing all of the longitudinal modes of a cavity to operate in phase. This process (termed *mode-locking*) can generate pulses as short as 11 femtoseconds (see Section 11.5).

6.1.1 A Qualitative Description of Relaxation Oscillations

Perhaps the simplest example of transient laser operation is a process called relaxation oscillation. Consider a solid-state laser material (such as Nd:YAG) at a time immediately after the pump source is turned on. Initially, there are negligible photons in the laser cavity and negligible atoms in state $|2\rangle$. The activation of the pump source causes a linear build up of excited atoms in state $|2\rangle$. At some point, the population in state $|2\rangle$ becomes larger than in state $|1\rangle$ and laser action is possible. However, since there are negligible photons in the cavity, the population in state $|2\rangle$ will overshoot the threshold for laser operation ($N_{2,\text{thres}}$) because no stimulated radiation yet exists in the cavity to pull the inversion down to threshold.

The large inversion means the cavity gain is extremely high, and so the photon population in the cavity begins to increase rapidly. The photon population increases much more rapidly than would normally be possible under steady-state conditions. This rapid increase in the photon population means that the upper state population N_2 will deplete quickly and the depletion rate may exceed the pumping rate. If this occurs, then the population in the upper state N_2 will begin to drop and may drop below the threshold value $N_{2,\text{thres}}$. When this happens, the gain in the cavity is less than 1 and the existing laser oscillation in the cavity begins to decrease quickly. However, the reduction in the *photon density* in the cavity will then permit an increase in the population N_2 and the process will repeat. The result is the distinctive series of relaxation oscillation spikes (illustrated in Figure 6.1).

Relaxation oscillations were a well-studied phenomenon of the maser era.[1] The first relaxation oscillations in lasers were observed in 1960 by Collins et al.[2] and by Maiman.[3] The first theoretical description of laser relaxation oscillations was given by Hellwarth in 1961.[4]

---

[1]C. L. Tang, *J. Appl. Phys.* 34:2935 (1963).

[2]R. J. Collins, D. F. Nelson, A. L. Schawlow, W. Bond, C. G. B. Garrett, and W. Kaiser, *Phys. Rev. Lett.* 5:303 (1960).

[3]T. H. Maiman, paper ATC1, presented at the 1960 annual meeting of the Optical Society of America, Boston, Massachusetts, 1960.

[4]R. W. Hellwarth, *Phys. Rev. Lett.* 6:9 (1961).

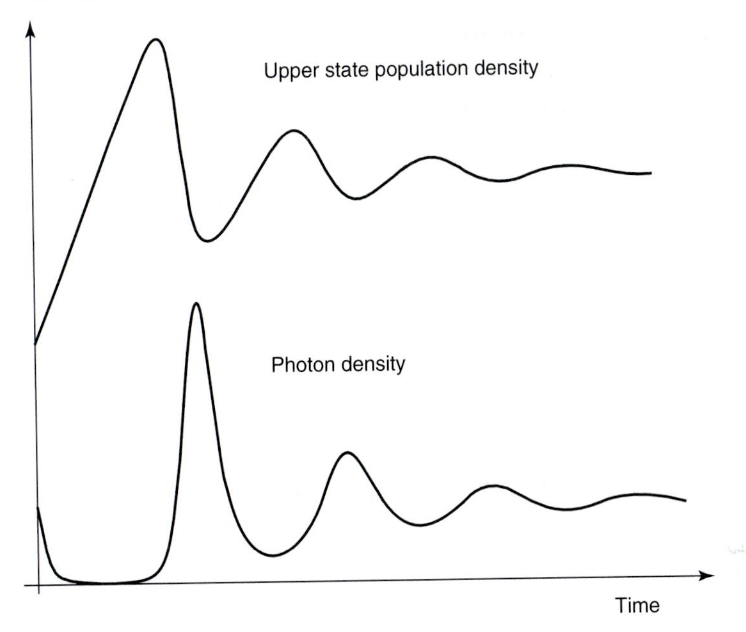

Figure 6.1 Relaxation oscillations occur in lasers when the upper state lifetime is significantly greater than the lifetime of a photon in the cavity. Relaxation oscillations occur in both the upper state population density and in the cavity photon density.

6.1.2 Numerical Modeling of Relaxation Oscillations

Modeling relaxation oscillations requires solving the laser rate equations for transient operation. This is *not* a trivial task! However, programs such as Mathematica do provide numerical techniques to aid in this process.

Consider the four-state rate equations for the two laser state populations (as discussed in Chapter 5),

$$\frac{dN_1}{dt} = P_1 - \left(R_{\text{trans}} + \frac{1}{\tau_1}\right) N_1 + \left(R_{\text{trans}} + \frac{1}{\tau_{21}}\right) N_2 \tag{6.1}$$

$$\frac{dN_2}{dt} = P_2 - \left(R_{\text{trans}} + \frac{1}{\tau_2}\right) N_2 + (R_{\text{trans}})\, N_1 \tag{6.2}$$

where

$$R_{\text{trans}} = \frac{\sigma_{12} I_\nu}{(h\nu)_{12}} \tag{6.3}$$

represents the transition rate due to laser action. I_ν is the intensity of the laser beam in the cavity, σ_{12} is the cross-section between states $|1\rangle$ and $|2\rangle$, $1/\tau_2 = 1/\tau_{20} + 1/\tau_{21}$ is used as the total relaxation lifetime from state $|2\rangle$, $1/\tau_1 = 1/\tau_{10}$ is used as the total relaxation lifetime from state $|1\rangle$, and the degeneracies of both states are assumed to be 1 so that $R_{\text{trans},12} = R_{\text{trans},21} = R_{\text{trans}}$.

Now, in the steady-state condition, the intensity I_ν is constant, and so it is not necessary to include a rate equation for I_ν. However, for the transient case, I_ν also varies, and so an additional rate equation is necessary. Since the equations for population density are in terms of atoms/cm^3, it is convenient to express the missing rate equation for I_ν in terms of the photon density ϕ in photons/cm^3 where ϕ is defined by[5]

$$\phi = \frac{I_\nu}{(h\nu)_{12}\, c}. \tag{6.4}$$

So, writing the rate equations for the state populations in terms of the photon density ϕ gives

$$\frac{dN_1}{dt} = P_1 + \frac{N_2}{\tau_{21}} - \frac{N_1}{\tau_1} + \sigma_{12} c\phi\,(N_2 - N_1) \tag{6.5}$$

$$\frac{dN_2}{dt} = P_2 - \frac{N_2}{\tau_2} - \sigma_{12} c\phi\,(N_2 - N_1) \tag{6.6}$$

where $\sigma_{12} c\phi = R_{\text{trans}}$.

The third equation for the photon density is then written as

$$\frac{d\phi}{dt} = \sigma_{12} c\phi\,(N_2 - N_1) - \frac{\phi}{\tau_{\text{cav}}} \tag{6.7}$$

where $\sigma_{12} c\phi\,(N_2 - N_1)$ is the rate at which photons are added to the optical wave by stimulated emission and ϕ/τ_{cav} is the rate at which photons leave the cavity via the output coupler.

The parameter τ_{cav} can be thought of as the decay time constant for the radiation in a passive resonator including both transmission through the mirrors and losses due to internal α mechanisms such as scattering or internal absorption.

To derive an expression for τ_{cav}, notice that the fractional loss per round trip can be expressed as

$$F_l = 1 - R_F R_B e^{-2\alpha L} \tag{6.8}$$

where R_F and R_B are the front and back mirror reflectances. The round-trip time of a photon in the cavity is $(2Ln/c_o)$. Thus, the exponential-decay constant is the fractional loss per round trip divided by the round-trip time as

$$\frac{1}{\tau_{\text{cav}}} = \frac{F_l}{2Ln/c_o} = \frac{1 - R_F R_B e^{-2\alpha L}}{2Ln/c_o} \tag{6.9}$$

[5]Walter Koechner, *Solid State Laser Engineering*, 3d ed. (Berlin: Springer-Verlag, 1992), p. 106.

where c_o/n is the speed of light in the laser material. The term $1 - R_F R_B e^{-2\alpha L}$ is small, and using the approximation $1 - x \approx -\ln(x)$ (for small $1 - x$) this term can be written as $-\ln(R_F R_B e^{-2\alpha L}) = -\ln(R_F R_B) + 2\alpha L$. Thus, τ_{cav} is approximated by[6]

$$\tau_{\text{cav}} = \frac{2nL}{c_o\left(-\ln\left(R_F R_B\right) + 2\alpha L\right)} \tag{6.10}$$

where L is the length and α is the loss parameter. (Notice that if the gain material does not fill the entire cavity, then the optical length nL needs to be replaced by $n_c L_c + n_g L_g$, where $n_c L_c$ is the optical length of that portion of the cavity without the gain material and $n_g L_g$ is the optical length of the gain material.)

Equations (6.5), (6.6), and (6.7) represent a system of coupled nonlinear differential equations. Such equations are common in physical situations, but are exceptionally difficult to solve. Even by computer, the solution of these equations is complex, as numerical techniques require careful attention to boundary conditions and to the rapid rates of changes of the various parameters.

Several approximations can be made to these equations to simplify the problem. In particular, for most four-state laser materials, the lower state is carefully selected to "drain" quickly (possess a very fast relaxation time) to other states. Thus, the population N_1 can be assumed negligible, an assumption that eliminates one parameter and one equation. Under these conditions, the system simplifies to a pair of coupled nonlinear differential equations,

$$\frac{dN_2}{dt} = P_2 - \frac{N_2}{\tau_2} - \sigma_{12} c \left(\phi N_2\right) \tag{6.11}$$

$$\frac{d\phi}{dt} = -\frac{\phi}{\tau_{\text{cav}}} + \sigma_{12} c \left(\phi N_2\right). \tag{6.12}$$

Assigning simpler constants to these equations aids in numerical modeling. Setting $A = P_2$, $B = 1/\tau_2$, $C = \sigma_{12} c$, and $D = 1/\tau_{\text{cav}}$ gives the mathematical system,

$$\frac{dN_2}{dt} = x' = A - Bx - Cyx \tag{6.13}$$

$$\frac{d\phi}{dt} = y' = -Dy + Cyx. \tag{6.14}$$

Example 6.1

Ignore the physical origin of the variables A, B, C, and D and use Mathematica to solve $x' = A - Bx - Cyx$ and $y' = -Dy + Cyx$ for x and y as a function of time. Include a plot of $x(t)$, and $y(t)$, and a parametric plot of x versus y with t as a parameter. Set $A = 30$, $B = 0.01$, $C = 3$, and $D = 30$. Assume the initial conditions on x and y to be 1.

---

[6] Amnon Yariv, *Optical Electronics*, 4th ed. (Philadelphia, PA: Saunders College Publishing, 1991), p. 177.

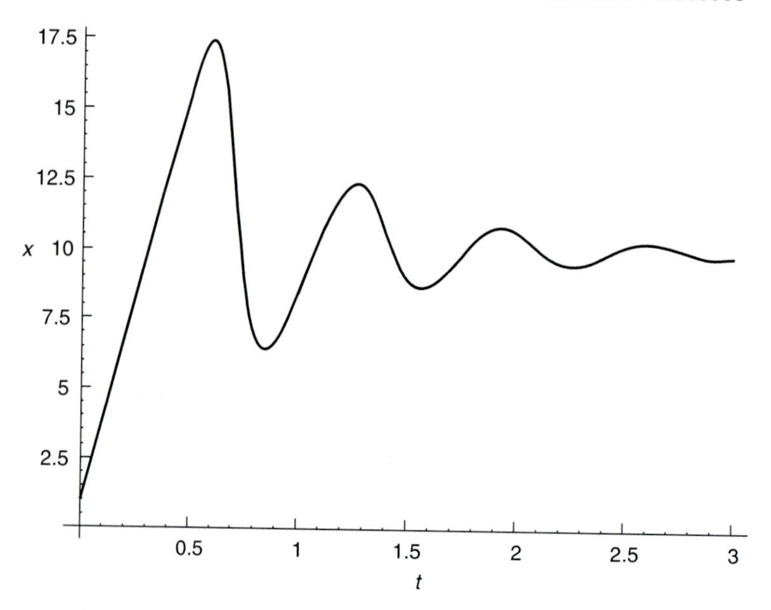

Figure 6.2 The Mathematica solution for the upper state population x as a function of time. The system is a test system where $A = 30$, $B = 0.01$, $C = 3$, and $D = 30$. The initial conditions on x and y are 1.

Solution

```
In[1]:=
eqone = x'[t] ==30 - 0.01 x[t] - 3 x[t] y[t];
eqtwo = y'[t] == 0 - 30 y[t] + 3 x[t] y[t];
tsolt = NDSolve[
{eqone, eqtwo, x[0]==1, y[0]==1},
{x[t], y[t]}, {t,0,20}];
Plot[{x[t]} /. tsolt, {t,0,3}, Compiled -> False]
Plot[{y[t]} /. tsolt, {t,0,3}, Compiled -> False]
ParametricPlot[{x[t],y[t]} /. tsolt, {t,0,20}, Compiled ->False]
```

Figure 6.2 illustrates the upper state population x as a function of time. The rapid linear buildup to a population much above threshold can be clearly seen in the illustration. Also notice the rapid oscillations that converge to the threshold value for the state population. Figure 6.3 illustrates the photon density y as a function of time. Here, a series of rapid pulses converges to the steady-state photon density. Figure 6.4 simultaneously illustrates both the upper state population x and the photon density y as they converge on their equilibrium point. Plotting nonlinear equations parametrically is very insightful, as the existence of a spiral (such as this) indicates both a final stable value, and an oscillatory path to obtain it.

Example 6.2

Now consider the physical origin of the variables B, C, and D. To assist in numerical modeling, establish a set of units that give normalized values for the variables B, C, and D. Assume that $L = 50$ cm, $R_1 = 0.65$, $R_2 = 1.0$, $\alpha = 0.002$ cm^{-1}, $n = 1.8$, $\sigma_{12} = 6 \cdot 10^{-19}$ cm^2, and

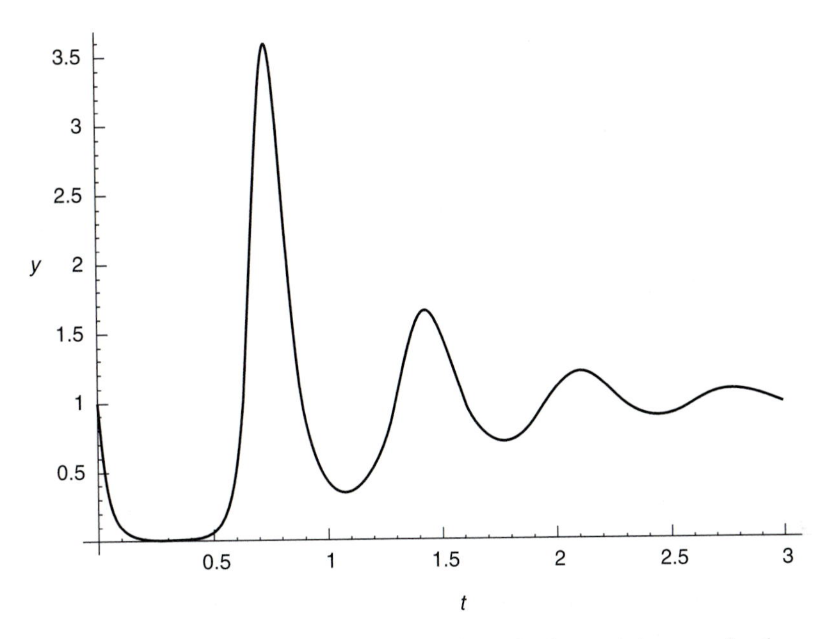

Figure 6.3 The Mathematica solution for the photon density population y as a function of time. The system is a test system where $A = 30$, $B = 0.0$, $C = 3$, and $D = 30$. The initial conditions on x and y are 1.

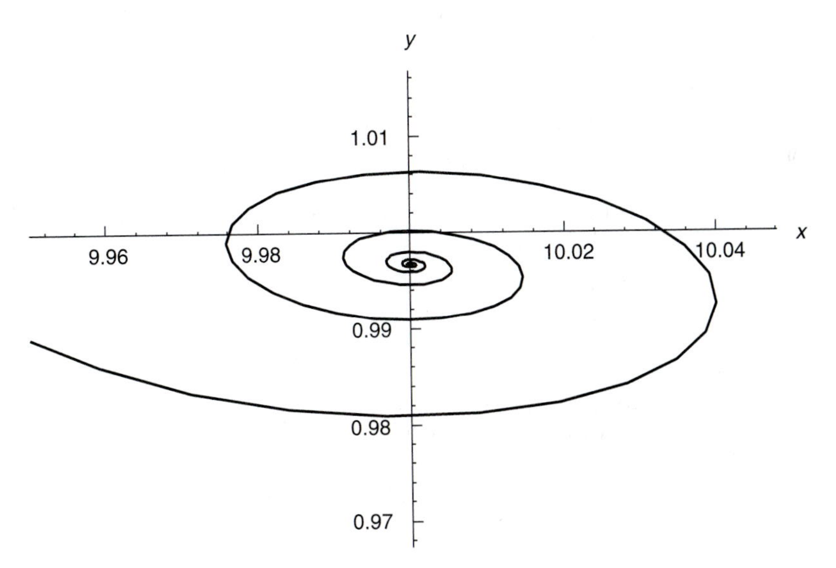

Figure 6.4 A parametric plot of the Mathematica solution for the upper state population x against photon density population y. The system is a test system where $A = 30$, $B = 0.01$, $C = 3$ and $D = 30$. The initial conditions on x and y are 1.

$\tau_2 = 230\ \mu s$. Also assume that most of the cavity is the gain region (thus $Ln = 50$ cm $\cdot$ 1.8) and that the loss is averaged over the entire cavity (thus $2\alpha L = 2 \cdot 0.002$ cm$^{-1} \cdot$ 50 cm).

Solution. A good normalizing value is 10^{15} atoms/cm^3 per nanosecond. Defining 10^{15} atoms/cm^3 as a *norm*, then evaluating B, C, and D in these units gives[7]

$$B = \frac{1}{\tau_2} = \frac{1}{230\mu s} = 4.348 \cdot 10^{-6}\ \text{ns}^{-1} \tag{6.15}$$

$$C = \frac{\sigma_{12} c_o}{n} = \frac{6 \cdot 10^{-19}\ \text{cm}^2 \cdot c_o}{1.8} = 0.01\ \text{norm}^{-1}\ \text{ns}^{-1} \tag{6.16}$$

$$\tau_{\text{cav}} = \frac{2Ln}{c_o\,(-\ln(R_1 R_2) + 2\alpha L)} \tag{6.17}$$

$$\tau_{\text{cav}} = \frac{2 \cdot 50\ \text{cm}\ \cdot 1.8}{c_o\left(-\ln(0.65 \cdot 1.0) + 2 \cdot 0.002\ \text{cm}^{-1} \cdot 50\ \text{cm}\right)} = 9.519\ \text{ns} \tag{6.18}$$

$$D = \frac{1}{\tau_{\text{cav}}} = \frac{1}{9.519\ \text{ns}} = 0.105\ \text{ns}^{-1}. \tag{6.19}$$

Example 6.3

Use Mathematica to model the relaxation oscillation behavior for a physical system. Assume that $B = 0.000004$ ns^{-1} and $C = 0.01$ norm^{-1} ns^{-1}. Set the range of A to be from 0.05 to 0.20 norm ns^{-1}. Set the range of D to be from 0.05 to 0.20 ns^{-1}. Create a plot of $N_2(t)$ and $\phi(t)$. Assume the initial conditions on N_2 and on ϕ to be $1 \cdot 10^{12}$ atoms/cm^3 (0.001 norm).

Solution

```
In[1]:=
Clear[sol]
sol[s_]:=Module[{tsolt,N,P,t,eqone,eqtwo},

eqone = N'[t] == s/20 - 0.000004 N[t] - 0.01 N[t] P[t];
eqtwo = P'[t] == 0 - s/20 P[t] + 0.01 N[t] P[t];
tsolt = NDSolve[
{eqone, eqtwo, N[0]==0.001, P[0]==0.001},
{N[t], P[t]}, {t,0,1000}];
Plot[{N[t]} /. tsolt, {t,0,1000}, Compiled -> False,
DisplayFunction -> Identity]]

graphs=Table[sol[qq], {qq, 1, 4, 1}
Show [graphs, AxesOrigin -> {0,0},
AxesLabel -> {``t - ns'', ``N2 - 1E15 cm^-3''},
PlotRange->All, DisplayFunction->$DisplayFunction]
```

[7]The use of normalized values is necessary for most current (1997) numerical differential equation solvers (Mathematica, MathCad, etc.) as a consequence of the way the programs select mesh points. As modeling technology advances, it is anticipated that this normalization step will become unnecessary.

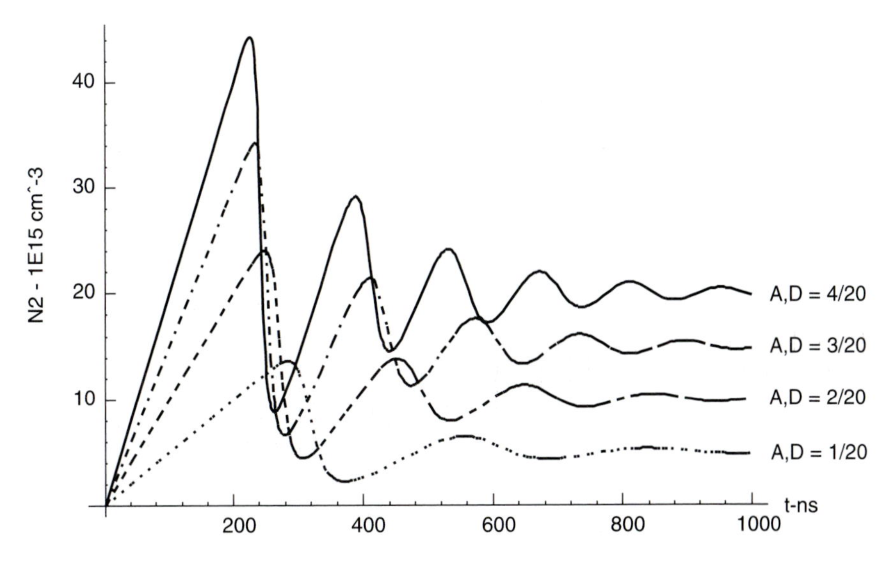

Figure 6.5 The Mathematica solution for the upper state population N_2 as a function of time and pumping. The system has $B = 0.000004$ ns^{-1} and $C = 0.01$ norm^{-1} ns^{-1}. The range of A is from 0.05 to 0.20 norm ns^{-1}. The range of D is from 0.05 to 0.20 ns^{-1}. The initial conditions on N_2 and on ϕ are $1 \cdot 10^{12}$ atoms/cm^3 (0.001 norm).

The above process yields the plot given in Figure 6.5. The process can be repeated with the substitution of `P[t]` for `N[t]` in the plot statement, and words ``` ``Photon density'' ``` for ``` ``N2'' ``` in the AxesLabel specification. This gives the plot in Figure 6.6.

Figure 6.5 illustrates the upper state population as a function of time and pumping. Notice for smaller values of pumping/extraction, the oscillations are small, and the system rapidly moves to a relatively low equilibrium value. As the pumping/extraction increases, the magnitude and number of oscillations of the upper state population increase. The final equilibrium value of the upper state population also increases.

Figure 6.6 illustrates the photon density as a function of time and pumping. Notice that an increase in the pumping/extraction term results in a narrower pulse width and faster oscillations. Also notice that the final equilibrium value is essentially the same for all cases.

6.1.3 Analytical Treatment of Relaxation Oscillations

Although it is nearly impossible to analytically evaluate the coupled nonlinear differential equations describing the upper state population and photon density, it *is* possible to obtain an approximate analytical answer using a simple *perturbative* technique. In perturbative techniques such as this one, the answer is assumed to be composed of a large background term and a small oscillating term.

To demonstrate this technique, recall the rate Equations (6.5) and (6.7) for the upper state population N_2 and the photon density ϕ

$$\frac{dN_2}{dt} = P_2 - \frac{N_2}{\tau_2} - \sigma_{12} c\, (\phi N_2) \tag{6.20}$$

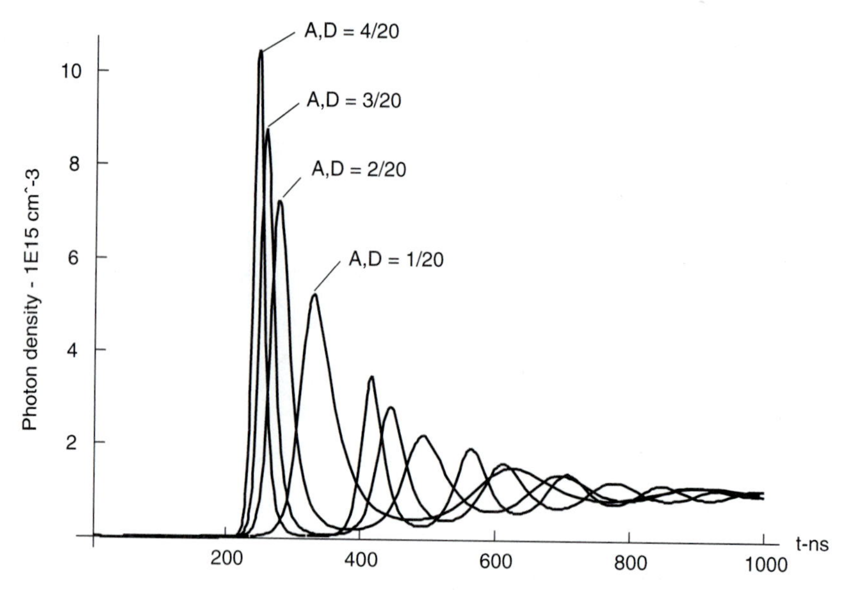

Figure 6.6 The Mathematica solution for the photon density ϕ as a function of time and pumping. The system has $B = 0.000004\ \text{ns}^{-1}$ and $C = 0.01\ \text{norm}^{-1}\ \text{ns}^{-1}$. The range of A is from 0.05 to 0.20 norm ns^{-1}. The range of D is from 0.05 to 0.20 ns^{-1}. The initial conditions on N_2 and on ϕ are $1 \cdot 10^{12}$ atoms/cm^3 (0.001 norm).

$$\frac{d\phi}{dt} = -\frac{\phi}{\tau_{\text{cav}}} + \sigma_{12}c\,(\phi N_2)\,. \tag{6.21}$$

Assume that the laser is operating near threshold, and that the fluctuations in the upper state population N_2 and the photon density ϕ are relatively small perturbations around the mean threshold densities. Under these conditions, N_2 and ϕ may be written as the sum of a large constant term (N_{2o}, ϕ_o) and a small varying term (ΔN_2, $\Delta\phi$),

$$N_2 = N_{2o} + \Delta N_2 \tag{6.22}$$

$$\phi = \phi_o + \Delta\phi. \tag{6.23}$$

It is possible to obtain an algebraic expression for the constant terms N_{2o} and ϕ_o by setting Equations (6.20) and (6.21) equal to 0 and solving for N_2 and ϕ. Doing so gives

$$N_{2o} = \frac{1}{\sigma_{12}c\tau_{\text{cav}}} \tag{6.24}$$

$$\phi_o = \frac{\sigma_{12}cP_2\tau_{\text{cav}}\tau_2 - 1}{\sigma_{12}c\tau_2}. \tag{6.25}$$

Notice that at threshold, $\phi_o = 0$. So, the threshold pumping density $P_{2\text{thres}}$ can be evaluated as

$$P_{2\text{thres}} = \frac{1}{\sigma_{12}c\tau_{\text{cav}}\tau_2}. \tag{6.26}$$

It is common to define a term r called *times-over-threshold*, which is the ratio of the pump power P_2 to the threshold power $P_{2\text{thres}}$ as

$$r \triangleq \frac{P_2}{P_{2\text{thres}}}. \tag{6.27}$$

Thus, P_2 can be expressed as

$$P_2 = r P_{2\text{thres}} = \frac{r}{\sigma_{12} c \tau_{\text{cav}} \tau_2}. \tag{6.28}$$

Substituting Equation (6.22) and (6.23) into the differential Equation (6.20) gives a new form for the differential equation,

$$\frac{d\left(N_{2o} + \Delta N_2\right)}{dt} = P_2 - \frac{\left(N_{2o} + \Delta N_2\right)}{\tau_2} - \sigma_{12} c \left(\phi_o + \Delta\phi\right)\left(N_{2o} + \Delta N_2\right). \tag{6.29}$$

Multiplying this out yields

$$\frac{dN_{2o}}{dt} + \frac{d\Delta N_2}{dt} = P_2 - \frac{N_{2o}}{\tau_2} - \frac{\Delta N_2}{\tau_2} - \sigma_{12} c \left(\phi_o N_{2o} + \Delta\phi N_{2o} + \phi_o \Delta N_2 + \Delta\phi \Delta N_2\right). \tag{6.30}$$

Assuming that the perturbations are small, the derivative term dN_{2o}/dt as well as the product term $\Delta\phi\Delta N_2$, can be neglected. Substituting the relations (6.24), (6.25), and (6.28) into (6.30) and simplifying gives

$$\frac{d\Delta N_2}{dt} = \left(\frac{-r}{\tau_2}\right)\Delta N_2 - \left(\frac{1}{\tau_{\text{cav}}}\right)\Delta\phi. \tag{6.31}$$

Performing the same operation on Equation (6.21) yields

$$\frac{d\Delta\phi}{dt} = \left(\frac{r}{\tau_2} - \frac{1}{\tau_2}\right)\Delta N_2. \tag{6.32}$$

Assume time-varying exponential solutions of the form

$$\Delta N_2 = \Delta N_{2o} e^{st} \tag{6.33}$$

and

$$\Delta\phi = \Delta\phi_o e^{st}. \tag{6.34}$$

Substituting these into Equation (6.31) gives

$$\left(-s - \frac{r}{\tau_2}\right)\Delta N_{2o} + \left(\frac{-1}{\tau_{\text{cav}}}\right)\Delta\phi_o = 0. \tag{6.35}$$

Performing the same operation on Equation (6.21) yields

$$\left(\frac{r}{\tau_2} - \frac{1}{\tau_2}\right)\Delta N_{2o} + (-s)\,\Delta\phi_o = 0. \tag{6.36}$$

Set the determinant of the system of Equations (6.35) and (6.36) equal to zero, and solve the resulting quadratic characteristic equation. The result is

$$s = \frac{-r}{2\tau_2} \pm \sqrt{\left(\frac{r}{2\tau_2}\right)^2 - \frac{r-1}{\tau_{cav}\tau_2}}. \tag{6.37}$$

Recall that the solutions were of the form e^{st}. Thus, the term under the square root is the term that can generate oscillatory behavior. If this term is negative, then the square root yields an imaginary number, and the behavior is damped oscillatory. The oscillations decay with a time constant of $r/2\tau_2$.

Therefore, to observe relaxation oscillations

$$\left(\frac{r}{2\tau_2}\right)^2 < \frac{r-1}{\tau_{cav}\tau_2} \tag{6.38}$$

$$\tau_2 > \frac{\tau_{cav}r^2}{4\,(r-1)} \approx \frac{\tau_{cav}r}{4}. \tag{6.39}$$

Thus, the upper state lifetime τ_2 must be greater than 1/4 of the product of the cavity lifetime and the pumping times-over-threshold. This condition is commonly met in solid-state laser materials such as ruby, Nd:YAG, and Nd:glass.

Example 6.4

Consider a solid-state laser material such as Nd:YAG, with a cavity such that $\tau_{cav} = 5$ ns. To accentuate the results, assume that the cavity is excessively overpumped with $r = 100$. Assume the upper state lifetime is 230 μs. Calculate the exponential decay constant and the relaxation oscillation frequency. Plot N_2 as a function of time.

Solution. The decay constant is the first term in Equation (6.37). Evaluating

$$\frac{1}{t_d} = \frac{-r}{2\tau_2} = \frac{-100}{2 \cdot 230\ \mu s} = -0.217\ \mu s^{-1}. \tag{6.40}$$

The oscillation frequency is the second term in Equation (6.37). Evaluating

$$j\omega_R = \sqrt{\left(\frac{r}{2\tau_2}\right)^2 - \frac{r-1}{\tau_{cav}\tau_2}} = \sqrt{\left(\frac{100}{2 \cdot 230\ \mu s}\right)^2 - \frac{100-1}{5\ \text{ns} \cdot 230\ \mu s}} \tag{6.41}$$

$$j\omega_R = j \cdot 9.276 \cdot 10^6\ \text{rad/sec}. \tag{6.42}$$

Thus,

$$f_R = \frac{\omega_R}{2\pi} = \frac{9.276 \cdot 10^6\ \text{rad/sec}}{2\pi} = 1.476\ \text{MHz}. \tag{6.43}$$

Figure 6.7 illustrates $\Delta N_2 = \Delta N_{2o} e^{st}$, assuming that $s = -0.217\ \mu s^{-1} + j \cdot 9.276 \cdot 10^6$ rad/sec and that $\Delta N_{2o} = 1$ norm. (As an aside, notice that this is a higher oscillation frequency than would be typical of a solid-state laser system such as Nd:YAG. Typical oscillation frequencies for solid-state laser systems will be on the order of tens of kilohertz, due to r being on the order of 2 to 5. The reason for using a large r in this example is to accentuate the decay constant for illustrative purposes.)

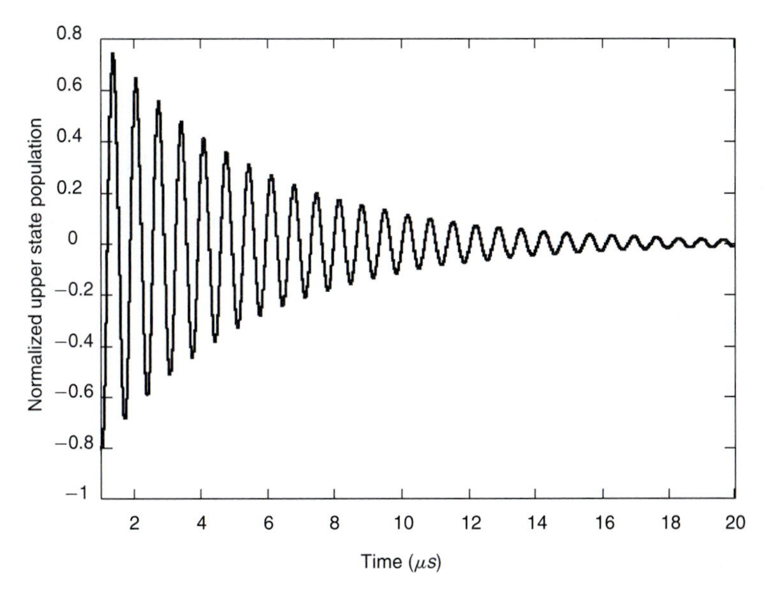

Figure 6.7 A plot of $\Delta N_2 = \Delta N_{2o} e^{st}$ assuming that $s = -0.217\ \mu\text{s}^{-1} + j \cdot 9.276 \cdot 10^6$ rad/sec and that $\Delta N_{2o} = 1$ norm.

Example 6.5

Consider the example where the relaxation oscillations are modeled using Mathematica. Recalculate the exponential decay constant and relaxation oscillation frequency using the analytical approach. Plot N_2 as a function of time and compare this with Figure 6.5 for the case of $D = 0.20$ ns^{-1}.

Solution. The upper state lifetime τ_2 can be extracted from B as

$$\tau_2 = \frac{1}{B} = \frac{1}{0.000004\ \text{ns}^{-1}} = 250\ \mu\text{s}. \tag{6.44}$$

The cavity lifetime τ_{cav} can be extracted from D as

$$\tau_{\text{cav}} = \frac{1}{D} = \frac{1}{0.20\ \text{ns}^{-1}} = 5\ \text{ns}. \tag{6.45}$$

The pumping density P_2 can be extracted from A as

$$P_2 = A = 0.20\ \text{norm ns}^{-1}. \tag{6.46}$$

The threshold density $P_{2\text{thres}}$ can be evaluated from C as

$$P_{2\text{thres}} = \frac{1}{\sigma_{12} c \tau_{\text{cav}} \tau_2} = \frac{1}{C \tau_{\text{cav}} \tau_2} = \frac{1}{(0.01\ \text{norm}^{-1}\text{ns}^{-1})(5\ \text{ns})(250\ \mu\text{s})} \tag{6.47}$$

$$P_{2\text{thres}} = 8 \cdot 10^{-5}\ \text{norm ns}^{-1}. \tag{6.48}$$

The pumping times-over-threshold can be extracted by taking the ratio of P_2 to $P_{2\text{thres}}$ as

$$r = \frac{P_2}{P_{2\text{thres}}} = \frac{0.20 \text{ norm ns}^{-1}}{8 \cdot 10^{-5} \text{ norm ns}^{-1}} = 2500. \tag{6.49}$$

(Again, the time constant in these examples has been made larger than is typical in a practical laser for the purpose of illustration.)

The decay constant is the first term in Equation (6.37). Evaluating

$$\frac{1}{t_d} = \frac{-r}{2\tau_2} = \frac{-2500}{2 \cdot 250\ \mu\text{s}} = -5\ \mu\text{s}^{-1}. \tag{6.50}$$

The oscillation frequency is the second term in Equation (6.37). Evaluating

$$j\omega_R = \sqrt{\left(\frac{r}{2\tau_2}\right)^2 - \frac{r-1}{\tau_{\text{cav}}\tau_2}} = \sqrt{\left(\frac{2500}{2 \cdot 250\ \mu\text{s}}\right)^2 - \frac{2500-1}{5 \text{ ns} \cdot 250\ \mu\text{s}}} \tag{6.51}$$

$$j\omega_R = j \cdot 4.443 \cdot 10^7 \text{ rad/sec.} \tag{6.52}$$

Thus,

$$f_R = \frac{\omega_R}{2\pi} = \frac{4.443 \cdot 10^7 \text{ rad/sec}}{2\pi} = 7.701 \text{ MHz.} \tag{6.53}$$

Figure 6.8 illustrates $\Delta N_2 = \Delta N_{2o} e^{st}$, assuming that $s = -5\ \mu\text{s}^{-1} + j \cdot 4.443 \cdot 10^7$ rad/sec and that $\Delta N_{2o} = 1$ norm.

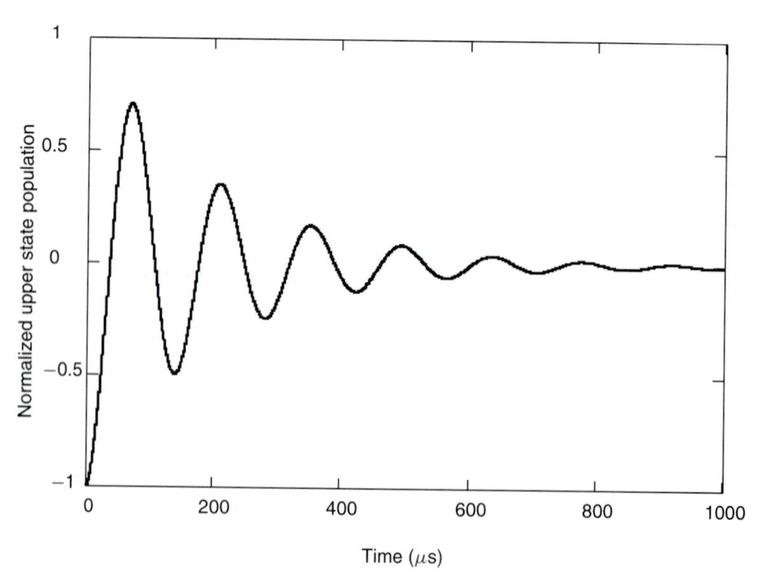

Figure 6.8 A plot of $\Delta N_2 = \Delta N_{2o} e^{st}$, assuming that $s = -5\ \mu\text{s}^{-1} + j \cdot 4.443 \cdot 10^7$ rad/sec and that $\Delta N_{2o} = 1$ norm. Note the similarities to the Mathematica plot for the case of $D = 0.20\ \text{ns}^{-1}$.

6.2 Q-SWITCHING

6.2.1 A Qualitative Description of Q-Switching

In a direct sense, Q-switching is artificially inducing relaxation oscillations in order to increase the amount of peak output power in a pulse. Q-switching is accomplished by placing something in the cavity that makes the cavity opaque (reduces the Q), so that laser action cannot occur. Since there is no laser action to reduce the population inversion, then the threshold inversion can increase to a level substantially higher than in the steady-state. If the cavity is suddenly returned to a normal transmissive condition, the increased population inversion will result in an increase in the gain and a dramatic increase in the photon density in the cavity. It is no surprise that the early experimental demonstrations of Q-switching referred to the process as creation of "Giant Pulses."[8,9]

However, in spite of the similarities between relaxation oscillations and Q-switching, several differences must be kept in mind. The initial conditions in Q-switching are that N_2 is at a relatively high value and ϕ is quite small. (Recall that for relaxation oscillations, it is typical for initial conditions on both N_2 and ϕ to be at relatively low values.) In Q-switching, the pump rate is only of importance to establish the high initial population of N_2. The pump can usually be neglected during the pulse formation. (Recall that for relaxation oscillations, the pump is a critical driver to the process.)

The possibility of Q-switching lasers was first proposed by Hellwarth in 1961.[10] The first experimental observation of Q-switched pulse behavior was made by McClung and Hellwarth in 1962 using an electrooptic Q-switch in a ruby laser.[11] A complete theoretical treatment was completed by Wagner and Lengyel in 1963[12] and a simplified version was given by Wang in the same year.[13] A more elaborate treatment including numerical modeling was performed by Fleck in 1970.[14]

6.2.2 Numerical Modeling of Q-Switching

The model for Q-switching uses the same pair of coupled nonlinear differential equations that are used to model relaxation oscillations. Recall that these are

$$\frac{dN_2}{dt} = P_2 - \frac{N_2}{\tau_2} - \sigma_{12} c\,(\phi N_2) \tag{6.54}$$

$$\frac{d\phi}{dt} = -\frac{\phi}{\tau_{cav}} + \sigma_{12} c\,(\phi N_2)\,. \tag{6.55}$$

---

[8] R. W. Hellwarth, *Advances in Quantum Electronics*, ed J. R. Singer (New York: Columbia University Press, 1961), pp. 334–341.

[9] F. J. McClung and R. W. Hellwarth, *J. Appl. Phys.* 33:828 (1962).

[10] R. W. Hellwarth, *Advances in Quantum Electronics*, ed J. R. Singer (New York: Columbia University Press, 1961), pp. 334–341.

[11] F. J. McClung and R. W. Hellwarth, *J. Appl. Phys.* 33:828 (1962).

[12] W. G. Wagner and B. A. Lengyel, *J. Appl. Phys.* 34:2040 (1963).

[13] C. C. Wang, *Proc. IEEE* 51:1767 (1963).

[14] J. A. Fleck, *Phys. Rev. B.* 1:84 (1970).

It is difficult to set up a modeling problem that presents a variable Q. However, it is simple to set P_2 equal to 0 and artificially increase the initial conditions for the upper state population N_2. This effectively sets time $t = 0$ in the numerical model to be the time at which the Q of the cavity was abruptly changed.

Example 6.6

Model the Q-switching behavior as a function of initial N_2 population for the situation of $P_2 = 0$. Assume that $B = 0.000004$ ns^{-1}, $C = 0.01$ norm^{-1} ns^{-1}, and that $D = 0.10$ norm ns^{-1}. Create a plot of $\phi(t)$. Assume the initial conditions on N_2 range from $1.5 \cdot 10^{16}$ atoms/cm^3 to $4.5 \cdot 10^{16}$ atoms/cm^3, (15 − 45 norm), and on ϕ are $1 \cdot 10^{12}$ atoms/cm^3 (0.001 norm).

Solution

```
In[1]:=
Clear[sol]
sol[s_]:=Module[{tsolt,N,P,t,eqone,eqtwo},

eqone = N'[t] == 0 - 0.000004 N[t] - 0.01 N[t] P[t];
eqtwo = P'[t] == 0 - 0.10 P[t] + 0.01 N[t] P[t];
tsolt = NDSolve[
{eqone, eqtwo, N[0]==s, P[0]==0.001},
{N[t], P[t]}, {t,0,200}];
Plot[{P[t]} /. tsolt, {t,0,200}, Compiled -> False,
DisplayFunction -> Identity]]

graphs=Table[sol[qq], {qq, 15, 45, 10}]
Show [graphs, AxesOrigin -> {0,0},
AxesLabel -> {``t - ns'', ``Photon density - 1E15 cm^-3''},
PlotRange->All, DisplayFunction->$DisplayFunction]
```

The above process yields the plot given in Figure 6.9. The process can be repeated with the substitution of `P[t]` for `N[t]` in the Plot statement, and the words `''N2''` for `''Photon density''` in the AxesLabel specification. This gives the plot in Figure 6.10.

Notice that the pulse width decreases rapidly with increasing initial upper state population density. The length of time between the initiation of the Q-switch pulse and the center of the peak also decreases with increasing upper state population density. Notice also that the upper state population is largely extracted for high initial upper state population densities, but a significant amount remains at lower initial upper state population densities.

6.2.3 Analytical Treatment of Q-Switching

Unlike relaxation oscillations, the boundary conditions on Q-switching make it possible to obtain a simplified analytical solution for the Q-switching parameters. The basic process is outlined as follows, and additional details can be found in Yariv[15] or Siegman.[16]

[15] Amnon Yariv, *Optical Electronics*, 4th ed. (Philadelphia, PA: Saunders College Publishing, 1991), pp. 205–12.

[16] Anthony Siegman, *Lasers* (Mill Valley, CA: University Science Books, 1986), Chapter 26.

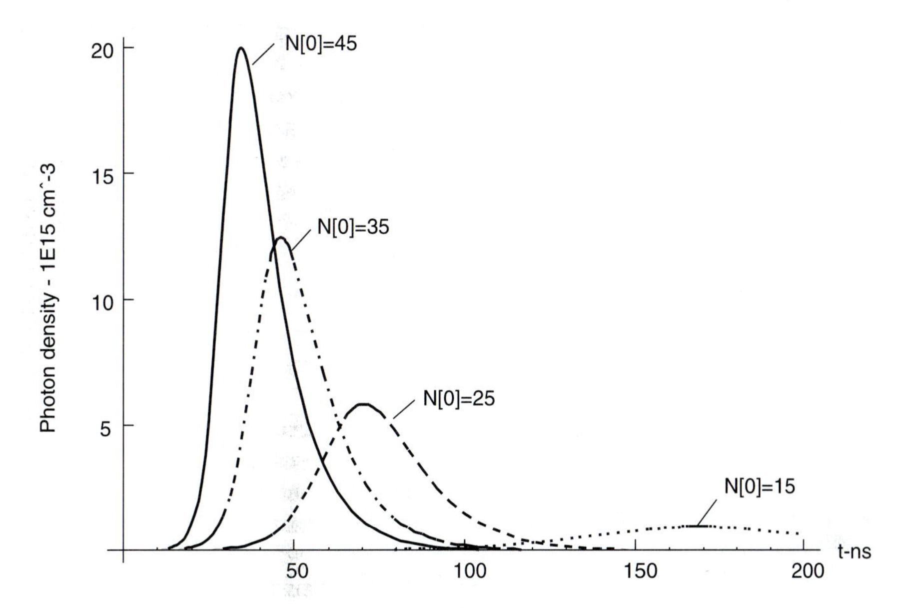

Figure 6.9 The Mathematica solution for the photon density ϕ in a Q-switched laser as a function of initial N_2 population for the case when $P_2 = 0$. In this case, $B = 0.000004$ ns^{-1}, $C = 0.01$ norm^{-1} ns^{-1}, and $D = 0.10$ norm ns^{-1}.

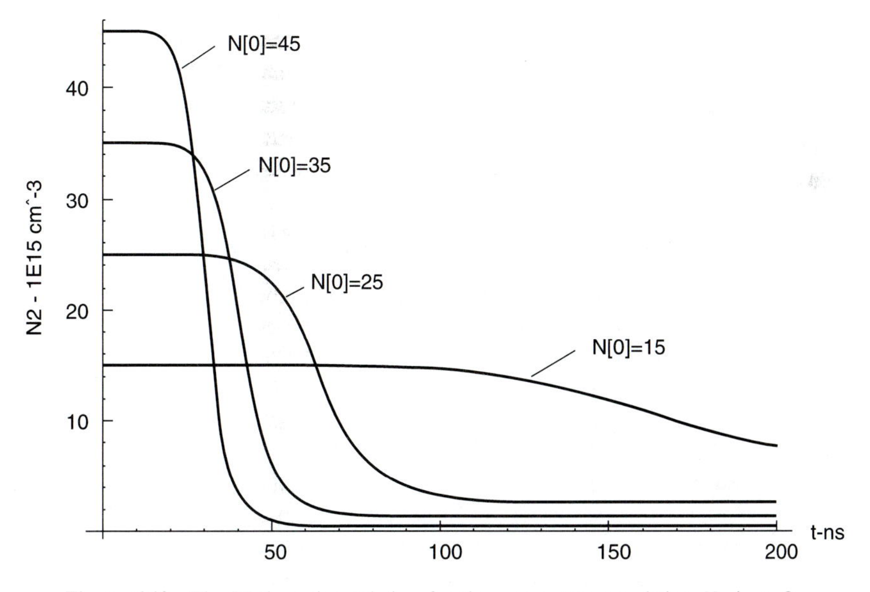

Figure 6.10 The Mathematica solution for the upper state population N_2 in a Q-switched laser as a function of initial N_2 population for the case when $P_2 = 0$. In this case, $B = 0.000004$ ns^{-1}, $C = 0.01$ norm^{-1} ns^{-1}, and $D = 0.10$ norm ns^{-1}.

Consider the simplified rate equations for a four-state laser system given in Equations (6.20) and (6.21),

$$\frac{dN_2}{dt} = P_2 - \frac{N_2}{\tau_2} - \sigma_{12}c\,(\phi N_2) \tag{6.56}$$

$$\frac{d\phi}{dt} = -\frac{\phi}{\tau_{\text{cav}}} + \sigma_{12}c\,(\phi N_2)\,. \tag{6.57}$$

Recall that the equilibrium values for N_{2o} and ϕ_o can be obtained by setting Equations (6.20) and (6.21) equal to 0 and solving for N_2 and ϕ. This problem is worked in Section 6.1, yielding Equations (6.24) and (6.25). Equation (6.24) is needed in this analysis also and is given by

$$N_{2o} = \frac{1}{\sigma_{12}c\tau_{\text{cav}}}. \tag{6.58}$$

For the Q-switching case, the pulse occurs quickly in comparison with the pumping, so the term P_2 may be neglected. Additionally, the spontaneous emission rate is usually small in comparison with the stimulated rate, so N_2/τ_2 may be neglected. Rewriting Equations (6.56) and (6.57) with these approximations in mind, and taking into account the value of N_{2o} expressed in Equation (6.58), gives

$$\frac{dN_2}{dt} = \frac{-N_2}{\tau_{\text{cav}}N_{2o}}\phi \tag{6.59}$$

and

$$\frac{d\phi}{dt} = \frac{\phi}{\tau_{\text{cav}}}\left(\frac{N_2}{N_{2o}} - 1\right). \tag{6.60}$$

Dividing Equation (6.60) by (6.59) yields

$$\frac{d\phi}{dN_2} = \left(\frac{N_{2o}}{N_2} - 1\right). \tag{6.61}$$

Integrating gives

$$\int_{\phi_i}^{\phi(t)} d\phi = \int_{N_{2i}}^{N_2(t)} \left(\frac{N_{2o}}{N_2} - 1\right) dN_2 \tag{6.62}$$

resulting in a photon density of

$$\phi(t) - \phi_i = N_{2o}\left(\ln\frac{N_2(t)}{N_{2i}}\right) - (N_2(t) - N_{2i})\,. \tag{6.63}$$

If the initial photon density is zero, then the final photon density can be written as

$$\phi_f = N_{2o}\left(\ln\frac{N_{2f}}{N_{2i}}\right) - \left(N_{2f} - N_{2i}\right). \tag{6.64}$$

Now, the output energy of the Q-switched pulse is related to the difference between the initial upper state population and the final upper state population as

$$E_Q = \left(N_{2i} - N_{2f}\right)h\nu. \tag{6.65}$$

The final upper state population can be obtained by setting Equation (6.64) equal to zero, and solving for $N_2(t) = N_{2f}$. This yields a transcendental equation for N_{2f},

$$\ln \frac{N_{2f}}{N_{2i}} = \frac{N_{2f} - N_{2i}}{N_{2o}}. \tag{6.66}$$

The peak photon density (ϕ_{peak}) can be obtained by noting that it occurs when $d\phi(t)/dt = 0$ and $N_2(t) = N_{2o}$. Under these conditions, Equation (6.63) may then be written as

$$\phi_{\text{peak}} = N_{2i} - N_{2o} - N_{2o} \ln (N_{2i}/N_{2o}) \tag{6.67}$$

leading to a peak power expression of

$$P_{\text{peak}} = \phi_{\text{peak}} \cdot \frac{h\nu}{\tau_{\text{cav}}} = \frac{h\nu}{\tau_{\text{cav}}} \left(N_{2i} - N_{2o} - N_{2o} \ln (N_{2i}/N_{2o})\right). \tag{6.68}$$

A good approximation to the pulse length Δt_p can then be obtained by taking the pulse energy, Equation (6.65), and dividing by the peak pulse power (Equation 6.68), to obtain:[17]

$$\Delta t_p = \frac{\left(N_{2i} - N_{2f}\right)}{N_{2i} - N_{2o} - N_{2o} \ln (N_{2i}/N_{2o})} \tau_{\text{cav}}. \tag{6.69}$$

Example 6.7

Consider the previous example, where Q-switching is modeled using Mathematica. Calculate the final upper state population N_{2f}, the final photon density ϕ_f, the energy of the pulse E_Q, the peak photon density of the pulse ϕ_{peak}, the peak power of the pulse P_{peak}, and the width of the pulse Δt_p. Assume that the laser is similar to Nd:YAG and possesses a wavelength of 1.064 μm. Compare the results with Figures 6.9 and 6.10 for the intermediate case where $N_{2i} = 25$ norm.

Solution. The upper state lifetime τ_2 can be extracted from B as

$$\tau_2 = \frac{1}{B} = \frac{1}{0.000004 \text{ ns}^{-1}} = 250 \ \mu\text{s}. \tag{6.70}$$

The cavity lifetime τ_{cav} can be extracted from D as

$$\tau_{\text{cav}} = \frac{1}{D} = \frac{1}{0.10 \text{ ns}^{-1}} = 10 \text{ ns}. \tag{6.71}$$

The threshold upper state population N_{2o} can be evaluated from C as

$$N_{2o} = \frac{1}{\sigma_{12} c \tau_{\text{cav}}} = \frac{1}{C\tau_{\text{cav}}} = \frac{1}{(0.01 \text{ norm}^{-1}\text{ns}^{-1})(10 \text{ ns})} \tag{6.72}$$

$$N_{2o} = 10 \text{ norm}. \tag{6.73}$$

The final upper state population N_{2f} can be obtained from Equation (6.66),

$$\ln \frac{N_{2f}}{N_{2i}} = \frac{N_{2f} - N_{2i}}{N_{2o}}. \tag{6.74}$$

Solving iteratively for the case of $N_{2i} = 25$ norm and $N_{2o} = 10$ norm gives $N_{2f} = 2.684$ norm.

[17] Walter Koechner, *Solid State Laser Engineering*, 3d ed. (Berlin: Springer-Verlag, 1992), p. 438.

The final photon density ϕ_f can be obtained from Equation (6.64),

$$\phi_f = N_{2o}\left(\ln\frac{N_{2f}}{N_{2i}}\right) - \left(N_{2f} - N_{2i}\right) \tag{6.75}$$

$$\phi_f = 10 \text{ norm}\left(\ln\frac{2.684 \text{ norm}}{25 \text{ norm}}\right) - (2.684 \text{ norm} - 25 \text{ norm}) = 0.00032 \text{ norm}. \tag{6.76}$$

The output energy of the Q-switched pulse E_Q can be obtained from Equation (6.65),

$$E_Q = \left(N_{2i} - N_{2f}\right) h \frac{c_o}{\lambda_o} \tag{6.77}$$

$$E_Q = (25 \text{ norm} - 2.684 \text{ norm})\, h \cdot \frac{c_o}{1.064\ \mu\text{m}} = 4.166 \text{ mJ cm}^{-3}. \tag{6.78}$$

The peak photon density ϕ_{peak} can be obtained from Equation (6.67),

$$\phi_{\text{peak}} = N_{2i} - N_{2o} - N_{2o} \ln\left(N_{2i}/N_{2o}\right) \tag{6.79}$$

$$\phi_{\text{peak}} = 25 \text{ norm} - 10 \text{ norm} - 10 \text{ norm} \cdot \ln(25 \text{ norm}/10 \text{ norm}) \tag{6.80}$$

$$\phi_{\text{peak}} = 5.837 \text{ norm}. \tag{6.81}$$

The peak power P_{peak} can be obtained from Equation (6.68),

$$P_{\text{peak}} = \phi_{\text{peak}} \cdot \frac{h\nu}{\tau_{\text{cav}}} \tag{6.82}$$

$$P_{\text{peak}} = 5.837 \text{ norm} \cdot h \cdot \frac{c_o}{1.064\ \mu\text{m}} \cdot \frac{1}{10 \text{ ns}} = 108.976 \text{ kW cm}^{-3}. \tag{6.83}$$

The pulse width Δt_p can be obtained from Equation (6.68),

$$\Delta t_p = \frac{\left(N_{2i} - N_{2f}\right)}{N_{2i} - N_{2o} - N_{2o} \ln\left(N_{2i}/N_{2o}\right)} \tau_{\text{cav}} \tag{6.84}$$

$$\Delta t_p = \frac{(25 \text{ norm} - 2.684 \text{ norm})}{25 \text{ norm} - 10 \text{ norm} - 10 \text{ norm} \cdot \ln(25 \text{ norm}/10 \text{ norm})} \cdot 10 \text{ ns} = 38.231 \text{ ns}. \tag{6.85}$$

Comparing the results to Figures 6.9 and 6.10 gives excellent agreement.

6.3 THE DESIGN OF Q-SWITCHES

The essential idea in designing a Q-switch for a laser system is to temporarily render the laser cavity opaque while the population inversion is building, then to rapidly switch the cavity to a transparent state to create the high power pulse. The transition from opaque to transparent should be swift, otherwise spurious pulses will be created. The extinction ratio ($I_{\text{max}}/I_{\text{min}}$) should be large, otherwise photons will leak through and reduce the maximum upper state population by prelasing. There are four major technologies used for Q-switches: mechanical, electrooptical, acoustical, and dye.

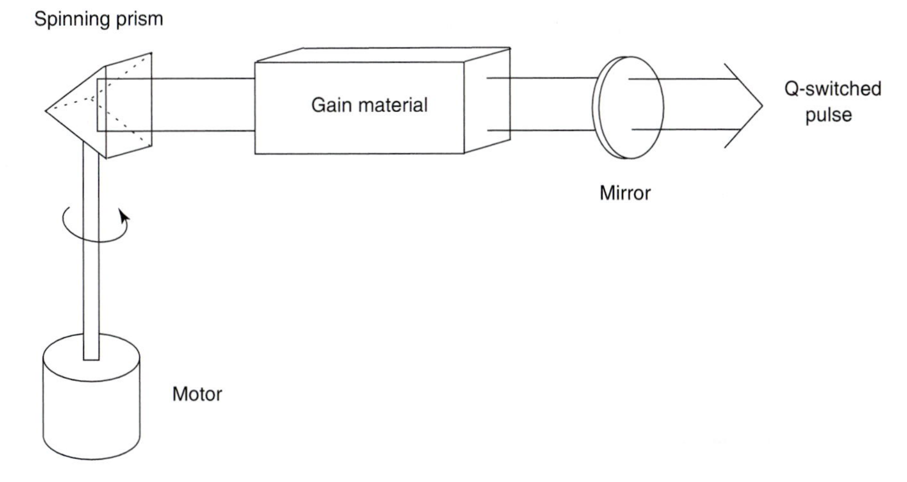

Figure 6.11 Roof prisms are often used in mechanical Q-switches to simplify alignment problems in the laser cavity.

6.3.1 Mechanical Q-Switches

A conceptually simple way to Q-switch a cavity is to incorporate a mechanical device within the cavity that blocks the laser beam. The first mechanical Q-switches used a rotating chopper, similar to that used in lock-in amplifiers.[18] However, choppers are slow and vibration-prone, and such techniques were soon abandoned in favor of rotating mirrors and prisms.[19] With rotating mirrors, the approach is to spin the mirror using a high-speed motor. Such designs usually incorporate a multisided mirror or multiplicative optical geometries so that several reflections are possible for each rotation.[20] However, rotating mirror Q-switches are prone to alignment difficulties because each face of the mirror must be aligned to within a fraction of a milliradian. Thus, roof prisms are often used as the rotating elements.[21,22] (See Figure 6.11.) As long as the roof of the prism is perpendicular to the axis of rotation, reflection is guaranteed at some angle of rotation.

In order to Q-switch successfully with a mechanical Q-switch, it is necessary to rotate at a very high speed (20,000 to 50,000 rpm) and to use multifaceted reflectors. Otherwise, the length of time between Q-switch pulses will be so long that multiple (and lower-power) pulses will be generated. A typical example would be a cw Nd:YAG laser with a twelve-sided prism rotating at 25,000 rpm to achieve a repetition rate of single pulses at 5 kHz.[23]

Mechanical Q-switches are the simplest and least expensive of the Q-switches. They have the additional advantages of polarization and wavelength insensitivity. However, the

---

[18]R. J. Collins and P. Kisliuk, *J. Appl. Phys.* 33:2009 (1962).

[19]R. C. Benson and M. R. Mirarchi, *IEEE Trans. Milit. Electr.* MIL-8:13 (1964).

[20]R. Daly and S. D. Sims, *Appl. Opt.* 3:1063 (1964).

[21]D. Findlay and A. F. Fray, *Opto-Electronics* 2:51 (1970).

[22]R. G. Smith and M. F. Galvin, *IEEE J. of Quan. Electron.* QE-3:406 (1967).

[23]E. J. Woodbury, *IEEE J. of Quan. Elect.* QE-3:509 (1967).

high rotational speeds mean that the devices are noisy and possess relatively short lifetimes. Furthermore, mechanical components are not robust in harsh environments.

6.3.2 Electrooptic Q-Switches

Basic operation of electrooptic Q-switches. *Electrooptic* Q-switches employ materials (such as KDP or $LiNbO_3$; see Table 6.1) that exhibit *birefringence* under an applied electric field. Birefringent crystals are crystals that possess two different indices of refraction in two orthogonal directions (see Section 8.3 for additional discussion on birefringence). An optical beam linearly polarized at 45 degrees to these axes and normal to the plane of these axes will be split into two orthogonally polarized components traveling in the same direction, but at different velocities. When the two components recombine at the end of the crystal, the result will be an elliptically polarized beam.

The most common arrangement for an electrooptic Q-switch is to place the Q-switch between a linear polarizer and the rear mirror. (See Figure 6.12.) The applied voltage across the Q-switch is chosen so as to cause a $\lambda/4$ difference in the phases of the emerging components. If linearly polarized light enters the crystal, then circularly polarized light will emerge. In other words, when voltage is applied to the Q-switch, it acts like a *quarter-wave plate* (see Section 8.3.2 for additional information on quarter-wave plates).

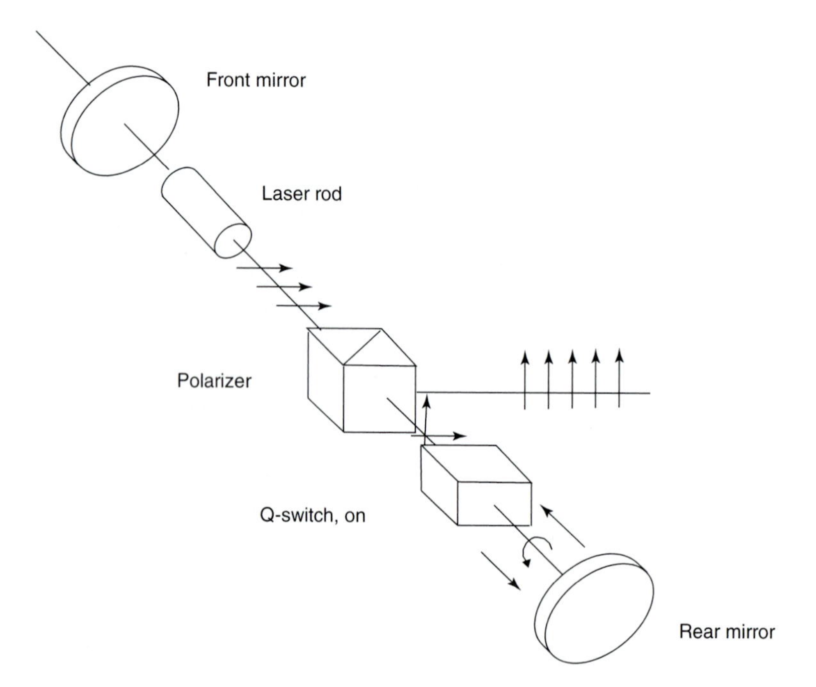

Figure 6.12 The most common arrangement for an electrooptic Q-switch is to place the Q-switch between a linear polarizer and the rear mirror. The applied voltage across the Q-switch is chosen so as to cause a $\lambda/4$ difference in the phases of the emerging components.

The sequence of events is as follows. Initially, the beam passes through the linear polarizer. It enters the electrooptic crystal and emerges as right circularly polarized light. After it reflects from the mirror, it converts into left circularly polarized light. When the beam passes through the electrooptic crystal, it emerges as linear polarized light but *perpendicular* to the direction of the original light polarization. (An alternative way to think about this is to observe that the $\lambda/4$ Q-switch plus the mirror reflection plus the $\lambda/4$ Q-switch again, acts like a $\lambda/2$- or *half-wave* plate. A $\lambda/2$-wave plate will convert linear polarization in one direction to linear polarization in the orthogonal direction.) The orthogonally polarized beam is then ejected from the cavity by the polarizer.

When the voltage is removed from the Q-switch, the crystal is no longer birefringent. Thus, the emerging beam from the crystal is unchanged and is not affected by the polarizer. Therefore, this Q-switch only produces a pulse when the voltage is off.

Another common arrangement for an electrooptic Q-switch is to place it between two crossed (polarizations rotated by 90 degrees or orthogonally polarized) linear polarizers. (See Figure 6.13.) The applied voltage across the Q-switch is chosen so as to cause a $\lambda/2$ difference in the phases of the emerging components. If linearly polarized light enters the crystal, then linearly polarized light rotated by 90 degrees will emerge. In other words, when voltage is applied to the Q-switch, it acts like a half-wave plate.

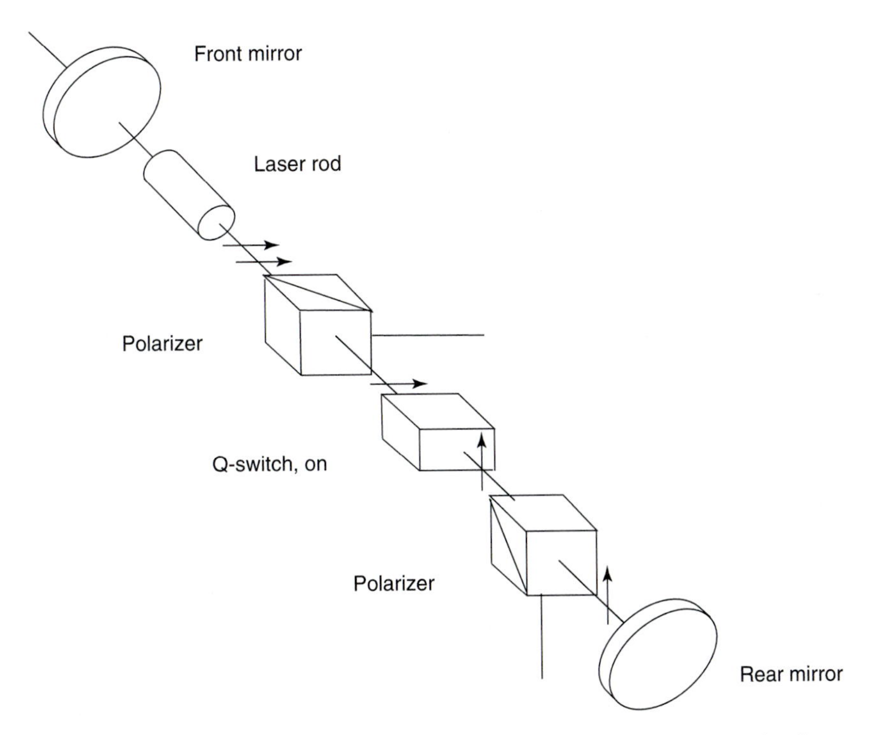

Figure 6.13 Another common arrangement for an electrooptic Q-switch is to place it between a two crossed linear polarizers. The applied voltage across the Q-switch is chosen so as to cause a $\lambda/2$ difference in the phases of the emerging components.

The sequence of events is as follows. Initially, the beam passes through the linear polarizer. It enters the electrooptic crystal and emerges as linear polarized light rotated by 90 degrees (orthogonal to the original polarization). It then can pass through the second polarizer. When the voltage is removed from the Q-switch, the crystal is no longer birefringent and the emerging beam from the crystal is unchanged. The beam has the same polarization as the original beam and is ejected from the cavity. Thus, this Q-switch only produces a pulse when the voltage is on.

Design of electrooptic Q-switches. One common configuration for a Q-switch is to apply the voltage longitudinally along the optic axis of the crystal and in the same direction as the light transmission (see Figure 6.14). When there is no voltage applied, the two orthogonal polarizations see the same index of refraction. However, when voltage is applied, the indices of refraction are no longer the same for both polarizations. Thus, an index ellipse is formed with its major and minor axes being the axes of induced birefringence. In this configuration, the propagation direction is along the z (optic) axis, the light is linearly polarized at 45 degrees to the axes of induced birefringence, and the voltage is applied along the z (optic) axis (see Section 8.3 for more information on birefringence). The half-wave voltage (the voltage required to induce a $\lambda/2$ change) for a $\bar{4}2m$ crystal[24] in this configuration is[25]

$$V_{1/2} = \frac{\lambda_o}{2n_o^3 r_{63}} \tag{6.86}$$

where n_o is the *ordinary index of refraction* in the material and r_{63} is one of the components of the electrooptical tensor (see Table 6.1). Notice that the half-wave voltage is independent of the dimensions of the crystal. However, fringing fields in longitudinal rectangular geometry Q-switches do adversely affect both the half-wave voltage and the extinction ratio. Thus, cylindrical geometry Q-switches are often employed.[26]

Another common configuration is to apply the voltage across (transverse) to the axis of light transmission in the laser (see Figure 6.15). In this configuration, the propagation direction is along one of the axes of induced birefringence, the light is linearly polarized at 45 degrees to the z-axis, and the voltage is applied along the z-axis. The half-wave voltage (the voltage required to induce a $\lambda/2$ change) for a $\bar{4}2m$ crystal cut in this configuration is[27]

$$V_{1/2} = \frac{\lambda_o d}{n_o^3 r_{63} l} \tag{6.87}$$

---

[24]Crystals are divided into seven major crystal systems defined by their symmetry properties. These classes are triclinic, monoclinic, orthorhombic, tetragonal, cubic, trigonal, and hexagonal. Each of these classes is further divided into point groups characterized by a number such as $\bar{4}3$m or 6. There are two common numbering systems for point groups, the Hermann-Maugin system (used more commonly by vendors) and the Schoenflies system. Boyd in *Nonlinear Optics* (San Diego, CA: Academic Press, 1992) gives a nice overview of point groups in Chapter 1 from a tensor viewpoint (Table 1.5.1 is especially informative). William Leonard and Thomas Martin in *Electronic Structure and Transport Properties of Materials* (Malabar, FL: Robert Krieger, 1987) also provide a good background in Chapter 5 on crystallographic notation.

[25]Walter Koechner, *Solid State Laser Engineering*, 3d ed. (Berlin: Springer-Verlag, 1992), p. 451.

[26]L. L. Steinmetz, T. W. Pouliot and B. C. Johnson, *Appl. Opt.* 12:1468 (1973).

[27]Walter Koechner, *Solid State Laser Engineering*, 3d ed. (Berlin: Springer-Verlag, 1992), p. 454.

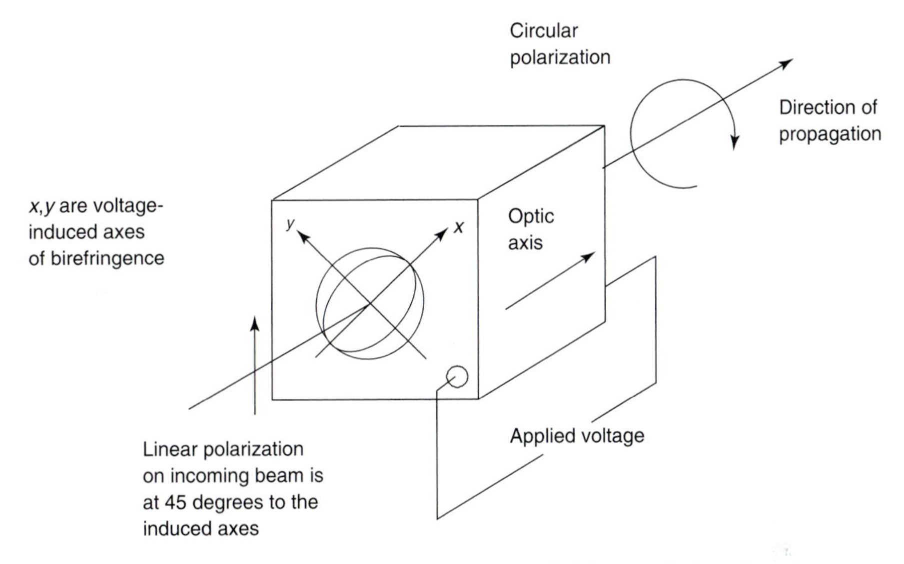

Figure 6.14 One common configuration for a Q-switch is to apply the voltage longitudinally along the optic axis of the crystal and in the same direction as the light transmission. In this configuration, the propagation direction is along the z (optic) axis, the light is linearly polarized at 45 deg to the axes of induced birefringence, and the voltage is applied along the z (optic) axis.

where d is the thickness between the electrodes and l is the length. Notice that the half-wave voltage is now dependent upon the dimensions of the crystal. This is of considerable advantage in reducing the value of the half-wave voltage.

Unfortunately, when the transverse geometry is used with KDP crystals, the propagation angle is not parallel with the optic axis. Thus, the ordinary and *extraordinary* rays propagate in different directions in the crystal (an undesirable phenomenon termed *walk-off*). As a consequence, the transverse configuration is rarely used in KDP.

However, lithium niobate can be operated in a transverse configuration where the propagation axis is parallel with the optic axis (see Figure 6.16). In this configuration,

TABLE 6.1 NONLINEAR CRYSTALS USED IN Q-SWITCHES

Material	Index of refraction	r_{63} (μm/V $\cdot 10^{-6}$)
Ammonium dihydrogen phosphate (ADP)	1.53	8.5
Potassium dihydrogen phosphate (KDP)	1.51	10.5
Deuterated KDP (KD*P)	1.51	26.4
Cesium dihydrogen arsenate (CDA)	1.57	18.6
Deuterated CDA (CD*A)	1.57	36.6
Lithium niobate	2.237	5.61 (r_{22})

Walter Koechner, *Solid State Laser Engineering*, 3d ed. (Berlin: Springer-Verlag, 1992), pp. 452, 455.

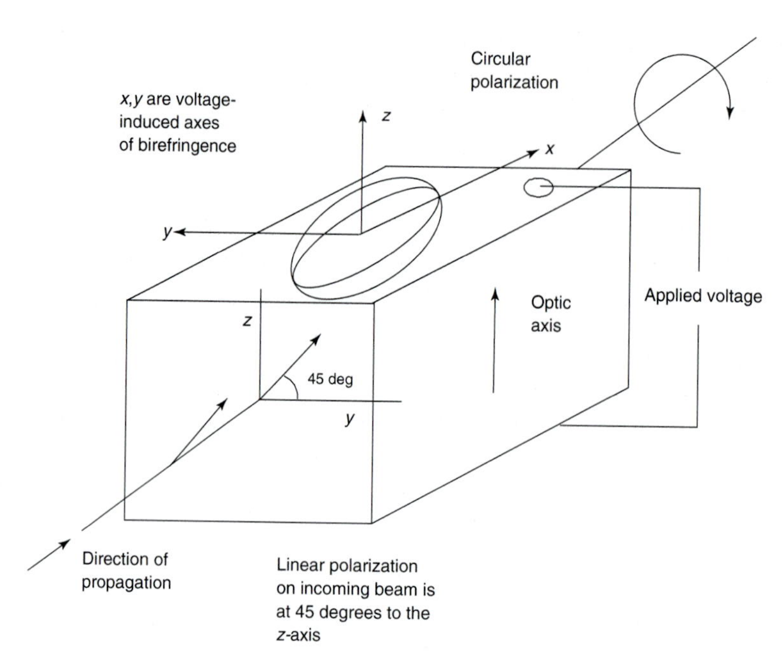

Figure 6.15 Another common configuration is to apply the voltage across (transverse) to the axis of light transmission in the laser. In this configuration, the propagation direction is along one of the axes of induced birefringence, the light is linearly polarized at 45 deg to the z-axis, and the voltage is applied along the z-axis.

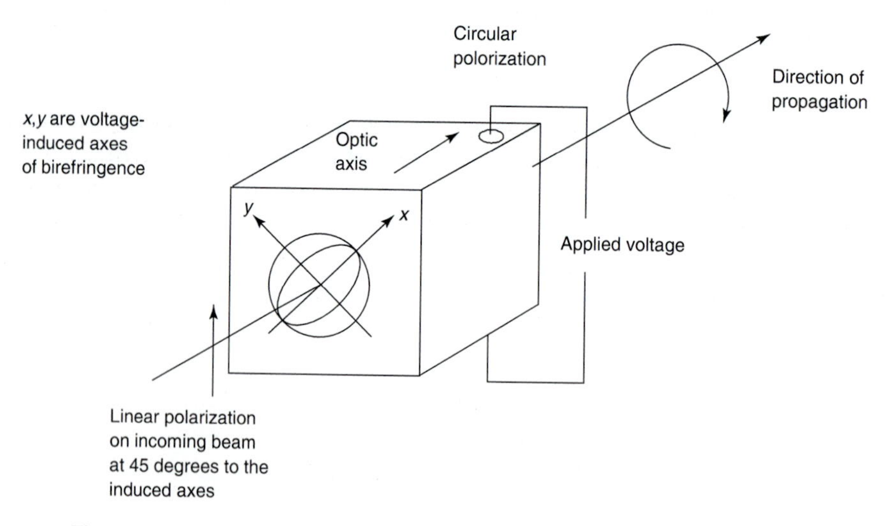

Figure 6.16 Lithium niobate can be operated in a transverse configuration where the propagation axis is parallel with the optic axis. In this configuration, the light is parallel with the optic axis, linearly polarized at 45 deg to the axes of induced birefringence, and the voltage is applied along the same direction as the linear polarization.

walk-off is not a problem. If the light is parallel with the optic axis, linearly polarized at 45 deg to the axes of induced birefringence, and the voltage is applied along the same direction as the linear polarization, then the half-wave voltage for a lithium niobate crystal is[28]

$$V_{1/2} = \frac{\lambda_o d}{2n_o^3 r_{22} l} \tag{6.88}$$

where r_{22} is one of the components of the electrooptical tensor (see Table 6.1).

Examples of various crystals used in electrooptic Q-switches are given in Table 6.1. The most useful crystals are deuterated KDP (KD*P) and deuterated CDA (CD*A) because they possess the highest r_{63} coefficients. However, virtually all commercial electrooptic Q-switches employ KD*P. Large strain-free optical quality KD*P crystals can be easily grown in water solution. The crystals are available in sizes up to 5 cm and can be readily cut and polished. Although KD*P crystals are *hygroscopic* (water-loving), this can be overcome by imaginative mounting techniques. One common solution is to submerge the crystal in an index-matching fluid, thus isolating it from humid room air. Other solutions include hermetically sealed mounts, molecular sieve water *getters* incorporated into the mount, and provisions for flowing dry air across the crystal.

Electrooptic Q-switches have the advantage of possessing extinction ratios I_{max}/I_{min} on the order of several hundred. Thus, they are well-suited for Q-switching high-gain lasers. However, they are usually fabricated from crystals which are hygroscopic and prone to damage by the laser beam. The polarization requirements mean that the laser beam must be polarized (undesirable in some applications). These Q-switches also require some cleverness in the design of the electrical system in order to switch a several kV electrical signal without transients. This combination of factors means that electrooptical Q-switches are typically used in high-peak power pulsed lasers, as well as in high-gain cw lasers.

Example 6.8

Calculate the half-wave voltages for KD*P and CD*A crystals assuming a longitudinal Q-switch configuration and an Nd:YAG laser with an operating wavelength of 1.064 μm. Calculate the half-wave voltage for lithium niobate assuming a transverse Q-switch configuration with a 1 cm square aperture, a length of 25 mm, and an operating wavelength of 1.064 μm.

Solution. The half-wave voltages for KD*P and CD*A crystals can be calculated from Equation (6.86) using the values for r_{63} from Table 6.1. The half-wave voltage for lithium niobate can be calculated from Equation (6.88) using the values for r_{22} from Table 6.1 as

$$V_{1/2(KD^*P)} = \frac{\lambda_o}{2n_o^3 r_{63}} = \frac{1.064\ \mu\text{m}}{2\cdot(1.51)^3(26.4\cdot 10^{-6}\ \mu\text{m/V})} = 5.853\text{ kV} \tag{6.89}$$

$$V_{1/2(CD^*A)} = \frac{\lambda_o}{2n_o^3 r_{63}} = \frac{1.064\ \mu\text{m}}{2\cdot(1.57)^3(36.6\cdot 10^{-6}\ \mu\text{m/V})} = 3.756\text{ kV} \tag{6.90}$$

$$V_{1/2(LN)} = \frac{\lambda_o d}{2n_o^3 r_{22} l} = \frac{1.064\ \mu\text{m}\cdot 1\text{ cm}}{2\cdot(2.237)^3(5.61\cdot 10^{-6}\ \mu\text{m/V})\cdot(25\text{ mm})} = 3.389\text{ kV}. \tag{6.91}$$

[28]Walter Koechner, *Solid State Laser Engineering*, 3d ed. (Berlin: Springer-Verlag, 1992), p. 455.

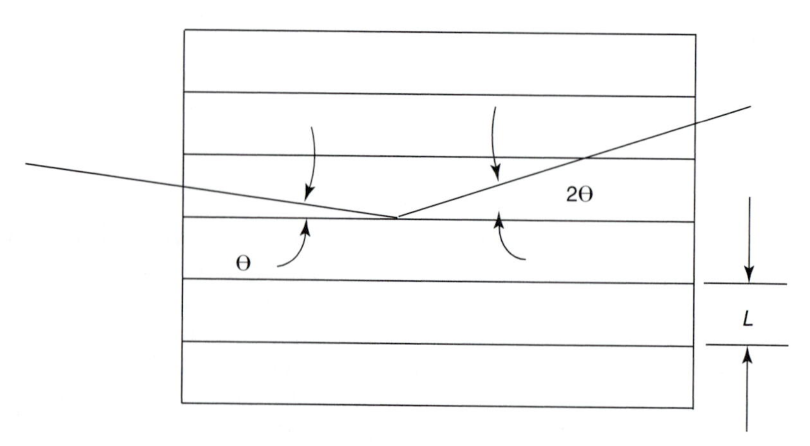

Figure 6.17 An acoustic standing wave can be created in a crystal by means of a transducer bonded to the crystal. The acoustic standing wave will generate a corresponding standing wave in the index of refraction. This index of refraction standing wave will interact with an incident beam like an optical grating.

6.3.3 Acousto-Optic Q-switches

Acousto-optic Q-switches employ materials (such as quartz) that exhibit a change in the index of refraction when the material is acoustically excited (*photoelastic effect*). The idea is to create an acoustic standing wave in the crystal by means of a transducer bonded to the crystal (see Figure 6.17). The acoustic standing wave will generate a corresponding standing wave in the index of refraction. This index of refraction standing wave will behave like an optical grating. With the appropriate choice of parameters, the optical wave can scatter from the induced index of refraction grating and be ejected from the cavity (see Figure 6.18).

Acousto-optic Q-switches are often operated in the Bragg scattering regime. In this regime, the interaction path is large and only the zeroth and first-order diffracted beams need to be considered. In the Bragg regime of operation, the acoustic grating is oriented at an angle of Θ with respect to the incoming light ray. The angle Θ is typically defined inside the acousto-optic modulator, and is given as[29]

$$\Theta = \sin^{-1}\left(\frac{\lambda_o}{2n\Lambda}\right) \tag{6.92}$$

where n is the index of refraction and Λ is the acoustical wavelength.

The scattering angle is 2Θ and the intensity of the scattered beam is given as[30]

$$\frac{I}{I_o} = \sin^2\left[\frac{\pi}{2}\left(\frac{2}{\lambda_o^2}\frac{l}{w}M_2 P_{ac}\right)^{1/2}\right] \tag{6.93}$$

where P_{ac} is the acoustical power, l and w are the length and width of the transducer, respectively, and M_2 is the acousto-optic figure of merit. (A typical value of M_2 for a

[29] Walter Koechner, *Solid State Laser Engineering*, 3d ed. (Berlin: Springer-Verlag, 1992), pp. 452, 468.

[30] Walter Koechner, *Solid State Laser Engineering*, 3d ed. (Berlin: Springer-Verlag, 1992), pp. 452, 468.

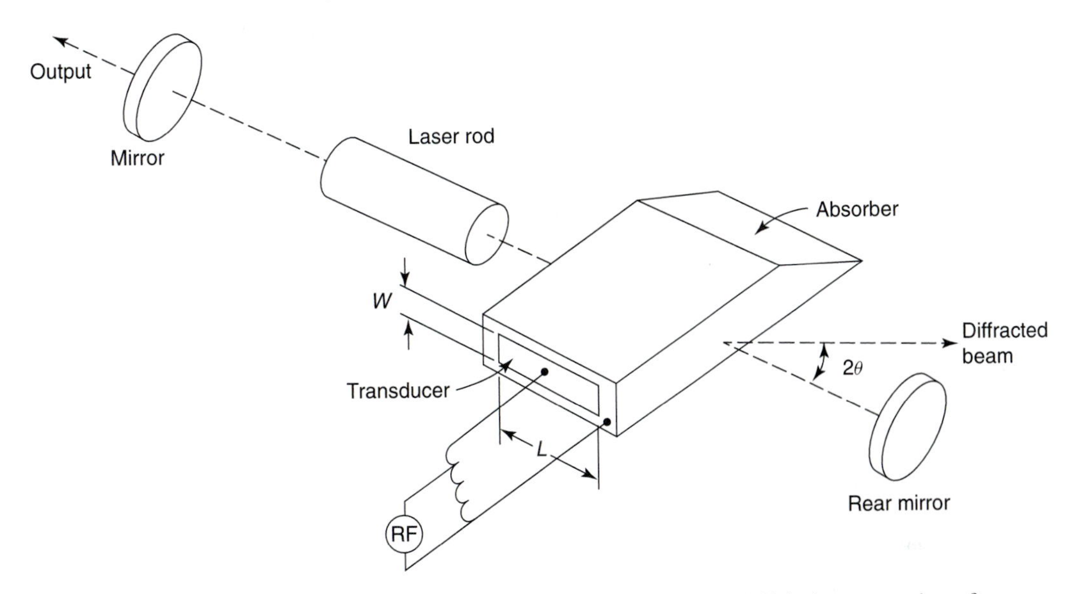

Figure 6.18 The acoustical standing wave can be used as a Q-switch if the laser system is configured so that the optical wave can scatter from the induced index of refraction grating and be ejected from the cavity. (From W. Koechner, *Solid State Laser Engineering*, 4th ed. (Berlin: Springer-Verlag, 1996), p. 484, Fig. 8.24, ©Springer-Verlag 1996. Reproduced with the permission of the author.)

longitudinal acoustic wave with the optical beam polarized parallel with the acoustic wave vector is $0.30 \cdot 10^{-18}$ sec^3/gram.)

Acousto-optical Q-switches have the advantage of being low-loss elements when not Q-switching. In contrast with the electrooptic crystals, acousto-optic crystals have high damage thresholds and are typically not hygroscopic. However, the extinction ratio is low for acousto-optical Q-switches (typically 20 to 30%) and the Q-switches generally require high-power RF power supplies at 20 to 60 W, 50 to 150 MHz. This combination of factors means that acousto-optic Q-switches are typically used in low-gain cw lasers. Their most common application is in continuous wave Nd:YAG Q-switched and mode-locked laser systems.

6.3.4 Saturable Absorber Dyes for Q-Switching

A *saturable absorber dye* can be used as a simple and economical Q-switch.[31] Such a dye exhibits a nonlinear absorption characteristic with respect to intensity. For low intensities, the dye is moderately absorbing and serves to reduce the cavity transmission. However when the intensity exceeds a certain critical intensity I_{dye}, the dye begins to bleach and the cavity becomes more transparent. Mathematically, the dye absorption is expressed as

$$\alpha_{\text{dye}} = \frac{\alpha_{\text{odye}}}{1 - I/I_{\text{dye}}}$$

[31] H. W. Mocker and R. J. Collins, *Appl. Phys. Lett.* 7:270 (1965).

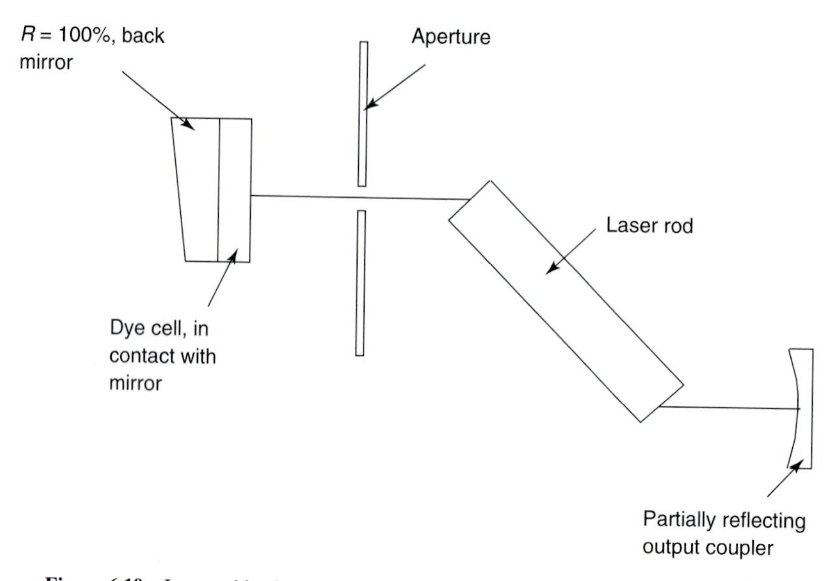

Figure 6.19 In saturable absorber dye Q-switches, the dye cell can be incorporated into the rear mirror to minimize reflective and scattering losses.

and the dye response time is given by

$$t_r = \frac{\tau_{\text{dye}}}{1 + I/I_{\text{dye}}}$$

where I is the intensity in the laser cavity, I_{dye} is the saturation threshold of the dye, $\alpha_{o\text{dye}}$ is the low-intensity absorption and τ_{dye} is the effective lifetime in the excited state. The saturation intensity I of a typical dye is given by

$$I_{\text{dye}} = \frac{h\nu}{\sigma_{21}\tau_{\text{dye}}}$$

where σ_{21} is the dye cross-section. Typical values are on the order of $\sigma_{21} = 5 \cdot 10^{-16}$ cm^2.

For a saturable absorber dye to be useful as a Q-switch, it needs to have a large cross-section (certainly larger than the laser cross-section) and a moderate effective lifetime. If the lifetime is too short, the laser may mode-lock (see Section 6.4 on dye mode-locking) rather than Q-switch. Since continuous wave lasers will typically not have enough intensity to effectively bleach the dye, the operation is usually limited to pulsed lasers.

In a saturable absorber dye Q-switch, the dye is usually placed in a dye cell immediately in front of the rear mirror (see Figure 6.19). In some systems, the dye cell is actually incorporated into the rear mirror to minimize reflective and scattering losses. Typical dye cells are 1 mm to 1 cm in length and the dye can be static or circulated. An interesting variation is to impregnate a plastic insert with the dye and place the insert in front of either the front or rear mirror.

Dye Q-switched lasers have the advantages of economy, simplicity, and emission of pulses with a relatively narrow linewidth. However, they suffer from timing jitter and

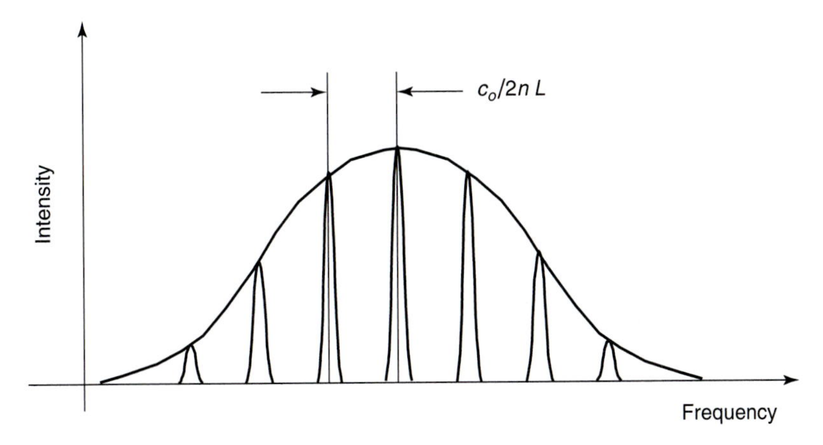

Figure 6.20 A laser cavity supports a comb of longitudinal modes separated by the frequency $c_o/2n\,L$.

absorption losses in the dye. Organic dyes are also very sensitive to UV light—both from the laser and from the ambient environment (such as room fluorescent lighting). In addition, most organic dyes have a refractive index variation with intensity, which creates a *self-focusing effect* where the spatial center of the pulse will see a higher index of refraction than the outer edges (see Section 7.5).

6.4 MODE-LOCKING

6.4.1 A Qualitative Description of Mode-Locking

Mode-locking is another technique to obtain temporal control over the laser beam. Mode-locking is achieved by controlling the relative phases and magnitudes of the various longitudinal modes in the laser. Thus, mode-locking techniques are most effective in lasers with numerous longitudinal modes running under the gain curve.

Recall that a laser cavity supports a comb of longitudinal modes separated by the frequency $c_o/2nL$ (see Figure 6.20). In a typical laser, each of these modes oscillates independently of all the others. The phases of the modes are randomly distributed between $-\pi$ and $+\pi$. The magnitudes of the modes do not follow a perfect Gaussian profile, but fluctuate in a Rayleigh distribution around some mean. The consequence of this randomness in the phase and magnitude characteristics of the longitudinal modes is that the laser output fluctuates randomly in time (see Figure 6.21).

Mode-locking is the process of forcing these oscillating longitudinal modes to maintain fixed phase relationships with each other. Depending on the nature of the phase relationships, mode-locking can produce streams of short pulses, FM modulated output, a spatially scanned laser beam, or a Gatling gun output (where single pulses are generated at different spatial locations on the target).

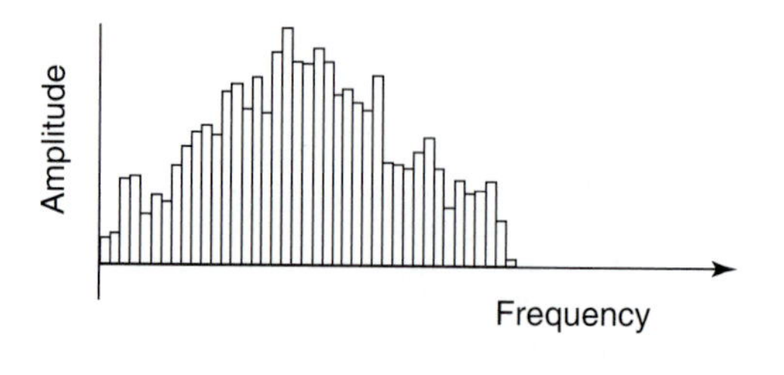

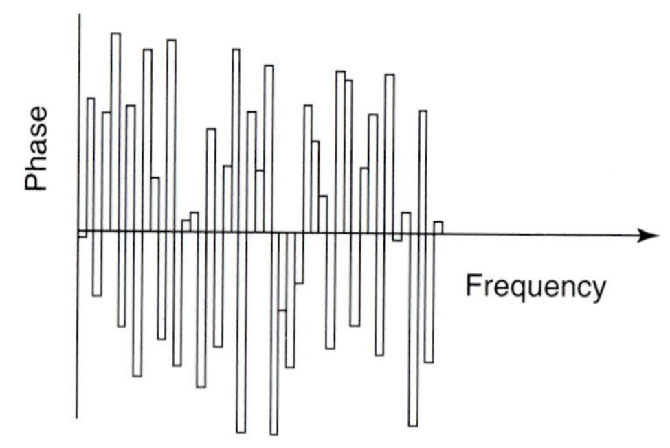

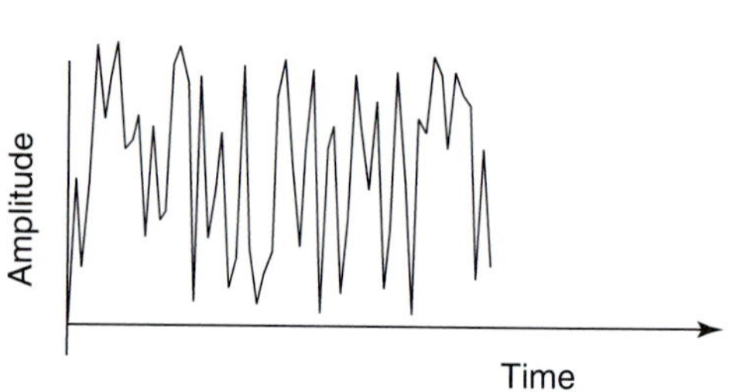

Figure 6.21 In a typical laser, each of the longitudinal modes oscillates independently of all the others. The phases of the modes are randomly distributed between $-\pi$ and $+\pi$. The magnitudes of the modes do not follow the perfect Gaussian profile. The consequence of this randomness in the phase and magnitude characteristics of the longitudinal modes is that the laser output fluctuates randomly in time.

Mode-locking was first observed experimentally in 1964 by Hargrove[32] in a HeNe laser mode-locked with an acousto-optic modulator. The FM mode-locking of a HeNe laser was demonstrated in 1964 by Harris and Targ,[33] using a HeNe laser with an internal phase modulator. In 1965 Corwell[34] reported HeNe laser experiments that demonstrated self-mode-locking due to nonlinearities in the HeNe gas. The argon ion laser was first mode-locked in 1965 by DeMaria and Stetser[35] and the ruby laser in 1965 by Deutsch.[36] Also in 1965, Mocker and Collins[37] first demonstrated both Q-switching and mode-locking using a saturable absorber. In 1966 DeMaria et al.[38] reported mode-locking Nd:glass with a loss modulator and soon afterwards with a saturable dye.[39] In 1966 Nd:YAG was mode-

---

[32] L. E. Hargrove, R. L. Fork, M. A. Pollack, *Appl. Phys. Lett.* 5:4 (1964).

[33] S. E. Harris, and R. Targ, *Appl. Phys. Lett.* 5:202 (1964).

[34] M. H. Corwell, *IEEE J. of Quan. Electron.* QE-1:12 (1965).

[35] A. J. DeMaria, and D. A. Stetser, *Appl. Phys. Lett.* 7:71 (1965).

[36] T. Deutsch, *Appl. Phys. Lett.* 7:80 (1965).

[37] H. W. Mocker and R. J. Collins, *Appl. Phys. Lett.* 7:270 (1965).

[38] A. J. DeMaria, C. M. Ferrar, and G. E. Danielson, *Appl. Phys. Lett.* 8:22 (1966).

[39] A. J. DeMaria, D. A. Stetser, and H. Heynau, *Appl. Phys. Lett.* 8:174 (1966).

locked by DiDomenico et al.[40] using an internal modulator. In 1968, mode-locking was reported in CO_2,[41,42] dye,[43,44] and semiconductor lasers.[45]

6.4.2 Analytical Description of Mode-Locking

At a given point inside the laser, the electromagnetic field can be expressed as the sum of the fields of all the various longitudinal modes. Thus, the field can be written as

$$E(t) = E_1 e^{j(\omega_o + 1\cdot\omega_{\text{FSR}})t+\theta_1} + E_2 e^{j(\omega_o+2\cdot\omega_{\text{FSR}})t+\theta_2} + \cdots \tag{6.94}$$

$$E(t) = \sum_p E_p e^{j[2\pi(\nu_o + p\cdot\nu_{\text{FSR}})t+\theta_p]} \tag{6.95}$$

where p is the mode number, ν_o is the fundamental laser frequency, and ν_{FSR} is the frequency spacing between longitudinal modes. The phase angle θ_p is the phase difference between modes, referenced to an arbitrary mode (such as the center mode) set at $\theta_p = 0$.

The sum of the various sinusoidal waves with different phases and different frequencies creates a complex beat pattern. This beat pattern repeats itself every $T = 2Ln/c_o$ seconds. Notice that T is the inverse of the frequency spacing of the modes $\nu_{\text{FSR}} = c_o/2Ln$, and represents a round-trip time for the cavity.

If all of the phase differences are set to a constant, then the random-looking temporal pattern changes to one with a well-defined pulse every $T = 2Ln/c_o$ seconds. The more longitudinal modes available in the summation, the narrower and more well-defined the pulse. The physical picture is that of a single small bundle of light traveling back and forth between the cavity mirrors and creating a stream of pulses at the output separated by the round-trip time of the cavity.

The most interesting aspect of the mode-locked pulse is its width. To estimate the width of the pulse, assume that all of the pulses have the same magnitude E_o, and set the relative phase differences to $\theta_p = 0$. Substituting these assumptions into Equation (6.95) yields

$$E(t) = \sum_p E_o e^{j[2\pi(\nu_o+p\cdot\nu_{\text{FSR}})t+\theta_p]} = E_o e^{j2\pi\nu_o t} \sum_p e^{j2\pi(k\cdot\nu_{\text{FSR}})t} \tag{6.96}$$

$$E(t) = E_o e^{j2\pi\nu_o t}\left(e^{j2\pi(1\cdot\nu_{\text{FSR}})t} + e^{j2\pi(2\cdot\nu_{\text{FSR}})t} + e^{j2\pi(3\cdot\nu_{\text{FSR}})t} + \cdots\right). \tag{6.97}$$

The series $\sum_p e^{p\cdot j2\pi\nu_{\text{FSR}}t}$ is a series with a known solution for a finite sum. Substituting the known solution in for $\sum_p e^{p\cdot j2\pi\nu_{\text{FSR}}t}$ yields

$$E(t) = E_o e^{j2\pi\nu_o t}\sum_p e^{j2\pi(p\cdot\nu_{\text{FSR}})t} = E_o e^{j2\pi\nu_o t}\left(\frac{\sin(M\cdot 2\pi\nu_{\text{FSR}}t/2)}{\sin(2\pi\nu_{\text{FSR}}t/2)}\right) \tag{6.98}$$

---

[40] M. DiDomenico, H. M. Marcos, J. E. Geusic, and R. E. Smith, *Appl. Phys. Lett.* 8:180 (1966).

[41] D. E. Caddes, L. M. Osterink, and R. Targ, *Appl. Phys. Lett.* 12:74 (1968).

[42] O. R. Wood and S. E. Schwarz, *Appl. Phys. Lett.* 12:263 (1968).

[43] W. Glenn, M. J. Brienza, and A. J. DeMaria, *Appl. Phys. Lett.* 12:54 (1968).

[44] W. Schmidt and F. P. Schafer, *Phys. Lett.* 26A:558 (1968).

[45] V. N. Morozov, V. V. Nikitin, and A. A. Sheronov, *JETP Lett.* 7:256 (1968).

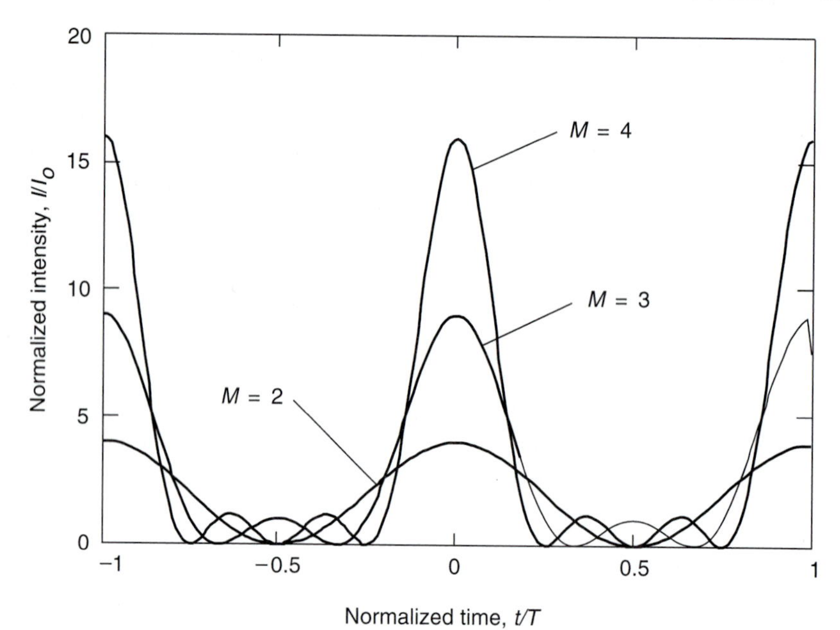

Figure 6.22 The normalized intensity of a mode-locked laser pulse plotted as a function of the normalized time t/T for the case of $M = 2$, 3, and 4 modes.

where M is the total number of modes.

The intensity of the laser is then given as[46]

$$I(t) = \frac{E(t) \cdot E(t)^*}{2\eta_o} = \frac{E_o^2}{2\eta_o}\left(\frac{\sin(M \cdot 2\pi\nu_{\text{FSR}}t/2)}{\sin(2\pi\nu_{\text{FSR}}t/2)}\right)^2 \tag{6.99}$$

where η_o is the impedance of free space given as $\eta_o = \sqrt{\mu_o/\epsilon_o} \approx 377$ ohm.

Example 6.9

Plot Equation (6.99) as a normalized function of t/T for the case of $M = 2$, 3, and 4 modes.

Solution. Equation (6.99) can be expressed as a normalized function of t/T,

$$\frac{I(t)}{I_o(t)} = \left(\frac{\sin(M\pi \cdot t/T)}{\sin(\pi \cdot t/T)}\right)^2. \tag{6.100}$$

This Equation is illustrated in Figure 6.22.

Equation (6.99) has some interesting properties. In particular, notice that the width (1/2 of the time from zero to zero) of the peak is approximately $\Delta t_p = T/M$. Recall that the number of modes M is roughly equal to the width of the gain curve $\Delta\nu$ divided by the frequency spacing of the modes $c_o/2nL$. Thus, an approximation to the pulse width Δt_p can be obtained as

$$\Delta t_p = \frac{1}{M} \cdot T = \frac{c_o}{\Delta\nu \cdot 2nL} \cdot \frac{2nL}{c_o} = \frac{1}{\Delta\nu}. \tag{6.101}$$

[46]Joseph T. Verdeyen, *Laser Electronics*, 2d ed. (Englewood Cliffs, NJ: Prentice Hall, 1989), p. 274.

Notice that this expression closely resembles the Heisenberg Uncertainty Principle.[47] Essentially, the product of the pulse length and frequency bandwidth is a constant.

A more elaborate approach to the analytical description of mode-locking incorporates both the Gaussian character of the pulse and a possible frequency chirp on the pulse. Assume that the electric field of the mode-locked pulse has a temporal profile given by

$$E(t) = \left(\frac{E_o}{2}\right) e^{-\kappa t^2} e^{j(\omega_o t + \beta t^2)} \tag{6.102}$$

where κ determines the Gaussian envelope of the pulse and β models the frequency shift (the chirp) during the pulse. The full-width half-maximum (FWHM) pulse width of this chirped pulse is given as

$$\Delta t_p = \left(\frac{2\ln 2}{\kappa}\right)^{1/2} \tag{6.103}$$

and the FWHM bandwidth is given by

$$\Delta \nu_p = \frac{1}{\pi}\left[2\ln 2\left(\frac{\kappa^2 + \beta^2}{\kappa}\right)\right]^{1/2}. \tag{6.104}$$

The time-bandwidth product for the pulse is especially interesting, and is given as

$$\Delta t_p \Delta \nu_p = \left(\frac{2\ln 2}{\pi}\right)\left[1 + \frac{\beta^2}{\kappa^2}\right]^{1/2}. \tag{6.105}$$

Notice that for $\beta = 0$ (no frequency chirp), the time-bandwidth product is

$$\Delta t_p \Delta \nu_p = \left(\frac{2\ln 2}{\pi}\right) = 0.441 \tag{6.106}$$

which is often termed a band-limited or Fourier-transform-limited pulse.

Example 6.10

Consider HeNe, Nd:YAG, and semiconductor diode lasers. Assume that the HeNe has a FWHM bandwidth of $\Delta\nu_p = 1500$ MHz; the Nd:YAG has $\Delta\nu_p = 300$ GHz, and the semiconductor diode laser has $\Delta\nu_p = 4$ THz. Calculate the Fourier-transform-limited pulse width assuming an unchirped pulse. Also calculate the physical length of the pulse.

Solution. Computing the Fourier-transform-limited pulse width requires that Equation (6.106) be solved for the various cases.

$$\Delta t_p = \frac{0.441}{\Delta \nu_p} = \frac{0.441}{1500\ \text{MHz}} = 294\ \text{ps} \tag{6.107}$$

$$\Delta t_p = \frac{0.441}{\Delta \nu_p} = \frac{0.441}{300\ \text{GHz}} = 1.47\ \text{ps} \tag{6.108}$$

$$\Delta t_p = \frac{0.441}{\Delta \nu_p} = \frac{0.441}{4\ \text{THz}} = 110.25\ \text{fs}. \tag{6.109}$$

---

[47] The Heisenberg Uncertainty Principle is discussed in most treatments of quantum mechanics. For an interesting view of the principle from its developer, consider W. Heisenberg, *The Physical Principles of the Quantum Theory* (Dover Publications, 1930). For an excellent overview and placement of the principle in the body of quantum theory, consider J. Gribbin, *Schrödinger's Kittens and the Search for Reality* (Boston, MA: Little Brown and Company, 1995).

Computing the physical length of the pulse can be accomplished by multiplying the Fourier-transform-limited pulse width by the speed of light c_o,

$$l_p = \frac{c_o \cdot 0.441}{1500 \text{ MHz}} = 8.814 \text{ cm} \tag{6.110}$$

$$l_p = \frac{c_o \cdot 0.441}{300 \text{ GHz}} = 0.441 \text{ mm} \tag{6.111}$$

$$l_p = \frac{c_o \cdot 0.441}{4 \text{ THz}} = 33.052 \ \mu\text{m}. \tag{6.112}$$

6.4.3 The Design of Mode-Locking Modulators

Mode-locking is used to create streams of short pulses by setting all of the longitudinal modes to a constant phase difference. When the problem is stated in this way, it sounds like a very hard thing to do! However, notice that if the longitudinal modes are all in phase, then the output from the laser is a series of short pulses separated by the round-trip cavity time $2nL/c_o$. Thus, the problem can be stated in reverse. In order to mode-lock a laser, it is necessary to create periods of low loss in the cavity that are synchronized with the $2nL/c_o$ round-trip time of the cavity. If the cavity can produce pulses separated by $2nL/c_o$, then the longitudinal modes are running in phase.

Although the final goal of mode-locking a laser is somewhat different from that of Q-switching; the technologies employed are very similar. Active mode-locking techniques include both electrooptic and acousto-optic modulators. Passive mode-locking is often accomplished with saturable absorber dyes.

Active mode locking. Electrooptic and acousto-optic elements similar to those described for Q-switching (Section 6.2) can also be used for mode-locking. The basic idea is to place an element in the cavity whose loss is externally modulated. Because the mode-locking frequency is set by an external tuned source, this type of mode-locking is typically termed *active*.

The active mode-locking element can be constructed in an AM configuration where the amplitude of the loss is synchronized to the $2nL/c_o$ round-trip time of the cavity (see Figure 6.23). For an AM modulator, the single-pass transmission function is given by[48]

$$\Upsilon(t) = \cos(\delta_{\text{AM}} \sin(2\pi f_m t)). \tag{6.113}$$

If the modulator driving frequency f_m is set at half the axial mode spacing ($f_m = c_o/4nL$), then the transmission function $\Upsilon(t)$ has a maximum every round-trip time of the laser $2nL/c_o$. Thus, AM modulators are driven at one-half the frequency of the axial mode spacing. The laser pulse then occurs once each AM cycle, when the loss is minimized. Under these conditions of AM modulation, the pulse width is given by[49]

$$\Delta t_{p(\text{AM})} = \frac{(\gamma L_g)^{1/4}}{2\sqrt{\delta_{\text{AM}}}} \left(\frac{1}{f_m \Delta\nu} \right)^{1/2} \tag{6.114}$$

---

[48] Walter Koechner, *Solid State Laser Engineering*, 3d ed. (Berlin: Springer-Verlag, 1992), pp. 452, 498.

[49] A. E. Siegman and D. J. Kuizenga, *Opto-Electronics* 6:43 (1974).

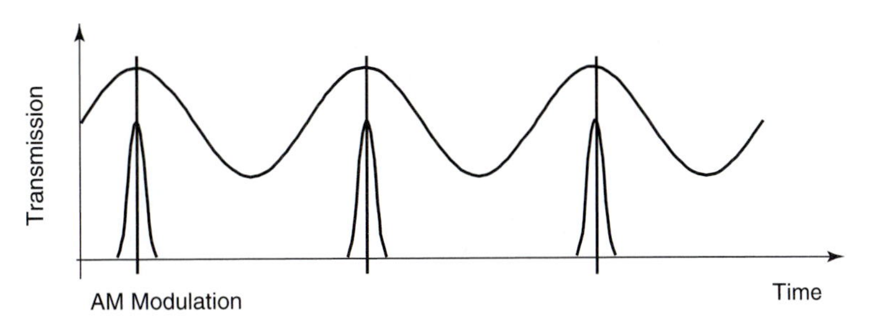

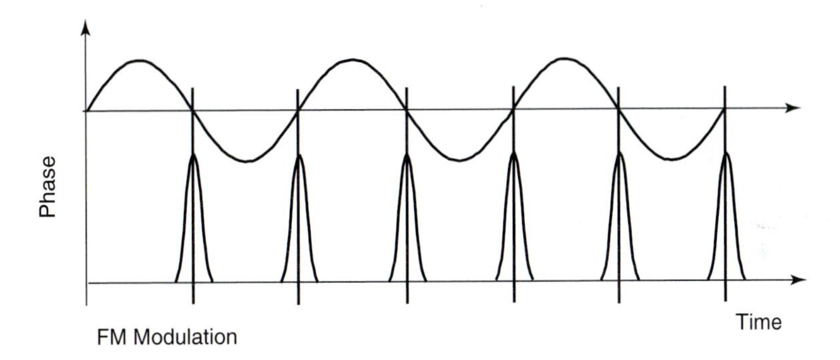

Figure 6.23 An active mode-locking element can be constructed in an AM configuration where the amplitude of the loss is synchronized to the $2nL/c_o$ round-trip time of the cavity. In the AM case, the laser pulse then occurs once each modulation cycle when the transmission is maximized. Alternatively, the mode-locking element can be constructed in an FM configuration where the modulator varies the frequency of the signal. In the FM case, the laser pulse then occurs twice each modulation cycle when the frequency shift is zero.

where γ is the saturated gain coefficient, L_g is the length of the gain material, δ_{AM} is the modulation index, $f_m = c_o/4nL$, and $\Delta\nu$ is the frequency width of the gain curve. (Notice that AM modulation does not produce a chirp on the pulse.)

The mode-locking element can also be constructed in an FM configuration where the modulator varies the frequency of the signal. The loss results from the frequency walking-off the $c_o/2nL$ stable frequencies of the cavity. The laser pulse then occurs twice each modulation cycle at the place where the frequency shift is zero (see Figure 6.23).

For an FM modulator, the single-pass transmission function is given by[50]

$$\Upsilon(t) = \exp(\pm j\,\delta_{\text{FM}}(2\pi f_m)^2 t^2) \tag{6.115}$$

where δ_{FM} is the peak phase retardation through the modulator and $f_m = c_o/2nL$. Under these conditions of FM modulation, the pulse width is given by[51]

$$\Delta t_{p(\text{FM})} = \frac{(\gamma L_g)^{1/4}}{2\,(\delta_{\text{FM}})^{1/4}}\left(\frac{1}{f_m\,\Delta\nu}\right)^{1/2}. \tag{6.116}$$

[50] Walter Koechner, *Solid State Laser Engineering*, 3d ed. (Berlin: Springer-Verlag, 1992), pp. 452, 499.

[51] A. E. Siegman and D. J. Kuizenga, *Appl. Phys. Lett.* 14:171 (1969).

The FM modulated pulse will be chirped with a frequency shift β given as[52]

$$\beta = \pi^2 \left(\frac{\delta_{\text{FM}}}{4\gamma L_g} \right)^{1/2} f_m \Delta\nu. \tag{6.117}$$

Lithium niobate is frequently used as both an AM and FM modulator material. For the AM case, the direction of light propagation is along the z (optic) axis, and the modulating field is applied along the x direction. The light is polarized along the x-axis. The effective value of the peak phase retardation δ_{AM} is given as[53]

$$\delta_{\text{AM}} = \sin^2 \left(\frac{\pi r_{22} n_o^3 V_o a}{\lambda_o d} \right) \tag{6.118}$$

where a is the length of the crystal in the z direction, r_{22} is one of coefficients of the electrooptic tensor ($5.61 \cdot 10^{-6}$ μm/V for lithium niobate), n_o is the ordinary index (2.24 at 1.064 μm for lithium niobate), V_o is the amplitude of the voltage, and d is the length of the crystal in the x direction.

For the FM case, the direction of light propagation is in the x direction and the modulating field is applied along the z direction. The light is polarized along the z-axis. The effective value of the peak phase retardation δ_{FM} is given as[54]

$$\delta_{\text{FM}} = \left[\cos\left(\frac{\pi Z_o}{L} \right) \text{sinc}\left(\frac{\pi a}{2L} \right) \right] \left(\frac{\pi r_{33} n_e^3 V_o a}{\lambda_o d} \right) \tag{6.119}$$

where Z_o is the distance of the modulator from the end mirror, L is the optical length of the laser cavity, a is the length of the crystal in the x direction, r_{33} is one of coefficients of the electrooptic tensor ($30.8 \cdot 10^{-6}$ μm/V for lithium niobate), n_e is the extraordinary index (2.16 at 1.064 μm for lithium niobate), V_o is the amplitude of the voltage, and d is the length of the crystal in the z direction.

Saturable absorber dyes for mode locking. Saturable absorber dyes with short excitation lifetimes can be used as mode-locking dyes. This technique is often called *passive* mode-locking. It is necessary for the dye to have: 1) a saturable absorption at the laser line, 2) a linewidth greater than the gain curve of the laser, and 3) a recovery time shorter than the round-trip time constant of the cavity. Ideally, the intensity at which the dye completely saturates should be near the maximum intensity of the output pulses.

An elegant computer simulation originally devised by Fleck[55] demonstrates the evolution of the mode-locked pulse starting from optical noise in a saturable absorber mode-locked system (see Figure 6.24). Initially, as the laser begins to lase, the output intensity in time consists of a random series of spikes. Now some of the spikes will be bigger than others, and they will begin to saturate the dye. Thus, the system loss on these spikes will be less than on the others. Therefore, the intense spikes will begin to grow and the less intense spikes will start to shrink. Now, of the intense spikes that are growing, some of

[52]Walter Koechner, *Solid State Laser Engineering*, 3d ed. (Berlin: Springer-Verlag, 1992), pp. 452, 500.

[53]Walter Koechner, *Solid State Laser Engineering*, 3d ed. (Berlin: Springer-Verlag, 1992), pp. 452, 502.

[54]Walter Koechner, *Solid State Laser Engineering*, 3d ed. (Berlin: Springer-Verlag, 1992), pp. 452, 501.

[55]J. A. Fleck, *Phys. Rev. B* 1:84 (1970).

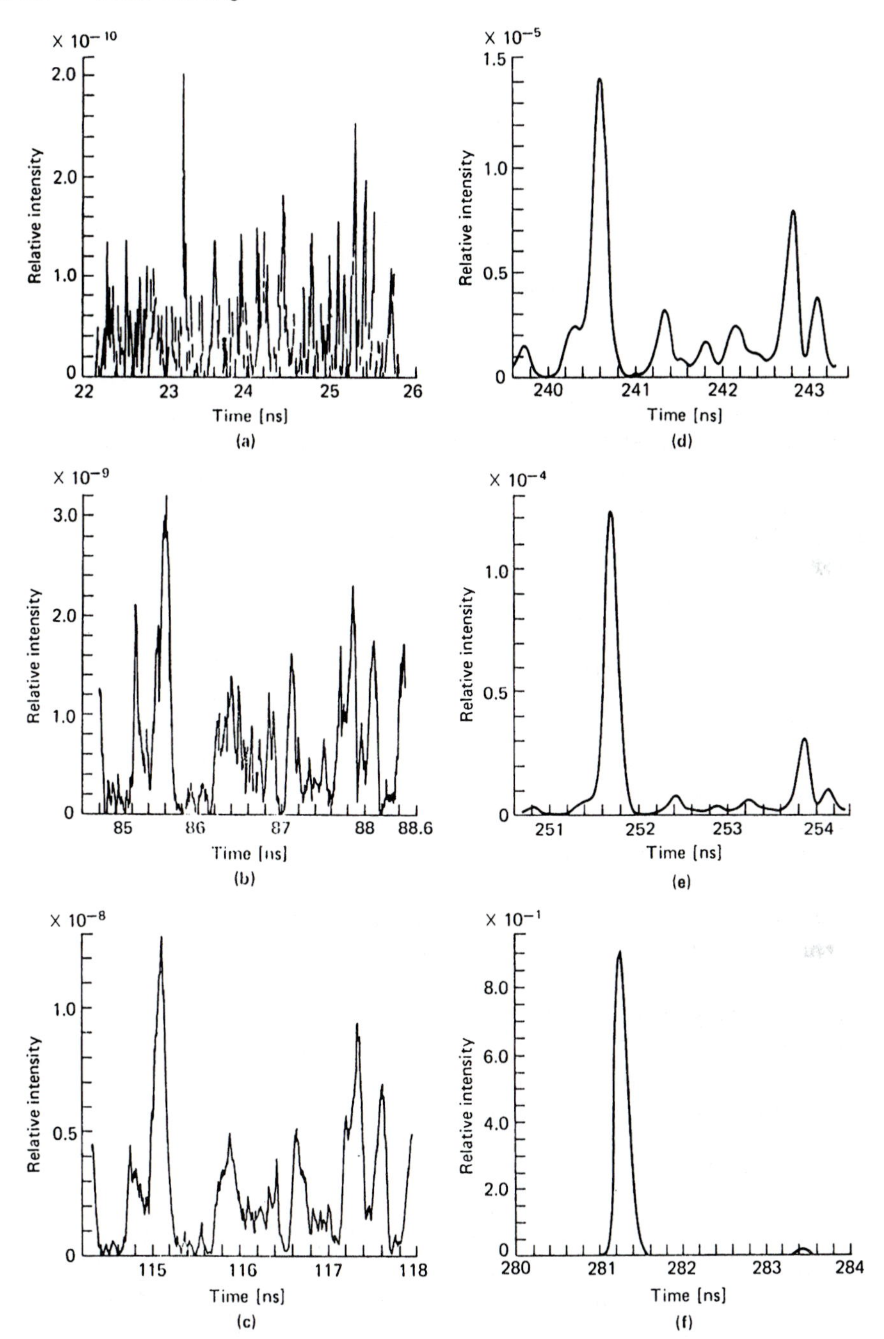

Figure 6.24 This elegant computer simulation demonstrates the evolution of the mode-locked pulse from noise in a saturable absorber mode-locked system. (From J. A. Fleck, *Phys. Rev. B* 1:84 (1970), Figures 3A, 3C, 3D, 3F, 3G, 3H. Reprinted with the permission of the author.)

them will be in phase with each other, and these will begin to grow much faster than the out-of-phase spikes. Relatively rapidly, the output will organize itself into a series of very intense phase-locked spikes separated by $2nL/c_o$ in time.

SYMBOLS USED IN THE CHAPTER

E_i, E_j: State energies (J or eV)
N_3, N_2, N_1: State population densities (atoms/cm^3)
g_3, g_2, g_1: State degeneracies (the number of states at the same energy) (unitless)
ν: Frequency (Hz)
τ_{12}, τ_{13}, τ_{23}: Thermal transition time (sec)
τ_1, τ_{10}, τ_2, τ_{20}, τ_{21}, τ_{31}, τ_{32}: Spontaneous transition time (relaxation lifetime) (sec)
R_{trans}, $R_{\text{trans},12}$, $R_{\text{trans},21}$: Stimulated transition probabilities due to laser action (1/sec)
P_1, P_2: External pumping density to states (four-state approach) (cm^{-3} sec^{-1})
σ_{ij}: Cross-section between states i and j (cm^2 or m^2)
I_ν: Intensity in the cavity due to laser action (watts/cm^2)
L: Laser resonator length (cm or m)
n: Real index of refraction (unitless)
R_F, R_B: Front and back mirror reflectances, between 0 and 1 (unitless)
T_F, T_B: Front and back mirror transmittances, between 0 and 1 (unitless)
γ_o: Unsaturated gain coefficient (cm^{-1} or m^{-1})
I_{sat}: Saturation intensity (watts/cm^2)
α: Loss coefficient (cm^{-1} or m^{-1})
ϕ: Photon density (atoms/cm^3)
τ_{cav}: Cavity relaxation time (sec)
F_l: Loss per round trip (unitless)
L_c: Length of the resonator minus the length of the gain material (cm or m)
L_g: Length of the gain material (cm or m)
n_c: Index of refraction of the resonator without the gain material (unitless)
n_g: Index of refraction of the gain material (unitless)
norm: Unit for simplifying numerical modeling (10^{15} atoms/cm^3)
A: Normalized variable representing upper state pumping (norm ns^{-1})
B: Normalized variable representing the inverse of the upper state lifetime (ns^{-1})
C: Normalized variable representing the stimulated process (norm^{-1} ns^{-1})
D: Normalized variable representing the inverse of the cavity lifetime (ns^{-1})
ΔN_2, $\Delta\phi$: Perturbed upper state and photon density populations (atoms/cm^3)
N_{2o}, ϕ_o: Constant or threshold upper state and photon densities (atoms/cm^3)
$P_{2\text{thres}}$: Threshold pumping density (norm ns^{-1})

r: Times-over-threshold (unitless)
t_d: Decay time constant (sec)
ω_R: Relaxation oscillation frequency (rad/sec)
f_R: Relaxation oscillation frequency (Hz)
ϕ_i, ϕ_f, ϕ_{peak}: Initial, final, and peak photon densities (atoms/cm^3)
N_{2i}, N_{2f}: Initial and final upper state population densities (atoms/cm^3)
E_Q: Energy of the Q-switched pulse (joules)
P_{peak}: Peak power density of the Q-switched pulse (watts/cm^2)
Δt_p: Temporal width of the Q-switched or mode-locked pulse (sec)
λ_o: Free space wavelength of light (m)
n_o, n_e: Ordinary and extraordinary indices of refraction (unitless)
r_{63}, r_{22}, r_{33}: Important elements of various electrooptic tensors (m/volt or μm/volts)
$V_{1/2}$: Voltage to induce a half-wave polarization change in a crystal (volts)
d, l, w, a: Various dimensions of modulator crystals (cm or m)
Θ: Internal orientation angle for an acousto-optic modulator (deg or rad)
Λ: Wavelength of an acoustic wave in an acousto-optic modulator (cm or m)
P_{ac}: Acoustic power (watts)
M_2: Acousto-optic figure of merit (sec^3/kg or sec^3/g)
I_{dye}: Dye saturation intensity (watts/cm^2)
α_{dye}, $\alpha_{o\text{dye}}$: Saturated and unsaturated dye absorptions (cm^{-1})
t_r: Dye response time (sec)
$E(t)$, E_o: Electric field as a function of time, and the electric field at $t = 0$ (V/cm)
ν_o: Center frequency of a transition (Hz)
ω_o: Center frequency of a transition (rad/sec)
ν_{FSR}: Frequency spacing between longitudinal modes (Hz)
ω_{FSR}: Frequency spacing between longitudinal modes (rad/sec)
θ_p: Phase different between modes (deg or rad)
$T = c_o/2nL$: Round-trip time of the laser cavity (sec)
M: Number of modes (unitless integer)
η_o: Impedance of free space (ohm)
$I(t)$, $I_o(t)$: Various light intensities (watts/cm^2)
Δt_p: Frequency width of the gain curve (Hz)
$\Delta \nu$: Frequency width of the gain curve (Hz)
κ: Gaussian pulse envelope (sec^{-2})
β: Frequency chirp (sec^{-2})
$\Delta \nu_p$: Bandwidth of the mode-locked pulse (Hz)
l_p: Spatial pulse length (cm or m)
$\Upsilon(t)$: Pulse transmission function (unitless)

f_m: Modulator frequency (Hz)

δ_{AM}, δ_{FM}: AM or FM modulation index (unitless)

γ: Saturated gain coefficient (cm^{-1} or m^{-1})

L_g: Length of the gain material (cm^{-1} or m^{-1})

V_o: Modulator signal amplitude (volts)

a: Crystal length in the z direction (cm or m)

r_{22}, r_{33}: Electro-optic tensor components (μm/V)

Z_o: Distance of the modulator from the end mirror (cm or m)

EXERCISES

Relaxation oscillations

6.1 Consider the simultaneous set of nonlinear equations,

$$x'[t] = 20 - 0.02x[t] - 3x[t]y[t] \tag{6.120}$$

$$y'[t] = 0 - 20y[t] + 3x[t]y[t] \tag{6.121}$$

Use Mathematica (or a similar robust numerical differential-equation solver) to solve these equations over the range of $t = 0$ to $t = 20$. Plot x as a function of t, y as a function of t, and create a parametric plot of x versus y.

6.2 Consider the simultaneous set of nonlinear equations,

$$x'[t] = 20 - 0.02x[t] - 3x[t]y[t] \tag{6.122}$$

$$y'[t] = 0 - 20y[t] + 3x[t]y[t] \tag{6.123}$$

The 20 in the first equation can be considered to be a pumping term. The 20 in the second equation can be considered to be an extraction term. In this example, the pumping and extraction are equal. Use Mathematica (or a similar robust numerical differential-equation solver) to explore the behavior of the system when the pumping does not match the extraction.

6.3 Consider a Nd:glass laser with a cross-section of $8 \cdot 10^{-19}$ cm^2, an index of refraction of 1.5, a length of 35 cm, a front reflector of $R = 75\%$, and a high reflecting back mirror. Assume the upper state lifetime is 300 microseconds and that the system is pumped at six times over threshold. Assume that the loss per unit length is $\alpha = 0.0001$ cm^{-1}. Calculate the exponential decay rate and frequency of the relaxation oscillations. Plot the relaxation oscillations versus time from 0 to 200 microseconds.

6.4 Consider a Nd:YAG laser with a cross-section of $2 \cdot 10^{-19}$ cm^2, an index of refraction of 1.8, a length of 15 cm, a front reflector of $R = 90\%$, and a high reflecting back mirror. Assume the upper state lifetime is 230 microseconds and that the system is pumped at ten times over threshold. Assume that the loss per unit length is $\alpha = 0.005$ cm^{-1}. Use Mathematica (or a similar robust numerical differential-equation solver) to explore the relaxation oscillation behavior of this laser as a function of the *pumping*.

6.5 Consider a Nd:YAG laser with a cross-section of $2 \cdot 10^{-19}$ cm^2, an index of refraction of 1.8, a length of 15 cm, a front reflector of $R = 90\%$, and a high reflecting back mirror. Assume the upper state lifetime is 230 microseconds and that the system is pumped at ten times over threshold. Assume that the loss per unit length is $\alpha = 0.005$ cm^{-1}. (This is the same as Exercise

6.4.) Use your results from Exercise 6.4 to pick a convenient pumping value. Use Mathematica (or a similar robust numerical differential-equation solver) to explore the relaxation oscillation behavior of this laser as a function of the *initial conditions* at this pumping value.

Q-switching

6.6 Consider a Nd:YAG laser with a cross-section of $2 \cdot 10^{-19}$ cm^2, an index of refraction of 1.8, a length of 15 cm, a front reflector of $R = 90\%$, and a high reflecting back mirror. Assume the upper state lifetime is 230 microseconds and that the system is pumped at ten times over threshold. Assume that the loss per unit length is $\alpha = 0.005$ cm^{-1}. (This is the same as Exercises 6.4 and 6.5.) Use Mathematica (or a similar robust numerical differential-equation solver) to explore the Q-switching behavior of this laser as a function of the *initial conditions on the upper state population.*

6.7 Consider a Nd:YAG laser with a cross-section of $2 \cdot 10^{-19}$ cm^2, an index of refraction of 1.8, a length of 15 cm, a front reflector of $R = 90\%$, and a high reflecting back mirror. Assume the upper state lifetime is 230 microseconds and that the system is pumped at ten times over threshold. Assume that the loss per unit length is $\alpha = 0.005$ cm^{-1}. (This is the same as Exercises 6.4 and 6.5.) Use the results from Exercise 6.6 to pick an initial condition for N_2 that demonstrates good Q-switching. Then use Mathematica (or a similar robust numerical differential-equation solver) to explore the behavior of the system as a function of the *initial conditions on the photon density.*

6.8 Consider a Cr:LiSAF laser with a cross-section of $2 \cdot 10^{-20}$ cm^2, an index of refraction of 1.6, a length of 15 cm, a front reflector of $R = 90\%$, and a high reflecting back mirror. Assume the upper state lifetime is 67 microseconds and that the system is pumped at six times over threshold. Assume that the loss per unit length is $\alpha = 0.005$ cm^{-1}. Use Mathematica (or a similar robust numerical differential-equation solver) to explore the Q-switching behavior of this laser as a function of the *initial conditions on the upper state population.* Pick a particular value of initial conditions that gives nicely Q-switched pulses and then check the computer results analytically.

6.9 Calculate and compare the half-wave voltages for a KDP Q-switch crystal versus a KD*P Q-switch crystal. Assume a longitudinal configuration and an operating wavelength of 1.064 microns. Compare the results.

6.10 Calculate the half-wave voltage for a lithium niobate Q-switch assuming a transverse configuration, an aperture of 5 mm, and a length of 1.5 cm. Assume the operating wavelength is 1.064 microns.

Mode-locking

6.11 (design) Create a numerical set of longitudinal modes spaced at $c_o/2nL$ and possessing a Gaussian intensity profile. Set all the phases equal and use a computer to sum the modes, using Equation (5.13), to obtain $E(t)$. Repeat with a more random intensity profile. Repeat with a random intensity profile and with a random phase profile. Discuss your results. Your answer should include:

(a) three computer plots, and

(b) a concise (a few sentences) discussion of the results.

6.12 Consider a mode-locked diode laser with 110 phase-locked modes. Consider a mode-locked Nd:YAG laser with 11 phase-locked modes. Use a computer program to create a plot that displays the ideal normalized pulse width for both lasers as a function of time. Discuss the differences between the lasers and speculate on various mode-locking applications that might use one or the other.

6.13 Consider a Ti:sapphire laser, a Cr:LiSAF laser, a Nd:glass laser, and a Nd:YAG laser. Assume that the FWHM frequency linewidth of the lasers is Ti:sapphire = 150 THz, Cr:LiSAF = 90 THz, Nd:glass = 7.5 THz, and Nd:YAG = 300 GHz. Calculate the Fourier-transform-limited pulse width assuming an unchirped pulse. Also calculate the physical length of the pulse. Discuss the differences between the laser pulses.

6.14 For an AM actively mode-locked modulator, the modulator driving frequency is set at half the axial mode spacing ($f = c_o/4nL$). Develop a simple argument to demonstrate that this will result in a transmission function that has a maximum at every round-trip time of the laser pulse in the cavity.

6.15 Consider a Cr:LiSAF system with an operating wavelength of 840 nm. Assume that a lithium niobate mode-locking crystal is used. The mode-locking crystal has a length of 20 mm and a width of 5 mm. The voltage applied is 300 volts. The modulator is located 1 cm from the end mirror. Assume the laser has a gain of 0.1 cm^{-1} and a frequency bandwidth of 90 THz. Calculate the effective value of the peak phase retardation for both AM and FM modulation. Calculate the observed pulse width for both AM and FM modulation.

7

Introduction to Nonlinear Optics

Objectives

- To describe the physical origins for nonlinear processes.
- To qualitatively describe the second harmonic generation process.
- To compute the conversion efficiency for second harmonic generation using both the low conversion efficiency approximation and the depletion approximation.
- To compute phase matching angles for Type I negative uniaxial crystals.
- To compute phase matching angles for arbitrary crystals.
- To compute the walk-off angle.
- To qualitatively describe the optical parametric amplification process.
- To qualitatively describe the stimulated Raman scattering process.
- To calculate the wavelengths for Stokes and anti-Stokes lines resulting from a particular Raman transition.
- To qualitatively describe the whole beam and small scale self-focusing processes.
- To compute the critical power for whole beam self-focusing.
- To compute the self-focusing distance for both whole beam and small scale self-focusing.
- To compare and contrast the properties of various common nonlinear crystals.

The nonlinear optical processes are among the most fascinating effects that can be produced in lasers. These processes are almost magical, as they permit light of one color to be converted into light of a different color. They are also important commercially and scientifically, as they make it possible to access parts of the spectrum that are difficult to obtain in other ways.

Nonlinear effects are most easily obtained using high-intensity laser beams. Thus, mode-locked Ti:sapphire lasers and Nd:YAG lasers are among the more common lasers used to generate nonlinear effects.

The field of nonlinear optics is complex and multifaceted. As a consequence, the study of nonlinear optics is often delayed until the graduate years. However, nonlinear optical elements are common in the industrial and scientific workplace and may be encountered early in an engineer's career. Thus, the goal of this chapter is to provide a review of the various nonlinear processes commonly encountered in the workplace. A more detailed approach can be obtained from the number of excellent texts available in the area of nonlinear optics[1,2] as well as the various review articles.[3]

7.1 NONLINEAR POLARIZABILITY

The science of nonlinear optics arises from the nonlinear nature of the polarizability in real materials. Recall the macroscopic vector form of Maxwell's equations:

$$\vec{\nabla} \cdot \vec{\mathcal{D}} = \rho \tag{7.1}$$

$$\vec{\nabla} \cdot \vec{\mathcal{B}} = 0 \tag{7.2}$$

$$\vec{\nabla} \times \vec{\mathcal{E}} = -\frac{\partial \vec{\mathcal{B}}}{\partial t} \tag{7.3}$$

$$\vec{\nabla} \times \vec{\mathcal{H}} = \vec{\mathcal{J}} + \frac{\partial \vec{\mathcal{D}}}{\partial t}. \tag{7.4}$$

$\vec{\mathcal{P}}$ is the electric polarization vector and is given by

$$\vec{\mathcal{P}} = \epsilon_o \chi \vec{\mathcal{E}} \tag{7.5}$$

where χ is the complex susceptibility and where $\vec{\mathcal{P}}$ at a point is defined as the net electric dipole moment per unit volume in a small volume around that point.

In Equation (7.5) $\vec{\mathcal{P}}$ has been assumed to be a linear function of $\vec{\mathcal{E}}$. However, there are many useful optical materials for which this is not the case (examples include potassium dihydrogen phosphate, KDP; lithium niobate, $LiNbO_3$; lithium iodate, LiIO; and KTP, $KTiOPO_4$).

---

[1] Robert W. Boyd, *Nonlinear Optics* (San Diego, CA: Academic Press, 1992).

[2] Amnon Yariv and Pochi Yeh, *Optical Waves in Crystals* (New York: John Wiley and Sons, 1984).

[3] R. W. Minck, R. W. Terhune, and C. C. Wang, *Proc. IEEE* 54:1357 (1966).

For a one-dimensional case, the polarizability $\vec{\mathcal{P}}$ can be expanded in terms of the power of $\vec{\mathcal{E}}$ as

$$P(t) = \epsilon_o \chi^{(1)} E(t) + \chi^{(2)} E^2(t) + \chi^{(3)} E^3(t) \tag{7.6}$$

or in the vector case as

$$\vec{\mathcal{P}}(\omega_j) = \epsilon_o \sum_j \chi_{ij}^{(1)}(\omega_m) E_j(\omega_m) + \sum_{jk} \sum_{(mn)} \chi_{ijk}^{(2)}(\omega_m, \omega_n) E_j(\omega_m) E_k(\omega_n) \tag{7.7}$$

$$+ \sum_{jkl} \sum_{(mno)} \chi_{ijkl}^{(3)}(\omega_m, \omega_n, \omega_o) E_j(\omega_m) E_k(\omega_n) E_l(\omega_o) + \cdots \tag{7.8}$$

where $\chi_{ij}^{(1)}$ is a second-rank tensor (9 components, xx, xy, xz, yx, ...), where $\chi_{ijk}^{(2)}$ is a third-rank tensor (27 components, xxx, xxy, xxz, xyx, ...), and $\chi_{ijkl}^{(3)}$ is a fourth-rank tensor (81 components, $xxxx$, $xxxy$, $xxxz$, $xxyx$, ...).

All of the optics discussed so far in this book have been linear optics encompassed in the term $\epsilon_o \chi_{ij}^{(1)}(\omega_m) E_j(\omega_m)$. This term represents optical phenomena that are proportional to the electric field and are at the frequency of incoming wave.

The term $\chi_{ijk}^{(2)}(\omega_m, \omega_n) E_j(\omega_m) E_k(\omega_n)$ is responsible for all of the two-wave effects. This includes second harmonic generation (two fields at ω to make one at 2ω) and parametric oscillation (one field at ω_1 and one field at ω_2 to create fields at $\omega_1 - \omega_2$ and $\omega_1 + \omega_2$). This also includes the Pockels effect (change of index of refraction with applied electric field) because if one wave is taken to be a DC field across the crystal ($\omega = 0$), then the index of refraction becomes a function of the DC field.

The term $\chi_{ijkl}^{(3)}(\omega_m, \omega_n, \omega_o) E_j(\omega_m) E_k(\omega_n) E_l(\omega_o)$ is responsible for all of the three-wave effects. This includes third harmonic generation (three fields at ω to make one at 3ω) and various combinations of three fields to produce sum and difference frequencies as in four-wave mixing. This also includes the Kerr effect (where the index of refraction is a function of the input light intensity). The imaginary component of $\chi_{ijkl}^{(3)}(\omega_m, \omega_n, \omega_o)$ is responsible for Raman, Brillouin, and Rayleigh scattering, as well as for two-photon absorption.

Although each of the above effects is important in laser design, this discussion will be confined to those effects commonly used to change the operating wavelength of a laser (second harmonic generation, parametric amplification, and stimulated Raman scattering), as well as to those effects important in optical damage (the Kerr effect).

7.2 SECOND HARMONIC GENERATION

Second harmonic generation is a $\chi_{ijk}^{(2)}(\omega_m, \omega_n)$ process where a wave at frequency ω is converted into one at frequency 2ω. The most common application is in the conversion of near-infrared light (such as the 1.064 μm of an Nd:YAG laser) into visible light (such as green at 532 nm). Second harmonic conversion finds application in photochemistry and in various spectroscopy applications, because many physical and chemical interactions are more efficient at lower wavelengths. Doubling the frequency of a near-infrared laser (for example, from 1.064 μm to 532 nm) is often a significantly more economical source of green light than using a laser that lases in the green (such as an argon-ion laser).

Optical second harmonic generation was first proposed by Khokhlov in 1961[4] and demonstrated by Franken in 1961.[5] In 1962, Armstrong et al. wrote their seminal paper "Interactions Between Light Waves in a Nonlinear Dielectric," which included optical second harmonic generation as one of the many analyzed cases.[6] In 1962, Maker et al.[7] first observed the oscillation of second harmonic generation with crystal length. The concepts of phase matching were proposed in 1962 by Kleinman,[8] Giordmaine,[9] and Maker et al.[10] The use of birefringent crystals was suggested in 1962 by Kleinman[11] and by Bloembergen and Pershan,[12] discussed by Boyd in 1965,[13] and reviewed by Hobden in 1967.[14] An excellent review article on second harmonic generation is that by Akhmanov.[15]

7.2.1 The Process of Conversion

Consider an electromagnetic field at the fundamental frequency ω_1, which is propagating through a crystal. If the crystal is a noncentrosymmetric crystal for which $\chi_{ijk}^{(2)}$ is not zero, then a polarization wave will be produced in the crystal at twice the frequency of the incident wave. This polarization wave will then transfer energy to an electromagnetic wave that is also at twice the frequency of the incident wave. The energy transmitted to the second harmonic wave will reach a maximum after some distance l_c and then the second harmonic wave will begin to transfer energy by the reverse process back to the first harmonic wave.

This distance l_c is given by[16]

$$l_c = \frac{\lambda_1}{4\left(n_2 - n_1\right)} \tag{7.9}$$

where n_2 is the real index of refraction at 2ω and n_1 is the real index of refraction at ω.

As the wave propagates through the crystal, the original energy in the incident electromagnetic field will simply transfer back and forth between the primary and second harmonic waves with a period of $2l_c$. The process is analogous to a pair of coupled pendulums where one pendulum swings, followed by the other (see Figure 7.1).

---

[4] R. V. Khokhlov, *Radiotek. Electron.* 6:1116 (1961).

[5] P. A. Franken, A. E. Hill, C. W. Peters, and G. Weinreich, *Phys. Rev. Lett.* 7:118 (1961).

[6] J. A. Armstrong, N. Bloembergen, J. Ducuing, and P. S. Pershan, *Phys. Rev.* 127:1918 (1962).

[7] P. D. Maker, R. W. Terhune, N. Nisenoff, and C. M. Savage, *Phys. Rev. Lett.* 8:21 (1962).

[8] D. A. Kleinman, *Phys. Rev.* 128:1761 (1962).

[9] J. A. Giordmaine, *Phys. Rev. Lett.* 8:19 (1962).

[10] P. D. Maker, R. W. Terhune, N. Nisenoff, and C. M. Savage, *Phys. Rev. Lett.* 8:21 (1962).

[11] D. A. Kleinman, *Phys. Rev.* 128:1761 (1962).

[12] N. Bloembergen and P. S. Pershan, *Phys. Rev.* 128:606 (1962).

[13] G. D. Boyd, A. Ashkin, J. M. Dziedzie, and D. A. Kleinman, *Phys. Rev.* 137:A1305 (1965).

[14] M. V. Hobden, *J. Appl. Phys.* 38:4365 (1967).

[15] S. A. Akhmanov, A. I. Kovrygin, and A. P. Sukhorukov, "Optical Harmonic Generation and Optical Frequency Multipliers" in *Quantum Electronics: A Treatise*, ed Herbert Rabin and C. L. Tang (New York: Academic Press, 1975), pp. 476–587.

[16] This definition is not universal. It is also common to see this characteristic length defined as the period of oscillation, rather than one-half the period of oscillation. In this case $l_c = \lambda_1/2\left(n_2 - n_1\right)$.

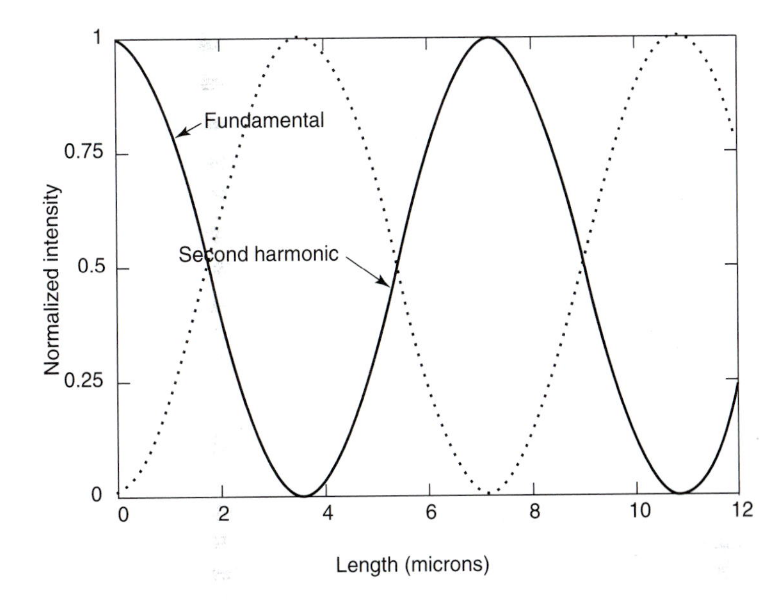

Figure 7.1 As a wave propagates through a second harmonic generation crystal, the original energy in the incident electromagnetic field transfers back and forth between the primary and second harmonic waves with a period of $2l_c$. The process is analogous to a pair of coupled pendulums where first one pendulum swings and then the other.

The fundamental equations describing second harmonic generation are the coupled differential equations,

$$\frac{dE_1}{dz} = -j\eta_1\omega_1 d_{\text{eff}} E_2 E_1^* e^{-j\Delta kz} \tag{7.10}$$

$$\frac{dE_2}{dz} = -j\eta_2\omega_2 d_{\text{eff}} E_1 E_1 e^{-j\Delta kz} \tag{7.11}$$

where E_1 is the electric field of the fundamental, E_2 is the electric field of the second harmonic, $\eta_x = 377/n_x$ in ohms, d_{eff} is the nonlinear coefficient of the nonlinear tensor in units of As/V^2, and $\Delta k = 2k_1 - k_2$.

The ratio between the maximum power at 2ω to the maximum power at ω can be obtained by solving these coupled equations. This general solution was first developed by Armstrong et al. in 1962.[17] However, although a general solution exists, the solution is mathematically complicated due to the necessity to express the solution in the form of elliptic integrals. Tutorial discussions of the general solution are given by Boyd[18] and by Yariv and Yeh.[19]

Typically, simplified forms of the general solution are used for important limiting cases. For example, if the input fundamental wave is assumed to be depleted (and if Δk

---

[17] J. A. Armstrong, N. Bloembergen, J. Ducuing, and P. S. Pershan, *Phys. Rev.* 127:1918 (1962).

[18] Robert W. Boyd, *Nonlinear Optics* (San Diego, CA: Academic Press, 1992).

[19] Amnon Yariv and Pochi Yeh, *Optical Waves in Crystals* (New York: John Wiley and Sons, 1984).

and the input second harmonic wave are both assumed to be small), then the conversion efficiency is given by[20]

$$\frac{P_{2\omega}}{P_\omega} = \tanh^2\left[L\sqrt{2\eta_o^3\omega_1^2 d_{\text{eff}}^2}\left(\frac{P_\omega}{A}\right)^{1/2}\right] \tag{7.12}$$

where $\eta_o = 377/n_o$ in ohms, d_{eff} is the nonlinear coefficient of the nonlinear tensor in units of As/V^2 and L is the total length.

An alternative expression appears in Koechner[21] for the case of finite Δk,

$$\frac{P_{2\omega}}{P_\omega} = \tanh^2\left[L\sqrt{2\eta_o^3\omega_1^2 d_{\text{eff}}^2}\left(\frac{P_\omega}{A}\right)^{1/2}\frac{\sin(\Delta k L/2)}{\Delta k L/2}\right] \tag{7.13}$$

where

$$\Delta k = \frac{4\pi(n_1 - n_2)}{\lambda_1}. \tag{7.14}$$

If the input fundamental wave is assumed to be undepleted (the low conversion efficiency approximation), then it is possible to use the simpler expression,[22,23]

$$\frac{P_{2\omega}}{P_\omega} = 2\eta_o^3\omega_1^2 d_{\text{eff}}^2 L^2\left(\frac{P_\omega}{A}\right)\left(\frac{\sin(\Delta k L/2)}{\Delta k L/2}\right)^2. \tag{7.15}$$

For the low-conversion case expressed by Equation (7.15), the conversion efficiency is proportional to the power in the fundamental wave P_ω. This illustrates the more general observation that second harmonic conversion is most effective in lasers with high-intensity beams. Also notice that the sinusoidal term $\sin(\Delta k L/2)$ is maximized when the indices n_1 and n_2 are as close as possible (a process called *phase matching*). Unfortunately, in most real materials the indices n_2 and n_1 are different and so conversion efficiencies are limited to the range of $1 \cdot 10^{-5}\%$ to $1 \cdot 10^{-3}\%$. However, birefringent crystals (see Section 8.3) can be used to minimize the difference between n_1 and n_2, thus permitting second harmonic generation efficiencies on the order of 30 to 70%.

Example 7.1

Use Equation (7.12) to plot the conversion efficiency for a 1 cm long lithium niobate crystal in a Type I phase-matching configuration (see Section 7.2.2) as a function of input power density. Assume $n_o = 2.2323$. Construct the plot from 0 to 100 MW/cm^2. The d coefficients for lithium niobate are

$$d_{22} = 3 \text{ pm/V}$$
$$d_{31} = -5 \text{ pm/V}$$
$$d_{33} = -33 \text{ pm/V}.$$

[20] Amnon Yariv and Pochi Yeh, *Optical Waves in Crystals* (New York: John Wiley and Sons, 1984), p. 528, with the approximation that $\eta_1 \approx \eta_2 \approx \eta_o$.

[21] Walter Koechner, *Solid State Laser Engineering*, 3d ed. (Berlin: Springer-Verlag, 1992), p. 511. The equation in Koechner references N. Bloembergen, *Nonlinear Optics* (New York: W. A. Benjamin, Inc., 1965).

[22] Walter Koechner, *Solid State Laser Engineering*, 3d ed. (Berlin: Springer-Verlag, 1992), p. 511.

[23] Amnon Yariv and Pochi Yeh, *Optical Waves in Crystals* (New York: John Wiley and Sons, 1984), p. 520.

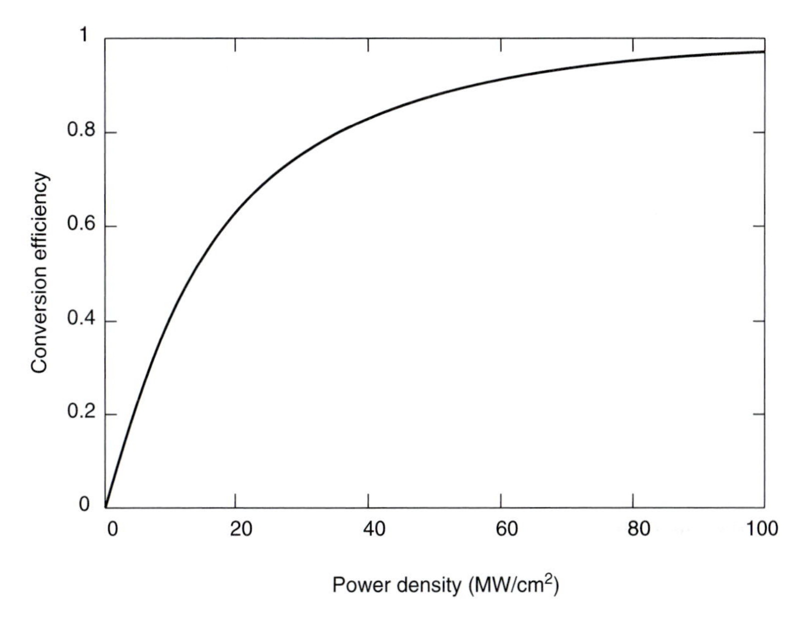

Figure 7.2 The conversion efficiency (calculated using the depletion approximation) for a 1 cm long lithium niobate crystal in a Type I phase-matching configuration as a function of input power density.

Solution. The Appendix (A.7) outlines the general procedure for evaluating d_{eff}. For a Type I phase match under optimal conditions, $d_{\mathrm{eff}} = d_{31}$.

To convert from mV to As/volt², d_{eff} must be multiplied by ε_o,

$$d_{\mathrm{eff}}(\mathrm{As/volt}^2) = d_{\mathrm{eff}} \cdot \varepsilon_o = d_{31} \cdot \varepsilon_o = -4.427 \cdot 10^{-23} \ \mathrm{As/volt}^2.$$

Evaluating Equation (7.12) for the d_{eff} above and the other given parameters yields the graph displayed in Figure 7.2. Notice that the conversion efficiency approaches 100% only for very large input powers.

Example 7.2

Use Equation (7.12) to plot the conversion efficiency for a lithium niobate crystal in a Type I configuration (with 20 MW/cm² input power density) as a function of crystal length. Assume $n_o = 2.2323$. Construct the plot from 0 to 3.5 cm.

Solution. Evaluating Equation (7.12) for the d_{eff} in Example 7.1 and for the parameters given yields the graph displayed in Figure 7.3.

Example 7.3

Use Equation (7.15) to plot the conversion efficiency for a lithium niobate crystal in a Type I configuration (with 40 MW/cm² input power density) as a function of crystal length. Plot the curves for three values of $n_1 - n_2$: 0.02, 0.01 and 0.005. Construct the plot from 0 to 40 μm. Assume $n_o = 2.2323$.

Solution. Evaluating Equation (7.15) for the previous d_{eff} and for the parameters given yields the graph displayed in Figure 7.4.

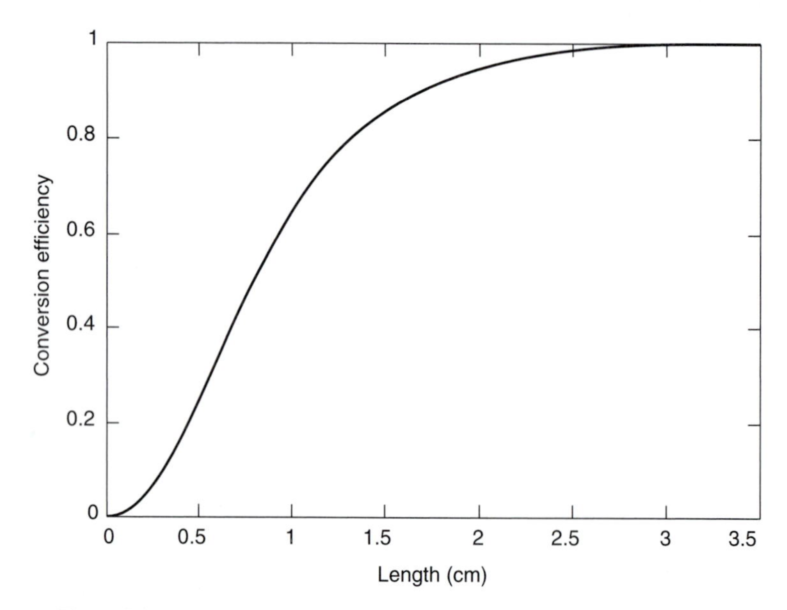

Figure 7.3 The conversion efficiency (calculated using the depletion approximation) for a lithium niobate crystal in a Type I configuration (with 20 MW/cm^2 input power density) plotted as a function of crystal length.

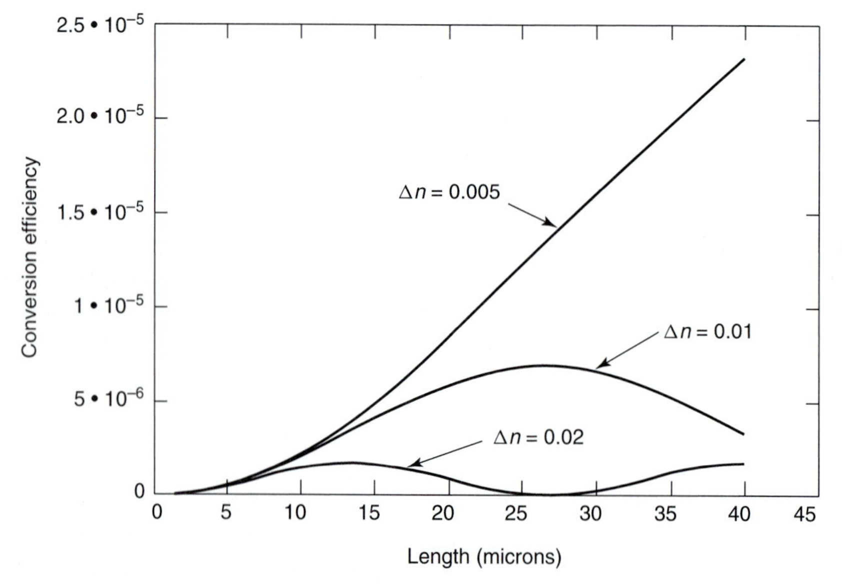

Figure 7.4 The conversion efficiency (using the low conversion approximation) for a lithium niobate crystal in a Type I configuration (with 40 MW/cm^2 input power density) as a function of crystal length. The three values of $n_1 - n_2$ are 0.02, 0.01, and 0.005.

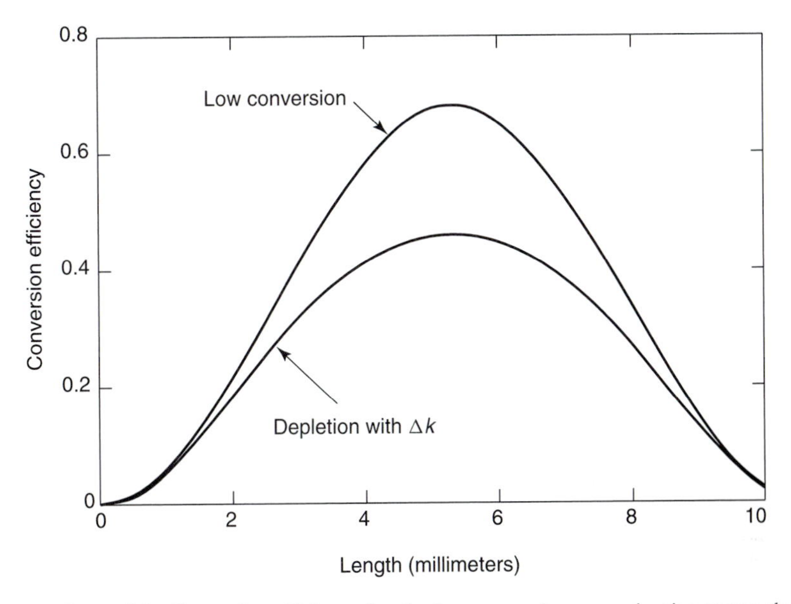

Figure 7.5 Conversion efficiency for the low conversion approximation versus the depletion approximation for the case of an initial power density of 100 MW/cm^2 and $n_1 - n_2 = 5 \cdot 10^{-5}$.

Example 7.4

Compare Equation (7.15) to Equation (7.13) for the case of an initial power density of 100 MW/cm^2 and $n_1 - n_2 = 5 \cdot 10^{-5}$. Construct the plot from 0 to 10 mm. Assume $n_o = 2.2323$.

Solution. Evaluating Equation (7.15) and Equation (7.13) for the d_{eff} in Example 7.1 and the parameters given yields the graph displayed in Figure 7.5.

7.2.2 Phase Matching

The purpose of phase matching is to set the value of the index at the fundamental frequency equal to the value of the index at the second harmonic frequency so that $\Delta k = 0$. Unfortunately, for most regions of the spectrum, indices of refraction increase in normal materials as the frequency increases. However, in birefringent materials, different indices exist for the different polarizations.[24] Thus, it is possible to phase match in birefringent materials by suitable choice of the polarizations.

If the material is positive uniaxial, then the extraordinary index will always be larger than the ordinary index and phase matching can occur at $n_e^{\omega} = n_o^{2\omega}$. (See Figure 7.6.) If the material is negative uniaxial, then $n_o^{\omega} = n_e^{2\omega}$. (See Section 8.3 for a discussion of positive and negative uniaxial crystals, and of ordinary and extraordinary indices of refraction.)

The polarization of the wave is important in phase matching. The outgoing wave at 2ω will be polarized in only one direction. However, the incoming wave may be one

[24] A clever method for measuring the indices of refraction of uniaxial crystals by measurement of the rings created from SHG using an unfocused beam is given in R. Trebino, *Appl. Opt.* 20:2090 (1981).

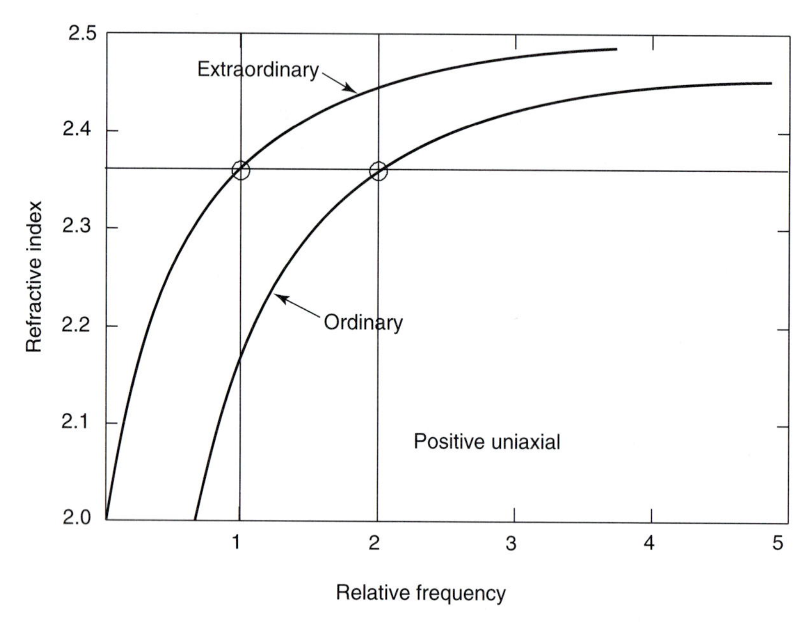

Figure 7.6 If a birefringent material is positive uniaxial, then the extraordinary index will always be larger than the ordinary index and phase matching can occur at $n_e^{\omega} = n_o^{2\omega}$.

polarization, or may be formed by the sum of two orthogonal polarizations. The condition where there is a single polarization for the incoming wave is called Type I phase matching. The condition where there are two orthogonal polarizations for the incoming wave is called Type II phase matching.

The direction of the polarization is often characterized by its orientation relative to the ordinary or extraordinary indices of the crystal. Thus, a wave polarized so as to experience the ordinary index n_o is termed the ordinary or o wave. Similarly, a wave polarized so as to experience the extraordinary index n_e is termed the extraordinary or e wave. With these conventions, the phase matching conditions are often abbreviated by notations such as "oeo" (ordinary + extraordinary = ordinary) and can be summarized as follows:

	Positive uniaxial	Negative uniaxial
Type I	$n_e^{\omega} + n_e^{\omega} = n_o^{2\omega}$ (eeo)	$n_o^{\omega} + n_o^{\omega} = n_e^{2\omega}$ (ooe)
Type II	$n_e^{\omega} + n_o^{\omega} = n_o^{2\omega}$ (eoo and oeo)	$n_o^{\omega} + n_e^{\omega} = n_e^{2\omega}$ (oee and eoe)

Robert W. Boyd, *Nonlinear Optics* (San Diego, CA: Academic Press, 1992), p. 88.

The indicatrix (see Section 8.3) for a uniaxial crystal is an ellipsoid with identical indices on the x and y axis (n_o) and a different index on the z-axis (n_e). Consider a beam propagating through the crystal at an angle of Θ as measured from the optic axis and at an angle Φ as measured from the x-axis. A cut perpendicular to the direction of propagation and through the center of the indicatrix will create an ellipse. The directions of the major and minor axes of the ellipse determine the two polarization directions, and the lengths of the axes determine the value of the index of refraction in that direction. Notice that since

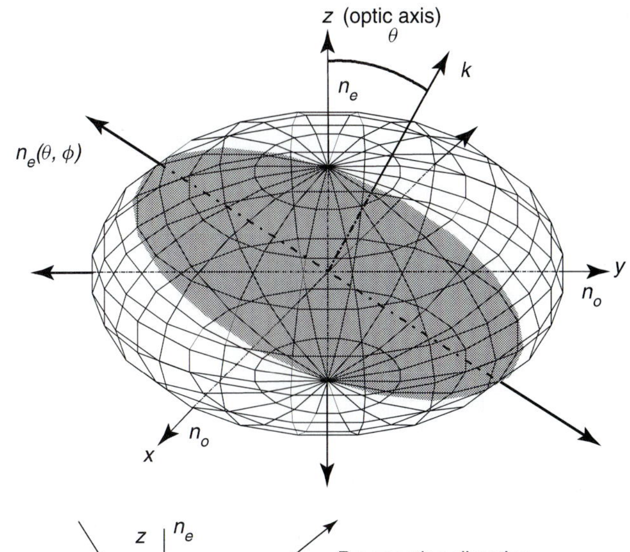

Figure 7.7 The indicatrix for a uniaxial crystal is an ellipsoid with identical indices on the x- and y-axis (n_o) and a different index on the z-axis (n_e).

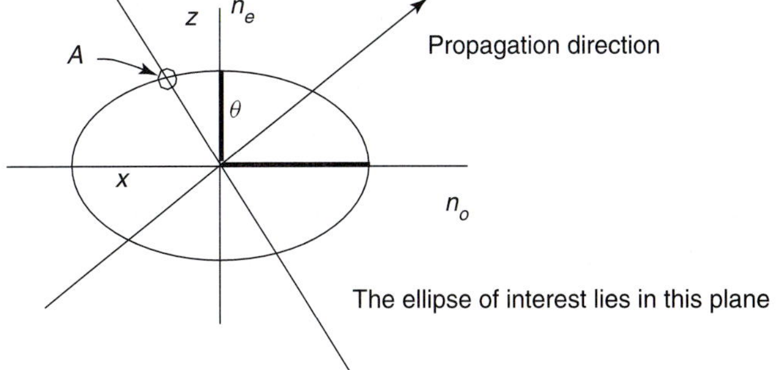

Figure 7.8 Computing the index $n_e(\Theta)$ requires computing the distance to the point A on the ellipse formed by the ordinary and extraordinary axes.

uniaxial crystals have the same index on both the x- and y-axes, the result is independent of Φ. (See Figure 7.7).

The most straightforward way to phase match a birefringent crystal is to change the angle at which the wave propagates through a crystal. Consider a beam propagating through the crystal at an angle of Θ as measured from the optic axis. For Type I phase matching in a negative uniaxial crystal, $n_e^{2\omega}(\Theta) = n_o^{\omega}$. For Type II phase matching in a negative uniaxial crystal, $n_e^{2\omega}(\Theta) = \frac{1}{2}\left(n_e^{\omega}(\Theta) + n_o^{\omega}\right)$.

Computing the index $n_e(\Theta)$ requires computing the distance to the point A on the ellipse formed by the ordinary and extraordinary axes (see Figure 7.8),[25]

$$\frac{1}{n_e(\Theta)^2} = \frac{\sin^2\Theta}{{n_e}^2} + \frac{\cos^2\Theta}{{n_o}^2}. \tag{7.16}$$

For Type I phase matching in a negative uniaxial crystal, $n_e^{2\omega}(\Theta) = n_o^{\omega}$, so

$$\frac{1}{(n_o^{\omega})^2} = \frac{\sin^2\Theta}{(n_e^{2\omega})^2} + \frac{\cos^2\Theta}{(n_o^{2\omega})^2} \tag{7.17}$$

[25]Robert W. Boyd, *Nonlinear Optics* (San Diego, CA: Academic Press, 1992), p. 88.

TABLE 7.1 SUMMARY OF FORMULAS FOR COMPUTING PHASE MATCHING ANGLES

Negative uniaxial	Positive uniaxial
$\tan^2 \Theta_{ooe} = (1-U)/(W-1)$	$\tan^2 \Theta_{eeo} = (1-U)/(U-S)$
$\tan^2 \Theta_{eoe} = (1-U)/(W-R)$	$\tan^2 \Theta_{oeo} = (1-V)/(V-Y)$
$\tan^2 \Theta_{oee} = (1-U)/(W-Q)$	$\tan^2 \Theta_{eoo} = (1-T)/(T-Z)$

where

$U = (A+B)^2/C^2$; $W = (A+B)^2/F^2$
$R = (A+B)^2/(D+B)^2$; $Q = (A+B)^2/(A+E)^2$;
$S = (A+B)^2/(D+E)^2$
$V = B^2/(C-A)^2$; $T = A^2/(C-B)^2$
$Y = B^2/E^2$; $Z = A^2/D^2$
$A = n_o^{\omega}/\lambda_1$; $B = n_o^{\omega}/\lambda_1$; $C = n_o^{2\omega}/\lambda_2$
$D = n_e^{\omega}/\lambda_1$; $E = n_e^{\omega}/\lambda_1$; $F = n_e^{2\omega}/\lambda_2$

V. G. Dmitriev, G. G. Gurzadyan, and D. N. Nikogosyan, *Handbook of Nonlinear Optical Crystals* (Berlin: Springer-Verlag, 1991), p. 13.

or

$$\sin^2 \Theta = \frac{(n_o^{\omega})^{-2} - (n_o^{2\omega})^{-2}}{(n_e^{2\omega})^{-2} - (n_o^{2\omega})^{-2}}. \tag{7.18}$$

Type I phase matching in negative uniaxial crystals is the most straightforward of the various possibilities. The others are summarized in Table 7.1, using the "ooe" nomenclature convention.

Unfortunately, whenever Θ has a value that is not 0 or 90 degrees, then the ordinary and extraordinary rays walk-off from each other as the beam transverses the crystal. This walk-off angle is given by[26]

$$\rho(\Theta) = \pm \tan^{-1}\left[\left(\frac{n_o}{n_e}\right)^2 \cdot \tan\Theta\right] \mp \Theta \tag{7.19}$$

where the upper signs refer to a negative uniaxial crystal and the lower signs refer to a positive uniaxial crystal.

For second harmonic generation, the walk-off angle ρ is given in terms of the phase matching angle as

$$\tan\rho = \frac{(n_o^{\omega})^2}{2}\left(\frac{1}{(n_e^{2\omega})^2} - \frac{1}{(n_o^{2\omega})^2}\right)\sin 2\Theta \tag{7.20}$$

and the walk-off distance is given approximately by

$$l_w \approx \frac{A}{\rho} \tag{7.21}$$

where A is the area of the beam.

[26] V. G. Dmitriev, G. G. Gurzadyan, and D. N. Nikogosyan, *Handbook of Nonlinear Optical Crystals* (Berlin: Springer-Verlag, 1991), p. 8.

It is possible to adjust the temperature of certain crystals so that the phase matching angle Θ is exactly 90 degrees. In these crystals, the linear walk-off (given above) will be zero and the restriction on crystal length will be significantly improved. The most popular application of this effect is in Type I lithium niobate ($LiNbO_3$) where the phase-matching angle is 90 degrees at a temperature of 47°C.

Example 7.5

Compute the phase matching and walk-off angles for lithium niobate in a Type I configuration at 25°C. Assume that the wavelength of the input beam is 1.064 μm and that the wavelength of the second harmonic beam is 532 nm. Assume that $n_o^\omega = 2.234$, $n_o^{2\omega} = 2.3251$, $n_e^\omega = 2.1554$, $n_e^{2\omega} = 2.233$ at 25°C.

Solution. The phase matching angle is computed using Equation (7.18), rewritten slightly,

$$\Theta = \sin^{-1}\left(\sqrt{\frac{(n_o^\omega)^{-2} - (n_o^{2\omega})^{-2}}{(n_e^{2\omega})^{-2} - (n_o^{2\omega})^{-2}}}\right) = 83.837 \text{ deg.}$$

Notice that this is consistent with 90-degree phase matching at 47 degrees.

The walk-off angle is computed using Equation (7.20), rewritten slightly,

$$\rho = \tan^{-1}\left(\frac{(n_o^\omega)^2}{2}\left(\frac{1}{(n_e^{2\omega})^2} - \frac{1}{(n_o^{2\omega})^2}\right)\sin 2\Theta\right) = 0.475 \text{ deg.} \tag{7.22}$$

Example 7.6

Compute the phase matching and walk-off angles for KDP in a Type II configuration at 25°C. Assume that the wavelength of the input beam is 1.064 μm and that the wavelength of the doubled beam is 532 nm. Assume that $n_o^\omega = 1.4928$, $n_o^{2\omega} = 1.5085$, $n_e^\omega = 1.4555$, and $n_e^{2\omega} = 1.4690$ at 25°C.

Solution. The Type II configuration is an "eoe" or "oee" configuration and KDP is a negative uniaxial crystal. Thus, the relevant equations from Table 7.1 are

$$\tan^2 \Theta_{eoe} = (1 - U)/(W - R) \tag{7.23}$$

and

$$\tan^2 \Theta_{oee} = (1 - U)/(W - Q) \tag{7.24}$$

(either one will work). Using Equation (7.23) and rewriting slightly gives

$$\Theta_{eoe} = \tan^{-1}\left(\sqrt{(1 - U)/(W - R)}\right) \tag{7.25}$$

where $U = (A + B)^2/C^2$; $W = (A + B)^2/F^2$; $R = (A + B)^2/(D + B)^2$; $A = B = n_o^\omega/\lambda_1$; $C = n_o^{2\omega}/\lambda_2$; $D = E = n_e^\omega/\lambda_1$; and $F = n_e^{2\omega}/\lambda_2$. Evaluating the equation gives

$$\Theta_{eoe} = \tan^{-1}\left(\sqrt{(1 - U)/(W - R)}\right) = 59.469 \text{ deg.} \tag{7.26}$$

The walk-off angle is computed using Equation (7.20), rewritten slightly,

$$\rho = \tan^{-1}\left(\frac{(n_o^\omega)^2}{2}\left(\frac{1}{(n_e^{2\omega})^2} - \frac{1}{(n_o^{2\omega})^2}\right)\sin 2\Theta\right) = 1.338 \text{ deg.} \tag{7.27}$$

7.2.3 Design Techniques for Frequency-Doubling Laser Beams

There are two major ways to double the frequency of an operating laser system. If the laser is a high-gain laser with a relatively high output intensity, then the second harmonic generation crystal is placed outside the laser cavity (see Figure 7.9). Typical conversion efficiencies for a 10 J pulse with a pulsewidth of 1 ns are 40 to 50%. KDP and KTP in angle- or temperature-tuned configurations are common choices for the nonlinear crystal.

It is also possible to frequency-double a low-intensity cw laser by placing the doubling crystal within the laser cavity (see Figure 7.10). The basic idea is to use two mirrors that are both nearly 100% reflecting at the primary wavelength. However, one mirror is transmissive

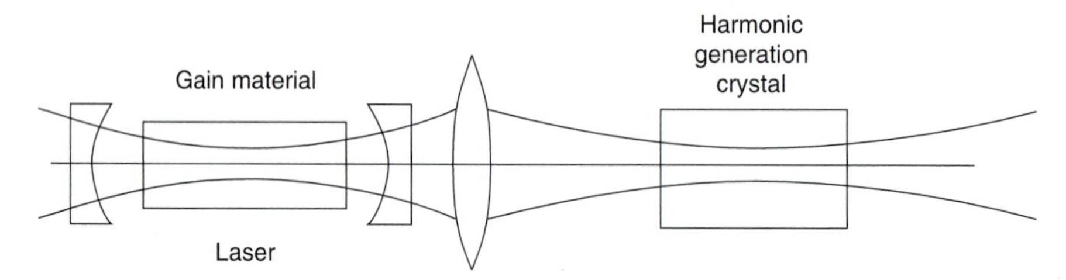

Figure 7.9 A very convenient geometry for second harmonic generation with high-intensity lasers is to place the nonlinear crystal outside the laser cavity.

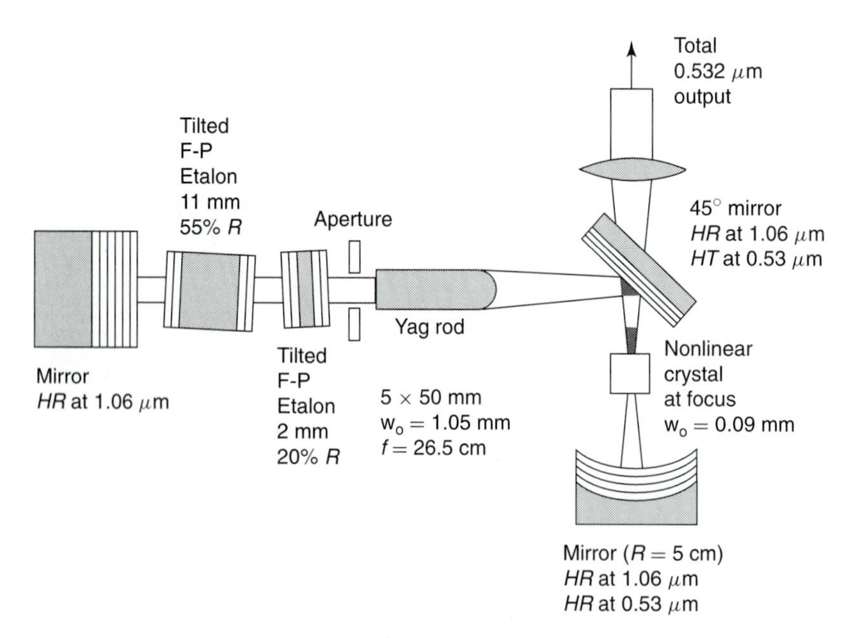

Figure 7.10 It is possible to double a low-intensity laser by placing the doubling crystal within the laser cavity. (From W. Culshaw, J. Kannelaud, and J. Peterson, "Efficient Frequency Doubled Single Frequency Nd:YAG Laser," *IEEE J. of Quantum Electron.* QE-10:253 (1974). © 1974 IEEE.

at the second harmonic wavelength. Optimal operation occurs when the efficiency of the second harmonic generation is the same as the optimal output coupling (including losses).[27] Under these circumstances, the doubling efficiency can be nearly 100%.

The first demonstration of intracavity doubling was made by Wright in 1963[28] and is a special case of resonant optical harmonic generation as described by Ashkin in 1966.[29] The initial demonstration of intracavity doubling was followed in rapid succession by other demonstrations using a variety of crystals including barium sodium nitrate ($Ba_2NaNb_5O_{15}$),[30] lithium niobate ($LiNbO_3$),[31] lithium iodate ($LiIO_3$),[32] and, more recently, KTP ($KTiOPO_4$).[33] Intracavity doubling is especially popular in diode-pumped Nd:YAGs, as it permits an efficient solid-state laser source in the green.

However, in spite of the potentially large conversion efficiency, there are a number of problems associated with intracavity doubling:

1. Absorption of the laser beam in the crystal will heat the crystal, thus changing the phase matching conditions and causing the optical beam to fluctuate in spatial location.
2. Nonlinear effects in the crystal will cause instabilities in multimode systems due to nonlinear mode coupling (the "green problem").[34,35,36]
3. The high nonlinear coefficient required means that materials with high nonlinear coefficients such as $LiNbO_3$ or $LiIO_3$ must be used. Unfortunately, the internal cavity intensities can be high and these crystals are more subject to damage than crystals such as KDP.

7.3 OPTICAL PARAMETRIC OSCILLATORS

Parametric oscillation is a $\chi^{(2)}_{ijk}$ process where two input beams at the frequency ω_1 and ω_2 generate a beam at $\omega_1 - \omega_2$. By using a birefringent crystal and tuning the refractive index (by means of angle, temperature, or electric field), the operating wavelength of the parametric oscillator can be tuned. Parametric oscillators have been demonstrated throughout the visible and infrared spectrum. They operate with high conversion efficiencies from the pump beam (ideally 100% of the pump beam is converted to the signal and idler beams) and relatively low pump power thresholds. Their solid-state construction and broad tunability mean that they commonly are used in applications such as spectroscopy and remote sensing.

---

[27] R. Smith, *IEEE J. of Quan. Electron.* QE-6:215 (1970).

[28] J. K. Wright, *Proc. IEEE* 51:1663 (1963).

[29] A. Ashkin, G. D. Boyd and J. M. Dziedzic, *IEEE J. Quan. Electron.* QE-2:109 (1966).

[30] J. E. Geusic, J. Levinstein, S. Singh, R. G. Smith, and L. G. Van Uitert, *Appl. Phys. Lett.* 12:306 (1968).

[31] T. R. Gurski, *Appl. Phys. Lett.* 15:5 (1969).

[32] U. Deserno and G. Nath, *Phys. Lett.* 30A:483 (1969).

[33] Y. S. Liu, D. Dentz, and R. Belt, *Opt. Lett.* 9:76 (1984).

[34] T. J. Baer, *Opt. Soc. Am. B* 3:1175 (1986).

[35] G. E. James, E. M. Harrell, C. Bracikowski, K. Wiesenfeld, and R. Roy, *Opt. Lett.* 15:1141 (1990); and G. E. James, E. M. Harrell, and R. Roy, *Phys. Rev. A* 41:2778 (1990).

[36] M. Oka and S. Kubota, *Opt. Lett.* 13:805 (1988).

In 1962, Armstrong et al. wrote their seminal paper "Interactions between Light Waves in a Nonlinear Dielectric," which included optical parametric amplification as one of many discussed cases.[37] In 1965, Wang and Racette[38] first observed parametric gain in an optical nonlinear system, and Giordmaine and Miller[39] demonstrated the first optical parametric oscillator using $LiNbO_3$. In 1966, Akhmanov et al.[40] first demonstrated an optical parametric oscillator in KDP. In 1966, Giordmaine and Miller[41] extended the $LiNbO_3$ optical parametric oscillator into the visible (0.73 μm to 1.93μm). Also in 1966, Boyd and Askin[42] theoretically proposed the possibility of cw optical parametric oscillation in $LiNbO_3$. In 1968, Smith et al. obtained cw optical parametric oscillation in $Ba_2NaNb_5O_{15}$[43] and Byer et al. obtained cw visible optical parametric oscillation in $LiNbO_3$.[44] Excellent review articles on optical parametric oscillators include Harris,[45] Byer,[46] and, more recently, Barnes.[47]

The output of an optical parametric oscillator (OPO) is similar to that of a laser. It is highly monochromatic and exhibits laser speckle. The spectrum is formed of one or several longitudinal modes. The transverse mode spectrum is often $TEM_{0,0}$ and propagates with Gaussian-like properties.

OPOs differ from a traditional laser in their sensitivity to the optical characteristics of the pump radiation. In a traditional laser, the pump source serves only to populate the higher-level energy states. As long as the pump source adequately populates the upper laser states, a traditional laser's output is independent of the spatial or frequency characteristics of the pump radiation. On the other hand, an OPO is very sensitive to the phase coherence between the three frequencies, and to the spectral and spatial character of the pump pulse. For example, if the longitudinal mode spacing of the idler beam is set equal to that of the pumping laser, then all of the modes in the pumping laser act together to produce gain for a single signal frequency mode.[48] Additionally, if the longitudinal mode spacings of all three beams are equal, then the peak pump power (rather than average pump power) will drive the oscillator.[49]

---

[37]J. A. Armstrong, N. Bloembergen, J. Ducuing, and P. S. Pershan, *Phys. Rev.* 127:1918 (1962).

[38]C. C. Wang and C. W. Racette, *Appl. Phys. Lett.* 6:169 (1965).

[39]J. A. Giordmaine and R. C. Miller, *Phys. Rev. Lett.* 14:973 (1965).

[40]S. A. Akhmanov, A. I. Kovrigin, V. A. Kolosov, A. S. Piskarskas, V. V. Fadeev, and R. V. Khokhlov, *JETP Lett.* 3:241 (1966).

[41]J. A. Giordmaine and R. C. Miller, *Appl. Phys. Lett.* 9:298 (1966).

[42]G. D. Boyd and A. Ashkin, *Phys. Rev.* 146:187 (1966).

[43]R. G. Smith, J. E. Geusic, H. J. Levinstein, J. J. Rubin, S. Singh, and L. G. Van Uitert, *Appl. Phys. Lett.* 12:308 (1968); also R. G. Smith, J. E. Geusic, H. J. Levinstein, S. Singh, and L. G. Van Uitert, *J. Appl. Phys.* 39:4030 (1968).

[44]Robert L. Byer, "Parametric Fluorescence and Optical Parametric Oscillation," Ph.D. dissertation, Stanford Univ., Stanford, CA, 1968; and also Byer, R. L., *Optical Spectra*, September 1970, p. 42.

[45]S. E. Harris, *Proc. IEEE* 57:2096 (1969).

[46]Robert L. Byer, "Optical Parametric Oscillators," in *Quantum Electronics: A Treatise*, ed Herbert Rabin and C. L. Tang (New York: Academic Press, 1975), pp. 588–694.

[47]Norman P. Barnes, "Optical Parametric Oscillators," in *Tunable Lasers Handbook*, ed F. J. Duarte (San Diego, CA: Academic Press, 1995), pp. 293–348.

[48]S. E. Harris, *IEEE J. Quan. Electron.* QE-2:701 (1966).

[49]S. E. Harris, *IEEE J. Quan. Electron.* QE-3:205 (1967).

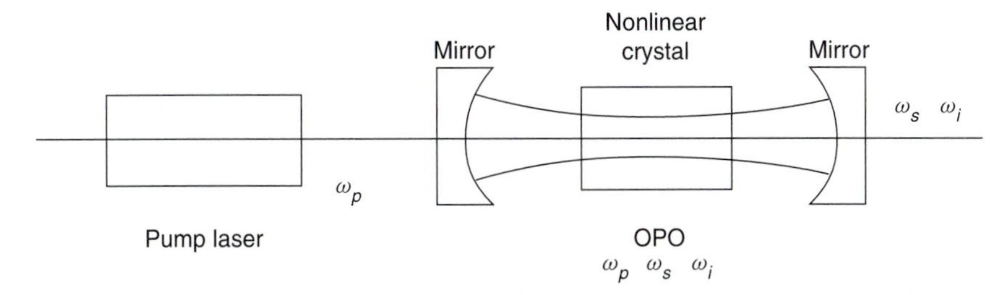

Figure 7.11 Optical parametric oscillators are constructed with a transmitting back mirror at the pump wavelength and either a reflective or transmitting front mirror at the pump wavelength. The front mirror can be transmitting for the signal wave, idler wave, or both.

Consider a crystal pumped by an electromagnetic wave at the pump frequency ω_p. Assume that another wave is incident on the crystal at the signal frequency ω_s. The interaction of these two waves will generate a polarization wave in the crystal at the difference frequency $\omega_i = \omega_p - \omega_s$ of the incident waves. This polarization wave will then transfer energy to an electromagnetic wave that is also at the idler frequency ω_i of the incident waves. Now, the interaction of the waves at ω_i and ω_p will generate a polarization wave in the crystal at the difference frequency $\omega_p - \omega_i = \omega_s$ of the incident waves. This will reinforce the signal wave at ω_s. Thus, the signal wave at ω_s and the idler wave at ω_i will both be amplified. For this process to have significant amplification, the three waves must simultaneously satisfy the frequency condition $\omega_p = \omega_s + \omega_i$ and the phase-matching condition

$$\frac{n_p}{\lambda_p} = \frac{n_s}{\lambda_s} + \frac{n_i}{\lambda_i}. \tag{7.28}$$

In practice, parametric oscillators are constructed by providing a pump wave at the wavelength λ_p. The resonator is constructed with a transmitting back mirror at the pump wavelength and either a reflective or transmitting front mirror at the pump wavelength (see Figure 7.11). In addition, the resonator can provide feedback at either the signal or idler frequency (singly resonant) or at both the signal and idler frequencies (doubly resonant). Although doubly resonant OPOs are more difficult to construct due to stability issues, they have the significant advantage of providing lower-oscillation thresholds.

Tuning the wavelength of the parametric oscillator is accomplished by changing the indices of refraction of the nonlinear crystal with respect to the beam. One common method is to change the angle of the crystal with respect to the cavity (or the pump beam[50]). Typical tuning curves for this process for lithium niobate are given in Figure 7.12. Another method is to change the temperature of the crystal. Typical tuning curves for this process for lithium niobate are given in Figure 7.13. It is also possible to tune the wavelength by using electrooptical crystals. The electrooptic method is much quicker, but provides a substantially smaller tuning range. For example, lithium niobate ($LiNbO_3$) has a tuning range of about

[50]In this method, the pump beam is introduced from the side (not colinear with the beam direction) and the angle is changed by an optical element such as a prism. Falk and Murray first demonstrated this method of angle tuning in J. Falk and J. E. Murray, *Appl. Phys. Lett.* 14:245 (1969).

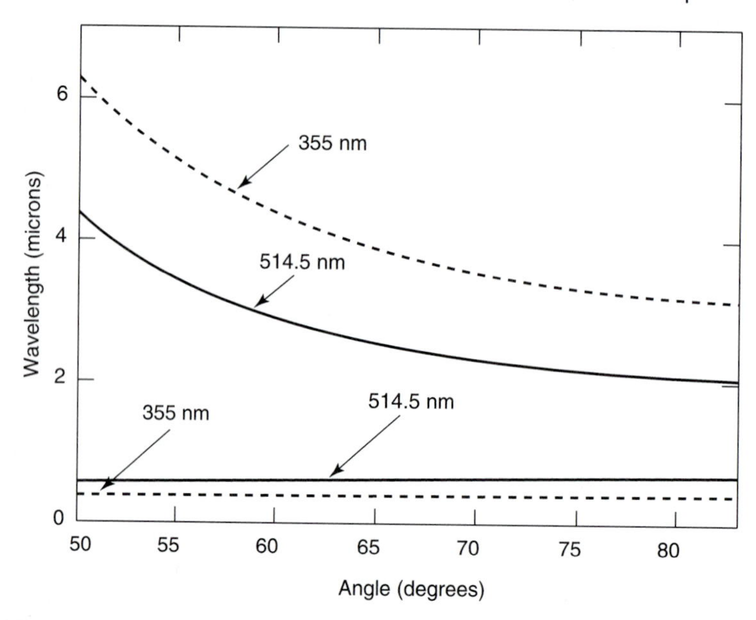

Figure 7.12 Optical parametric oscillators can be tuned by changing the angle of the crystal with respect to the cavity. These tuning curves are calculated for lithium niobate with a 355 nm and 514.5 nm pump.

6.7 angstroms per applied kV with electrooptic tuning, as contrasted to approximately 700 angstroms per angular degree for angle tuning.[51]

The fundamental coupled equations for parametric amplification are similar to those of second harmonic generation. Following the procedure of Harris,[52] these equations can be written as

$$\frac{dE_s}{dz} = -j\eta_s\omega_s d_{\text{eff}} E_p E_i^* e^{-j\Delta kz} \tag{7.29}$$

$$\frac{dE_i}{dz} = -j\eta_i\omega_i d_{\text{eff}} E_p E_s^* e^{-j\Delta kz} \tag{7.30}$$

$$\frac{dE_p}{dz} = -j\eta_p\omega_p d_{\text{eff}} E_s E_i e^{-j\Delta kz} \tag{7.31}$$

where E_s is the electric field of the signal, E_i is the electric field of the idler, E_p is the electric field of the pump, $\eta_x = 377/n_x$ in ohms, d_{eff} is the nonlinear coefficient of the nonlinear tensor in units of As/V^2, and $\Delta k = k_p - k_s - k_i$.

The ratio between the signal and idler powers can be obtained by solving these coupled equations. This general solution was first developed by Armstrong et al. in 1962.[53] As with second harmonic generation, the general solution is mathematically complex due to the

[51] S. E. Harris, *Proc. IEEE* 57:2096 (1969).

[52] S. E. Harris, *Proc. IEEE* 57:2096 (1969).

[53] J. A. Armstrong, N. Bloembergen, J. Ducuing, and P. S. Pershan, *Phys. Rev.* 127:1918 (1962).

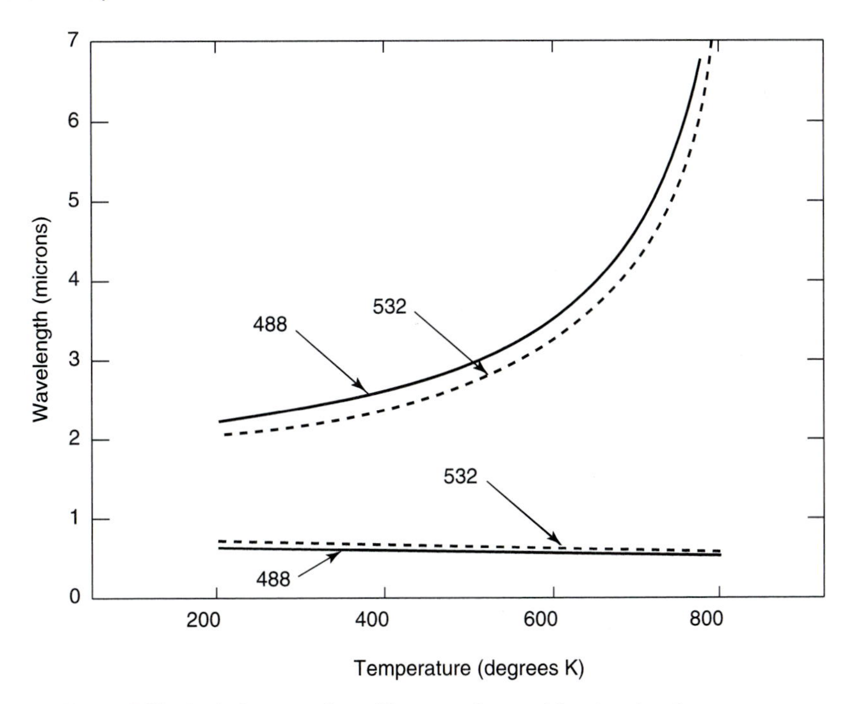

Figure 7.13 Optical parametric oscillators can be tuned by changing the temperature of the crystal. These tuning curves are calculated for lithium niobate with a 488 nm and 532 nm pump.

necessity of expressing the solution in the form of elliptic integrals. Tutorial discussions of the general solution are given by Boyd[54] and by Yariv and Yeh.[55]

As with second harmonic generation, simplified forms of the general solution are used for important limiting cases. For example, if the pump wave is assumed to be undepleted, then the signal and idler fields as a function of length are given as

$$\begin{aligned} E_s(L) &= E_s(0)\exp\left(-j\frac{\Delta k L}{2}\right)\left[\cosh sL + j\frac{\Delta k}{2s}\sinh sL\right] \\ &\quad -j\frac{\kappa_s}{s}E_i^*(0)\exp\left(-j\frac{\Delta k L}{2}\right)[\sinh sL] \end{aligned} \tag{7.32}$$

$$\begin{aligned} E_i(L) &= E_i(0)\exp\left(-j\frac{\Delta k L}{2}\right)\left[\cosh sL + j\frac{\Delta k}{2s}\sinh sL\right] \\ &\quad -j\frac{\kappa_i}{s}E_s^*(0)\exp\left(-j\frac{\Delta k L}{2}\right)[\sinh sL] \end{aligned} \tag{7.33}$$

where $\kappa_s = \eta_s\omega_s d_{\text{eff}}E_p$, $\kappa_i = \eta_i\omega_i d_{\text{eff}}E_p$, $\Gamma^2 = \kappa_s \cdot \kappa_i$, and $s = \sqrt{\Gamma^2 - \Delta k^2/4}$.

---

[54] Robert W. Boyd, *Nonlinear Optics* (San Diego, CA: Academic Press, 1992).

[55] Amnon Yariv and Pochi Yeh, *Optical Waves in Crystals* (New York: John Wiley and Sons, 1984).

The single pass gain for $E_i(0) = 0$ is given by

$$G = \Gamma^2 L^2 \frac{\sinh^2\left(\sqrt{\Gamma^2 - \Delta k^2/2} \cdot L\right)}{\left(\Gamma^2 - \Delta k^2/4\right) L^2} \tag{7.34}$$

and, for $\Delta k \approx 0$,

$$G = \sinh^2(\Gamma L) \approx \Gamma^2 L^2 = 2\omega_s \omega_i \eta_s \eta_i \eta_p \left|d_{\text{eff}}^2\right| L^2 \left(P_p/A\right) \tag{7.35}$$

where P_p is the intensity of the pump beam.

Taking the complex conjugates of Equations (7.29) and (7.30), and multiplying Equations (7.29), (7.30), and (7.31) by E_s/η_s, E_s/η_s, and E_s/η_s, respectively, gives the rate equations modeling the power flow in an optical parametric amplifier

$$\frac{1}{\omega_s}\frac{d}{dz}\left(\frac{|E_s|^2}{2\eta_s}\right) = \frac{1}{\omega_i}\frac{d}{dz}\left(\frac{|E_i|^2}{2\eta_i}\right) = -\frac{1}{\omega_p}\frac{d}{dz}\left(\frac{|E_p|^2}{2\eta_p}\right). \tag{7.36}$$

7.4 STIMULATED RAMAN SCATTERING

Second harmonic generation can be considered to be a photon summing method, one where two photons at frequency ω sum to make one photon at 2ω. Similarly, optical parametric amplification can be considered to be a photon differencing method, one where the photon ω_p splits into two photons at $\omega_p = \omega_i + \omega_s$. In neither case do the energetic properties of the nonlinear material enter into the process.

Raman scattering is different from second harmonic generation and optical parametric amplification in that the characteristic Raman absorption state energy of the material is a part of the process. Raman scattering is a $\chi_{ijk}^{(3)}$ process, where sum and difference frequencies are generated between a pump source and characteristic Raman absorption states in a material. The resulting photons are produced at frequencies $\omega_P \pm m\omega_R$, where ω_P is the pump frequency, ω_R is the characteristic Raman absorption state frequency, and m is an integer (see Figure 7.14).

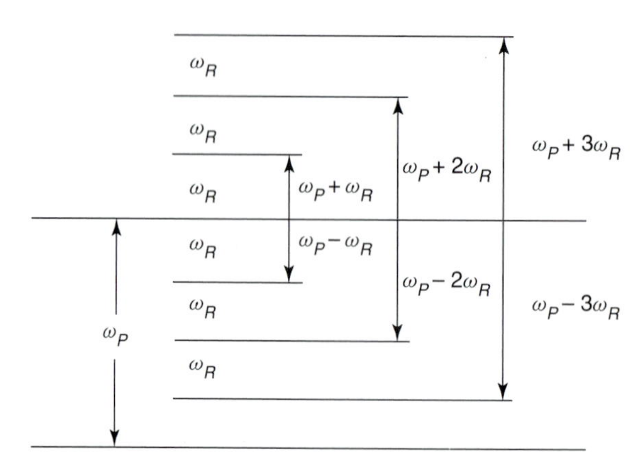

Figure 7.14 Raman scattering is a $\chi_{ijk}^{(3)}$ process where sum and difference frequencies are generated between a pump source and characteristic Raman absorption states in a material. The resulting photons are produced at frequencies $\omega_P \pm m\omega_R$, where ω_P is the pump frequency, ω_R is the characteristic Raman absorption state frequency, and m is an integer.

Spontaneous Raman scattering was first observed by C. V. Raman in 1928, and the study of spontaneous Raman scattering was a well-developed field of spectroscopy at the time the laser was invented. However, spontaneous Raman scattering is a very weak process. Under excitation by a laser beam, a more efficient Raman process called stimulated Raman scattering can occur. Stimulated Raman scattering differs from spontaneous Raman scattering both in the efficiency of conversion and in the emission pattern. Spontaneous Raman radiation is emitted nearly isotropically, while stimulated Raman radiation is emitted in a narrow cone aligned with the pump beam.

Stimulated Raman scattering was first observed by Woodbury and Ng in 1962[56] and was theoretically explained by Eckhardt et al. that same year.[57] Additional experiments in 1963 by Hellworth[58] and Eckhardt[59] demonstrated the stimulated Raman effect in a variety of materials. The process was further analyzed by Terhune,[60] Garmire et al.,[61] and Hellworth in 1963;[62] and by Bloembergen and Shen in 1964[63] and 1965.[64] Excellent review articles have been published by Kaiser and Maier,[65] Penzkofer et al.,[66] and Raymer and Walmsley.[67] Tutorial discussions of the general solution are given by Bloembergen,[68] Boyd,[69] and Yariv.[70]

The stimulated Raman process has gain on the order of a few dB per centimeters in a Raman active material. There are two major sets of emission lines available in Raman scattering. The Stokes lines consist of a series of intense lines at the Stokes frequencies,

$$\omega_P - \omega_R, \quad \omega_P - 2\omega_R, \quad \omega_P - 3\omega_R, \quad \ldots. \tag{7.37}$$

where ω_P is the pump transition frequency and ω_R is the Raman transition frequency. Stokes scattering usually occurs on the axis of the beam and is the primary process used in Raman lasers.

It is also possible to obtain Raman wavelength conversion using anti-Stokes scattering. Anti-Stokes scattering consists of a series of intense lines at the anti-Stokes frequencies,

$$\omega_P + \omega_R, \quad \omega_P + 2\omega_R, \quad \omega_P + 3\omega_R, \quad \ldots. \tag{7.38}$$

---

[56] E. J. Woodbury and W. K. Ng, *Proc. IRE* 50:2367 (1962).

[57] G. Eckhardt, R. W. Hellwarth, F. J. McClung, S. E. Schwarz, D. Weiner, and E. J. Woodbury, *Phys. Rev. Lett.* 9:455 (1962).

[58] R. W. Hellwarth, *Applied Optics* 2:847 (1963).

[59] G. Eckhardt, D. P. Bortfeld, and M. Geller, *Appl. Phys. Lett.* 3:36 (1963).

[60] R. W. Terhune, *Solid State Design* 4:38 (1963).

[61] E. Garmire, F. Pandarese, and C. H. Townes, *Phys. Rev. Lett.* 11:160 (1963).

[62] R. W. Hellwarth, *Phys. Rev.* 130:1850 (1963).

[63] N. Bloembergen and Y. R. Shen, *Phys. Rev.* 133:A37 (1964).

[64] Y. R. Shen and N. Bloembergen, *Phys. Rev.* 137:A1787 (1965).

[65] W. Kaiser and M. Maier, in *Laser Handbook*, edited by F. T. Arecchi and E. O. Schulz-DuBois (Amsterdam: North-Holland, 1972).

[66] A. Penzkofer, A. Laubereau, and W. Kaiser, *Prog. Quantum Electronics* 6:55 (1979).

[67] M. G. Raymer and I. A. Walmsley, in *Progress in Optics, Vol. 28*, E. Wolf (Amsterdam: North-Holland, 1990).

[68] N. Bloembergen, *Nonlinear Optics* (New York: W.A. Benjamin, Inc., 1965).

[69] Robert W. Boyd, *Nonlinear Optics* (San Diego, CA: Academic Press, 1992).

[70] Amnon Yariv, *Quantum Electronics*, 2d ed. (New York: John Wiley and Sons, 1975).

TABLE 7.2 RAMAN FREQUENCIES AND SCATTERING CROSS-SECTIONS

Material	State (cm^{-1})	$d\sigma/d\Omega$ (cm^2/sr)
H_2	4155	$8.1 \cdot 10^{-31}$
CH_4	2914	$3.0 \cdot 10^{-30}$
N_2	2330	$3.7 \cdot 10^{-31}$
HF	3962	$4.8 \cdot 10^{-31}$

where ω_R is the frequency of the Raman transition. Anti-Stokes scattering occurs at the angle to the beam that matches the phase and is thus usually observed as an off-axis process.

The fundamental coupled equations for stimulated Raman scattering are similar to those of optical parametric amplification. Following the notation of Koechner,[71] these equations can be written as

$$\frac{\partial E_p}{\partial z} = -\frac{\omega_p}{2cn_p}\chi_R'' \, |E_{st}|^2 \, E_p \tag{7.39}$$

$$\frac{\partial E_{st}}{\partial z} = -\frac{\omega_{st}}{2cn_{st}}\chi_R'' \, |E_p|^2 \, E_{st} \tag{7.40}$$

where E_{st} is the electric field of the first Stokes signal; E_p is the electric field of the pump; n_p and n_{st} are the indices of refraction at the pump and Stokes waves, respectively; ω_p and ω_{st} are the angular frequencies at the pump and Stokes waves, respectively; and χ_R'' is the peak value of the imaginary part of the Raman susceptibility.

For a constant pump wave, the Stokes intensity grows exponentially, with a gain coefficient given by

$$\gamma_s = \frac{\omega_{st}\chi_R'' \, |E_p|^2}{cn_{st}} = \frac{4\pi \chi_R'' I_p}{\lambda_{st} n_{st} n_p \varepsilon_o c} = \frac{\lambda_p \lambda_{st}^2 N (d\sigma/d\Omega) \, I_p}{c\pi \, \Delta\nu_R n_{st}^2 \hbar} \tag{7.41}$$

where λ_{st} is the free space wavelength of the Stokes wave, I_p is the pump intensity, $d\sigma/d\Omega$ is the spontaneous Raman scattering cross-section (cm^2/steradian), N is the number density of scattering centers, and $\Delta\nu_R$ is the FWHM Raman frequency linewidth.

The maximum theoretical conversion efficiency of a Raman laser is given as

$$\eta = \frac{\nu_P - \nu_R}{\nu_P} \tag{7.42}$$

where ν_P is the pump frequency and ν_R is the frequency of the Raman state. Typical Raman frequencies and Raman scattering cross-sections are given in Table 7.2.

A common Raman scattering demonstration is to focus an intense beam inside a cell containing a Raman material.[72] The center of the beam will convert using the Stokes frequencies and the outside of the beam will convert using the anti-Stokes frequencies (see Figure 7.15). If the starting beam is a red laser, the result will be a colorful bull's-eye where

[71] Walter Koechner, *Solid State Laser Engineering*, 3d ed. (Berlin: Springer-Verlag, 1992).

[72] This experimental configuration was first demonstrated by R. W. Terhune in *Solid State Design* 4:38 (1963).

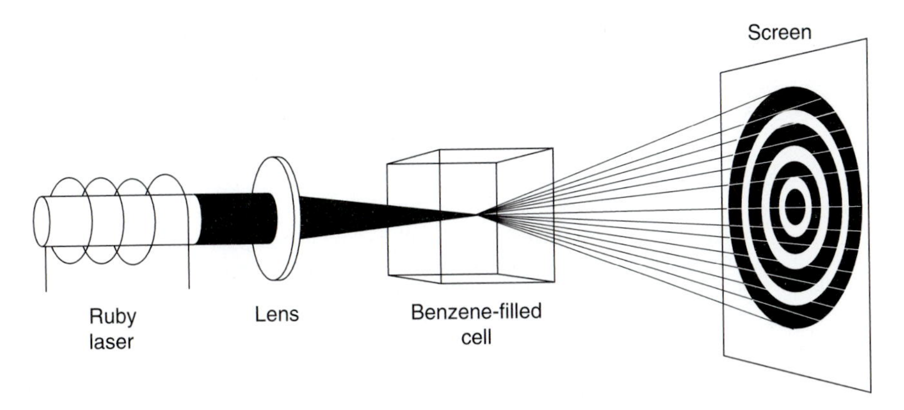

Figure 7.15 A common Raman demonstration is to focus an intense beam inside a cell containing a Raman material. The center of the beam will convert using the Stokes frequencies and the outside of the beam will convert using the anti-Stokes frequencies. (From Amnon Yariv, *Quantum Electronics*, 2d ed. Copyright ©1975 John Wiley & Sons. Reprinted by permission of John Wiley & Sons, Inc.)

the interior rings vary from red to IR generated by the Stokes frequencies, and the exterior rings range from green to blue generated by the anti-Stokes frequencies.[73]

A variation on this experimental arrangement is to construct a high-pressure Raman gas cell (as shown in Figure 7.16). This type of single-pass cell has the advantage of simplicity of alignment and operation. However, the lack of a resonance condition means that other nonlinear processes (such as stimulated Brillouin scattering and optical distortions of the beam) may effect the output. Additionally, the single-pass cell will produce all of the Stokes and anti-Stokes lines. If the user is only planning to use the first Stokes line, then the other lines constitute unnecessary loss. Typical fill gases for a single pass cell are hydrogen, deuterium, and CH_4; fill pressures are on the order of 300 to 1200 psi. Typical conversion efficiencies are 20-50% into the first Stokes line, but only a few percent into the other lines.

Another option is to construct the Raman cell within a resonator, as illustrated in Figure 7.17. By using selective mirrors at each end of the resonator, the desired Raman line can be selectively amplified. Although the first Stokes line is the usual line that is chosen, it is possible to operate Raman lasers on other lines by suitable choice of the reflective optics.

Example 7.7

Compute the wavelengths of the Stokes and anti-Stokes lines, assuming a Nd:YAG laser pump at $\lambda = 1.064\ \mu$m and a high-pressure external hydrogen cell.

Solution. The values for the state transition of a Raman line are usually given in wavenumbers ($1/\lambda$). The reason is that this permits the states to be treated like energies. Thus, it is easy to obtain the various Raman energy states by simple addition or subtraction of the wavenumber. The final answer is then inverted to obtain the wavelength.

[73] An outstanding color photograph of this effect can be seen in the title page of Amnon Yariv's text, *Quantum Electronics*, 2d ed. (New York: John Wiley and Sons, 1975).

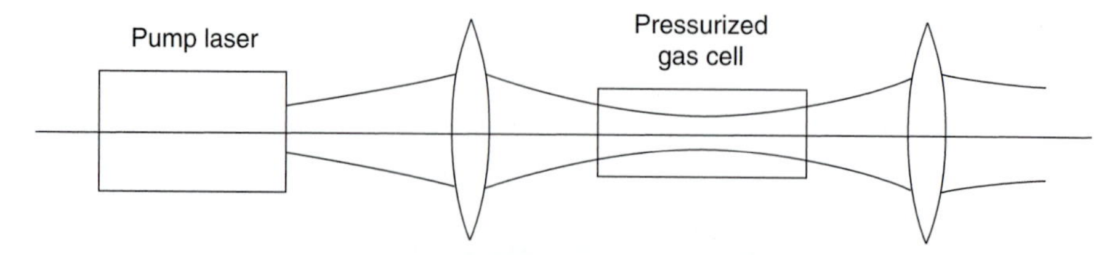

Figure 7.16 A single-pass Raman cell has the advantage of simplicity of alignment and operation. However, the lack of a resonance condition means that other nonlinear processes (such as stimulated Brillouin scattering and optical distortions of the beam) may affect the output.

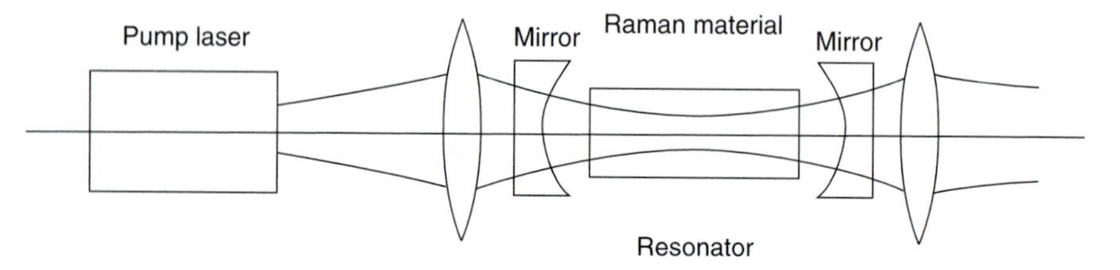

Figure 7.17 A desired raman line can be amplified in a Raman laser configuration by using selective mirrors at each end of the resonator.

The wavenumber for the Nd:YAG laser is

$$wn_p = \frac{1}{\lambda} = 9398.5 \text{ cm}^{-1}$$

and the first Stokes transition is simply

$$S_1 = wn_p - wn_R = 9398.5 \text{ cm}^{-1} - 4155 \text{ cm}^{-1}$$

$$\frac{1}{S_1} = 1.907 \ \mu\text{m}.$$

Similarly, the second Stokes transition is

$$S_2 = wn_p - 2 \cdot wn_R = 9398.5 \text{ cm}^{-1} - 2 \cdot 4155 \text{ cm}^{-1}$$

$$\frac{1}{S_2} = 9.187 \ \mu\text{m}.$$

There is no third transition because there is no remaining pump energy.

For the anti-Stokes lines, the same procedure holds, but with a sign difference,

$$AS_1 = wn_p + wn_R = 9398.5 \text{ cm}^{-1} + 4155 \text{ cm}^{-1}$$

$$\frac{1}{AS_1} = 738 \text{ nm}.$$

Similarly, the second anti-Stokes transition is

$$AS_2 = wn_p + 2 \cdot wn_R = 9398.5 \text{ cm}^{-1} + 2 \cdot 4155 \text{ cm}^{-1}$$

$$\frac{1}{AS_2} = 565 \text{ nm}$$

and so on. There is no end to the possible anti-Stokes transitions. (Although, in practice, anything beyond the fourth transition has negligible power!)

TABLE 7.3 VALUE OF N_2

Material	n_o	n_2 (cm^2/watt)
Fused silica	1.456	$2.72 \cdot 10^{-16}$
Ruby	1.75	$3.2 \cdot 10^{-16}$
BK-7	1.517	$3.85 \cdot 10^{-16}$
YAG	1.829	$8.945 \cdot 10^{-16}$
LASF-7	1.914	$1.296 \cdot 10^{-15}$

7.5 SELF-FOCUSING AND OPTICAL DAMAGE

Self-focusing is a $\chi_{ijk}^{(3)}$ process where the light intensity causes an increase in the index of refraction. This increase in the index of refraction causes a lensing effect where the beam is focused more tightly. The more tightly focused beam has a higher intensity and thus causes increased focusing. If the focusing is severe enough to compensate for beam diffraction, then the laser beam will focus smaller and smaller until the intensity at the focal spot is sufficient to burn a hole in the optical material. A good overview of the problems of self-focusing in large laser systems can be obtained from Brown.[74] For additional information on optical damage, see Weber.[75]

For historical reasons, the nonlinear index of refraction for self-focusing is termed n_2 and for the expression

$$\Delta n = n_2 |E|^2 \tag{7.43}$$

is defined as

$$\Delta n_2 = \frac{2\pi}{n_o} \left(\chi_{1122}^3 + \chi_{1212}^3 + \chi_{1221}^3 \right) . \tag{7.44}$$

Values of n_2 for various materials are given in Table 7.3.

Whole beam self-focusing occurs when the change in the index of refraction generated by the laser beam is sufficient to overcome diffraction. The critical power P_{crit} is the power level for which self-focusing just overcomes diffraction. P_{crit} (for n_2 in cm^2/watt) is given for Gaussian beams by[76]

$$P_{\text{crit}} = \frac{\pi (0.61)^2 \lambda_o^2}{8 n_o^3 n_2} . \tag{7.45}$$

At this critical power the beam neither focuses nor expands. Thus, the condition is termed self-trapping. If the power is greater than the critical power, the beam will focus. For a Gaussian beam, this focusing distance is given by[77]

$$z_{\text{self-focus}} = \frac{2 w_o^2}{0.61 \lambda_o} \frac{1}{\sqrt{P / P_{\text{crit}}}} \tag{7.46}$$

where w_o is the initial beam waist of the Gaussian beam.

[74] David C. Brown, *The Physics of High Peak Power Nd:Glass Laser Systems* (Berlin: Springer-Verlag, 1980).

[75] Marvin J. Weber, ed, *Handbook of Laser Science and Technology, Vol. III, Optical Materials* (Boca Raton, FL: CRC Press, Inc., 1982).

[76] E. L. Dawes and J. H. Marburger, *Phys. Rev.* 179:862 (1969).

[77] W. G. Wagner, H. A. Haus, and J. H. Marburger, *Phys. Rev.* 175:256 (1968).

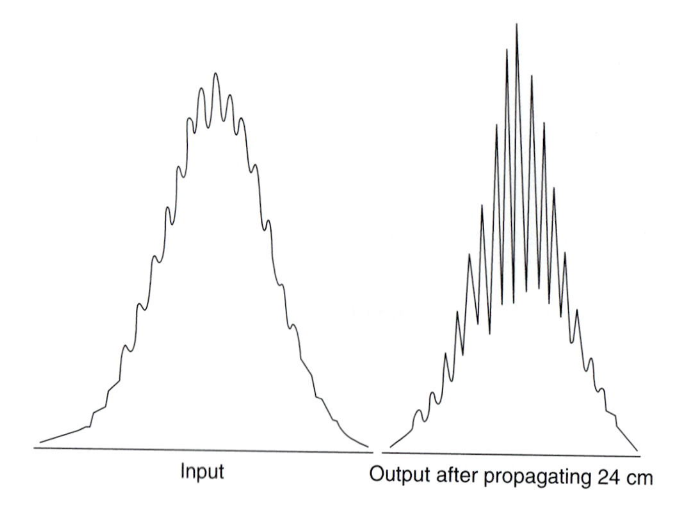

Figure 7.18 Small-scale diffraction effects (such as those generated by dust and clipping from optical surfaces) will produce interference fringes in the optical beam. These interference fringes can be quite intense and are typically subject to self-focusing effects long before the whole beam focuses. (Reprinted with permission from E. S. Bliss, D. R. Speck, J. F. Holzrichter, J. H. Erkkila, and A. J. Glass, *Appl. Phys. Lett.* 25:448 (1974). Copyright 1974 American Institute of Physics.)

In practice, this phenomena can be easily demonstrated by sending a high-intensity laser beam (such as a pulsed mode-locked Nd:YAG) through air. Air has a small nonlinear index n_2 and the beam will focus down to a spot several feet away from the laser. The intensity will exceed the dielectric breakdown of air and the air will pop into a plasma. This will scatter the beam and decrease the intensity. However the scattering is reasonably coherent and the expanded beam will continue to propagate. As it propagates, the beam will again be focused by the nonlinear constant of air and again it will focus down to a spot and pop into a plasma. This process will repeat three or four times as the laser beam propagates across the room. Eventually the process will terminate because the incoherent scattering will drop the intensity below the critical power.[78]

In large laser systems another self-focusing problem appears, termed small-scale self-focusing. The essential problem is that laser beams are not spatially uniform. Small-scale diffraction effects (such as those generated by dust and clipping from optical surfaces) tend to produce interference fringes in the optical beam. These interference fringes can be intense and are typically subject to self-focusing effects long before the whole beam focuses (see Figure 7.18).

For an initial perturbation $\delta = \Delta E/E$ the self-focusing length is given by[79]

$$z_{\text{self-focus}} = \left(\frac{\lambda_o}{2\pi n_2 I_p}\right) \ln\left(\frac{3}{\delta}\right) \tag{7.47}$$

where I_p is the beam intensity (watt/cm^2).

Small scale self-focusing can be reduced by means of a technique called spatial filtering. As is well-known, the image at the focal point of a lens represents the Fourier transform of the image entering the lens. Thus, if the image at the lens contains a high-frequency component, this will appear as widely separated spots at the focal point. If the focused laser beam is directed through a small dimension pinhole, these high-frequency components can

[78] A picture of this effect is given in Figure 13 of C. G. Young, *Proc. IEEE* 57:1267 (1969).

[79] A. J. Campillo, S. L. Shapiro, and B. R. Suydam, *Appl. Phys. Lett.* 23:628 (1973); and *Appl. Phys. Lett.* 24:178 (1974).

be stripped from the beam. The unfortunate side-effect of spatial filtering is that it imposes the Fourier transform of a round circle (which is an Airy disk) on the laser beam. Thus, the propagating laser beam is no longer Gaussian.

Example 7.8

Calculate the critical power for self-focusing for Nd:YAG at $\lambda = 1.064\ \mu$m.

Solution. The n_2 index for YAG from Table 7.3 is $8.945 \cdot 10^{-16}$ cm^2/watt and $n_o = 1.829$. Evaluating the critical power from Equation (7.45) gives

$$P_{\text{crit}} = \frac{\pi (0.61)^2 \lambda_o^2}{8 n_o^3 n_2} = 0.320 \text{ MW} \tag{7.48}$$

Example 7.9

Assuming the critical power calculated in Example 7.8, calculate the large- and small-scale focusing distances. Assume a Gaussian beam with an input beam waist of 1 mm and an input power of 10 MW. For the small-scale case, assume a 10% initial perturbation ($\delta = 0.1$).

Solution. For a Gaussian beam, the large-scale focusing distance is given by Equation (7.46) as

$$z_{\text{self-focus}} = \frac{2w_o^2}{0.61\lambda_o} \frac{1}{\sqrt{P/P_{\text{crit}}}} = 53.57 \text{ cm} \tag{7.49}$$

and for small-scale self-focusing by Equation (7.47) as

$$z_{\text{self-focus}} = \left(\frac{\lambda_o}{2\pi n_2 I}\right) \ln\left(\frac{3}{\delta}\right) = 202.29 \text{ cm.} \tag{7.50}$$

7.6 NONLINEAR CRYSTALS

A number of crystals are used in nonlinear optics. An excellent reference to these crystals is the *Handbook of Nonlinear Optical Crystals* by V. G. Dmitriev.[80] The data sheets from the various crystal manufacturers are also helpful. Only the major characteristics of the most important crystals are summarized below.

7.6.1 Major Crystals

Potassium dihydrogen phosphate (KDP) family. This group[81,82,83,84] of negative uniaxial crystals includes KDP (KH_2PO_4), ADP ($NH_4H_2PO_4$), and CDA (CsH_2AsO_4). The crystals are all $\bar{4}2$m and possess tetragonal symmetry.[85] Deuterated forms of the crystals

[80] V. G. Dmitriev, G. G. Gurzadyan, and D. N. Nikogosyan, *Handbook of Nonlinear Optical Crystals* (Berlin: Springer-Verlag, 1991).

[81] V. S. Suvarow and A. S. Sonin, *Sov. Phys. Crys.* 11:711 (1967).

[82] T. R. Sliker and S. R. Burlage, *J. Appl. Phys.* 34:1837 (1963).

[83] D. Eimerl, *IEEE J. Quantum Electron.* QE-23:575 (1987).

[84] R. C. Eckardt, H. Masuda, Y. X. Fan, and R. L. Byer, *IEEE J. Quan. Electron.* QE-26:922 (1990).

[85] Crystals are divided into seven major crystal systems defined by their lattice symmetry properties. (See Section 6.3.2, footnote 24.)

are also available and are indicated by the notation D*. The crystals are grown in water solution and therefore are exceptionally hygroscopic.[86] (It is virtually impossible to use KDP-type crystals in the laboratory environment without a special housing.) In addition, they are fragile and sensitive to thermal shock as well as hairline fractures. However, the ease of crystal growth means that it is possible to obtain large crystals of very high optical quality. Most members of the KDP family have a high damage threshold for optically induced damage. Additionally, certain members of the family (CDA and CD*A) phase match at 90 degrees for the 1.06 μm transition of Nd:YAG. The isomorph KDP is an exceptionally popular crystal for doubling high-power large-aperture lasers due to its availability in large sizes at relatively low cost. KDP possesses an absorption band at 1 to 1.2 μm, so KD*P is preferred over KDP for Nd:YAG and Nd:glass lasers.

Lithium niobate ($LiNbO_3$). $LiNbO_3$[87,88] is a 3m negative uniaxial crystal with trigonal symmetry. $LiNbO_3$ crystals are grown from the melt by a Czochralski process. The crystals are hard and nonhygroscopic. In addition, they have a much higher nonlinear coefficient than members of the KDP family. $LiNbO_3$ has the additional advantage of possessing a phase matching angle of 90 degrees for 1.06 μm at a temperature of 47°C. However, $LiNbO_3$ has a significantly lower damage threshold than members of the KDP family. In addition, many crystals of $LiNbO_3$ are susceptible to a type of damage (often referred to a photorefractive damage) that alters the index of refraction within the crystal. Doping $LiNbO_3$ with MgO reduces the photorefractive damage and permits 90-degree phase matching at a temperature of 107°C.[89] $LiNbO_3$ is most commonly used as a doubling crystal for internally frequency-doubled Nd:YAG lasers, and for lower-power externally frequency-doubled Nd:YAG and Nd:glass lasers. (See Figure 7.19.)

Potassium titanyl phosphate (KTP). KTP ($KTiOPO_4$)[90,91,92,93] is a mm2 biaxial crystal with orthorhombic symmetry. KTP is a difficult crystal to grow, and is currently grown by hydrothermal and flux growth techniques. However, KTP possesses good optical properties, a large acceptance angle, large temperature acceptance, a large nonlinear coefficient, and high optical damage thresholds. It is a mechanically rugged and nonhygroscopic crystal. However KTP suffers from a cumulative photochemical degradation phenomena, termed *grey tracking*, caused by long-term exposure to the intense fundamental and second harmonic radiation. Although this photochemical effect can be reversed by operating the crystal at an elevated temperature, absorption in the crystal due to the grey tracking may

---

[86]Due to the hygroscopic nature of the crystals, they are often polished in solutions of ethylene glycol rather than in water.

[87]R. S. Weis and T. K. Gaylord, *Appl. Phys.* A37:191 (1985).

[88]A. Rauber, "Chemistry and Physics of Lithium Niobate," in *Current Topics in Materials Science*, Vol. 1, ed E. Kaldis (Amsterdam: North-Holland, 1978), pp. 481–601.

[89]D. A. Bryan et al., *Appl. Phys. Lett.* 44:847 (1984).

[90]K. Kato, *IEEE J. Quan. Electron.* QE-24:3 (1988).

[91]H. Vanherzeele, *Opt. Lett.* 14:728 (1989).

[92]J. Q. Yao and T. S. Fahlen, *J. Appl. Phys.* 55:65 (1984).

[93]R. C. Eckardt, H. Masuda, Y. X. Fan, and R. L. Byer, *IEEE J. Quan. Electron.* 26:922 (1990).

Figure 7.19 $LiNbO_3$ is commonly used as a doubling crystal for internally doubled Nd:YAG lasers, and for lower-power externally doubled Nd:YAG and Nd:glass lasers. (Photograph courtesy of Crystal Technology Inc., Palo Alto, CA)

damage the crystal beyond repair. KTP is a more recently developed crystal than KDP or $LiNbO_3$, but is emerging as one of the most popular frequency-doubling crystals for Nd:YAG and Nd:glass lasers. KTP is also finding application as an OPO material and in difference frequency applications (see Figure 7.20).

Beta-barium borate (BBO). BBO (β-BaB_2O_4)[94,95,96,97] is a 3m negative uniaxial crystal with trigonal symmetry. The crystal possesses a moderate nonlinear coefficient, high-temperature stability, and an outstanding damage threshold. It is mildly hygroscopic, typically requiring a special housing under laboratory conditions. The crystal's optical transmission also extends down to 200 nm, making it possible to perform multiphoton nonlinear processes into the blue and UV. However, BBO also possesses a very low angular tolerance, meaning that good alignment and high-quality optical beams are required for efficient nonlinear conversion. The large walk-off angle also results in beams that are elliptical. Due to its broad phase-matching region, BBO finds good application for harmonic generation in Ti:sapphire lasers. It also is finding application in broadly tunable OPOs and optical parametric amplifiers (OPAs). The exceptionally high IR transmittance allows high average power OPO and OPA operation with minimal thermal heating caused by the idler radiation.

7.6.2 Other Crystals Used in Nonlinear Optics

Lithium iodate ($LiIO_3$). $LiIO_3$ is a 6 crystal with hexagonal symmetry. $LiIO_3$ crystals are grown from water solution. The crystal is hygroscopic and significantly more fragile than $LiNbO_3$. In addition, this crystal has a rather low optical-damage threshold.

[94] X. D. Wang, P. Basseras, J. D. Miller, and H. Vanherzeele, *Appl. Phys. Lett.* 59:519 (1991).

[95] C. J. Vanderpoel, J. D. Bierlein, J. B. Brown, and S. Colak, *Appl. Phys. Lett.* 57:2074 (1990).

[96] Y. X. Fan, R. C. Eckhardt, R. L. Byer, C. Chen, and A. D. Jiang, *IEEE J. of Quantum Electron.* 25:1196 (1989).

[97] D. N. Nikogosyan, *Appl. Phys.* A52:359 (1991).

Figure 7.20 KTP is a more recently developed crystal than KDP or $LiNbO_3$, but is emerging as a common doubling crystal for Nd:YAG and Nd:glass lasers. KTP is also used as an OPO material and in difference-frequency applications. (Photograph courtesy of Crystal Technology Inc., Palo Alto, CA)

However, the crystal does have a high nonlinear coefficient and does not suffer from the photorefractive damage of $LiNbO_3$. $LiIO_3$ is most commonly used as a frequency-doubling crystal for internally doubled Nd:YAG lasers where $LiNbO_3$ has not proven successful.

Barium sodium niobate ($Ba_2NaNb_5O_{15}$). $Ba_2NaNb_5O_{15}$ is a mm2 biaxial crystal with orthorhombic symmetry. $Ba_2NaNb_5O_{15}$ or "Banana" crystals were a popular nonlinear material in the mid-1970s. However, the crystal is exceptionally difficult to grow with high optical quality and is no longer available commercially.

Lithium tri-borate (LBO). LBO is a mm2 biaxial crystal with orthorhombic symmetry. It is a mechanically hard, nonhygroscopic crystal with good chemical stability and UV transparency. It possesses a moderate nonlinear coefficient and good optical damage thresholds.

SYMBOLS USED IN THE CHAPTER:

σ: Conductivity (Ω-m)
ρ: Charge density (C/m^3)[98]
$\vec{\mathcal{D}}$: Electric displacement vector (C/m^2)
$\vec{\mathcal{B}}$: Magnetic induction vector (Webers/m^2)
$\vec{\mathcal{E}}$: Electric field (V/cm)

[98] Notice that ρ finds common application as *both* the charge density and as the walk-off angle. This is usually not a problem, as the two usages rarely conflict in the same application, and they have different units.

$\vec{\mathcal{H}}$: Magnetic intensity vector (A/m)
$\vec{\mathcal{J}}$: Current density (A/m^2)
$\vec{\mathcal{P}}$: Electric polarization vector (C/m^2)
$\vec{\mathcal{M}}$: Magnetic polarization vector (A/m)
χ: Complex susceptibility (m/V)
$\chi_{ij}^{(1)}$: Second-rank susceptibility tensor (m/V)
$\chi_{ijk}^{(2)}$: Third-rank susceptibility tensor (m/V)
$\chi_{ijkl}^{(3)}$: Fourth-rank susceptibility tensor (m/V)
ω_m, ω_n, ω_o: Various angular frequencies (rad/sec)
E_j, E_k, E_l: Various scalar electric fields (V/m)
l_c: Second harmonic coherence length (cm or m)
λ_1: Free space wavelength of light at the fundamental (m)
λ_2: Free space wavelength of light for the second wave (m)
n_1, n_2: Indices of refraction for the fundamental and second wave[99]
E_1, E_2: Electric fields for the fundamental and second wave (V/m)
η_x: Characteristic impedance for the x wave (ohms)
d_{eff}: Effective nonlinear coefficient (As/V^2)
P_ω, $P_{2\omega}$: Power in the fundamental or second harmonic (watts)
A: Area of the beam (cm^2 or m^2)
n_o^ω, $n_o^{2\omega}$: Ordinary indices of refraction for the fundamental and second harmonic (unitless)
n_e^ω, $n_e^{2\omega}$: Extraordinary indices of refraction for the fundamental and second harmonic (unitless)
Θ: Angle of propagation as measured from the optic axis (degrees or radians)
Φ: Angle of propagation as measured from the x-axis (degrees or radians)
ρ: Walk-off angle (degrees or radians)
l_w: Walk-off distance (cm or m)
λ_p, λ_s, λ_i: Free space wavelength of light for the pump, signal and idler (cm or m)
ω_p, ω_s, ω_i: Free space angular frequencies of light for the pump, signal and idler (rad/sec)
n_p, n_s, n_i: Indices of refraction for the pump, signal and idler (unitless)
E_p, E_s, E_i: Electric fields for the pump, signal and idler (V/m)
P_p, P_s, P_i: Intensities for the pump, signal and idler (watts)
η_p, η_s, η_i: Characteristic impedances for the pump, signal and idler (ohms)
G: Single-pass gain
ω_P, ω_R: Free space angular frequencies of light for the pump and Raman line (rad/sec)
λ_p, λ_{st}: Free space wavelength of light for the pump and first Stokes wave (cm or m)
ω_p, ω_{st}: Free space angular frequencies of light for the pump and first Stokes wave (rad/sec)

---

[99]Notice that n_2 finds common application as *both* the index at the second harmonic (for SHG) and as the nonlinear index for self-focusing. This is usually not a problem as the two usages have different units.

n_p, n_{st}: Indices of refraction for the pump and first Stokes wave

E_p, E_{st}: Electric fields for the pump and first Stokes wave (V/m)

E_p, E_{st}: Electric fields for the pump and first Stokes wave (V/m)

γ_s: Gain of the first Stokes wave (cm^{-1})

χ_R'': Peak value of the imaginary part of the Raman susceptibility (m/V)

I_p: Pump intensity (watts/cm^2)

$d\sigma/d\Omega$: Spontaneous Raman scattering cross section (cm^2/steradian)

N: Number density of scattering centers (centers/cm^3)

$\Delta\nu_R$: Full-width, half-maximum (FWHM) Raman frequency linewidth (Hz)

η: Peak conversion efficiency for Raman shifting

wn_p: Pump wavenumber (cm^{-1})

wn_R: Raman state wavenumber (cm^{-1})

n_2: Nonlinear index for self-focusing[100] (cm^2/watt)

P_{crit}: Power level for which whole-beam self-focusing overcomes diffraction (watts)

$z_{\text{self-focus}}$: Self-focusing distance (cm or m)

w_o: Initial beam waist for an arbitrary Gaussian beam (cm or m)

P: Initial power for an arbitrary Gaussian beam (watts)

δ: Initial perturbation for small signal self-focusing (unitless)

EXERCISES

Second harmonic generation

7.1 Calculate the second harmonic generation energy transfer distance l_c for KD*P and for lithium niobate assuming Type I phase matching at room temperature with a phase-matching angle of 90 degrees. Assume the input fundamental wavelength is 1.064 μm. For room temperature, assume the ordinary index for KD*P at 1.064 μm is 1.4928 and at 532 nm is 1.5085, and that the extraordinary index for KDP at 1.064 μm is 1.4555 and at 532 nm is 1.469. For room temperature, assume the ordinary index for lithium niobate at 1.064 μm is 2.235 and at 532 nm is 2.3251, and that the extraordinary index for lithium niobate at 1.064 μm is 2.1554 and at 532 nm is 2.233. Both crystals are negative uniaxial.

7.2 Use the depletion approximation Equation (7.12) to plot the conversion efficiency as a function of *intensity* for KD*P in a Type II configuration. Assume the crystal is 2 cm long and is phase-matched at an angle of $\theta = 50$ degrees and $\phi = 90$ degrees. Plot input intensities of 1 to 500 MW/cm^2. Comment on the effect of intensity on the second harmonic conversion efficiency.

7.3 Use the depletion approximation Equation (7.12) to plot the conversion efficiency as a function of *length* for KD*P in a Type II configuration. Assume the input intensity is 100 MW/cm^2 and that the crystal is phase matched at an angle of $\theta = 50$ degrees and $\phi = 90$ degrees. Plot lengths of 1 to 10 cm. Comment on the effect of length on the second harmonic conversion efficiency.

---

[100]Notice that n_2 is commonly used for both the second harmonic real index of refraction (unitless) and for the nonlinear focusing index (cm^2/watt). Both context and the difference in units can be used to distinguish these.

7.4 Observe the effects of small phase mismatches on the second harmonic conversion efficiency by plotting the conversion efficiency as a function of length using for KD*P in a Type II configuration using the low-conversion approximation Equation (7.15). Assume the input intensity is 100 MW/cm^2 and that the crystal is phase matched at an angle of $\theta = 50$ degrees and $\phi = 90$ degrees. Plot for Δn values of 0.001, 0.002, and 0.005. Set the plot range from 0 to 500 μm. Comment on the effects of small phase mismatches on the second harmonic conversion efficiency.

7.5 Compare Equations (7.12) and (7.15) by plotting the conversion efficiency as a function of length using for KD*P in a Type II configuration. Assume the input intensity is 100 MW/cm^2 and that the crystal is phase matched at an angle of $\theta = 50$ degrees and $\phi = 90$ degrees. Set the plot range from 0 to 500 μm. Assume $\Delta n = 0.00005$. Discuss the differences between the two equations and develop a rule of thumb characterizing when it is appropriate to use Equation (7.15).

7.6 Calculate the phase matching and walk-off angles for beta barium borate (BBO) in a Type I configuration at 25°C. Assume that the input beam is 1.064 μm, and that the doubled beam is 532 nm. Assume that the ordinary index for the fundamental at 25°C is 1.6551 and for the second harmonic is 1.675. Assume that the extraordinary index for the fundamental at 25°C is 1.5426 and that the extraordinary index for the second harmonic is 1.5555.

7.7 Calculate the phase matching and walk-off angles for both Type I and Type II phase matching for AD*P at 25 degrees C. Assume that the input beam is 1.064 μm, and that the doubled beam is 532 nm. Assume that the ordinary index for the fundamental at 25°C is 1.5049 and for the second harmonic is 1.5210. Assume that the extraordinary index for the fundamental at 25°C is 1.4659 and that the extraordinary index for the second harmonic is 1.4778. (AD*P is a member of the KDP family and is a 42m uniaxial material.)

7.8 Calculate the phase matching and walk-off angles for CD*A at 25°C. Calculate the phase matching angles for both Type I and Type II phase matching. Assume that the input beam is 1.064 μm, and that the doubled beam is 532 nm. Assume that the ordinary index for the fundamental at 25°C is 1.5499 and for the second harmonic is 1.5692. Assume that the extraordinary index for the fundamental at 25°C is 1.5341 and that the extraordinary index for the second harmonic is 1.5496. (CD*A is a member of the KDP family and is a 42m uniaxial material.)

Optical parametric oscillators

7.9 (design) There is some good-natured controversy on whether optical parametric oscillators should really be included in a book on lasers. Compare and contrast optical parametric oscillators to traditional lasers. Your answer should include a list of similarities and differences between optical parametric oscillators and conventional lasers.

Stimulated Raman scattering

7.10 Assume that your laboratory has a Nd:YAG laser that can run both at the fundamental and at the second harmonic (1.064 μm and 532 nm). Assume you purchase a Raman gas cell, which you operate on high-pressure gas mixtures of either hydrogen or nitrogen. How many different visible coherent light wavelengths can you conceivably obtain and what are their values? (Assume the visible spectrum extends from 400 to 760 nm.)

Self-focusing

7.11 Calculate the critical power for large-scale self-focusing in fused silica and compare it with that of BK-7. Assume a beam is propagating though each material and calculate the distance to the large-scale self-focus for an input power of 50 MW and a minimum beam waist of 2 mm. Assume the wavelength is the second harmonic of Nd:YAG at 532 nm.

7.12 Calculate the critical power for large-scale self-focusing in a ruby. Assume a beam is propagating though the ruby and calculate the distance to the large-scale self-focus for an input power of 100 MW and a minimum beam waist of 1 mm. Calculate the distance to the small-scale self-focus assuming that the initial perturbation is 30%. Assume the wavelength is the wavelength of the ruby laser at 694.3 nm.

Nonlinear crystals

7.13 (design) Contact a few vendors and obtain data sheets for several major nonlinear crystals. (You may want to coordinate with your classmates to avoid having 30 people call the same vendor!) Create a table comparing the various optical and mechanical properties of these crystals. Your answer should include

(a) your comparison table, and

(b) a concise (a few sentences) description of which material you would pick for the intended application and why.

[Possible vendors include Litton/Airtron, Union Carbide, Crystal Technology, and Casix for crystals. Check the *Photonics Buyers Guide to Products and Manufacturers* (Pittsfield, MA: Laurin Publishing Co., Inc.) for additional vendors.]

8

Supportive Technologies

Objectives

- To qualitatively describe the reasons that multilayer dielectric optical coatings are used for anti-reflectance and high-reflectance laser coatings.
- To qualitatively describe the process by which transmittance and reflectance are calculated for multilayer dielectric optical films.
- To qualitatively describe the physics underlying the design of a quarter-wavelength anti-reflectance coating.
- To compute the index of refraction for an optimal quarter-wavelength anti-reflectance coating.
- To qualitatively describe the physics underlying the design of a multilayer $(HL)^N$ high-reflectance coating.
- To compute the peak reflectance and width of the high-reflectance zone for a simplified N-layer quarter-wave $(n_H d_H = n_L d_L = \lambda_o/4)$ high-reflectance coating.
- To describe and sketch indicatrices for positive and negative uniaxial crystals.
- To describe and sketch the process of using the indicatrix to determine the directions and indices of refraction of the ordinary and extraordinary polarizations for light interacting with a crystal at an angle of θ from the z-axis and ϕ from the x-axis.
- To qualitatively describe the operation of quarter- and half-wave plates.
- To calculate the thickness of first and multiple order, quarter- and half-wave plates.
- To describe the use of a quarter-wave plate to reduce back reflections in laser systems.
- To qualitatively describe the operation of thermal detectors such as thermopiles, bolometers, and pyroelectric detectors.
- To qualitatively describe the operation and use of photoelectric detectors.

- To qualitatively describe the operation and use of photomultiplier tubes.
- To qualitatively describe the operation and use of a photoconductor.
- To compute the current generated in a photoconductor for a given incident laser power.
- To qualitatively describe the operation and use of junction photodetectors such as PN photodiodes, PIN photodiodes, avalanche photodiodes, and phototransistors.
- To distinguish between photoconductive and photovoltaic operation of a photodiode and to be able to compute the appropriate current or voltage.
- To qualitatively describe the use of a MOS capacitor as a photoconducting device.

8.1 INTRODUCTION

Designing a laser system involves integrating a number of different technologies. Some of these (such as power supply design) are generic technologies used in many industries. Others (such as birefringent filter design; see Section 11.4.2) are technologies used almost exclusively by the laser industry.

This chapter presents three enabling technologies that are critical to the design of laser systems. Section 8.2 discusses multilayer dielectric films used for the manufacture of anti-reflection coatings, high-reflection coatings, beam-splitters, and interference filters. Section 8.3 reviews the properties of birefringent crystals as used for the manufacture of polarizers, harmonic generators, Q-switches, mode-lockers, and broadband filters. Section 8.4 summarizes the common photodetector technologies used with modern laser systems.

8.2 MULTILAYER DIELECTRIC FILMS

Multilayer dielectric films are used to construct high-reflecting mirrors, narrow-bandwidth interference filters, anti-reflection coatings, and dielectric polarizers. Although a complete treatment is outside the scope of this book, a basic treatment that permits calculation of the key features is presented in the following section. This treatment follows the method and notation presented in Born and Wolf, *Principles of Optics*.[1] Other excellent treatments of multilayer films can be found in optics reference texts.[2,3,4,5,6]

---

[1]Max Born and Emil Wolf, *Principles of Optics,* 6th ed. (Oxford, U.K.: Pergamon Press, 1980), pp. 57–61.

[2]J. A. Dobrowolski, "Optical Properties of Films and Coatings," in *Handbook of Optics, Vol. I, Fundamentals, Techniques, and Design*, ed Michael Bass (New York: McGraw Hill, 1995), pp. 42-1 to 42-130.

[3]J. A. Dobrowolski, "Coatings and Filters," in *Handbook of Optics,* eds W. G. Driscoll and W. Vaughan (New York: McGraw-Hill, 1978), pp. 8-1 to 8-124.

[4]H. A. McCleod, *Thin Film Optical Filters* (New York: McGraw-Hill, 1986).

[5]J. D. Rancourt, *Optical Thin Film Users' Handbook* (New York: Macmillan, 1987).

[6]A. Thelen, *Design of Optical Interference Filter Coatings* (New York: McGraw-Hill, 1988).

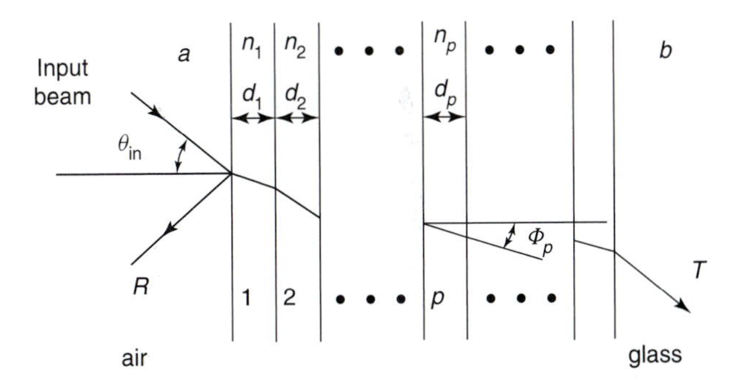

Figure 8.1 General configuration of a multilayer film. The wave is incident in material *a* (typically air) and exits in material *b* (typically glass). The angle ϕ_p is the angle (as measured from the normal) of the transmitted beam in the *p*th film.

8.2.1 The Fundamentals of Multilayer Film Theory

Imagine a situation where an electromagnetic wave is incident upon a stack of thin dielectric films (see Figure 8.1). The wave is incident in material *a* (typically air) and exits in material *b* (typically glass). Assume that z direction is vertical to the film and that the electromagnetic wave is linearly polarized. The angle ϕ_p is defined as the angle (as measured from the normal) of the transmitted beam in the *p*th film. This angle is related to the incident angle θ_{in} (as measured from the normal) for the first film by Snell's law as

$$\phi_{\text{first}} = \sin^{-1}\left(\frac{n_a}{n_1} \cdot \sin\theta_{\text{in}}\right) \tag{8.1}$$

where n_a is the index of air and n_1 the index of the first film layer. Additional films have

$$\phi_p = \sin^{-1}\left(\frac{n_p}{n_{p+1}} \cdot \sin\phi_{p-1}\right) \tag{8.2}$$

and the last film has

$$\phi_{\text{out}} = \sin^{-1}\left(\frac{n_p}{n_b} \cdot \sin\phi_{p-1}\right) \tag{8.3}$$

where n_b is the index of the substrate. (See Appendix A.5.)

The *p*th film in the multilayer is characterized by a matrix $\mathcal{M}$, which is given as[7]

$$\mathcal{M} = \begin{bmatrix} \cos\delta_p & \frac{-j}{u_p}\sin\delta_p \\ -ju_p\sin\delta_p & \cos\delta_p \end{bmatrix} = \begin{bmatrix} m_{11} & m_{12} \\ m_{21} & m_{22} \end{bmatrix} \tag{8.4}$$

where

$$\delta_p = \frac{2\pi}{\lambda_o}\left(n_p d_p \cos\phi_p\right) \tag{8.5}$$

and the quantity $n_p d_p \cos\phi_p$ can be considered to be the optical thickness of the layer at the angle ϕ_p.

[7]This is from Max Born and Emil Wolf, *Principles of Optics,* 6th ed. (Oxford, U.K.: Pergamon Press, 1980), p. 58. It differs from J. A. Dobrowolski, "Optical Properties of Films and Coatings," in *Handbook of Optics, Vol. I, Fundamentals, Techniques, and Design*, ed Michael Bass (New York: McGraw Hill, 1995), p. 42-11. Dobrowolski is missing the minus sign on the cross term.

Now, the effective index of refraction u_p for the film layers is

$$u_p = n_p \cos\phi_p \quad \text{for polarization } \perp \text{ to the plane of incident (TE polarization)} \tag{8.6}$$

$$u_p = \frac{1}{n_p}\cos\phi_p \quad \text{for polarization } \parallel \text{ to the plane of incident (TM polarization)} \tag{8.7}$$

Of course, there is also a u_p for the entering wave,

$$u_{\text{in}} = n_a \cos\theta_{\text{in}} \quad \text{for polarization } \perp \text{ to the plane of incident (TE polarization)} \tag{8.8}$$

$$u_{\text{in}} = \frac{1}{n_a}\cos\theta_{\text{in}} \quad \text{for polarization } \parallel \text{ to the plane of incident (TM polarization)} \tag{8.9}$$

and similarly for the leaving wave (with the substitution of n_b for n_a and ϕ_{out} for θ_{in}).

If a material is absorbing, then the index of refraction n_p must be given in its complex form as

$$n_p = \overline{n_p} + j\kappa_p \tag{8.10}$$

where $\overline{n_p}$ is the real part of the index and κ_p is the extinction coefficient.

A complete multilayer system is composed of many multilayer components. The characteristic matrix for the complete system is the product of the characteristic matrices of the various component layers as

$$\mathcal{M} = \mathcal{M}_1 \cdot \mathcal{M}_2 \cdot \mathcal{M}_3 \cdot \cdots \cdot \mathcal{M}_l \tag{8.11}$$

where each $\mathcal{M}_p$ represents a different layer.

Notice that the matrices are multiplied together *in the order in which they are encountered* (rather than in the reverse order, as is the case with lens matrices). The order reversal in the case of a lens matrix system is because the solution matrix from a ray matrix calculation is referenced from the output ray, while the solution matrix from a film matrix calculation is referenced from the starting ray.

The field (amplitude) reflection and transmission coefficients for the perpendicular polarization case are given by[8]

$$r_\perp = \frac{(m_{11\perp} + u_{\text{out}\perp} \cdot m_{12\perp}) \cdot u_{\text{in}\perp} - (m_{21\perp} + u_{\text{out}\perp} \cdot m_{22\perp})}{(m_{11\perp} + u_{\text{out}\perp} \cdot m_{12\perp}) \cdot u_{\text{in}\perp} + (m_{21\perp} + u_{\text{out}\perp} \cdot m_{22\perp})} \tag{8.12}$$

and

$$t_\perp = \frac{2 \cdot u_{\text{in}\perp}}{(m_{11\perp} + u_{\text{out}\perp} \cdot m_{12\perp}) \cdot u_{\text{in}\perp} + (m_{21\perp} + u_{\text{out}\perp} \cdot m_{22\perp})} \tag{8.13}$$

and the intensity transmission and reflection coefficients are given by

$$T = \frac{u_{\text{out}\perp}}{u_{\text{in}\perp}} |t_\perp|^2 \tag{8.14}$$

and

$$R = |r_\perp|^2 \tag{8.15}$$

[8]Max Born and Emil Wolf, *Principles of Optics,* 6th ed. (Oxford, U.K.: Pergamon Press, 1980), p. 60.

where the notation of $|x|^2$ is taken to mean the product of a complex number with its complex conjugate as $|x|^2 = x \cdot x^*$. Similar relationships exist for the parallel case with the substitution of $\parallel$ for $\perp$ in Equations (8.12) and (8.13). (Note that in the parallel or TM case, r and t represent the reflection and transmission of the magnetic field rather than electric field components.)

For obliquely nonpolarized radiation[9]

$$T = \frac{1}{2}\left(T_{\parallel} + T_{\perp}\right) \tag{8.16}$$

and

$$R = \frac{1}{2}\left(R_{\parallel} + R_{\perp}\right). \tag{8.17}$$

The phase changes φ on reflection and transmission are given as

$$\varphi_R = Arg\,(r)$$

and

$$\varphi_T = Arg\,(t) \tag{8.18}$$

where the absorption of the multilayer structure can be calculated from

$$A = 1 - T - R. \tag{8.19}$$

8.2.2 Anti-Reflection Coatings from Multilayer Films

Anti-reflection (AR) coatings are essential to the design of laser systems. A single glass surface typically has a 4% reflection loss at normal incidence. It does not take too many glass surfaces within a laser cavity to make the laser so lossy that it cannot lase. Additionally, spurious reflections from uncoated glass surfaces may cause secondary lasing or amplification of back reflections.

Visualize a laser beam incident upon a piece of glass at normal incidence (see Figure 8.2). Assume that the glass is coated with a single thin film layer. Also assume that neither the glass nor the thin film are lossy. Under these conditions

$$\theta_{\text{in}} = \phi_p = 0 \tag{8.20}$$

$$\delta_p = \frac{2\pi n_p d_p}{\lambda_o} \tag{8.21}$$

and

$$u_p = n_p \quad \text{for both polarizations.} \tag{8.22}$$

The pth film in the multilayer is characterized by a matrix $\mathcal{M}$, which is given as

$$\mathcal{M} = \begin{bmatrix} \cos\left(\frac{2\pi n_p d_p}{\lambda_o}\right) & \frac{-j}{n_p}\sin\left(\frac{2\pi n_p d_p}{\lambda_o}\right) \\ -jn_p \sin\left(\frac{2\pi n_p d_p}{\lambda_o}\right) & \cos\left(\frac{2\pi n_p d_p}{\lambda_o}\right) \end{bmatrix} = \begin{bmatrix} m_{11} & m_{12} \\ m_{21} & m_{22} \end{bmatrix}. \tag{8.23}$$

[9] J. A. Dobrowolski, "Optical Properties of Films and Coatings," in *Handbook of Optics, Vol. I, Fundamentals, Techniques, and Design*, ed Michael Bass (New York: McGraw Hill, 1995), p. 42-12.

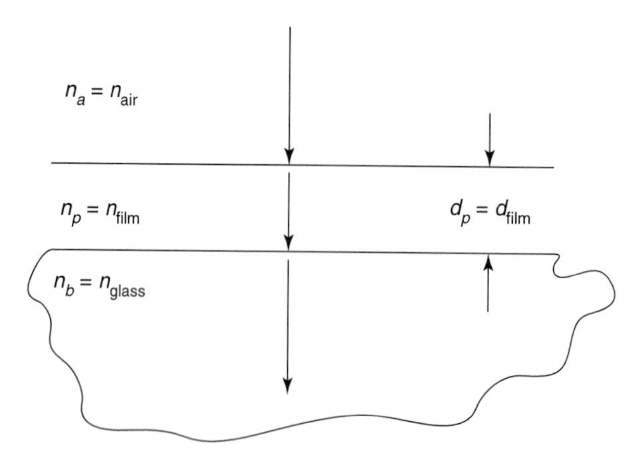

Figure 8.2 Consider a laser beam incident upon a thin film deposited on a piece of glass at normal incidence.

Notice that if

$$\frac{2\pi n_p d_p}{\lambda_o} = \frac{\pi}{2} \tag{8.24}$$

that is, if the thickness is an integral multiple of a quarter-wavelength

$$d_p = \frac{\lambda_o}{4 \cdot n_p} \tag{8.25}$$

then the sin terms go to unity and the cos terms go to zero. This leaves

$$\mathcal{M} = \begin{bmatrix} 0 & \frac{-j}{n_p} \\ -jn_p & 0 \end{bmatrix} = \begin{bmatrix} m_{11} & m_{12} \\ m_{21} & m_{22} \end{bmatrix}. \tag{8.26}$$

Assuming that there is only one film, then the $\mathcal{M}$ coefficients substitute directly into Equation (8.12) as

$$r = \frac{\left(0 + n_b \cos\theta_{\text{out}} \cdot \frac{-j}{n_p}\right) \cdot n_a \cos\theta_{\text{in}} - \left(-jn_p + n_b \cos\theta_{\text{out}} \cdot 0\right)}{\left(0 + n_b \cos\theta_{\text{out}} \cdot \frac{-j}{n_p}\right) \cdot n_a \cos\theta_{\text{in}} + \left(-jn_p + n_b \cos\theta_{\text{out}} \cdot 0\right)} \tag{8.27}$$

and

$$r = \frac{\left(n_b \cos\theta_{\text{out}} \cdot \frac{-j}{n_p}\right) \cdot n_a \cos\theta_{\text{in}} - \left(-jn_p\right)}{\left(n_b \cos\theta_{\text{out}} \cdot \frac{-j}{n_p}\right) \cdot n_a \cos\theta_{\text{in}} + \left(-jn_p\right)}. \tag{8.28}$$

Assuming normal incidence, $\cos\theta_{\text{out}} = 1$ and $\cos\theta_{\text{in}} = 1$ so[10]

$$r = \frac{\left(n_b \cdot \frac{-j}{n_p}\right) \cdot n_a - \left(-jn_p\right)}{\left(n_b \cdot \frac{-j}{n_p}\right) \cdot n_a + \left(-jn_p\right)} = \frac{n_b \cdot n_a - \left(n_p\right)^2}{n_b \cdot n_a + \left(n_p\right)^2} \tag{8.29}$$

[10] Max Born and Emil Wolf, *Principles of Optics,* 6th ed. (Oxford, U.K.: Pergamon Press, 1980), p. 63.

and

$$R = |r|^2 = \frac{\left(n_a n_b - n_p^2\right)^2}{\left(n_a n_b + n_p^2\right)^2}. \tag{8.30}$$

Thus, if the reflection is to equal zero,[11]

$$\left(n_a n_b - n_p^2\right) = 0 \tag{8.31}$$

or

$$n_a n_b = n_p^2 \tag{8.32}$$

or

$$n_p = \sqrt{n_a n_b}. \tag{8.33}$$

Example 8.1

Assume that an argon-ion laser at $\lambda = 514.5$ nm is incident on a glass surface of index 1.5. What is the optimal coating index? How thick is the anti-reflection coating? What if the material were not glass but Nd:YAG? What is the optimal coating index?

Solution. The film index of refraction is given by Equation (8.33) as

$$n_{\text{film}} = \sqrt{n_{\text{air}} n_{\text{glass}}}. \tag{8.34}$$

For glass of index $n_{\text{glass}} = 1.5$, this would mean a film index of

$$n_{\text{film}} = \sqrt{1 \cdot 1.5} = 1.225. \tag{8.35}$$

The film thickness is given by Equation (8.25) as

$$d_{\text{film}} = \frac{\lambda_o}{4 \cdot n_{\text{film}}} = \frac{514.5 \cdot 10^{-9}\ \text{m}}{4 \cdot 1.5} = 105.022 \cdot 10^{-9}\ \text{m}. \tag{8.36}$$

Although there are thin film materials with indices near 1.2 (sol-gels for example) these materials are typically not rugged enough for most laser applications. MgF_2 films (with an index of 1.38 at 550 nm) are almost universally used for single-layer anti-reflection coatings. Although such films give a maximum reflection of approximately 2% when used on glasses such a BK-7, they are rugged nonhygroscopic films with good damage thresholds.

Notice that as the material index increases, the MgF_2 film becomes closer to the optimal. For Nd:YAG with $n_{\text{YAG}} = 1.82$ the optimal film index would be

$$n_{\text{film}} = \sqrt{1 \cdot 1.82} = 1.349. \tag{8.37}$$

Single film quarter-wavelength MgF_2 anti-reflection coatings are still the most popular anti-reflection coatings used in the laser industry. However, for specialized applications, it is possible to obtain lower reflectances and broader bandwidths by use of multiple-laser anti-reflection coatings.

Various combinations of 2 to 4 dielectric materials in 10 to 50 layers can produce anti-reflection coatings with bandwidths of hundreds of nanometers and reflectances less than 1%. Many such coatings are available; however, most are proprietary to various optical houses (see Figure 8.3). As would be expected, such complex multilayer coatings are significantly more expensive (and often have lower optical-damage thresholds) than the more conventional single-layer MgF_2 anti-reflection coatings.

[11]Max Born and Emil Wolf, *Principles of Optics,* 6th ed. (Oxford, U.K.: Pergamon Press, 1980), p. 64.

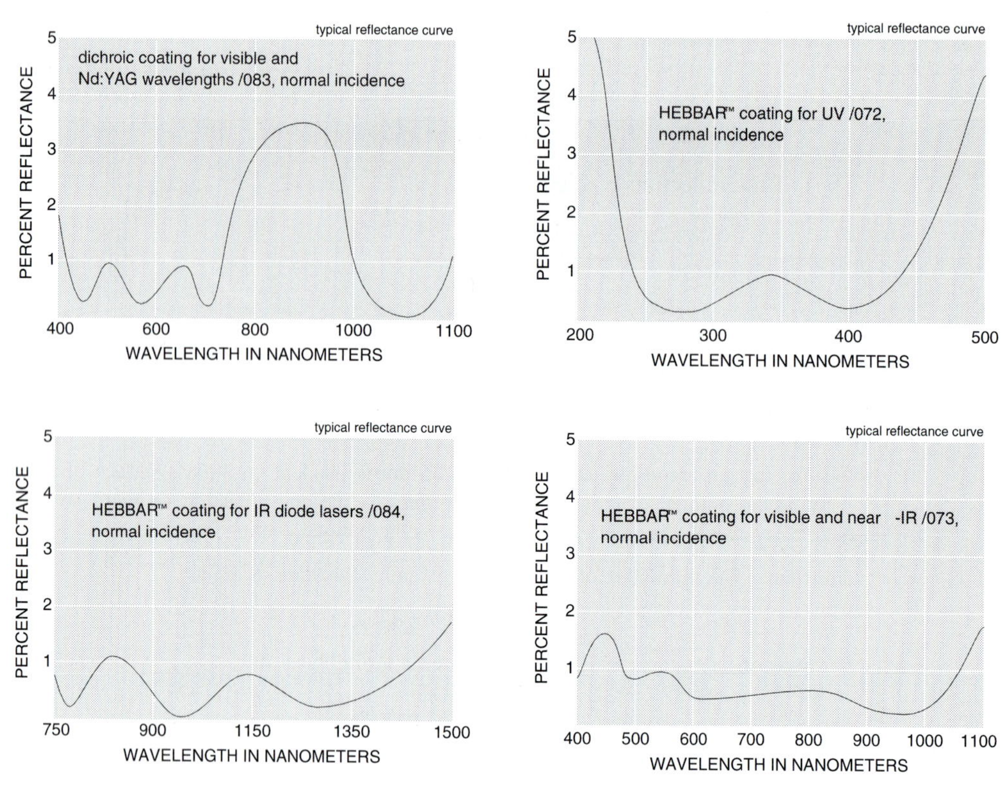

Figure 8.3 Various combinations of materials can produce anti-reflection coatings with bandwidths of hundreds of nanometers and reflectances near 1%. (Courtesy of Melles Griot)

8.2.3 High-Reflectance Coatings from Multilayer Films

An uninitiated outsider might imagine a laser mirror as looking somewhat like a bathroom mirror (perhaps cleaner!), but certainly metallic and probably coated on the back side. However, laser mirrors rarely look like this. Instead of being formed of metals, laser mirrors are generally constructed of dielectric multilayers and are usually coated on the front surface. To the human eye, they often look translucent or transparent. (Since this can often lead to them being installed backwards, a tiny arrow is usually drawn on the side of the mirror that points to the front coated surface.)

Such multilayer high-reflectance coatings are the backbone of the laser industry. Notice that even the most reflective metals are only about 90% reflective (see Figure 8.4). If laser engineers were limited to metal mirrors, cavity losses would be significantly higher, and mirror damage thresholds would be much lower.

The simplest high-reflectance dielectric multilayer is composed of N pairs of alternating high- and low-index materials as

$$\text{air}...HL..HL..HL..HL...\text{glass} = \text{air}\,(HL)^N\,\text{glass} \tag{8.38}$$

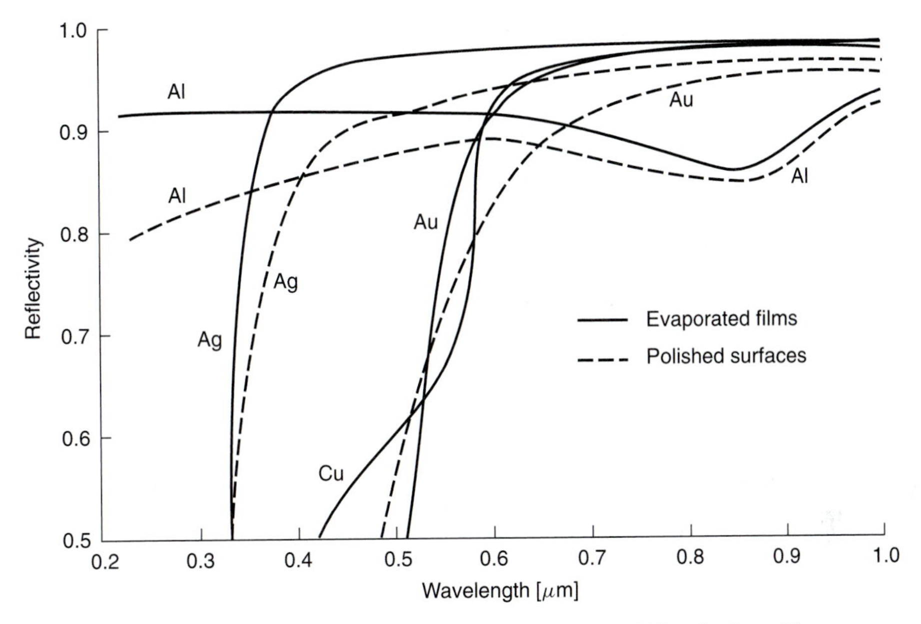

Figure 8.4 Even the most reflective metals are only about 90% reflective. (From W. Koechner, *Solid State Laser Engineering*, 4th ed. (Berlin: Springer-Verlag, 1996), Figure 6.87, p. 383. ©Springer-Verlag 1966. Reproduced with permission of the author.)

It is often the case that these layers are of identical optical thicknesses and the optical thicknesses are chosen to be a quarter-wavelength as

$$n_H d_H = n_L d_L = \frac{\lambda_o}{4}. \tag{8.39}$$

Under these conditions, the reflectance profile will consist of a series of peaks, as illustrated in Figure 8.5. The first peak reflectance occurs at

$$n_H d_H + n_L d_L = \frac{\lambda_o}{2} \tag{8.40}$$

and subsequent reflectances for N layers are given by

$$N\,(n_H d_H + n_L d_L) = p \cdot \frac{\lambda_o}{2} \quad \text{where } p = 2, 3, 4 \ldots \tag{8.41}$$

The highest reflectances are obtained when $n_H d_H$ and $n_L d_L$ are equal and an odd multiple of $\lambda_o/4$. Under these conditions, the peak reflectance for N layers is given by[12]

$$R_{\max} = \left[\frac{n_{\text{air}}/n_{\text{glass}} - (n_H/n_L)^{2N}}{n_{\text{air}}/n_{\text{glass}} + (n_H/n_L)^{2N}}\right]^2 \tag{8.42}$$

and the peak reflectance increases dramatically with the number of layers (see Figure 8.6).

[12] J. A. Dobrowolski, "Optical Properties of Films and Coatings," in *Handbook of Optics, Vol. I, Fundamentals, Techniques, and Design*, ed Michael Bass (New York: McGraw Hill, 1995), Chapter 42, p. 42-36.

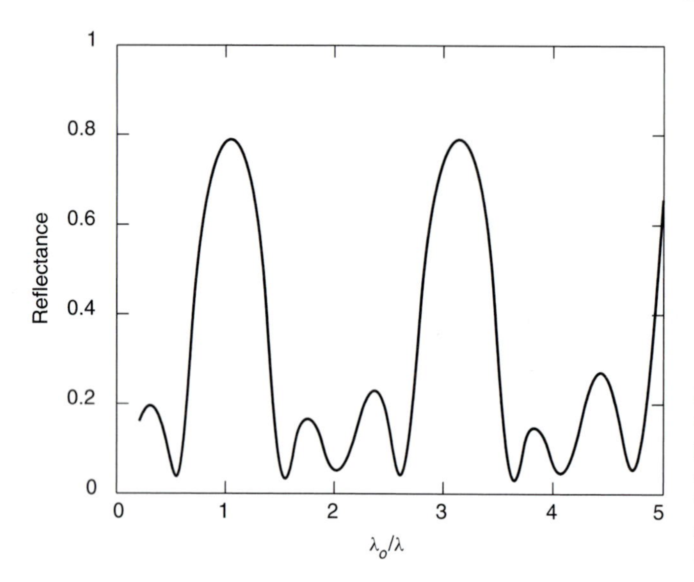

Figure 8.5 The reflectance profile of a dielectric multilayer consists of a series of peaks. This was calculated for a two-layer quarter-wavelength high-reflectance film, where $\lambda_o = 632.8$ nm, $n_a = 1.0$, $n_b = 1.52$, $n_H = 2.36$, $n_L = 1.38$, and where the incident angle is $\theta = 30$ deg.

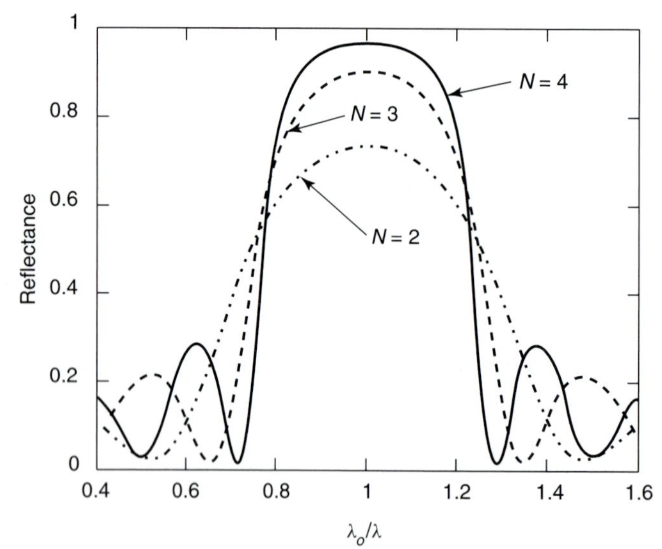

Figure 8.6 The peak reflectance increases dramatically with the number of layers.

The width of the high reflectance zone is greatest when $n_H d_H = n_L d_L = \lambda_o/4$ and is given by[13]

$$\frac{\Delta\lambda}{\lambda} = \frac{4}{\pi}\sin^{-1}\left(\frac{1 - n_H/n_L}{1 + n_H/n_L}\right). \tag{8.43}$$

The various combinations of materials, the ordering of high- and low-index layers, and the choice of number of layers can produce high-reflectivity coatings with a variety of

[13] J. A. Dobrowolski, "Optical Properties of Films and Coatings," *Handbook of Optics, Vol. I, Fundamentals, Techniques, and Design*, ed Michael Bass (New York: McGraw Hill, 1995), Chapter 42, p. 42-38.

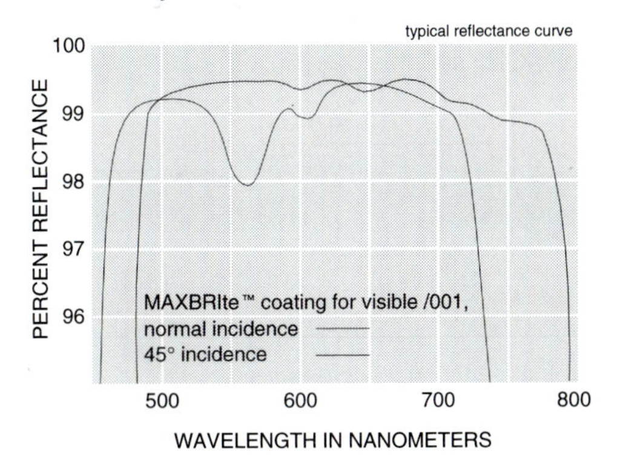

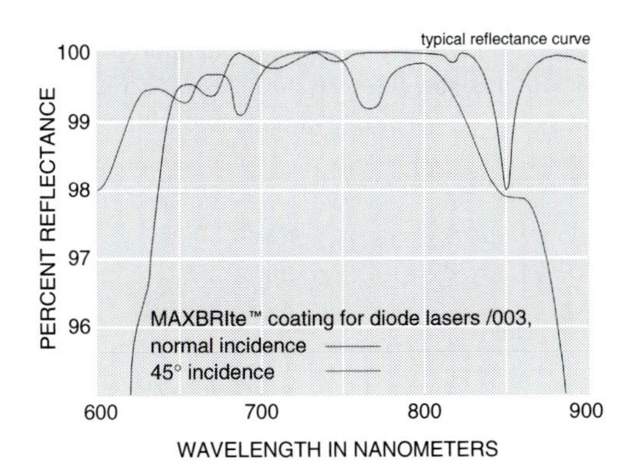

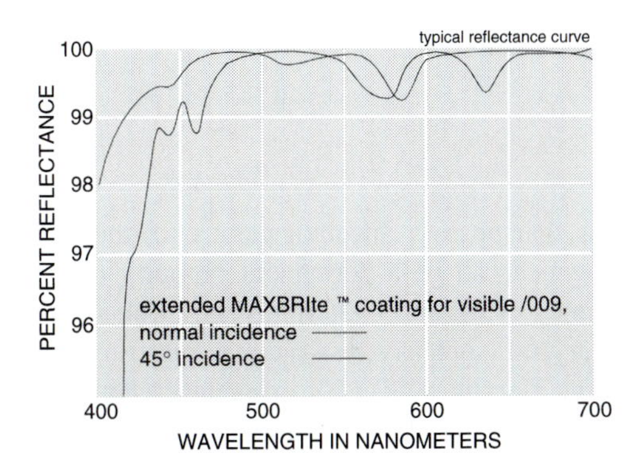

Figure 8.7 Various combinations of materials can produce high reflectivity coatings with a variety of bandwidths and reflectances. (Courtesy of Melles Griot)

bandwidths and reflectances. As with the AR coatings, many coatings are available, but most are proprietary to the various optical houses (see Figure 8.7).

Example 8.2

What is the maximum reflectance of a two-period dielectric multilayer of the form air $\cdot$ $(HL)^N$ $\cdot$ glass assuming air has an index of 1.0, the glass has an index of 1.520, the high-index layer is 2.360 and the low-index layer is 1.380? What if it is increased to five periods? Ten periods?

Solution. This uses Equation (8.42) as

$$R_{\max} = \left[\frac{1/1.52 - (2.36/1.38)^{2N}}{1/1.52 + (2.36/1.38)^{2N}}\right]^2. \tag{8.44}$$

Evaluating for the various cases yields

N	$R_{\max}$
2	0.734711
5	0.987776
10	0.999943

Example 8.3

What is the maximum reflectance of a two-period dielectric multilayer of the form air $\cdot$ $(LH)^N$ $\cdot$ glass assuming air has an index of 1.0, the glass has an index of 1.520, the high-index layer is 2.360 and the low-index layer is 1.380? What if it is increased to five periods? Ten periods?

Solution. This uses Equation 8.42 as

$$R_{\max} = \left[\frac{1/1.52 - (1.38/2.36)^{2N}}{1/1.52 + (1.38/2.36)^{2N}}\right]^2. \tag{8.45}$$

Evaluating for the various cases yields

N	$R_{\max}$
2	0.487498
5	0.971983
10	0.999867

8.3 BIREFRINGENT CRYSTALS

Birefringent crystals are crystals that possess different values for the indices of refraction in different directions. Birefringence can be a permanent property of a material (such as the birefringence displayed by mica or calcite). Alternatively, birefringence can arise from some force being applied to the crystal (such as stress-induced birefringence or birefringence arising from the application of an electric field). Birefringent crystals are used to make a number of important optical components, including polarizers, waveplates, Q-switches, and mode-lockers.

8.3.1 Positive and Negative Uniaxial Crystals

Birefringent crystals can be obtained in both uniaxial and biaxial configurations. Birefringent crystals (such as calcite) that possess two indices of refraction (i.e., the x and y directions have the same index, but the z direction has a different index) are termed *uniaxial*. Birefringent crystals (such as mica) that possess three indices of refraction (all three crystal directions x, y and z have different indices) are termed *biaxial*. Uniaxial crystals are the

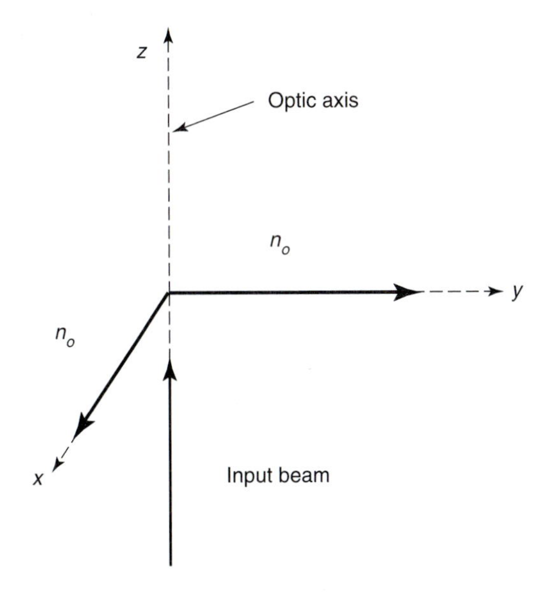

Figure 8.8 The x- and y-axes in a uniaxial birefringent material. A beam propagating through the crystal along the optic axis will split into two orthogonal polarizations, both of which have the same index of refraction n_o. The direction of these two polarizations defines the x- and y-axes.

most commonly used crystals in the laser field. Thus, this discussion will be confined to positive and negative uniaxial crystals.

If a beam of light is passed through a uniaxial crystal in any direction, the beam will split into two orthogonal polarizations where the plane of these polarizations is perpendicular to the propagation direction of the beam. For each of these two orthogonal polarizations, the crystal will possess a distinct index of refraction. However, there exists one direction in the crystal for which the index of refraction for both of the orthogonal polarizations is the same value. This direction is called the *optic axis* and the value of the index of refraction is n_o (called the *ordinary* index of refraction).

Consider a beam propagating through the crystal along the optic axis. The beam splits into two orthogonal polarizations, both of which experience the same index of refraction n_o. The direction of these two polarizations defines the x- and y-axes (see Figure 8.8).

Now, consider a beam propagating through the crystal along the x-axis. Again, the beam will split into two orthogonal polarizations. The one along the y-axis will experience the index of refraction n_o and the one along the optic axis (the z-axis) will experience the value n_e (called the *extraordinary* index of refraction). If $n_e > n_o$ the crystal is said to be *positive uniaxial* and if $n_e < n_o$ the crystal is *negative uniaxial* (see Figure 8.9).

The relationship between the index of refraction and the direction of propagation can be visualized by a graphical structure called the *indicatrix*.[14] For uniaxial crystals, the indicatrix is an ellipsoid of revolution where the optic axis is the axis of rotation. If

---

[14]*Indicatrix* is the most common name for this ellipsoid. Other names include *optical indicatrix*, *indicatrix ellipsoid*, and *index ellipsoid*.

Unfortunately, there is another common geometrical structure where the indices for wave propagation in a given direction are plotted on that direction. Since a propagating wave will experience two indices (one for each orthogonal polarization direction), this structure is quite different in appearance, as it consists of two concentric surfaces (picture a chicken egg inside of a goose egg!). This alternative structure is generally termed the *normal surface* or *index surface*. To add to the confusion, some authors also term this alternate structure an indicatrix.

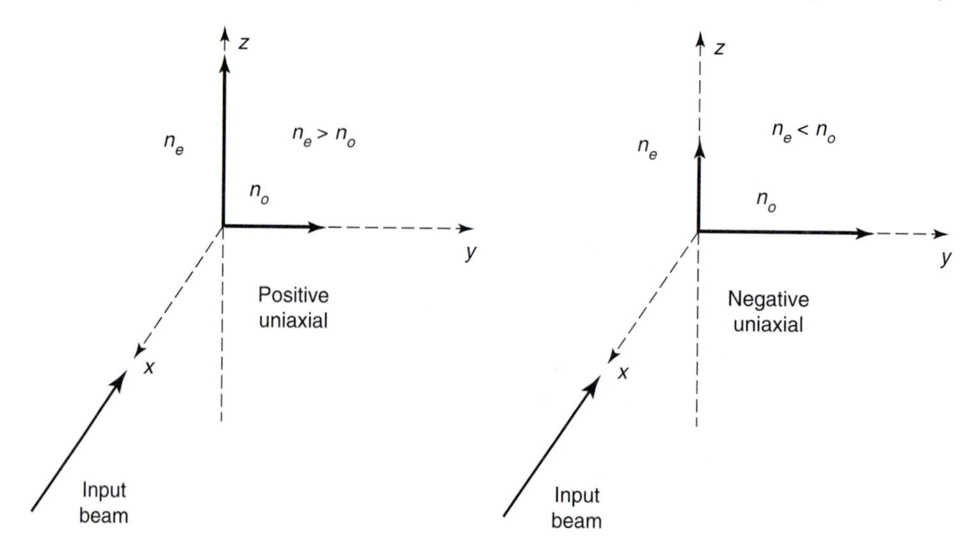

Figure 8.9 Consider a beam propagating through the crystal along the x-axis. The beam will split into two orthogonal polarizations. The one along the y-axis will have the index of refraction n_o and the one along the optic axis (the z-axis) will have the value n_e. If $n_e > n_o$ the crystal is said to be positive uniaxial and if $n_e < n_o$ the crystal is negative uniaxial.

the crystal is positive uniaxial, the indicatrix looks like a football standing on its tip (see Figure 8.10). If the crystal is negative uniaxial, the indicatrix looks like a flying saucer (see Figure 8.11).

Consider a beam propagating through the crystal at an angle of Θ as measured from the optic axis and at an angle Φ as measured from the x-axis. A cut perpendicular to the direction of propagation and through the center of the indicatrix will create an ellipse. The major and minor axes of the ellipse determine the two polarization directions, and the length of the axes determine the value of the index of refraction in that direction (see Figures 8.12 and 8.13).

8.3.2 Wave Plates from Birefringent Crystals

In laser engineering, it is common to need to change the polarization angle of linearly polarized light, or to turn linearly polarized light into circularly polarized light. These operations are performed with optical elements called *wave plates* or retardation plates. Wave plates function by resolving a light wave into orthogonal polarization components and then retarding (or advancing) the phase of one with respect to the other.

A quarter-wave plate will change the relative phase of the two polarization components by $\pm\pi/2$. Thus, a linearly polarized wave of the form

$$\vec{\mathcal{E}}(z,t) = E_x\hat{x}\cos(\omega t - kz) + E_y\hat{y}\cos(\omega t - kz) \tag{8.46}$$

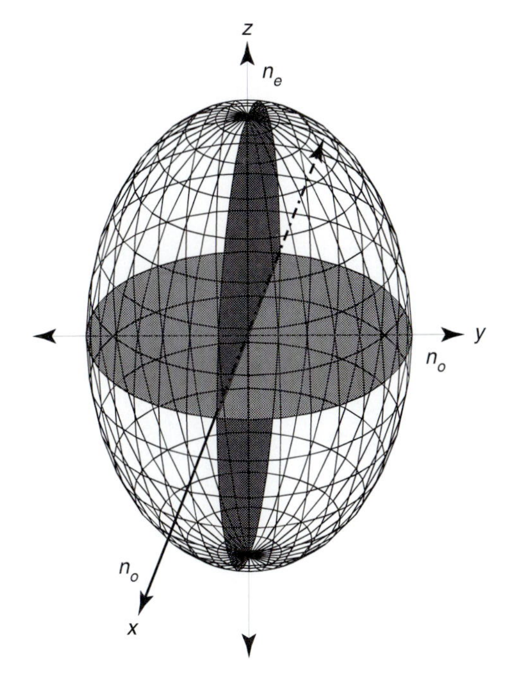

Figure 8.10 If a crystal is positive uniaxial, the indicatrix looks like a football standing on its tip.

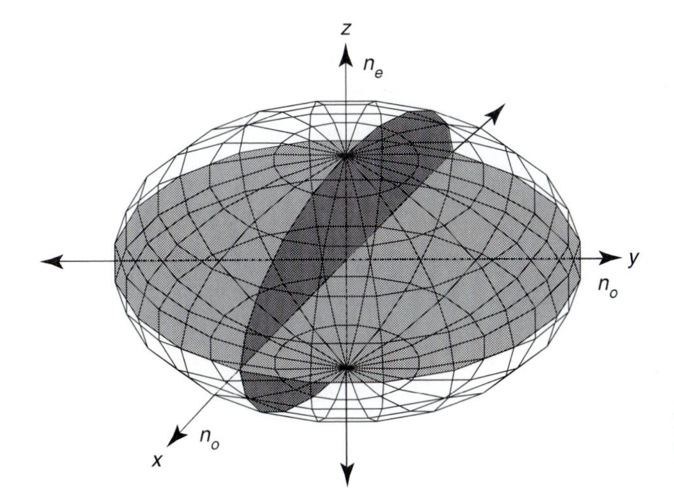

Figure 8.11 If a crystal is negative uniaxial, the indicatrix looks like a flying saucer.

will emerge from the quarter-wave plate as

$$\vec{\mathcal{E}}(z,t) = E_x\hat{x}\cos(\omega t - kz) + E_y\hat{y}\cos(\omega t - kz \pm \pi/2) = E_x\hat{x}\cos(\omega t - kz) \mp E_y\hat{y}\sin(\omega t - kz). \tag{8.47}$$

where ω is the frequency and k is the wavenumber. Therefore, a quarter-wave plate converts linearly polarized light into circularly polarized light (see Figure 8.14).

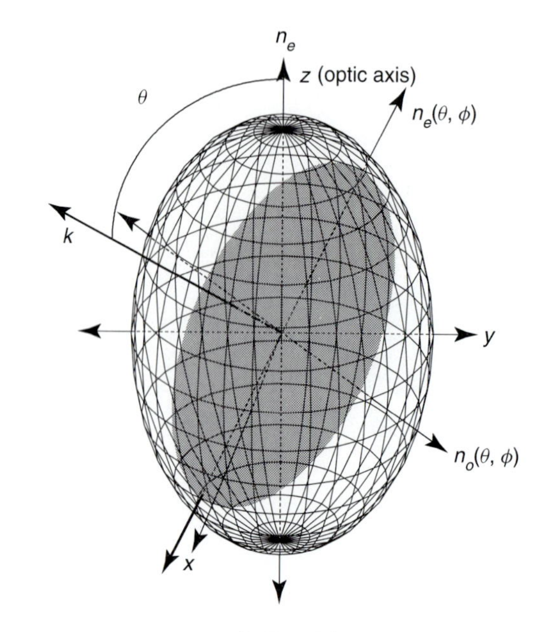

Figure 8.12 Consider a beam propagating through a positive uniaxial crystal at an angle of Θ as measured from the optic axis z and at an angle Φ as measured from the x-axis. A cut perpendicular to the direction of propagation and through the center of the indicatrix will create an ellipse. The major and minor axes of the ellipse determine the two polarization directions, and the length of the axes determine the value of the index of refraction in that direction.

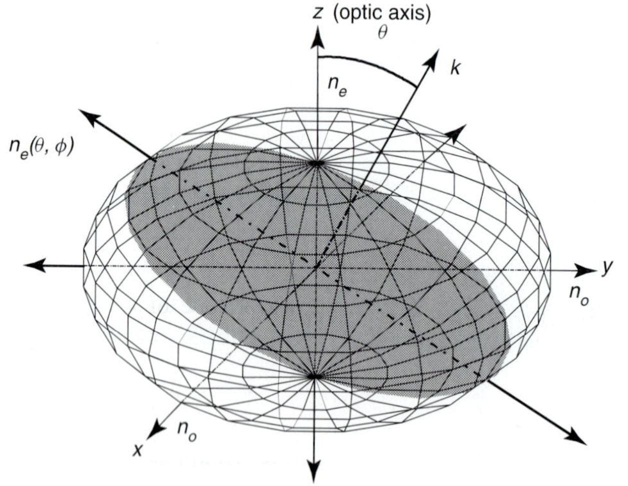

Figure 8.13 Consider a beam propagating through a negative uniaxial crystal at an angle of Θ as measured from the optic axis z and at an angle Φ as measured from the x-axis. A cut perpendicular to the direction of propagation and through the center of the indicatrix will create an ellipse. The major and minor axes of the ellipse determine the two polarization directions, and the length of the axes determine the value of the index of refraction in that direction.

A half-wave plate will change the relative phase of the two polarization components by $\pm\pi$. Thus, a linearly polarized wave of the form

$$\vec{\mathcal{E}}(z,t) = E_x\hat{x}\cos(\omega t - kz) + E_y\hat{y}\cos(\omega t - kz) \tag{8.48}$$

will emerge from the half-wave plate as

$$\vec{\mathcal{E}}(z,t) = E_x\hat{x}\cos(\omega t - kz) + E_y\hat{y}\cos(\omega t - kz \pm \pi) = E_x\hat{x}\cos(\omega t - kz) - E_y\hat{y}\cos(\omega t - kz). \tag{8.49}$$

Therefore, a half-wave plate converts linearly polarized light into linearly polarized light that is orthogonal to the original polarization (see Figure 8.15).

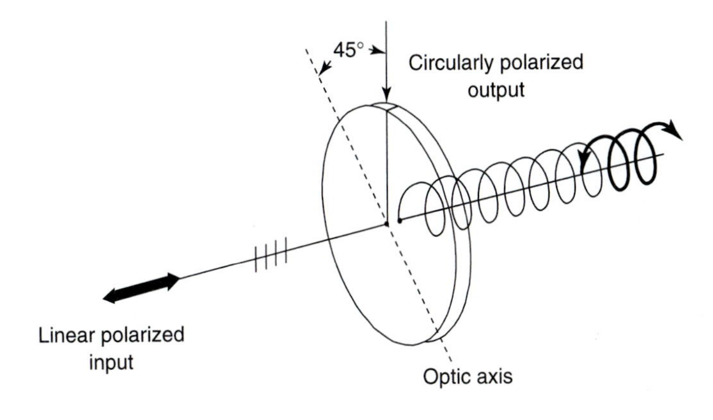

Figure 8.14 A quarter-wave plate will change the relative phase of the two polarization components by $\pm\pi/2$. Therefore, a quarter-wave plate converts linearly polarized light into circularly polarized light. (Courtesy of Melles Griot)

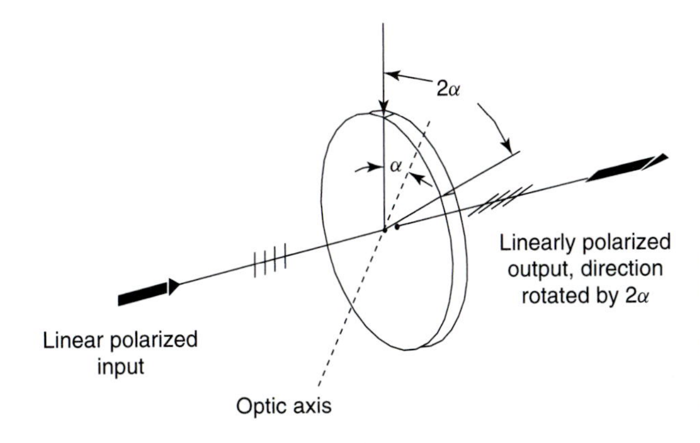

Figure 8.15 For a half-wave plate, if the input light beam has a polarization that is α degrees from the optic axis, the emerging beam will also be linearly polarized, but with a direction that is $-\alpha$ degrees from the optic axis. (Courtesy of Melles Griot)

Using wave plates to minimize back reflections. Quarter and half-wave plates find a number of applications in laser systems. However, one of the most fascinating uses of wave plates is to prevent stray light, which has reflected off various optical surfaces, from propagating backward in a laser amplifier system.

Consider the YAG laser amplifier depicted in Figure 8.16. The YAG laser beam passes through a dielectric polarizer, which sets the polarization direction as horizontal relative to the surface of the optical table. Immediately prior to the laser amplifier, a quarter-wave plate is used to convert the beam to right circularly polarized light. The majority of the right circularly polarized laser light then travels through the amplifier. At the end of the amplifier, another quarter-wave plate converts the light back into linearly polarized light, but now polarized vertically to the table top.

However, notice the fate of the right circularly polarized light reflected back from the end of the YAG amplifier. When the light reflects off the YAG rod, its propagation direction will change, but not its direction of rotation. Thus, it will now be left circularly polarized light, traveling backwards through the quarter-wave plate. When it emerges from the quarter-wave plate, it will be vertically polarized with respect to the table top. In other words, it now has a polarization direction *orthogonal* to the incoming beam. When this back-reflected light reaches the polarizer, it will be ejected from the optical system.

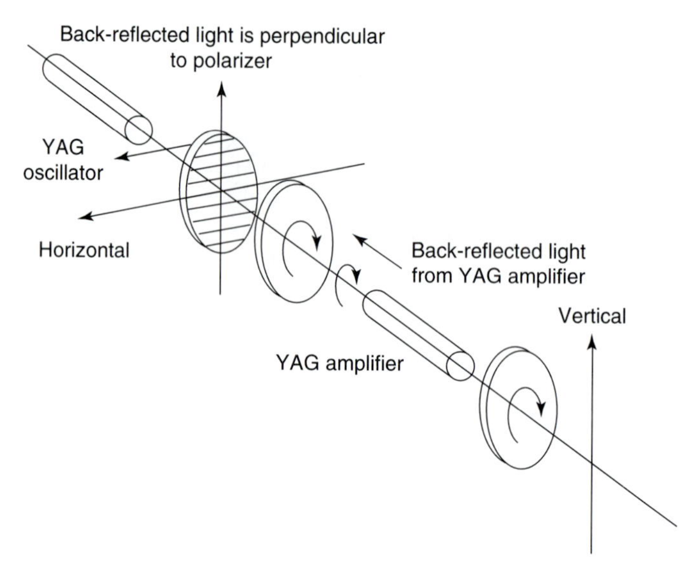

Figure 8.16 One of the most fascinating uses of wave plates is to prevent stray light that has reflected off various optical surfaces from propagating backward in a laser amplifier system.

Manufacturing wave plates. One common way to manufacture a wave plate is to use a birefringent material, such as quartz or calcite. The material can be cut so that the optic axis lies in the plane of the wave plate (see Figure 8.17). If the input light beam has a polarization that is 45 degrees to the optic axis (and normal to the face of the wave plate), then it will be split into two equal components, one traveling subject to an index of n_o and one traveling subject to an index of n_e. The relative phase difference between the two orthogonal polarizations is then given by[15]

$$\Delta\phi = \frac{2\pi}{\lambda_o} d_w \left(|n_o - n_e| \right) \tag{8.50}$$

where d_w is the thickness of the wave plate.

Recall that the relative phase difference should be $\Delta\phi = \pm\pi$ for a half-wave plate and $\Delta\phi = \pm\pi/2$ for a quarter-wave plate. However, it is difficult to fabricate a wave plate such that $\Delta\phi = \pm\pi$ or $\pm\pi/2$ because the wave plate is quite thin (a wave plate fabricated with $\Delta\phi = \pm\pi$ or $\pm\pi/2$ is termed a *first-order wave plate*). Thus, many practical wave plates are fabricated with many multiples of 2π plus the critical $\pm\pi$ or $\pm\pi/2$ (a wave plate fabricated with many multiples of 2π plus the critical $\pm\pi$ or $\pm\pi/2$ is termed a *multiple-order wave plate*).

For a quarter-wave plate (see Figure 8.14), notice that the input beam must be at 45 degrees to the optic axis for the resulting light to be circularly polarized. If the angle is not 45 degrees, then the two vector components will be unequal and the resulting beam will be elliptically polarized.[16] For a half-wave plate (see Figure 8.15), if the input light beam has

[15] Eugene Hecht, *Optics,* 2d ed. (Reading MA: Addison-Wesley, 1987), p. 300.

[16] Melles Griot catalog, 1995–6, p. 12-28.

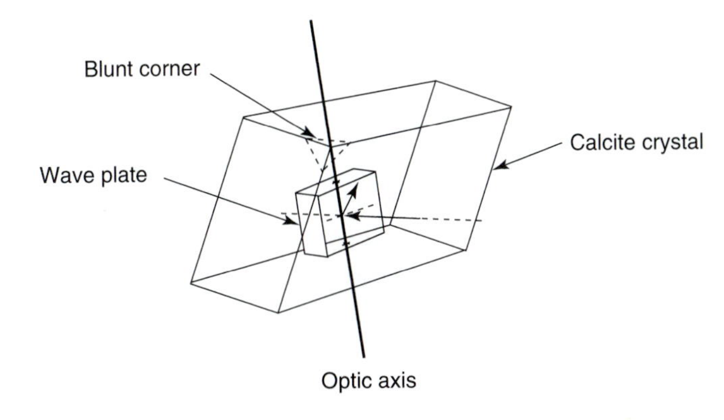

Figure 8.17 The wave plate must be cut in exactly the right direction relative to the optical axis of calcite to obtain quarter- or half-wave plate behavior. (Adapted from E. Hecht, *Optics*, 2d ed., Figure 8.45, page 301. ©1987, 1974 Addison-Wesley Publishing Company, Inc. Reprinted by permission of Addison-Wesley Longman, Inc.)

a polarization that is α degrees from the optic axis, then the emerging beam will also be linearly polarized, but with a direction that is -α degrees from the optic axis. In other words, the half-wave plate has produced a total angular rotation of the linearly polarized light by 2α degrees. If the initial light is polarized at 45 degrees from the optic axis, the resulting polarization will be orthogonal to the initial polarization.[17] (An example of a typical quartz data sheet is given in Figure 8.18.)

Example 8.4

Consider a wave plate fabricated from quartz ($n_o = 1.5443$ and $n_e = 1.5534$). How thick must the wave plate be in order to be a half-wave plate possessing a π phase shift for a 632.8 nm laser?

Solution. This can be obtained from Equation (8.50) as

$$\Delta\phi = \frac{2\pi}{\lambda_o} d_w \left(|n_o - n_e|\right)$$

Therefore

$$d_w = \frac{\Delta\phi}{\left(|n_o - n_e|\right)} \frac{\lambda_o}{2\pi} \tag{8.51}$$

$$d_w = \frac{\Delta\phi}{\left(|n_o - n_e|\right)} \frac{\lambda_o}{2\pi} = \frac{\pi}{\left(|1.5443 - 1.5534|\right)} \frac{632.8 \cdot 10^{-9}\ \text{m}}{2\pi} = 34.76 \cdot 10^{-6}\ \text{m} \tag{8.52}$$

Notice that 34.76 μm is extremely thin! Thus, first-order wave plates must be cleverly mounted (usually between windows of another material) in order to avoid mechanical damage.

Example 8.5

Consider a wave plate fabricated from quartz ($n_o = 1.5443$ and $n_e = 1.5534$) that is 0.9214 mm thick and designed to operate at 632.8 nm. Assume that the input light beam has a polarization that is 45 degrees to the optic axis. Is this a quarter- or half-wave plate?

Solution. In order to ascertain if the wave plate is a quarter- or half-wave, it is necessary to determine the relative phase difference between the n_o and n_e axes. This can be obtained from

[17] Melles Griot catalog, 1995–6, p. 12-28.

Quartz Retardation Plates

These are quarter- or half-wave first-order or multiple-order plates, suitable for high- and low-power laser applications. Multiple-order quartz retarders are assembled from a single crystalline plate, while first-order retarders are assembled from pairs of optically contacted crystalline quartz plates whose effects partially counterbalance each other. These component plates have orthogonal optic axis directions, so that the roles of the ordinary and extraordinary rays are interchanged in passing from one plate to the other. The retardation of first-order plates is essentially invariant with temperature, while the retardation of multiple-order plates is slightly temperature dependent. The temperature coefficient of retardance (phase difference between O- and E-rays at emergence) in multiple-order quartz retarders is approximately 0.07 nm/°C. The net retardation in first-order plates is a function of the thickness difference between the two plates. The net retardations are almost temperature invariant because both plates are of very nearly equal thickness. Both plates expand and contract essentially the same amount, and so almost cancel each other's effect.

SPECIFICATIONS:
QUARTZ RETARDATION PLATES

Wavelength: Specified by customer

Wavelength Range: 200 nm to 2300 nm

Retardation Tolerance: $\lambda/500$

Wavefront Distortion: $\lambda/10$ peak-to-peak

Parallelism: 0.5 arc seconds

Material: Crystal quartz

Optic Axis: Normal to facet on circumference of retarder

Surface Quality: 20–10 scratch and dig

AR Coatings:

For multiple order plates, see suffix system

For first order plates, coatings can be supplied on special order only.

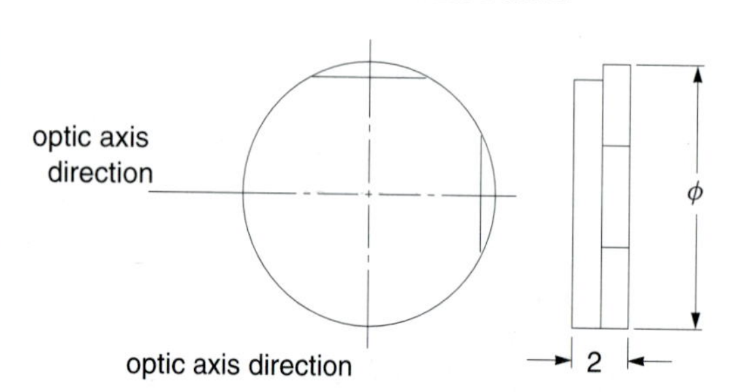

02 WRQ/ first order

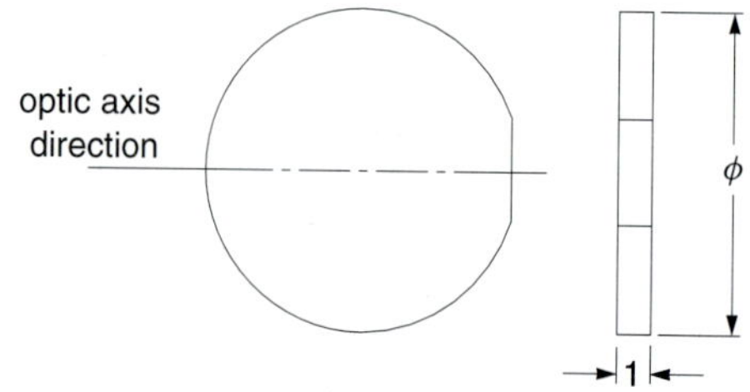

02 WRQ/ multiple order

Quartz Retardation Plates; First-Order

Diameter ϕ (mm)	PRODUCT NUMBER Retardation $\lambda/2$	$\lambda/4$
10	02WRQ023	02WRQ003
20	02WRQ027	02WRQ007
30	02WRQ031	02WRQ011

Please specify the wavelength when ordering these retardation plates (for example, 02WRQ003/632.8). A number of laser wavelengths are held in inventory for your convenience.

Quartz Retardation Plates; Multiple-Order

Diameter ϕ (mm)	PRODUCT NUMBER Retardation $\lambda/2$	$\lambda/4$
10	02WRQ021	02WRQ001
20	02WRQ025	02WRQ005
30	02WRQ029	02WRQ009

Please specify the wavelength when ordering these retardation plates (for example, 02WRQ003/632.8). A number of laser wavelengths are held in inventory for your convenience.

Figure 8.18 Quartz retardation plate data sheet. (Courtesy of Melles Griot)

Equation (8.50) as

$$\Delta\phi = \frac{2\pi}{\lambda_o} d_w \left(|n_o - n_e|\right) \tag{8.53}$$

$$\Delta\phi = \frac{2\pi}{632.8 \cdot 10^{-9}\ \text{m}} 0.9214 \cdot 10^{-3}\ \text{m} \left(|1.5443 - 1.5534|\right) = 13.25 \cdot 2\pi \tag{8.54}$$

$$\Delta\phi = 13.25 \cdot 2\pi = 13 \cdot 2\pi + \frac{\pi}{2} \tag{8.55}$$

So, this is a multiple-order quarter-wave plate.

8.4 PHOTODETECTORS

Quantitative measurement of light intensities using photodetectors is a critical part of any laser project. This section is only intended to give a broad overview of photodetector technologies. There are a number of references available on photodetectors for the reader seeking a more detailed approach.[18,19]

8.4.1 Thermal Detectors

Thermopiles, bolometers and pyroelectric detectors are the most common thermal detectors.

Thermopiles rely on thermocouples to measure the increase in temperature of an absorbing substance after it has been exposed to the laser beam. An example of a simple thermopile is a black disk with a thermocouple attached to the back. Thermopiles have the advantages of simplicity and insensitivity to wavelength. However, they suffer from a linked problem between response time and optical damage. A thin film thermopile responds quickly to incident light, but possesses a relatively low damage threshold. A large complex thermopile with many partially absorbing baffles has a high damage threshold, but possesses a very long response time.

Bolometers measure the change in resistance of a semiconductor or metal upon heating by the laser beam. An example of a simple bolometer is an enclosure containing a tangle of insulated wire (a rat's nest bolometer[20]). An example of a more complex bolometer is a high-T_c superconductor held at a temperature close to its critical temperature.[21] In both cases, the incident laser beam heats the material and the resulting change in resistivity is measured. Bolometers typically possess faster response times than thermopiles, but suffer from relatively low damage thresholds.

Pyroelectric detectors use pyroelectric crystals to measure the change in temperature of a material upon heating by the laser beam.[22] Pyroelectric crystals are materials that develop

[18] Michael Bass, ed, *Handbook of Optics, Vol. I, Fundamentals, Techniques, and Design* (New York: McGraw Hill, 1995), Part 5, pp. 15-1 to 23-37.

[19] S. M. Sze, *The Physics of Semiconductor Devices,* 2d ed. (New York: John Wiley and Sons, 1981), Chapter 7 (MIS diodes and CCDs) and Chapter 13 (Photodetectors).

[20] R. M. Baker, *Electronics* 36:36 (1963).

[21] P. W. Kruse, "High Tc Superconducting IR Detectors," in *Proc. SPIE* 1292: 108–177 (1990).

[22] E. H. Putley, "The Pyroelectric Detector," *Semiconductors and Semimetals*, ed R. K. Willardson and A. C. Beer (New York: Academic Press, 1970), pp. 259–285; and the update E. H. Putley, "The Pyroelectric Detector —

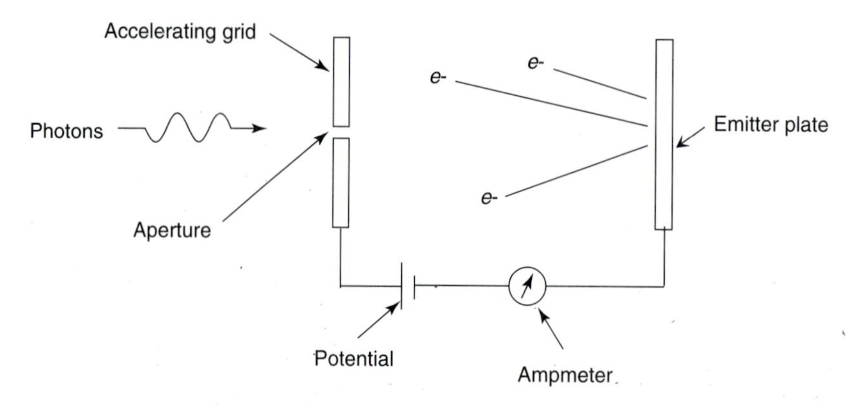

Figure 8.19 A simple photoelectric detector.

a change in surface charge as a function of a change in temperature. (The most familiar pyroelectric crystal is Rochelle salt.) Notice that this detection technique is a function of the *change* in temperature, not the total quantity of heat deposited. Thus, pyroelectric detector systems often use some sort of gate or shutter to create a temperature pulse on the pyroelectric detector. As a consequence, measurement times are swift and largely independent of beam power. Gated pyroelectric systems are also less likely to suffer from optical damage. An example of a common pyroelectric detector is a commercial baby thermometer that measures temperature by the IR emission from the baby's eardrum.

8.4.2 Photoelectric Detectors

The *photoelectric effect* is the creation of electrons by photons interacting with materials. The photoelectric effect can be used alone in simple photoelectric detectors or in combination with accelerating voltages for photomultiplier tubes.

Simple photoelectric detectors. A simple *photoelectric detector* can be constructed by setting up an accelerating grid at a potential V with a hole in it (see Figure 8.19). Light passes through the hole and is imaged on an emitter plate. Interaction of light with the material in the emitter plate will produce electrons, which will be attracted by the accelerating grid. The resulting current between the sensor plate and the grid can be easily measured.

If the accelerating grid is positive with respect to the emitter plate, then the electrons will be attracted to the accelerating grid. As the light intensity increases, more electrons are emitted and the measured electrical current increases.

If the accelerating grid is negative with respect to the emitter plate, then the electrons will be repelled from the grid. Thus, if an electron is emitted with a certain kinetic energy K_E, then there is a particular value of potential $V = K_E/q$ that will prevent the electron from reaching the accelerating grid. Therefore, for a population of electrons, the current at

An Update," *Semiconductors and Semimetals*, ed R. K. Willardson and A. C. Beer (New York: Academic Press, 1977), pp. 441–449.

each value of accelerating voltage represents the number of electrons of higher energy than that determined by the potential on the accelerating plate.

The relationship between the kinetic energy K_E and the light frequency ν can be written as

$$K_E = Vq = \frac{1}{2} m_o \mathrm{v}^2 = h\nu - W \tag{8.56}$$

where W is a property of the emitter plate material called the *work function*. Thus, the maximum wavelength that can be detected using a photoelectric detector is

$$\lambda = \frac{hc_o}{W}. \tag{8.57}$$

Photomultiplier tubes. A photoelectric detector has no intrinsic gain. Each photon (usually) produces a single electron. However, if the electron can be accelerated (for example, by a voltage differential) and then allowed to collide with other atoms, the single emitted electron can create a cascade of secondary electrons.

The most popular implementation of this effect is in a device called a *photomultiplier tube* (see Figure 8.20). A photon impacts on a photocathode and produces a single electron through the photoelectric effect. This electron is accelerated through an electric potential difference and impacts on the first dynode in the chain. The electron is now more energetic and creates a cascade of secondary electrons. These secondary electrons are accelerated through a potential difference to the second dynode, where the process is repeated. The cascade process is continued through additional dynodes until the desired current is obtained.

Photomultiplier tubes (PMTs) are astonishingly sensitive detectors. They are the only detectors that can be used for photon counting and are ubiquitous in low intensity measurement applications. However, the cascade process is very sensitive to dynode voltage and temperature. Thus, the gain of a PMT varies significantly with relatively minor variations in the environment.

8.4.3 Photoconductors

A *photoconductor* is a resistor where the resistance is a function of the incident light intensity. The simplest example of a photoconductor is a block of semiconductor material with two ohmic contacts (see Figure 8.21). The block is biased with a small voltage and the current

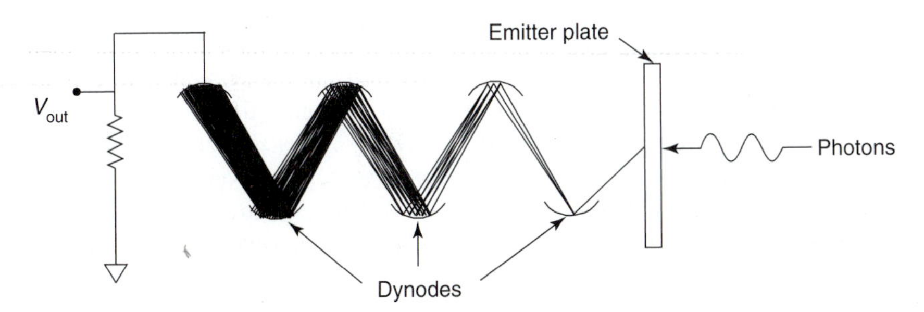

Figure 8.20 The photomultiplier tube.

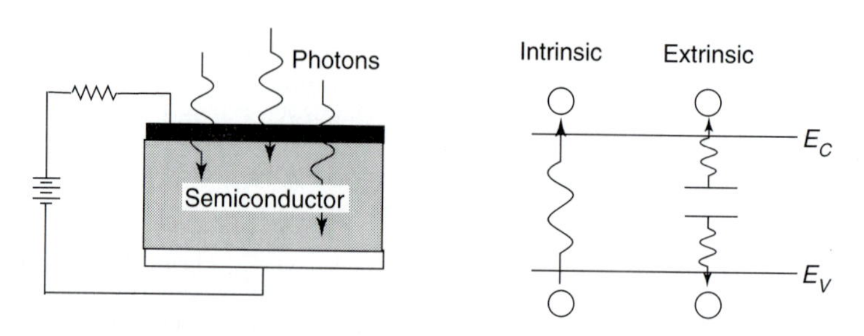

Figure 8.21 A simple photoconductor.

(or resistance) measured. A photon incident on this block of material can create an electron-hole pair by a band-to-band transition (intrinsic) or promote a carrier into the conduction or valence band from a state within the bandgap (extrinsic).[23] Either situation reduces the resistivity of the semiconductor.

Although the basic picture is simple (namely, a decrease in resistance due to carriers created by photon interaction with the semiconductor) there are a number of subtle effects that must be considered when using photoconductors. For example, the exponential absorption profile of light into the photodetector ($I = I_o e^{-\alpha L}$) means that the majority of the photons will be absorbed near the surface. Additionally, holes and electrons do not have the same mobility and can also be trapped by defect states in the material.

Assuming a material for which the mobilities of the holes and electrons are the same (and ignoring the other subtle effects!) the primary photocurrent that is created in a photodetector is a function of the input optical power P_{in} as

$$I_{\text{ph}} = q\left(\frac{\eta P_{\text{in}}}{h\nu}\right) \tag{8.58}$$

where η is the quantum efficiency (number of carriers produced per photon).

If the carriers recombine in the semiconductor with a time constant τ then the current between the electrodes is given as

$$I_p = q\left(\frac{\eta P_{\text{in}}}{h\nu}\right)\frac{\mu_n \tau \mathcal{E}}{L} = q\left(\frac{\eta P_{\text{in}}}{h\nu}\right)\left(\frac{\tau}{t_r}\right) \tag{8.59}$$

where $\mathcal{E}$ is the electric field, μ_n is the mobility of the carrier (assuming $\mu_n = \mu_p$ for this case), L is the length of the photoconductor, and t_r is the carrier transit time across the photoconductor.

Notice that the current between the electrodes is greater than the primary induced photocurrent by a factor of τ/t_r. Thus, the photoconductor has gain. In order to increase the gain, photoconductors should have a short interaction length, high carrier mobility, high electric field, and long recombination time τ. However (as might be expected) long recombination times lead to photoconductors that respond slowly. Attempting to increase the response speed while maintaining the gain requires very short lengths. Optimizing these varied parameters results in photoconductors fabricated on thin films with large widths and

[23] R. H. Bube, *Photoconductivity in Solids* (New York: John Wiley and Sons, 1962).

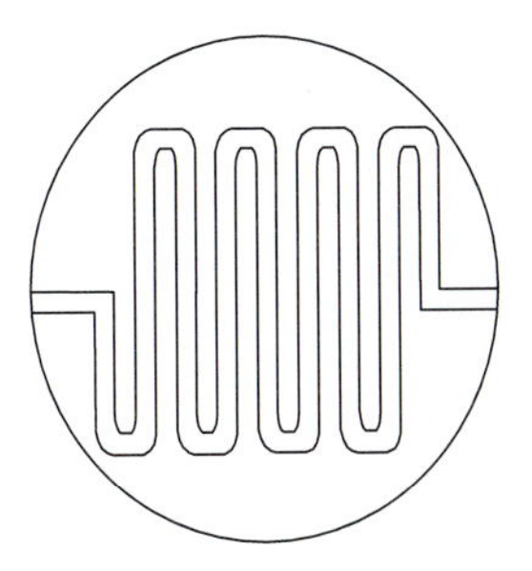

Figure 8.22 The characteristic interdigitated fingers of a simple photoconductor.

short lengths. The usual geometry is closely spaced interdigitated metal fingers deposited on thin films (see Figure 8.22).

The simplicity, ruggedness, and gain of photoconductors means that they are extremely useful devices. The majority of cheap "photodetectors" that you see in electronics shops are photoconductors (easily distinguished visually because of the characteristic interdigitated fingers). Cadmium sulfide (CdS) photoconductors are particularly common because they are one of the few photodetectors that can handle 110/220 volts.

8.4.4 Junction Photodetectors

PN-photodiodes, PIN-photodiodes, avalanche photodiodes, and phototransistors are the most common junction photodetectors.

PN-junction photodiodes. *PN-photodiodes* are PN-junctions with an optical window. Their operation can be understood by recalling the operation of a PN-junction diode (see Figure 8.23). An n-type material (with the Fermi level E_{Fn} near the conduction band edge) is mated to a p-type material (with the Fermi level E_{Fp} near the valence band edge). The electrons in the n-type material diffuse into the p-type material and recombine. The holes in the p-type material diffuse into the n-type material and recombine.

The positive ions left on the n-side of the junction and the negative ions left on the p-side of the junction form a depletion region where no free charges exist. Free charges entering the depletion region are swept to the appropriate side by the field produced by the ions.[24] Thus, holes entering the junction are swept to the p-side by the negative ions. Similarly, electrons entering the junction are swept to the n-side by the positive ions. The magnitude of the built-in field is such that (in equilibrium) the drift current produced by the field exactly opposes the carrier diffusion process.

If the diode is forward-biased (a positive voltage on the p-side) then diffusion of holes from the p-side and electrons from the n-side across the junction will be increased. The total current in the diode is then formed by the sum of the injected minority current across the PN-junction (that is, electrons on the p-side or holes on the n-side) plus the diffusing majority current which is recombining with the minority carriers.

Now, consider a forward-biased PN-junction exposed to light. A hole-electron pair is formed when light is incident on the depletion region. The hole will be swept toward the

[24]The current produced by carrier motion under the influence of an electric field is called the *drift current*.

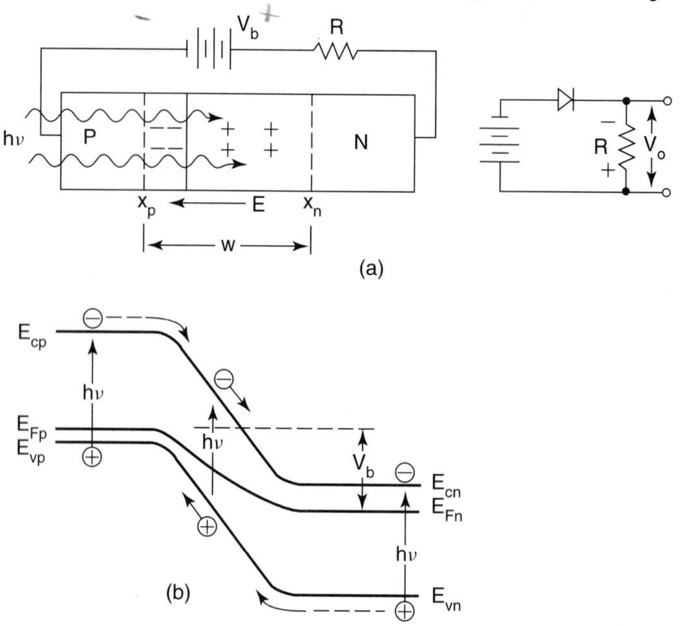

Figure 8.23 A PN-junction photodetector. (From LASER ELECTRONICS 2E by VERDEYEN, J.T. ©1989, Figure 16.5, p. 573. Adapted by permission of Prentice-Hall, Inc., Upper Saddle River, NJ.)

negative ions (p-side) and the electron will be swept toward the positive ions (n-side). This photo-induced current flows in the opposite direction of the normal forward biased current through the diode.

Notice that the production of current in the photodetector is intrinsically related to this separation of the optically created carriers by the built-in field. If the carriers are created too far from the depletion region, then they will recombine immediately and no current will flow. Therefore, only light incident within about one diffusion length of the depletion region will contribute to the photocurrent.

The general equation describing the current under these conditions is given as

$$I_{\text{diode}} = I_{\text{sat}}\left(e^{qV/kT} - 1\right) - I_{\text{light}} \tag{8.60}$$

where I_{diode} is the current through the diode, I_{sat} is the saturation current, V is the voltage across the diode, and

$$I_{\text{light}} = q\frac{\eta P_{\text{in}}}{h\nu} \tag{8.61}$$

where P_{in} is the input intensity and η is the quantum efficiency.

If the diode is open-circuited ($I_{\text{diode}} = 0$, no *net* current), then the optically-induced carrier current must be balanced by an equivalent diffusion current in the opposite direction in order to keep the total current at zero. Diodes can be operated as photodetectors in this open circuit mode. This is termed *photovoltaic* operation, and is used when dynamic range is more important than response speed. In the photovoltaic mode, I_{diode} in Equation (8.60) is set to zero and the equation is solved for the voltage as

$$V_{oc} = \frac{kT}{q} \ln \left(\frac{I_{\text{light}}}{I_{\text{sat}}} \right). \tag{8.62}$$

Notice that this equation gives a logarithmic variation in output voltage as a function of light intensity. This can be advantageous for certain experiments.

If the diode is reverse-biased (a positive voltage on the n-side), then diffusion of majority carriers (holes from the p-side and electrons from the n-side) will be dramatically reduced. Diodes can be operated as photodetectors in a reverse-biased mode (see Figure 8.23). This is termed *photoconductive* operation and is used when response speed or linearity is important. In the photoconductive mode, V is large, negative, and

$$I_{\text{diode}} = -I_{\text{sat}} - I_{\text{light}}. \tag{8.63}$$

If I_{sat} is small compared with I_{light}, then

$$I_{\text{diode}} = -I_{\text{light}} = -\frac{q\eta P_{\text{in}}}{h\nu}. \tag{8.64}$$

Notice that in this case the generated current is linear with respect to the incident light.

PIN-junction photodiodes. In a PN photodiode, only the light absorbed within one diffusion length of the depletion region is useful. If the light is absorbed further into the n or p regions, then the carriers recombine before they reach the junction.

The most straightforward way to increase the quantity of absorbed light is to make the depletion region larger. One of the easiest ways to make the depletion region larger is to add an intrinsic layer between the p and n regions. The thicker the intrinsic layer, the more optical carriers are created in the depletion region.

However, the response time for the photodetector is limited by the carrier transit time through the depletion layer. Thus the depletion layer thickness is usually kept to approximately $1/\alpha$ where α is the optical absorption at the wavelength of interest.

PIN diodes follow essentially the same mathematical model as PN-junction diodes. Because the PIN geometry provides significantly more control over the device characteristics, the majority of commercial photodiodes are actually PIN devices.

PN and PIN devices provide no gain. However, because they are easy to use and can be operated in either a photovoltaic or photoconductive mode, they are perhaps the most widely used of all photodetectors.

Avalanche photodiodes. If a PN- or PIN-junction diode is operated at a high value of reverse bias voltage, it is possible for the optical carriers to generate secondary carriers by an avalanche multiplication process.

Avalanche photodiodes are fast and can have high gains. However, they are touchy devices. The multiplicative gain is a very sensitive function of the bias voltage and temperature. The devices can also be noisy. Additionally, they are more difficult to fabricate due to the requirements for high-quality processing with low dislocation densities and uniform doping in order to sustain the high reverse biases without damage.

Avalanche photodiodes are generally used in applications requiring high frequency performance rather than gain. Applications requiring gain at lower frequencies are often better met by PMT devices (extremely low light applications) or by phototransistors.

Phototransistors. Virtually all semiconductor transistors can be operated in a photodetection mode. Thus, there are photodetecting BJT devices, MOSFET devices, and HBT devices. In a few cases, it is even possible to use minimally modified electronic devices as photodetectors (such as using an array of CMOS devices, with the package removed, as an array detector). However, commercial phototransistors intended for optical applications are usually modified significantly so as to enhance the optical absorption region.

Consider a basic bipolar junction transistor (BJT). A BJT can be considered to be two PN-junctions back-to-back (p-n-n-p or n-p-p-n). The middle junction (n-n or p-p) is termed the base, and the other two are termed the collector and emitter.

If both junctions are reverse-biased, negligible current flows, and the device is in cut-off (essentially an open circuit). If both junctions are forward-biased, the current is determined by the outside circuit, and the device is in saturation. If one junction is forward-biased, and the other is reverse-biased, the device is in active mode. In active mode, the current traveling through the device between the collector and emitter is a function of the current traveling through the base. For an NPN device, $I_c = \beta I_b$.

A transistor constructed for optical applications typically has no base lead. Instead, the physical area of the base is extended in order to enhance the carrier absorption. In a phototransistor, the base current I_b is produced by carriers created from absorbed photons. Thus, all common BJT circuits are applicable for phototransistors if I_b (the base current) is replaced by I_p (the photocurrent).

8.4.5 MOS Capacitor Devices

Metal oxide semiconductor (MOS) capacitors are very commonly used as photodetectors. The popular MOS charge-coupled device (CCD) arrays used in imaging are an excellent example of MOS photocapacitors.

The basic MOS capacitor (see Figure 8.24) is fabricated from a silicon substrate covered with a thin high-quality oxide. A gate conductor of metal or polysilicon is added on top of the oxide.

If the substrate material is p-type, then application of a positive gate voltage will deplete the holes from the oxide-semiconductor interface. This will expose the negative dopant ions and create a potential *well* at the oxide-semiconductor interface. This operating state (termed depletion) is a nonconducting mode, as the dopant atoms are immobile.

If light is incident on the material, then photons will be absorbed in the depletion layer and in the substrate. These photons will produce hole-electron pairs. Since there is a positive potential applied to the gate, the holes will drift away, and the electrons will move toward the gate. The electrons will then be trapped in the potential well at the oxide-semiconductor interface. The number of electrons in the well is proportional to the intensity of the light.

Usually, MOS capacitors are constructed in arrays of devices. Once the devices have been charged, the charge is then transferred down the devices by suitably manipulating the depth of the potential wells (altering the gate voltages) thereby forcing the charges to move

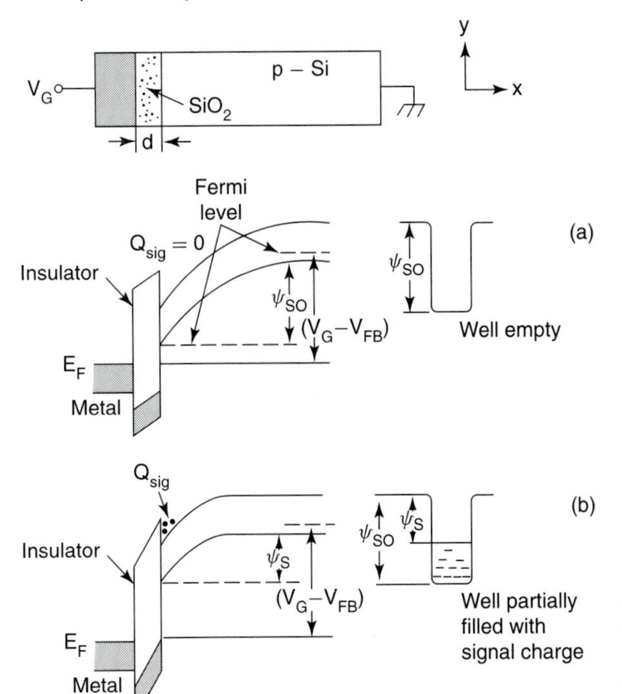

Figure 8.24 A simple MOS capacitor photodetector. (From D. F. Barbe, "Imaging Devices Using the Charge Coupled Concept," *Proc. IEEE* 63:38 (1975). ©1975 IEEE.)

from potential well to potential well. Thus, a "bucket brigade" is created by transferring the charges from device to device down the length of the array.

SYMBOLS USED IN THE CHAPTER

a, b: First and last materials in a thin dielectric film (generally air and glass)

n_a, n_b: Indices of refraction of the a and b materials in a thin dielectric film (unitless)

θ_{in}, θ_b: Incident and transmitted angles in the a and b materials in a thin dielectric film (rad or deg)

ϕ_{first}: Transmission angle in first film, air to glass interface (rad or deg)

ϕ_{out}: Transmission angle in last film, glass to substrate interface (rad or deg)

ϕ_p: Transmission angle in pth film layer (rad or deg)

δ_p: Optical thickness of the pth film layer (cm or m)

n_p: Real index of refraction in pth film layer (unitless)

$n_p = \overline{n_p} + j\kappa_p$: Real and imaginary parts of the refractive index n_p (unitless)

d_p: Thickness of pth film layer (cm or m)

$\mathcal{M}_p$: Characteristic matrix of pth film layer

$\perp$, $\|$: Subscripts representing the TE and TM waves

r, t, $r_\perp$, $r_\|$, $t_\perp$, $t_\|$: The electric field reflection and transmission coefficients (unitless)

R, T, A, $R_\perp$, $R_\parallel$, $T_\perp$, $T_\parallel$, $A_\perp$, $A_\parallel$: The intensity reflectance, transmittance, and absorptance coefficients (unitless)

m_{11}, m_{12}, m_{21}, and m_{22}: Matrix elements of the $\mathcal{M}_p$ matrix

u_{in}: Effective index of refraction for film to glass interface

u_{out}: Effective index of refraction for glass to substrate interface (unitless)

u_p: Effective index of refraction at ϕ_p (unitless)

φ_R, φ_T: Phase changes on reflection and transmission (deg or rad)

H, L: High and low index of refraction layers in a multilayer film (unitless)

n_H, n_L: Indices of refraction for the high- and low-index layers (unitless)

d_H, d_L: Thicknesses for the high- and low-index layers (cm or m)

N: Number of layers (or layer pairs) in a multilayer film (unitless)

R_{max}: Maximum reflectance in a multilayer film (unitless)

$\Delta\lambda/\lambda$: Wavelength bandwidth of a multilayer film (unitless)

n_o: Ordinary real index of refraction in a birefringent material (unitless)

n_e: Extraordinary real index of refraction in a birefringent material (unitless)

n: General real index of refraction (unitless)

Θ: Angle of propagation in a birefringent material as measured from the optic axis (deg or rad)

Φ: Angle of propagation in a birefringent material as measured from the x-axis (deg or rad)

$\vec{\mathcal{E}}$: Electric field (V/cm) (The vector components of $\vec{\mathcal{E}}$ are indicated by subscripts without the vector symbol, as $\vec{\mathcal{E}} \rightarrow E_x$, E_y, and E_z in the system indicated by the Cartesian unit vectors $\hat{x}$, $\hat{y}$, and $\hat{z}$.)

z: Spatial variable representing the direction of wave propagation (cm or m)

t: Time (sec)

ν: Frequency (Hz)

ω: Frequency (rad/sec)

λ: Material wavelength (cm or m)

λ_o: Free space wavelength (cm or m)

k: Analytic wavenumber $(2\pi/\lambda)$ (m^{-1} or cm^{-1})

$\Delta\phi$: Phase difference between n_o and n_e in a wave plate (rad)

d_w: Thickness of a wave plate (cm or m)

α: Angle of rotation of a wave plate from the optic axis (deg or rad)

K_E: Kinetic energy (eV)

W: Work function (eV)

V: Electric potential (volts)

I_{ph}: Primary photocurrent (amps or mA)

η: Quantum efficiency (unitless)

P_{in}: Input optical power (watt)

τ: Carrier recombination time (sec)

μ_n: Mobility (cm^2/V-sec)

L: General length (cm or m)

tr: Carrier transit time (sec)

I_{diode}: Diode current (amps or mA)

I_{sat}: Saturation current (amps or mA)

I_{light}: Light-induced current (amps or mA)

V_{oc}: Open-circuit voltage (V)

Part II

Design of Laser Systems

9
Conventional Gas Lasers

Objectives

- To summarize the sequence of historical events leading to the development of the modern commercial HeNe laser.
- To summarize the commercial applications of HeNe lasers.
- To describe the various energy states of the HeNe laser and to summarize how these states interact with each other.
- To describe the construction of a modern sealed HeNe laser.
- To describe the difference between a neutral and an ion laser.
- To summarize the sequence of historical events leading to the development of the modern commercial ion laser.
- To summarize the commercial applications of ion lasers.
- To describe the various energy states of krypton- and argon-ion lasers and to summarize how these states interact with each other.
- To describe the construction of a modern commercial argon-ion laser.

9.1 HeNe LASERS

9.1.1 History of HeNe Lasers

The first demonstration of laser action from HeNe was observed by Javan, Bennett and Herriott in 1961. The laser action was observed on various neon 2s to various neon 2p states. The observed wavelengths were 1.118 μm, 1.153 μm, 1.16 μm, 1.199 μm and 1.207 μm. (The various HeNe laser states are discussed in Section 9.1.3.)

Javan, Bennett and Herriott used an 80 cm long plasma tube, 1.5 cm in diameter, with a 10:1 fill at a total pressure of 1.1 mm Hg. A maximum power of 15 mW was observed on the 1.153 μm line. These initial experiments also dramatically demonstrated the low gain of the HeNe laser, as mirrors of 0.3% transmission were employed.[1] (The general principles of laser gain are discussed in Section 2.2 and the idea of gain saturation and choice of optimal mirrors for a given laser system are discussed in Chapter 5.)

The popular red 632.8 nm line was first observed in 1962 by White and Rigden.[2] They used a 120 cm long plasma tube, 7 mm in diameter, with a 10:1 fill at a total pressure of 0.7 mm Hg. A maximum power of 0.1 mW was observed on the 632.8 nm line. In the same experiment, they also made the first visual observations of transverse mode patterns in a laser. (Transverse modes are discussed in Section 4.2 and an experiment to observe transverse modes is outlined in Section A.8.1.)

Also in 1962, McFarlane, Patel, Bennett, and Faust observed five additional neon 2s to neon 2p transitions including 1.0798 μm, 1.0844 μm, 1.1143 μm, 1.139 μm, 1.1767 μm, and 1.5231 μm.[3] They used a 225 cm long plasma tube, 7 mm in diameter, with a 10:1 fill at a total pressure of 1 to 2 mm Hg. In 1963, White and Rigden first observed the 3.39 μm transition. In the same experiment they first explained the relationship between the 3.39 μm transition and gain suppression on the 632.8 nm line.[4] (See Section 9.1.3 for a detailed discussion on the relationship between the 3.39 μm transition and gain suppression on the 632.8 nm line.)

In almost all of the early experiments, some attempt was made to lase on the higher energy (shorter wavelength) neon 2s to 2p transitions. However, these attempts failed until 1971, when Perry obtained laser oscillation on the green 543.5 nm line. He used a 65 cm long plasma tube, 4 mm in diameter, with a 7:1 fill at 1 torr. A maximum power of 50 μW was observed on the 543.5 nm line. To align the laser, he used the method of first obtaining laser oscillation at 612.0 nm and then replacing the mirrors with 543.5 nm mirrors.[5]

The first HeNe lasers used a single long narrow glass tube as the primary skeleton of the laser. A second capillary tube was typically mounted inside the primary tube to provide a narrower bore diameter for the operating laser. Additional glass tubes were added to the side to hold electrodes, getters, and additional gas. Although the early HeNe publications do not provide illustrations of the tube geometries, an excellent window into the past is provided by two *Scientific American* articles written by C. L. Stong in 1964 and 1965.[6] A typical tube design of the era from Stong's articles is illustrated in Figure 9.1.

The neon 2p lower states ($2p_2$ to $2p_{10}$)[7] decay to the neon 1s states and the neon 1s states typically decay by wall collisions, not by radiative recombination. In practice, this means that the gain of the HeNe laser is inversely proportional to the diameter of the bore.

---

[1] A. Javan, W. R. Bennett, and D. R. Herriott, *Phys. Rev. Lett.* 6:106 (1961).

[2] A. D. White and J. D. Rigden, *Proc. IRE* 50:1697 (1962).

[3] R. A. McFarlane, C. K. N. Patel, W. R. Bennett, and W. L. Faust, *Proc. IRE* 50:2111 (1962).

[4] A. D. White and J. D. Rigden, *Appl. Phys. Lett.* 2:211 (1963).

[5] D. L. Perry, *IEEE J. of Quantum Electron.* QE-7:102 (1971).

[6] C. L. Stong, *Scientific American*, Sept.: 227–241 (1964); Dec.: 106–113 (1965).

[7] The nomenclature of laser lines is discussed in more detail in Section 9.1.3.

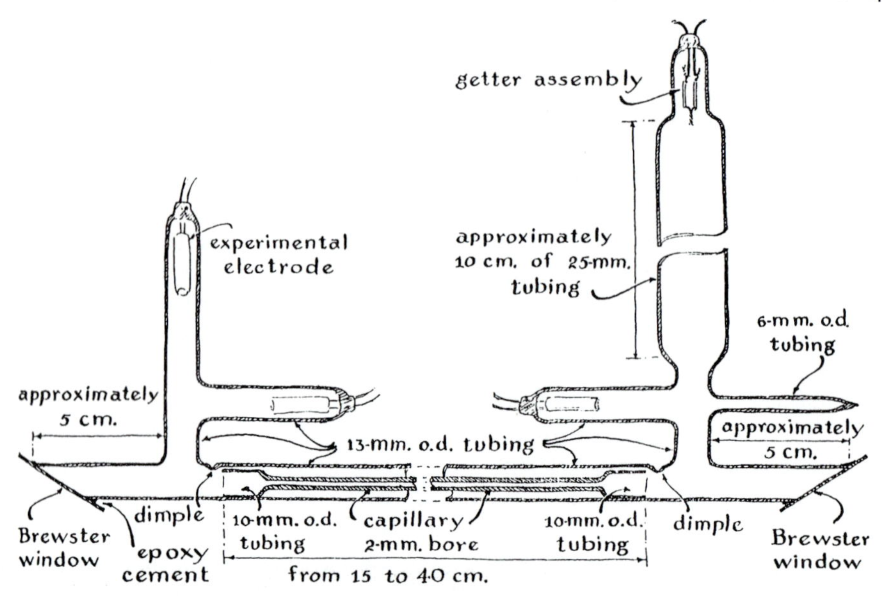

Figure 9.1 Although the early HeNe publications do not provide illustrations of the tube geometries, an excellent window into the past is provided by two *Scientific American* articles in "The Amateur Scientist" written by C. L. Stong in 1964 and 1965. A typical tube design of the era from Stong's articles is illustrated in here. (From "The Amateur Scientist," C. L. Stong, *Scientific American*, Dec., 1965, p. 112. © 1965, Scientific American, Inc. All rights reserved.)

For optimal power output, the relationship between the total gas fill pressure of the laser P (in torr) and the bore diameter D (in mm) is given by the empirical equation[8]

$$P \cdot D = C \tag{9.1}$$

where the constant C is reported with various values from 3.2[9] to 3.6.[10] (The importance of wall collisions in HeNe design is also discussed in Section 9.1.4.)

HeNe lasers will operate with diminishing intensity up to about twice the optimum pressure and down to about half of it. At pressures much lower than about 0.5 torr, the electrons acquire enough energy to damage the glass envelope by impact and to erode the electrodes. The metal vaporized in this manner condenses on the inside of the glass envelope and getters gas atoms, thus lowering the operating pressure even more.[11]

9.1.2 Applications for HeNe Lasers

HeNe lasers are widely used in such diverse commercial applications as supermarket barcode scanning; industrial alignment; laser document printing and scanning; laser surveying; machine vision; Doppler and interferometric velocity measurement; noncontact measurement

[8]John F. Ready, *Industrial Applications of Lasers* (New York: Academic Press, 1978), p. 71.

[9]John F. Ready, *Industrial Applications of Lasers* (New York: Academic Press, 1978), p. 71.

[10]C. L. Stong, *Scientific American*, Sept.: 227–241 (1964); Dec.: 106–113 (1965).

[11]C. L. Stong, *Scientific American*, Sept.: 232 (1954).

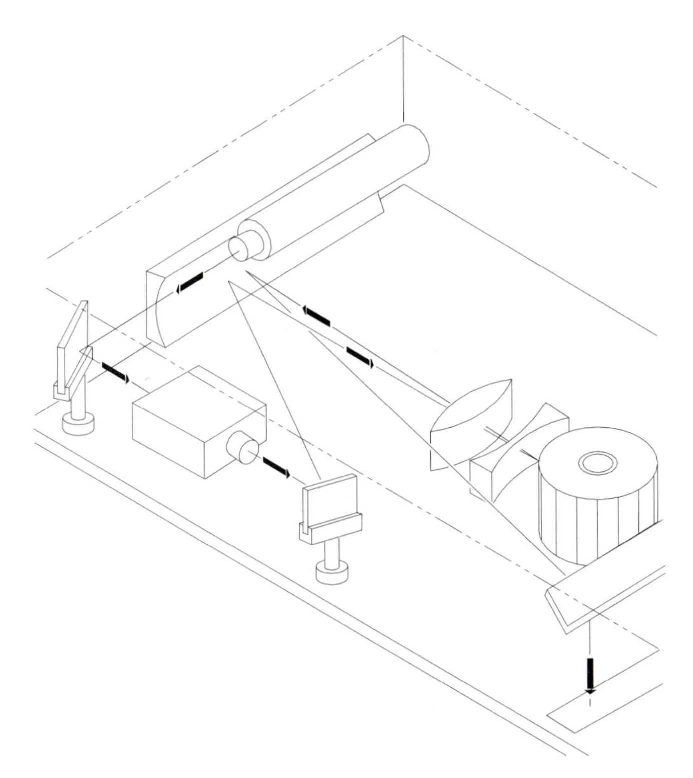

Figure 9.2 The most widely known application of HeNe lasers is in grocery store scanners. In these scanners, the beam is rastered in an $x - y$ pattern, using either multisided mirrors or resonant galvonometric scanners. (Courtesy of Melles Griot)

of thickness, surface quality, and surface index of refraction; and holography. In the medical world, HeNe lasers find application in laser-flow cytometry, laser-induced fluorescence, and as guide lasers for other technologies such as CAT and MRI scans. Many of the common commercial applications for HeNe lasers can be adapted into excellent laser design projects (see Appendix A.8).

The most widely known application of HeNe lasers is in grocery store scanners. In these scanners, the beam is rastered in an $x - y$ pattern using either multisided mirrors or resonant galvanometric scanners. The scan pattern is shaped to assure that the barcode can be read independent of its orientation. The scanned laser beam is reflected off the bright and dark areas on the barcode and received by a photodetector. The high spatial beam quality of the HeNe laser results in a large depth of field capability for the scanner (see Figure 9.2).

The excellent spatial beam quality of HeNe lasers has generated a number of applications that use the HeNe laser as a sophisticated "chalk line." One very common application is to use a rotating mirror to create a flat plane of laser light. This plane can be used to align anything from a drop ceiling to a parking lot (see Figure 9.3). Another application is the laser theodolite. A theodolite is constructed by bouncing a laser beam off a mirror attached to a large object and measuring the angular deflection of the reflected beam. Theodolites are used in applications ranging from land surveying to constructing airplane wings.

The excellent longitudinal mode quality of HeNe lasers means that HeNe lasers possess very large coherence lengths (typically from meters to kilometers). Lasers with large coherence lengths are ideal for interferometric applications such as interferometry and holography.

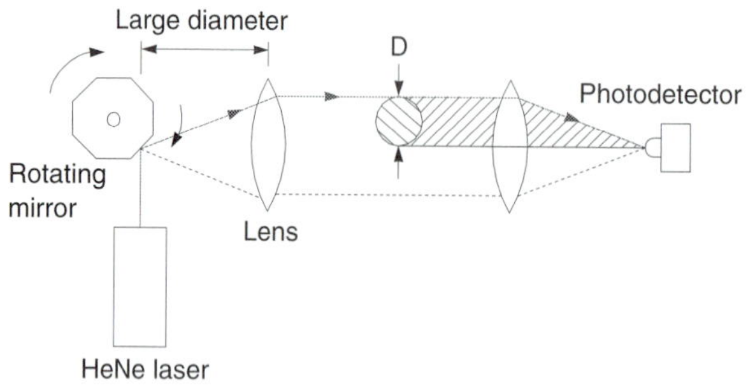

Figure 9.3 One very common application for HeNe lasers is to use a rotating mirror to create a flat plane of laser light. This flat plane can be used in a variety of applications. Here, the flat plane of light is used to measure the diameter of a rod. (Courtesy of Melles Griot)

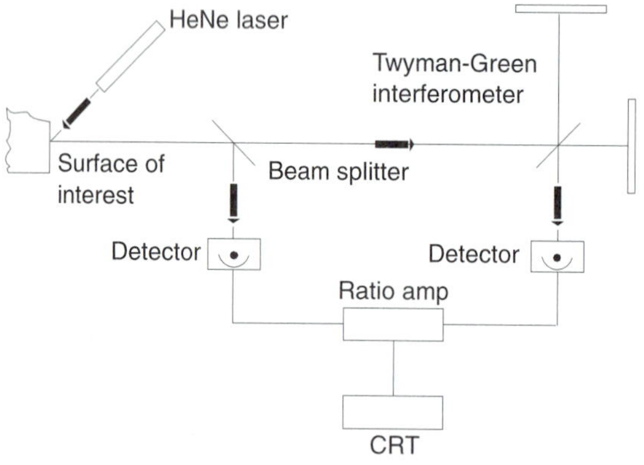

Figure 9.4 In a typical interferometric application, part of the beam is retained as the reference beam and part of the beam interacts with the sample. Interference between the reference beam and the sample beam provides phase information about the sample. (Courtesy of Melles Griot)

In a typical interferometric application, part of the beam is retained as the reference beam and part of the beam interacts with the sample. Interference between the reference beam and the sample beam provides phase information about the sample. Depending on the geometry of the setup, this can provide length information, velocity information, or both (see Figure 9.4).

Modifications on interferometric techniques frequently incorporate the scattered light from laser speckle. Variations in the scattered light profile can be used to detect velocity changes (noninvasive measurement of blood flow; see Figure 9.5) or can interfere with a primary source (measurement of combustive gas velocity).

Holographic applications typically rely on creating an interference pattern within an emulsion. In a typical two-beam transmission holography set-up, a reference beam interacts with a beam reflected off an object to produce the hologram. Variations on the technique include reflectance holography and real-time holography (see Figure 9.6).

The HeNe laser is often used in applications where a small signal needs to be isolated from a larger background. For example, in flow cytometry, the desired signal is a small signal obtained by fluorescence from a dye bound to proteins in the sample. This fluorescence signal can be extracted from the background by using a notch filter. (Other possibilities for isolating a small signal from a larger background include polarization and phase modulation.)

Another common application of HeNe lasers is the measurement of particle density and film thickness of semiconductor wafers. Undesirable particles on a surface interact with the laser beam and scatter light. In addition, the polarization characteristics of the reflected and refracted light can be used to determine the film thickness (see Figure 9.7).

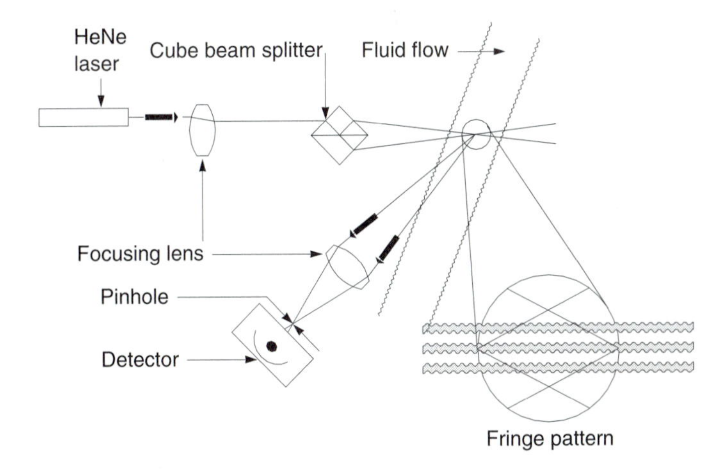

Figure 9.5 Variations in the the scattered light from laser speckle can be used to detect velocity changes in blood flow. (Courtesy of Melles Griot)

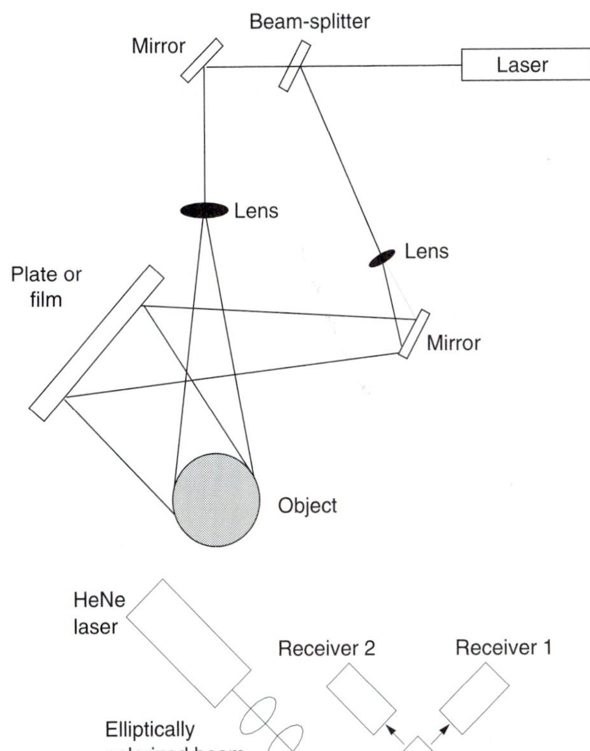

Figure 9.6 In a typical transmission holography set-up, a reference beam interacts with a beam reflected off an object to produce the hologram.

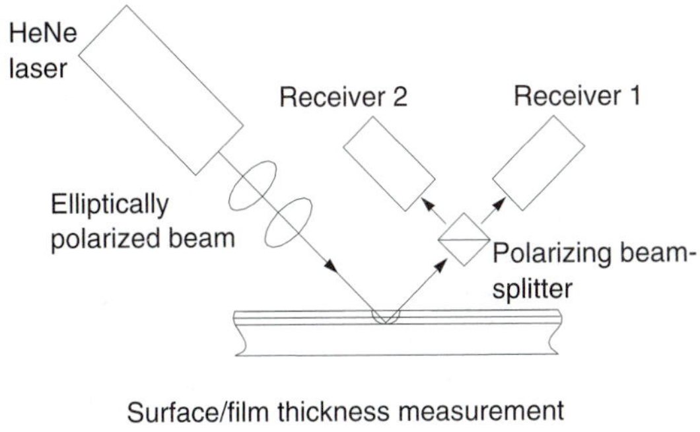

HeNe
laser
Receiver 2
Receiver 1
Elliptically
polarized beam
Polarizing beam-
splitter

Surface/film thickness measurement

Figure 9.7 HeNe lasers are used in the semiconductor industry to measure film thickness, surface cleanliness, defect structures, and impurities. (Courtesy of Melles Griot)

9.1.3 The HeNe Energy States

HeNe lasers lase on the various atomic transitions of neon. Although there are a large number of these transitions (see Figure 9.8), the transitions at 543.5 nm, 594.1 nm, 611.9 nm, 632.8 nm, 1.15 μm, 1.523 μm, and 3.39 μm are the ones typically used in commercial lasers.

Although HeNe lasers lase on the atomic transitions of neon, it is difficult to put neon in a tube and construct a laser. One reason is that the upper laser states of neon (the $3s_2$ and $2s_2$ states;[12] see Figure 9.8) are approximately 20 eV from the ground state. This large energy separation means that neon is extremely difficult to pump with electromagnetic radiation. Another reason is that the neon $3s_2$ and $2s_2$ states have very short lifetimes (less than 100 ns).[13] This means that they are difficult to pump with an electrical discharge.

One solution to the problems of pumping neon is to add helium to the neon gas mixture. Helium has two excited states (the 2^3S and 2^1S states), which are very close in energy to the $3s_2$ and $2s_2$ neon states. These helium excited states are much longer lived than the neon upper laser states (the 2^3S state has a lifetime of 5 μs and the 2^1S state has a lifetime of 100 μs[14]) and decay by resonant collisions with neon. These resonant collisions between helium and neon pump the $3s_2$ and $2s_2$ neon upper laser states. (It is interesting to note that a typical HeNe laser gas mixture is predominantly helium with less than 15% neon.)[15]

An electrical discharge is used to excite electrons from the ground state to the 20.6 eV helium 2^1S excited state. When the excited helium atoms collide with neon atoms, the helium potential energy (plus a little kinetic energy) excites electrons into the neon $3s_2$ upper laser state. Electron recombination from the neon $3s_2$ upper laser state to the neon $2p_4$ lower laser state produces the 632.8 nm red HeNe line. The transition from the neon $3s_2$ upper laser state to the neon $2p_6$ lower laser state produces the 611.9 nm orange HeNe line, the transition from the neon $3s_2$ upper laser state to the neon $2p_8$ lower laser state produces the 594.1 nm yellow HeNe line, and the transition from the neon $3s_2$ upper laser state to the neon $2p_{10}$ lower laser state produces the 543.5 nm green HeNe line.

However, there are some interesting subtleties. Notice that the electrical discharge will also populate the helium 2^3S excited state. This results in electrons being collisionally excited into the neon $2s_2$ upper laser state in the same fashion as with the helium 2^1S to neon $3s_2$ excitation. This then allows the 1.15 μm neon $2s_2$ to neon $2p_4$ transition to lase. However, both the 632.8 nm neon $3s_2$ to neon $2p_4$ and the 1.15 μm neon $2s_2$ to neon

---

[12] The notation $3s_2$ and $2s_2$ is an example of Russel-Saunders notation for energy levels. It is not necessary to understand the quantum mechanics behind the notation to understand the operation of the HeNe laser! The numbers can simply be assumed to be general labels for the states.

For the reader who has had quantum mechanics, the general Russel-Saunders notation is given as $Nl^k\ {}^{2S+1}L_J$ where N is the Bohr orbit number (where $S = 0$, $P = 1$, $D = 2$ ...), l is the angular momentum state of the last k active electrons (and the combination l^k is frequently omitted), S is the total spin angular momentum, L is the total orbital angular momentum quantum number, and $J = L + S$ is the magnitude of the total angular momentum. A more complete description of the notation is given by Joseph Verdeyen, in *Laser Electronics*, 2d ed. (Englewood Cliffs, NJ: Prentice-Hall, 1989), pp. 548–551.

[13] Amnon Yariv, *Optical Electronics*, 4th ed. (Philadelphia, PA: Saunders College Publishing, 1991), p. 240.

[14] Amnon Yariv, *Optical Electronics*, 4th ed. (Philadelphia, PA: Saunders College Publishing, 1991), p. 240.

[15] Melles Griot, 1995–6 catalog, p. 53-2.

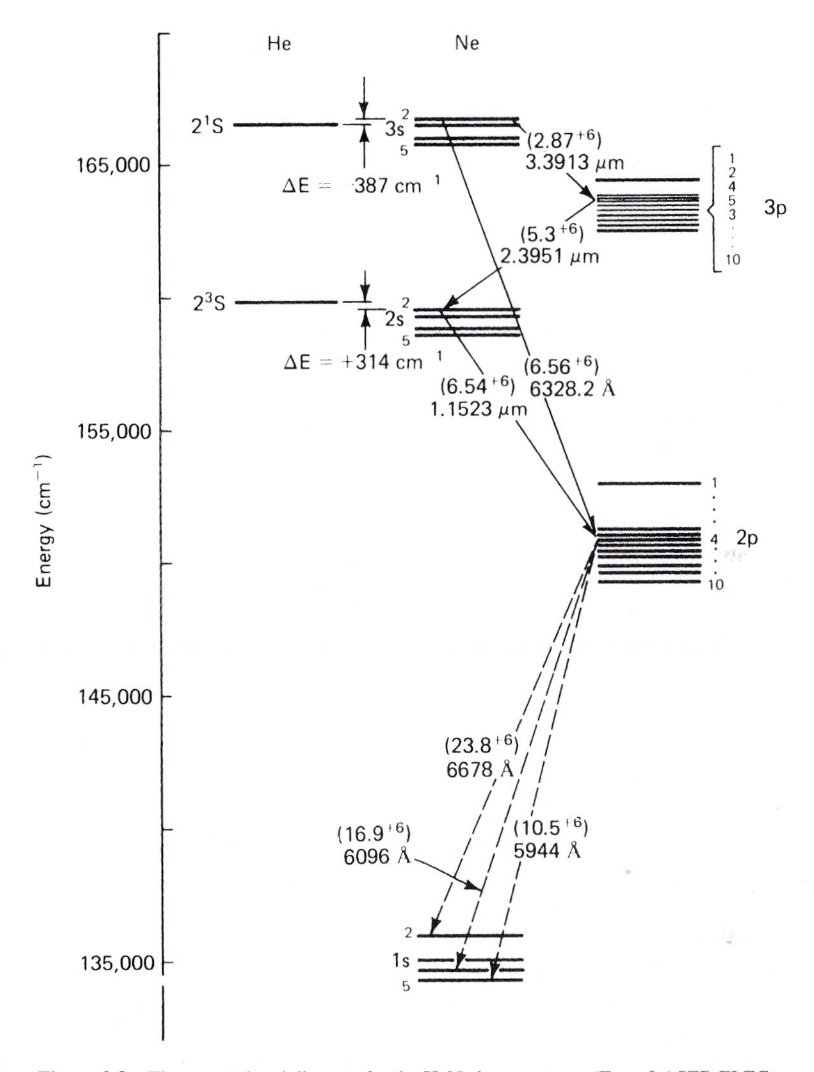

Figure 9.8 The energy band diagram for the HeNe laser system. (From LASER ELECTRONICS 2E. by VERDEYEN, J. T. ©1989, Figure 10.11, p. 326. Adapted by permission of Prentice-Hall, Inc., Upper Saddle River, NJ.)

$2p_4$ transitions share the same lower state. This makes it difficult to sustain a population inversion on the 632.8 nm transition if the 1.15 μm line is allowed to lase. Thus, in order to obtain laser action on the 632.8 nm transition, the 1.15 μm transition must be suppressed. This is typically accomplished by using a 1.15 μm absorber in the cavity and 1.15 μm transparent mirrors on the cavity. The same problem exists if operation is desired on the 1.15 μm line, because now the 632.8 nm line will be filling the desired lower laser state. Again, use of 632.8 nm transparent cavity mirrors and internal absorbers permits operation on the 1.15 μm line.

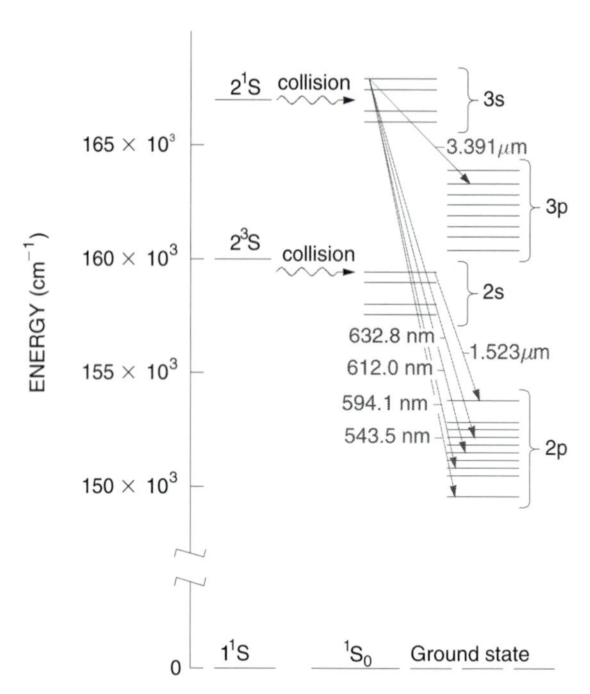

Figure 9.9 A detailed view of the energy band diagram for the HeNe laser system emphasizing the commercially available lines. (Courtesy of Melles Griot)

Another interesting subtlety involves the 632.8 nm neon $3s_2$ to neon $2p_4$ and the 3.39 μm neon $3s_2$ to neon $3p_4$ transitions. Both of these transitions share the same neon $3s_2$ upper laser state. However, the 3.39 μm neon $3s_2$ to neon $3p_4$ transition has the higher cross-section and will preferentially lase. Unfortunately, the gain on the 3.39 μm transition is quite high. Even when 3.39 μm transparent cavity mirrors and internal absorbers are used, it may not be possible to completely suppress this line. Thus, some HeNe lasers incorporate permanent magnets into the laser cavity design. These magnets create an inhomogeneous magnetic field that serves to broaden the 3.39 μm transition and reduce its gain.[16]

A similar situation exists with the 1.523 μm $2s_2$ to neon $2p_1$ transition. This transition shares an upper state with the 1.15 μm neon $2s_2$ to neon $2p_4$ transition. However, it does not share a lower state with any significant transition. Thus, operating on the 1.523 μm line simply requires suppressing the 1.15 μm transitions. A more complex situation exists between the 632.8 nm (red) neon $3s_2$ to neon $2p_4$, and the visible non-red transitions such as the 611.9 nm (orange) neon $3s_2$ to neon $2p_6$, 594.1 nm (yellow) neon $3s_2$ to neon $2p_8$, and 543.5 nm (green) neon $3s_2$ to neon $2p_{10}$ transitions. All of these visible transitions share the same upper state. Furthermore, although the lower states are different, electrons can easily cascade downward in energy to populate the neon $2p_6$, $2p_8$, and $2p_{10}$ lower states by recombination from the neon $2p_2$ or $2p_4$ states. Thus, lasing on any of the visible nonred

[16]John F. Ready, *Industrial Applications of Lasers* (New York: Academic Press, 1978), p. 69.

Figure 9.10 Melles Griot is the largest supplier of HeNe lasers in the world and all of their cylindrical lasers are produced at their facility in Carlsbad, CA. (Courtesy of Melles Griot)

HeNe transitions requires suppression of all the other visible transitions, plus the 3.39 μm and 1.15 μm transitions.

As yet another subtlety, the neon 2p lower states ($2p_2$ to $2p_{10}$) decay to the neon 1s states. The neon 1s states typically decay by wall collisions, not by radiative recombination. Thus, enhancing the number of collisions with the wall (for example, by reducing the size of the laser bore) can improve the overall laser performance.[17]

Interestingly enough, the key limiting factor in HeNe output power is not the complex relationships between the laser states, but rather the electrical discharge pumping of the neon itself. The characteristic red-orange lines that are associated with neon signs arise from the 2p to 1s transitions in neon. As the helium-neon mixture is excited by an electrical discharge, the 2p lower laser states fill up due to the 1s to 2p electrical excitation of neon. Squelching of the output power due to increasing 2p lower state populations is the largest factor limiting the output power of the HeNe laser.

9.1.4 Design of a Modern Commercial HeNe Laser

This section will highlight the design and construction of a modern family of cylindrical HeNe lasers produced by Melles Griot. Melles Griot is the largest supplier of HeNe lasers in the world and all of their cylindrical lasers are produced at their facility in Carlsbad, CA (see Figures 9.10 and 9.11).

A typical hard-sealed Melles Griot HeNe laser is illustrated in Figure 9.12. The laser action occurs between the two end mirrors and through an internal bore defined by a borosilicate capillary tube. The diameter of the capillary bore in HeNe lasers is a trade-off.

---

[17] Amnon Yariv, *Optical Electronics*, 4th ed. (Philadelphia, PA: Saunders College Publishing, 1991), p. 240.

Figure 9.11 Fabrication of a Melles Griot HeNe tube in their facility in Carlsbad, CA. (Courtesy of Melles Griot)

If the bore diameter is too narrow, diffraction losses in the laser beam will limit the output power and increase the beam divergence. If the bore diameter is too large, then higher order transverse modes will begin to lase and destroy the output mode purity. Typical bore diameters range from 1 to 3 mm.

The outside envelope of the plasma tube is also fabricated of borosilicate glass and is kept large in order to improve thermal and mechanical stability. The borosilicate glass selected by Melles Griot is a very hard glass with a leak rate of less than 0.01 torr/year. The precision capillary bore is firmly attached to the plasma tube at the anode end, and a unique metallic spring spider supports the cantilevered capillary tube at the cathode end. This results in greatly improved rotational stability of the laser, as gravitational sagging has been minimized (see Figure 9.13).

Melles Griot lasers are sealed with low-leakage hard seals (see Figure 9.14). Both the cathode-retaining endcap and the anode tube are made of Kovar. (Kovar is a nickel/iron/cobalt alloy whose thermal expansion coefficient is similar to that of borosilicate glass.) Prior to the sealing process, the cavity mirrors are bonded to the cathode and anode mirror cells. A special preformed ring of the frit sealing material is placed around the parts during initial assembly. Radio-frequency induction heating can be used to weld the Kovar mirror cell and the anode tube to the outer envelope of the plasma tube. During the heating process, the frit material flows fully into all of the component parts, creating a reliable long-lived seal. The mirror cells are cleverly designed to permit minor adjustments of laser alignment (using gentle deformation of cell material) after the sealing process is complete.

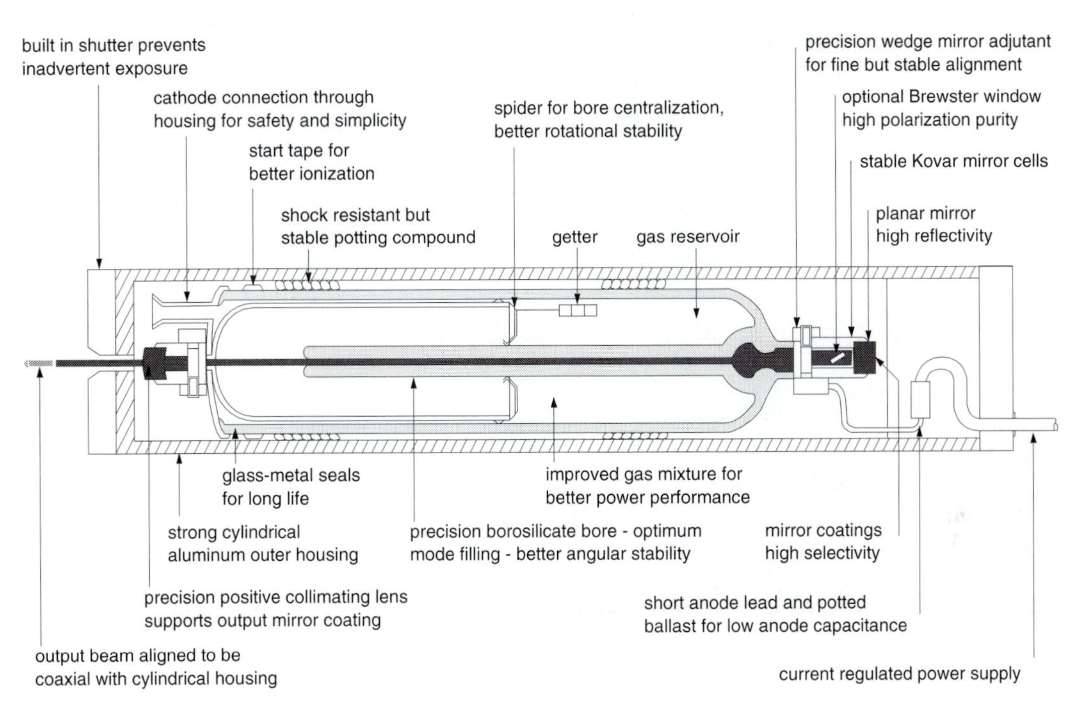

Figure 9.12 A typical hard-sealed Melles Griot HeNe laser plasma tube. (Courtesy of Melles Griot)

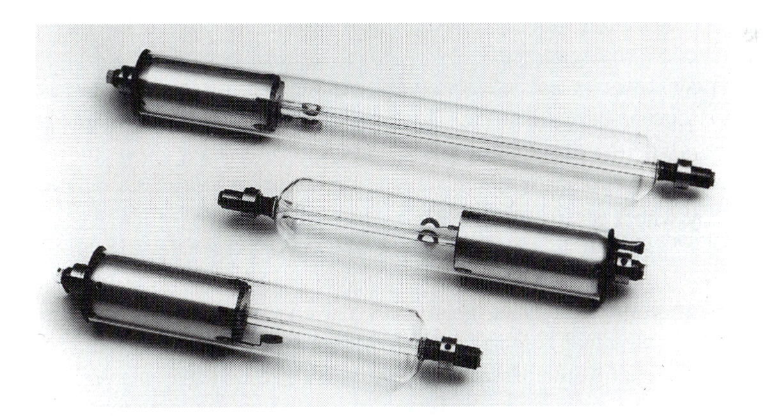

Figure 9.13 The outside envelope of the Melles Griot plasma tube is fabricated from borosilicate glass and is kept large in order to improve thermal and mechanical stability. (Courtesy of Melles Griot)

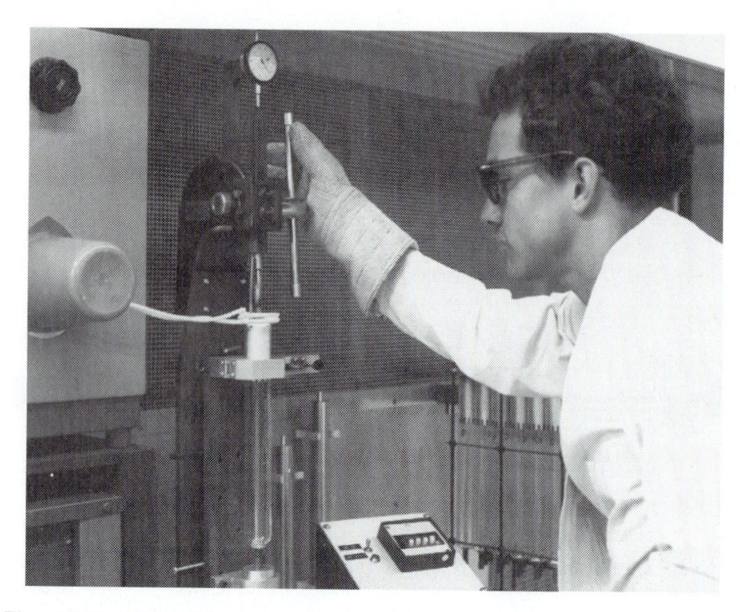

Figure 9.14 Melles Griot lasers are sealed with low-leakage hard seals. A special ring of the frit sealing material is placed around the parts during initial assembly. Radio-frequency induction heating is used to weld the Kovar endcap and the anode tube to the outer envelope of the plasma tube. (Courtesy of Melles Griot)

Cathode construction is also of importance in HeNe laser design.[18] Melles Griot lasers use deep coaxial cathode to assure a symmetrical discharge. The patented design also includes a hemispherical endcap to protect the nickel-iron endplate from bombardment by the plasma discharge.[19] Without this endcap, material from the end plate sputters on the optical surfaces and reduces the laser lifetime. Additionally, the hemispherical end design assures a more uniform current density distribution between the open end of the bore and the cathode (see Figure 9.15).

Like most gas lasers, HeNe tubes require one voltage input to create the plasma, and a second voltage source to sustain it. A high-voltage, low-current input is required to start the plasma. Once the plasma is formed, the required voltage drops to 1000 to 4000 volts and the laser output power is roughly linear with current until a saturation current is reached.[20] A typical 5 mW HeNe requires 8 to 12 kV to start the plasma, and 1800–2300 volts at 6.5 mA for DC operation.

A typical HeNe laser has a single pass gain on the order of 1%. Thus, typical cavity mirrors are a rear mirror at 99.9% and a front mirror at 98.5%. Most Melles Griot HeNe lasers use a modified hemispherical resonator configuration. The rear mirror is a flat mirror at the anode end of the tube. The front mirror is a concave mirror at the cathode end of the tube and has a typical radius of several meters. This combination produces a beam waist at

[18]L. Hall, *J. Appl. Phys.* 64:2630 (1988).

[19]U.S. Patent #4,311,969.

[20]John F. Ready, *Industrial Applications of Lasers* (New York: Academic Press, 1978), p. 71.

Figure 9.15 HeNe laser manufacturing requires a number of detailed and careful processes: (a) mirror inspection, (b) cleaning the outside envelope, (c) the gas fill process, and (d) operational testing. (Courtesy of Melles Griot)

the rear mirror of the laser, and a mildly divergent output beam. To reduce the divergence, the front mirror is fabricated on the concave surface of a glass meniscus lens. The output beam is refocused by this lens to a waist close to the physical exit of the laser beam from the laser housing.[21]

In a low-gain laser such as a HeNe, the gain is enhanced when adjacent longitudinal modes are orthogonally polarized. Thus, HeNe lasers without external polarization control will operate with adjacent longitudinal modes orthogonally polarized.

A simple HeNe laser with cylindrical symmetry has a randomly polarized output. Each individual cavity mode has a linear polarization at any one time. However, overall laser output is a time-varying mix of different polarization modes. Consequently, the beam appears to be randomly polarized.

Although the beam intensity of a randomly polarized laser is fairly constant, if the experiment or application involves polarization-dependent optics (such as beamsplitters) then large rapid amplitude fluctuations may be apparent. If this is not acceptable in the application, then a polarized HeNe laser should be used. Polarized HeNe lasers usually have an intracavity Brewster window that introduces sufficient loss in the s-plane so that only p-polarized output is produced.[22]

A typical HeNe laser has three longitudinal modes operating within the gain curve. As the laser length changes with small thermal variations, these modes will slowly sweep across the gain curve. Since the intensity of the longitudinal modes depends on their position in the gain curve, this means that most HeNe lasers will exhibit amplitude fluctuations as the longitudinal modes sweep across the gain curve. These fluctuations can be reduced by using a longer laser (more operating modes) or by protecting the laser from thermal variations.[23]

9.2 ARGON LASERS

It is possible to make both neutral and ion lasers. A neutral laser is one that lases between the various excited states of a neutral atom. (The helium-neon laser is an example of a neutral laser.) An ion laser is one that lases between the various excited states of various ion levels. Neutral and ion lasers are typically denoted by a Roman numeral equal to the ionization state plus one. For example, a neutral argon laser is Ar-I, a typical argon-ion laser is Ar-II, an ultraviolet (UV) argon laser is Ar-III, and so on.

Although a large number of ion lasers have been constructed, only a few have ever reached commercial potential. The most important of these are: Ar-II (conventional argon-ion), Kr-II (conventional krypton-ion), Ar-III (UV argon-ion), Kr-III (blue/UV krypton-ion), and HeCd-II. Cd-II, Se-II, Xe-IV, Zn-II, Ne-II, and Hg-II have received some interest, but have never reached commercial potential.

---

[21] Melles Griot, 1995–6 catalog, p. 53-11.

[22] D. J. Brangaccio, *The Rev. of Sci. Inst.* 33:921 (1962).

[23] Melles Griot, 1995–6 catalog, p. 53-12.

Argon-ion lasers are discussed in most modern laser texts. For a more detailed overview representative of the era, consider Bridges,[24,25] Bloom,[26] and Patel.[27]

9.2.1 History of Argon- and Krypton-Ion Lasers

The first ion laser was a mercury-ion laser, discovered by W. Earl Bell at Spectra-Physics in January 1964.[28]

In February 1964, William Bridges obtained laser action from ionized argon by modifying a mercury-ion laser.[29] In his initial argon-ion experiments,[30] Bridges used a 107 cm long (hot cathode) or 110 cm long (cold cathode) 4 mm diameter plasma tube with Brewster angle windows. Neon or helium was used as a buffer gas and typical pressures were a few mtorr of argon and 200 mtorr of the buffer gas. Dielectric coated mirrors (see Section 8.2) of approximately 1% transmission were used to obtain the laser lines. However, aluminized mirrors of approximately 10% transmission were also used successfully for the stronger lines. Average powers of 0.2 to 0.4 mW were obtained.

In March 1964, Guy Convert and his group at CSF in France[31] reported a new line in an Hg-Ar mixture, a line that was later found to be Ar-II.

Within a few months of the original discovery, Bridges had also obtained laser action in krypton and xenon.[32] At virtually the same time William Bennett et al. at Yale[33] and Guy Convert et al.[34] had independently operated pulsed argon-ion lasers. Bennett and his group used a 3 m long cavity with dielectric coated mirrors. The plasma tube was 5 mm in diameter and used Brewster angle windows. Mercury, neon, or helium was used as a buffer gas and typical pressures were 20 mtorr of argon and 200 mtorr of the buffer gas. Peak powers of 1 watt were obtained (on the 488.0 nm line) and approximately 10 watts (on the 514.5 nm line).

Later in 1964, Eugene Gordon, E. Labuda and W. Bridges achieved cw laser action in argon-ion (as well as in krypton and xenon).[35] Gordon's experiments used a selection of laser tubes that were shorter and narrower than the ones Bridges used. These included Tube A with a bore diameter of 1.9 mm and a length of 25 cm, Tube B with a bore diameter of 1.25 mm and a length of 10 cm, and Tube C with a bore diameter of 2.5 mm and a

---

[24] W. B. Bridges, A. N. Chester, A. S. Halsted, and J. V. Parker, *Proc. IEEE* 59:724 (1971).

[25] W. B. Bridges, "Atomic and Ion Lasers," in *Methods for Experimental Physics,* Vol. 15A, ed C. L. Tang, (New York: Academic Press, 1979).

[26] A. L. Bloom, *Gas Lasers* (New York: John Wiley and Sons, 1968); and A. L. Bloom, "Gas Lasers," *Proc. IEEE* 54:1262 (1966).

[27] C. K. N. Patel, "Gas Lasers," in *Lasers*, Vol. 2, ed A. K. Levine, (New York: Dekker, 1968), pp. 1–190.

[28] W. E. Bell, *Appl. Phys. Lett.* 4:34 (1964).

[29] William Bridges, "Ion Lasers: A Retrospective," in *Proc. of the Int. Conf. Lasers '81*, ed Carl B. Collins, (1981), pp. 1–2.

[30] W. B. Bridges, *Appl. Phys. Lett.* 4:128 (1964).

[31] G. Convert, M. Armand, and P. Martinot-LaGarde, *Comptes Rendus Acad. Sci.* 258:3259 (1964).

[32] W. B. Bridges, *Proc. IEEE.* 52:843 (1964).

[33] W. R. Bennett, J. W. Knutson, G. N. Mercer, and J. L. Detch, *Appl. Phys. Lett.* 4:180 (1964).

[34] G. Convert et al., *Comptes Rendus Acad. Sci.* 258:4467 (1964).

[35] E. I. Gordon, E. F. Labuda, and W. Bridges, *Appl. Phys. Lett.* 4:178 (1964).

length of 20 cm. The typical pressure was 0.45 torr of argon. Output powers ranged from a few milliwatts out each end (Tube B) to 80 mW out each end (Tube C). Air cooling of the argon-ion tubes was also demonstrated in this experiment.

Continuous wave argon-ion lasers became immediately attractive, not only because they lased in the green and blue, but also because of the high gain.[36,37] Relatively rapidly, Bridges produced a cw argon-ion laser with an output power of 150 mW[38] (which was significantly more power than the best performance obtained from HeNe). By October 1964, Roy Paananen and his team at Raytheon had achieved 4 watts output from the cw argon-ion (and this had doubled by January 1965[39]). As more researchers entered the field, even higher powers were achieved.[40,41] However, academic publications on argon-ion lasers dropped rapidly as the first systems entered the commercial marketplace. Hughes Aircraft and Raytheon began to market commercial argon-ion lasers in 1966, Spectra-Physics and RCA in 1967, and Coherent Radiation in 1968.

9.2.2 Applications for Argon- and Krypton-Ion Lasers

Ion lasers are one of the most mature laser technologies, with some 67% of the laser systems sold going to industrial applications and only about 33% being sold for scientific purposes.[42] Although there are a very large number of ion laser sources, only argon and krypton are of significant commercial interest.

Argon-ion lasers lase on several important lines in the green and blue but have the highest output power on the emerald green 514.5 nm and blue-green 488 nm lines. Krypton-ion lasers have a number of important lines in the red and yellow but have the highest output power on the 676.4 nm and 647.1 nm red lines. (See Section 9.2.3.) In spite of the poor efficiency of both argon-ion and krypton-ion lasers, these lasers have consistently maintained a good position in the laser market.[43]

Argon-ion lasers provide blue or green laser light of excellent beam quality with a narrow frequency linewidth. There are still no reliable commercial blue or green semiconductor diode lasers, so argon-ion lasers (unlike HeNe lasers) are not competing with semiconductor

[36] S. M. Jarrett and G. C. Barker, *J. Appl. Phys.* 39:4845 (1968).

[37] M. D. Sayers, *Phys. Lett.* 29A:591 (1969).

[38] W. B. Bridges Papers, Hughes Research Laboratories, Quarterly report for project 3232, Gaseous Lasers, 1964.

[39] *Electronics* 37:17 (1964); and *Raytheon News* 14:4 (1965).

[40] The early technical issues of argon-ion laser design are covered in W. B. Bridges, A. N. Chester, A. S. Halsted, and J. V. Parker, *Proc. IEEE* 59:724 (1971).

[41] A good historical overview of ion laser development is given by the interview with William Bridges by Jeff Hecht, "William Bridges," in *Laser Pioneers*, revised ed. (Boston, MA: Academic Press, 1992), pp. 205–226.

[42] K. Leggett, "Reports of Gas Lasers' Demise Have Been Greatly Exaggerated," *The Photonics Design and Applications Handbook 1996* (Pittsfield, MA: Laurin Publishing Co., 1996), p. H-256.

[43] One of the great argon-ion laser quotes is from William Bridges, "There have been many surprises over the past three decades, but the most surprising thing to me is that the argon ion laser is still with us!" and, later in the same reference, "How hard could it be to find something more practical than a fragile vacuum-tube-like monster that used 25 kW of three-phase electricity and several gallons per minute of cooling water to just produce a few watts of blue light?" Quoted in Jeff Hecht, *Laser Pioneers*, revised ed. (Boston, MA: Academic Press, 1992), p. 222.

diode lasers. However, argon-ion lasers *do* compete with laser diode-pumped solid-state lasers (see Section 10.5.7).

Argon-ion lasers operate in two major market niches. The first niche is small economical air-cooled argon-ion lasers (in the 20–150 mW output power range). The biggest use of these lasers is in the biological instrumentation business, for fluorescent cell sorters and DNA sequencers. Air-cooled argon-ion lasers are also used in printing, scanning, pointing, and interferometric applications.

The competition in this market primarily comes from green HeNe lasers and diode-pumped solid-state lasers. However, green HeNe lasers can only achieve power levels up to a few mW and high beam quality diode-pumped solid-state lasers are difficult to build as cheaply as small air-cooled argon-ion lasers. In addition (at least for biological applications) the original fluorescent dyes were specifically developed for the argon-ion laser wavelengths. Thus, air-cooled argon-ion lasers have a stable foothold in this market.

The second niche is large (expensive) water-cooled argon-ion lasers with output powers greater than 1 watt. These lasers find commercial applications which include printing, display technology, and medicine. Argon-ion lasers are (still) the only commercial laser source in the green and blue offering high continuous wave output power, outstanding beam quality, and good long-term performance.

Krypton-ion lasers provide red and yellow light of excellent beam quality with a narrow frequency linewidth. Krypton-ion lasers also operate on some blue transitions of commercial value. Since HeNe and red semiconductor diode lasers dominate the low power end of the red laser market, red krypton-ion lasers are used almost exclusively used for applications where high output power and excellent beam quality are required in the red or yellow. Green and blue krypton-ion lasers are used interchangeably with green and blue argon-ion lasers for applications requiring specific wavelengths.

One of the major commercial applications for the blue lines in both argon-ion and krypton-ion lasers is the mastering of compact disks (see Figure 9.16). Compact disks are manufactured by an injection molding process that requires creating an initial glass *master disk*. Creating the master disk begins by spinning photoresist on a polished glass plate. An argon-ion or krypton-ion laser writes the CD information into the photoresist. The photoresist is developed, leaving a pattern of *pits* and *land* in the photoresist. A thin layer of metal is deposited on the photoresist and is used as a conductive back plane to electroform a thicker metal backing. When the thick metal layer is removed from the glass master it is a negative of the CD and is termed a *father*. The electroforming process is repeated on the father to make positive plates, which are termed *mothers*, and repeated yet again to make negative plates from the mothers, which are termed *sons*. The sons are used as the molds for the CD injection-molding process.

Entertainment represents nearly 20% of the industrial ion laser market and is expanding due to increasing interest in laser light shows. The wide color selection and high output power make argon- and krypton-ion lasers the first choice for laser light shows and other laser display applications.

Green and blue argon-ion lasers also find application in microelectronics. This includes applications ranging from circuit board alteration to link-blowing. Deep blue, UV, and deep UV ion lasers are also finding application in lithography.

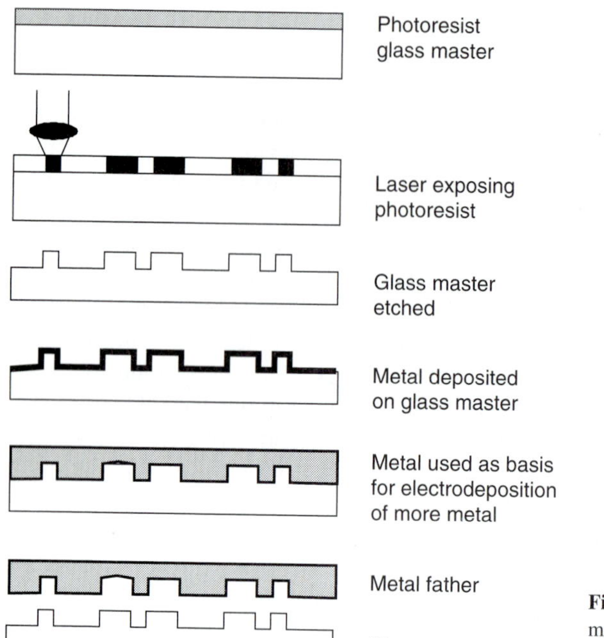

Figure 9.16 Compact disks are manufactured by using a laser-based lithography process.

Another fascinating application for blue and UV ion lasers is stereolithography. In this process, an ion laser writes a two-dimensional structure in resin. The resin begins to photocure where the ion laser has written the two-dimensional pattern. The depth of the resin bath is then changed, and the ion laser writes another layer. Thus, a three-dimensional structure is constructed layer by layer. Stereolithography is ideal for rapid prototyping applications, as it allows rapid creation of complex structures directly from computer graphics packages.

Argon-ion lasers are also used in medicine. One of the largest applications is in photocoagulation treatments for retinopathy. Retinopathy (retinal detachment) is seen in premature babies, diabetic patients, and in older patients with extreme myopia. Photocoagulation is a bit of a misnomer, as the argon-ion laser is used to heat (and intentionally damage) the retinal tissue. As the retinal tissue heals from the laser damage, it bonds more strongly to the detaching retina. It is the healing process (rather than the laser itself) that is therapeutic.

9.2.3 Argon and Krypton Laser States

The Ar-II spectrum is wonderfully rich. However, only the lines from the 4p to 4s bands are used commercially. These simplified lines are given in Figure 9.17.

The Kr-II spectrum is also quite rich and the simplified spectrum of the major commercial lines is given in Figure 9.18. Notice the similarities between the argon and krypton spectra (the 647.1 nm and 568.2 nm transitions in krypton correspond well to the 514.5 nm and 488.0 nm transitions in argon).

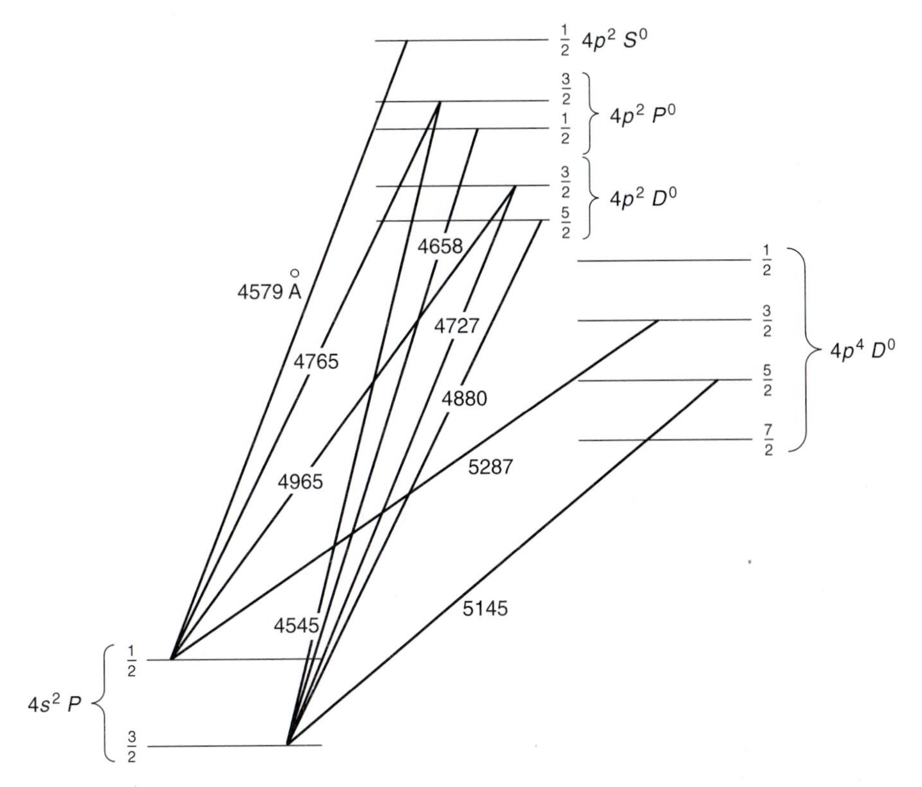

Figure 9.17 Commercial Ar-II argon-ion laser lines. (Reprinted with permission from W. B. Bridges, *Appl. Phys. Lett.*, 4:128, 1964, with a correction noted in *Appl. Phys. Lett.*, 5:39, 1964. © 1964 American Institute of Physics.)

The two highest power lines in argon are the 514.5 nm and the 488.0 nm lines. Although these lines share a lower state, they do not seem to compete as do similar lines in HeNe lasers. The lower state appears to drain rapidly enough (have a short enough spontaneous lifetime) to avoid bottlenecking when both lines are running. (This, by the way, contrasts with the analogous 647.1 nm and 568.2 nm lines in Kr-II. These lines also share a common lower laser level and *do* compete when both lines are running.)[44]

In Ar-II, the lower level depletes with 73 nm ultraviolet radiation. This contrasts quite significantly with HeNe, which depletes primarily with collisional interaction. Thus, the strong correlation of increasing laser power with decreasing bore diameter so significantly displayed by HeNe lasers is not an issue in argon-ion lasers.

However, argon-ion lasers do display a phenomena called radiation trapping. This is not a major effect in cw argon lasers, but can be quite dramatic in pulsed lasers due to the higher current densities. In radiation trapping the 73 nm emitted radiation from the lower laser states may be reabsorbed again and remitted and reabsorbed and so on (hence the name). The net effect is an increase in the lower state population. Under certain conditions

[44]W. Bridges, "Ionized Gas Lasers," in *CRC Handbook of Laser Science and Technology, Vol. II, Gas Lasers* (Boca Raton, FL: CRC Press, Inc., 1982), p. 196.

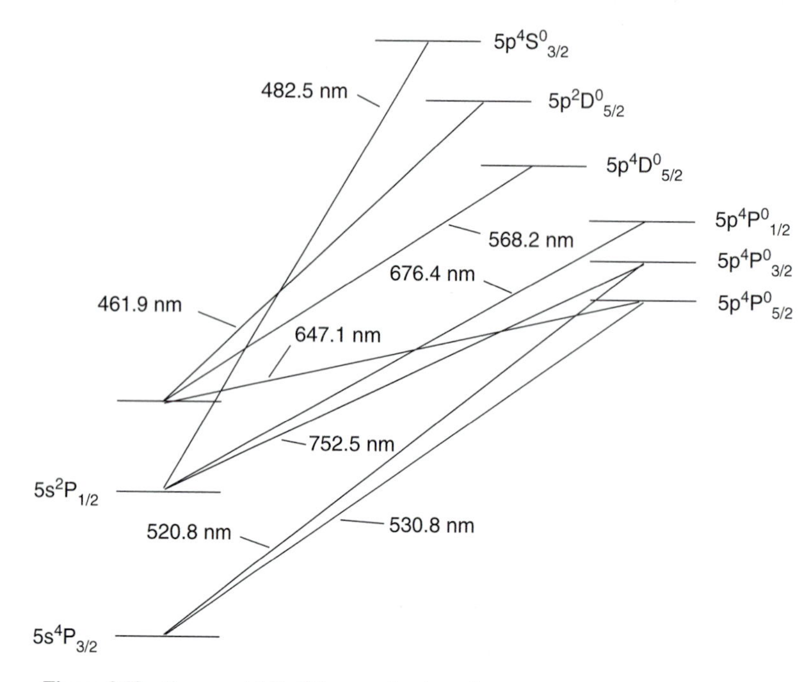

Figure 9.18 Commercial Kr-II krypton-ion laser lines. Notice the similarities between this and the argon spectra in figure 9.17, as the 647.1 nm and 568.2 nm transitions in krypton correspond well to the 514.5 nm and 488.0 nm transitions in argon.

the entire laser pulse can be "swallowed up" by radiation trapping[45] or, even more bizarre, plasma conditions can be created where radiation trapping kills the center part of the laser but allows the edges to continue to lase.[46]

Figure 9.19 illustrates the typical output power as a function of applied current for an argon-ion laser.[47] These curves were taken with laser mirrors of relatively constant reflectivity across the 520 nm to 4 μm range. However, the output power of any particular line could be enhanced by optimizing the reflectivity for the gain of that particular line (see Section 5.6).

9.2.4 Design of a Modern Commercial Argon-Ion Laser

This section will highlight the design and construction of the Beam-Lok argon-ion laser, which is produced by Spectra-Physics. The Beam-Lok is a stabilized single longitudinal mode argon-ion laser (single-mode laser operation is discussed in more detail in Chapter 3). A Spectra-Physics Beam-Lok argon-ion laser is illustrated in Figure 9.20. The laser consists of a plasma tube mounted on an external resonator structure. (An external resonator is not absolutely necessary for an argon-ion laser. Spectra-Physics also produces the Spinnaker line

[45]W. Bridges, *Proc. IEEE* 59:724 (1971).

[46]P. K. Cheo and H. G. Cooper, *Appl. Phys. Lett.* 6:1777 (1965).

[47]W. Bridges, "Ionized Gas Lasers," *CRC Handbook of Laser Science and Technology, Vol. II, Gas Lasers* (Boca Raton, FL: CRC Press, Inc., 1982), p. 247, Figure 2.20.

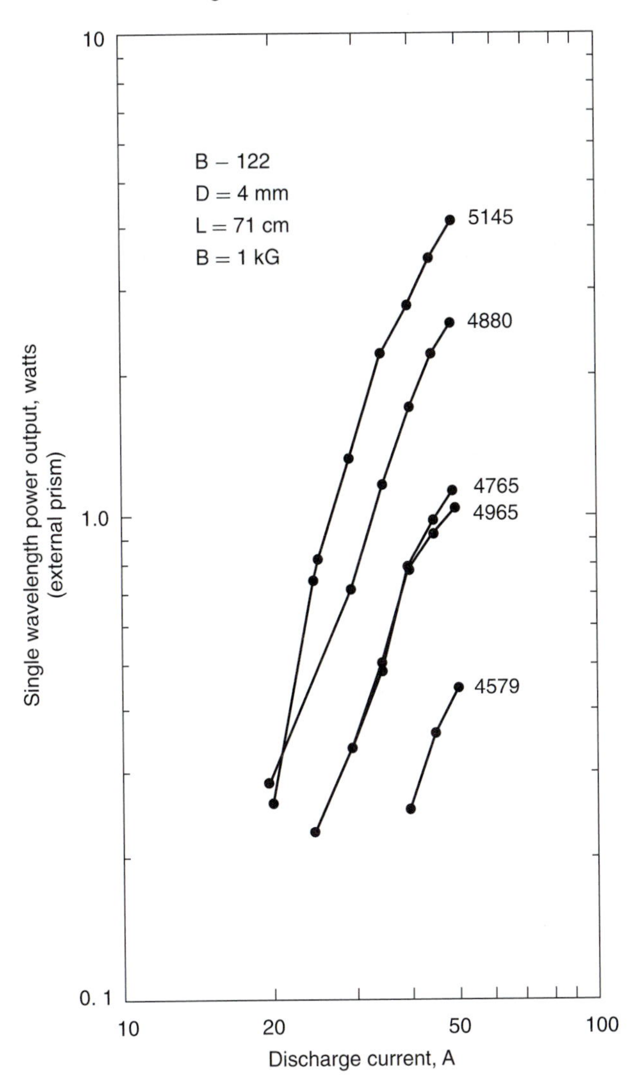

Figure 9.19 Typical output power as a function of applied current for an argon-ion laser. (Illustration from W. B. Bridges and A. S. Halsted, Gaseous Ion Laser Research Tech. Rep. AFAL-TR-67-89 (DDC No. AD-814897), Hughes Research Laboratories, Inc., Malibu, Calif., 1967. With permission of Hughes Research Laboratories, Inc.)

of argon-ion lasers, which incorporate hard-sealed mirrors similar to a HeNe laser. However, the Beam-Lok external resonator structure permits the use of mirrors, prisms, and etalons in order to tailor the beam wavelength and mode structure to the intended application.)

The plasma tube is the heart of an ion laser. The Spectra-Physics plasma tube is illustrated in Figure 9.21. The tube has two Brewster windows which are coated with a patented coating[48] to minimize degradation and color center formation generated by the vacuum ultraviolet radiation emitted by the plasma. The quartz windows are oriented so that the thermal expansion coefficient of the quartz window matches the thermal expansion coefficient of the endbell. The quartz windows are superpolished and attached (optically

[48]U.S. Patent #4,685,110.

Figure 9.20 A Spectra-Physics Beam-Lok argon-ion laser system. (Courtesy of Spectra-Physics, Mountain View, CA)

contacted with a post-bonding anneal) to the quartz-metal endbell.[49] This process requires no intermediate bonding material (such as a frit) and virtually eliminates window contamination. (Frits contain lead and the lead tends to migrate to the inside of the optical window.)

Each quartz window is individually inspected by placing the window inside of a UV laser source. The laser heats the window by UV absorption and the resulting window temperature is measured with an infrared pyrometer. Quartz windows not meeting a uniformity specification are rejected.

The laser bore of the Spectra-Physics laser is defined by thin tungsten disk segments separated by approximately 1/4 inch. Segments (rather than a continuous bore) are used so that the voltage drop is very small across any single segment. (Segmented construction does have the slight disadvantage that the plasma is not completely uniform between the segments.)

Tungsten is the material of choice for the bore, because it is a refractory material that is compatible with high-vacuum processing and possesses superior sputter resistance. Large copper disks are brazed to the tungsten bore segments in order to provide good heat transfer from the hot plasma to the ceramic wall. Molybdenum/manganese metalization brazed with a copper-silver eutectic is used to bond the copper disks to the ceramic sleeve. The molybdenum/manganese metalization is striped down the inside of the ceramic sleeve to achieve electrical contact without shorting from the anode to the cathode. (The striping pattern is achieved by depositing the metalization uniformly inside the tube and then removing any unwanted metal.)

In neutral ion lasers (such as HeNe lasers), the desired plasma is a weakly ionized plasma with current densities on the order of 10 to 200 mA/cm^2. In ion lasers, the plasma is strongly ionized, and current densities can rise to 10 to 200 A/cm^2. Strongly ionized plasmas of this type generate a great deal of heat, and so heat transfer is a critical part of argon-ion laser design. The Spectra-Physics design maximizes heat transfer from the plasma to the ceramic sleeve, while minimizing thermal stress and the subsequent microcracking of

[49]U.S. Patent #4,706,256.

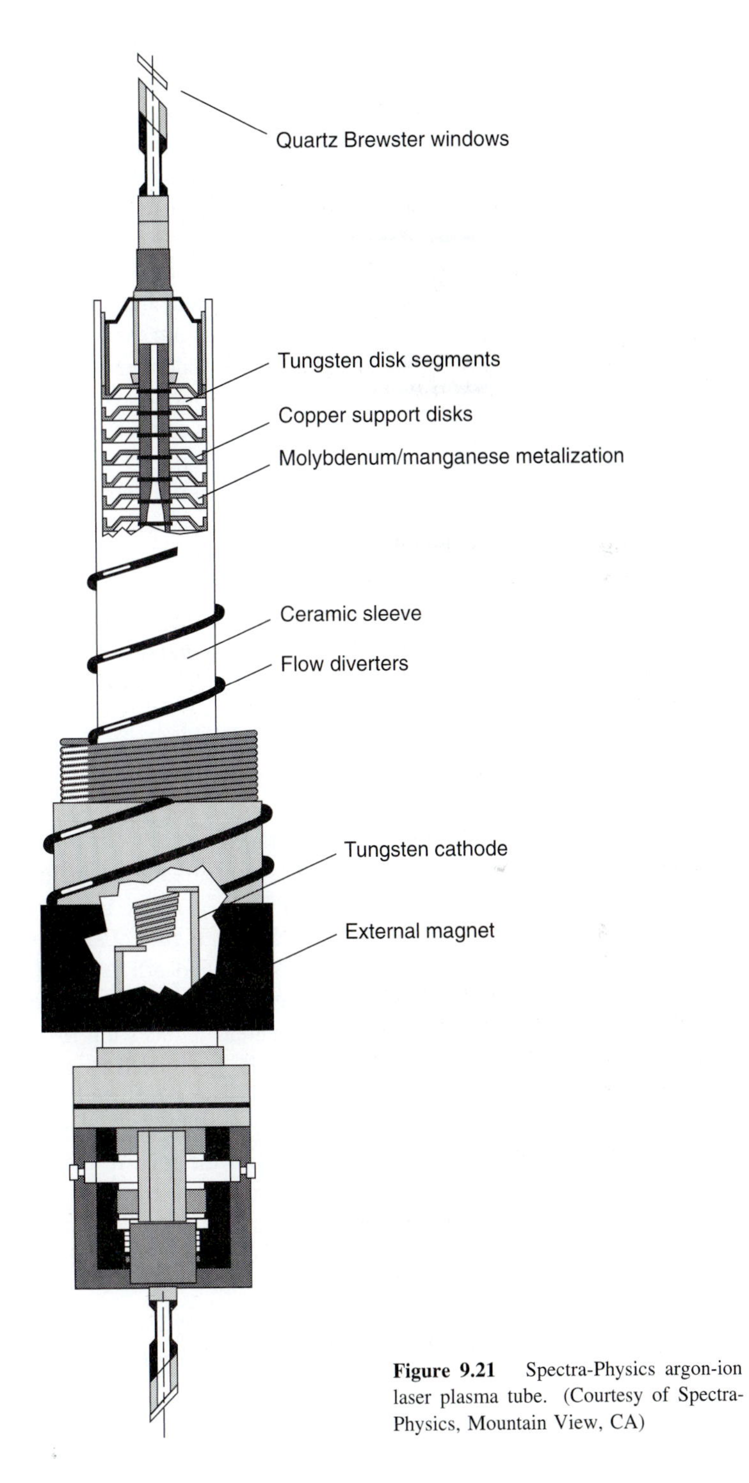

Figure 9.21 Spectra-Physics argon-ion laser plasma tube. (Courtesy of Spectra-Physics, Mountain View, CA)

the sleeve. The large diameter ceramic sleeve is water cooled and flow diverters are added to enhance heat transfer from the ceramic envelope to the cooling water.

A critical part of the design of the argon-ion laser tube is the cathode (electron emitter). Argon-ion laser cathode technology is a direct descendent of cathode technology from early electronic tubes. The cathode is a helical structure (not unlike a helical light bulb filament) made from a sintered tungsten matrix impregnated with a mixture of calcium, aluminum, barium, and strontium.

The tungsten matrix is a spongy form of tungsten made by taking tungsten powder and copper, sintering them, and dissolving out the copper. The matrix is approximately 80% tungsten and 20% cavities. The cavities are then impregnated with a material such as barium or strontium by soaking the matrix under pressure.

The barium monolayer (or fractional monolayer) reduces the work function of the tungsten surface so that the surface produces much more emission at a lower temperature. The barium continually evaporates off the surface. Thus, during operation, new layers of barium migrate to the surface from the materials stored in the matrix. (In new lasers, a bake-out and high-temperature firing step is necessary to create the first starting layer of free barium.)

A traditional problem with argon- and krypton-ion lasers is physical damage (usually by sputtering) of the walls confining the discharge, particularly at the cathode end of the tube. The damage is maximum in the *throat* region (the location where the tube necks down into the bore). Design of the throat area is critical to avoid lifetime reduction due to sputtering damage.

Anode laser design is not as critical as cathode design. The anode can be fabricated out of any thermally and electrically conductive material. Copper is a common choice for anode material.

An external magnet is used in the Spectra-Physics laser to create an axially directed magnetic field. Axial magnetic fields are often used in ion lasers as they demonstrably improve efficiency and output power.[50] In general, for any given diameter of tube, there is an optimal value of magnetic field (see Figure 9.22). In addition, the magnetic field can be contoured to enhance power output and efficiency. Interestingly enough, the exact mechanism by which the magnetic field improves the efficiency is not well understood. It is generally believed that the magnetic field reduces ion loss by slowing down the electron transport to the walls.[51] This theory is supported by research results that suggest that magnetic fields are not necessary in noble gas lasers with diameters larger than 1 cm.[52]

One of the more unusual features of high-current-density low-pressure ion lasers is the rapid pumping of gas from one end of the tube to the other.[53] Although both the ions and electrons gain equal momentum from the electric field, greater momentum is transferred to the wall by the ions than by the electrons. This pumping effect is so severe that early ion lasers shut themselves off after a few seconds of operation because all of the active gas

[50] First shown by I. Gorog and F. W. Spong, *Appl. Phys. Lett.* 9:61 (1966).

[51] C. E. Webb, "Ion Laser—Part I: The Radial Distribution of Excited Atoms and Ions in a Capillary Discharge in Argon," *Proc. Symp. Modern Optics* (New York: 1967).

[52] G. Herziger and W. Seelig, *Z. Phys.* 215:437 (1968).

[53] First discussed in detail by E. I. Gordon and E. F. Labuda, *Bell Sys. Tech. J.* 43:1827 (1964).

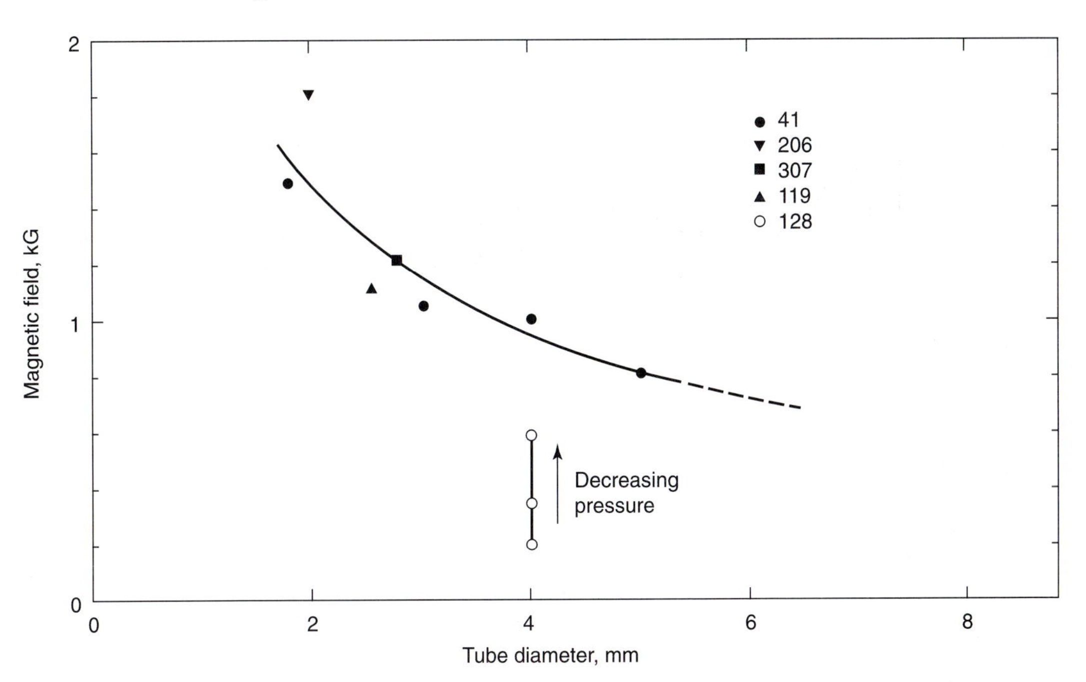

Figure 9.22 Axial magnetic fields are often used in ion lasers, as they demonstrably improve efficiency and output power. In general, for any given diameter of tube, there is an optimal value of magnetic field (typically 500–2000 G). (Reprinted with permission from W. B. Bridges, "Ionized Gas Lasers," in *Handbook of Laser Science and Technology, Volume II, Gas Lasers*, ed Marvin J. Weber (Boca Raton, FL: CRC Press LLC, 1982), p. 249. Copyright CRC Press, Boca Raton, FL, ©1982.)

ended up pumped to one end of the tube! In order to solve this problem, some sort of return path must be provided for the neutral atoms.

Early lasers simply provided a bypass path around the electrodes with a second glass tube. However, a bypass glass tube significantly reduces the robustness of the laser structure. The Spectra-Physics Beam-Lok Laser incorporates the modern approach of using several small off-axis holes in the copper support webs forming the tube. Not only is the gas returned through these holes, it is cooled by the heat transfer through the copper to the ceramic sleeve.

The output power of an ion laser is very sensitive to the gas pressure. However, sputtering during laser operation slowly consumes the gas. Thus, some sort of active fill system is required to replace the consumed gas. In the Spectra-Physics laser, the current-voltage characteristics of the laser during operation is used to determine the operating gas pressure. The laser incorporates an EPROM map of the current-voltage characteristics to determine when a fill sequence is necessary. When the tube pressure is low, an automatic fill is initiated from a high-pressure on-board reservoir. A two-valve system is used to meter the gas from the high-pressure reservoir into the laser. The first valve connects the high pressure reservoir to a small plenum and the second valve opens the plenum into the laser bore. This way, the quantity of new gas can be precisely controlled.

EXERCISES

HeNe lasers

9.1 (design) In the early HeNe laser experiments, very long (1 to 1.5 m) tubes with relatively small internal capillary bore diameters (1 to 3 mm) were used. Propose and describe three ways to align the front and rear mirrors on such long narrow lasers. Your answer should include a sketch (or sketches) and a brief written description of each proposed procedure.

9.2 (design) Consider the older style tube design given in Figure 9.1 and the more modern design given in Figure 9.12. Propose ten reasons why the more modern design has supplanted the older design.

9.3 (design) The raster display for a conventional (NTSC) television has 525 lines. The horizontal scan rate is 525*29.97 or 15,734 Hz. 63.6 μs are allocated per line. (Typically about 10 μs of this is devoted to the blanking line on the horizontal scan.) NTSC television is interlaced. This means that each frame is actually scanned twice. The first scan includes only the odd lines, the next scan includes only the even lines. With this method, the field rate is double the frame rate. Thus, NTSC systems have a field rate of 59.94 Hz. Assume you want to create a simple laser TV using a HeNe laser. Design an optical setup (you may model any required electronics by black boxes), incorporating two rotating multi-sided mirrors (and any additional optics that you feel is necessary), that is capable of simultaneously performing the required vertical and horizontal scans. Your answer should include:

(a) a sketch of your proposed design,

(b) a brief description of the design, and

(c) calculations which support that the design will perform to the given specification.

9.4 (design) One key feature of the Melles Griot laser design is that the mirror alignment is adjustable after the laser tube is hard-sealed by physically deforming the metal cell that holds the mirrors. List three advantages of this technique. List three disadvantages. Propose an alternative method for mounting and aligning hard-sealed mirrors. Your answer should include a sketch of the alternative method as well as a brief description.

9.5 (design) Assume that another major (high-gain) commercially viable laser line is discovered in HeNe. Assume this line runs between the $3p_4$ and the $2p_4$ states. Describe (in a fashion similar to Section 2.1) the possible effects that this new line would have on the other commercial HeNe laser lines.

9.6 (design) What property of the HeNe laser makes it difficult to construct a high power ($>$ 50 mW) HeNe? Propose three possible ways to circumvent this problem. Pick one of these as your favorite and expand on it in some detail. This expanded version should include physical calculations as necessary to support your idea.

Argon- and krypton-ion lasers

9.7 (design) Argon-ion lasers are notorious for being exceptionally inefficient and expensive to operate. For many years, it has been assumed that someday green and blue laser diode technology would advance sufficiently to replace the argon-ion laser. Do some library research (for example, check recent issues of trade journals such as *Laser Focus World* and *Photonics Spectra*, and recent issues of academic journals such as the *IEEE Journal of Quantum Electronics*, *Applied Physics Letters*, and *Optics Letters*) and determine the current state of the art in blue and green laser diodes. Do you think that blue and green lasers diodes will eventually supplant argon-ion lasers?

Your answer should include:

(a) a one- to two-paragraph description of the state of the art in blue and green laser diodes (including references!) and

(b) a one- to three-paragraph summary of your opinion (supported by facts, references and calculations) of the viability of blue and green laser diodes replacing argon-ion lasers.

9.8 (design) Consider the argon-ion laser states given in Figure 1.5. Speculate on which laser lines might conflict (for example, by suppressing gain). In other words, recreate the discussion of Section 9.1.3 using the argon-ion laser as the example.

9.9 (design) A critical part of the fabrication of the argon-ion laser is the bonding of the quartz windows to the metal endbell. Bonding quartz to metal is a difficult process. Why? Propose three different schemes for effectively bonding quartz to metal. Your answer should include:

(a) a few sentences describing why bonding quartz to metal is difficult,

(b) three simple sketches of your proposed schemes for bonding quartz to metal, and

(c) a few sentences on each approach (with calculations as appropriate) describing the advantages and disadvantages of the approach.

9.10 (design) Cathode design is a major part of effective gas laser design. Research other (nonlaser) types of cathode designs in the literature (for example, flashtubes, and electron tubes). Compare and contrast the cathode designs that you discover in your research. Your answer should include:

(a) a sketch (or photocopy) of the cathode designs that you located (and a reference for each!), and

(b) a few sentences for each comparing the various design features.

10

Conventional Solid-State Lasers

Objectives

- To summarize the sequence of historical events leading to the development of the modern commercial conventional solid-state laser.
- To summarize the commercial applications of conventional solid-state lasers.
- To describe the major laser host materials for conventional solid-state lasers.
- To describe the various energy states of the Nd:YAG laser and to summarize how these states interact with each other.
- To compare and contrast noble gas discharge lamp pumping versus semiconductor diode laser pumping for conventional solid-state lasers.
- To describe the design and construction of noble gas discharge lamp pumped conventional solid-state lasers. This includes the details of the noble gas discharge lamps, the power supplies, and the pump cavities.
- To describe the construction of a modern noble gas discharge lamp pumped Nd:YAG laser.
- To describe the design and construction of semiconductor diode laser pumped conventional solid-state lasers. This includes the details of the semiconductor pump lasers, the power supplies, and the pump cavities.
- To describe the construction of a modern semiconductor laser-pumped Nd:YAG laser.

10.1 HISTORY

One application that has driven a great deal of laser development is the idea of a directed beam weapon. From Buck Rogers to Captain Janeway, the concept of aiming a beam of energy at a target and vaporizing it has been very compelling to the military. In 1960, soon after the invention of the ruby laser, the Department of Defense (DOD) spent approximately $1.5 million on laser development. In 1961 the DOD spent $4 million on laser-related research, increasing to $12 million in 1962 and $19 to 24 million in 1963.[1] From 1960 to roughly 1980, the majority of academic research and much of industrial research was funded (directly or indirectly) by the military community.[2]

Very early in the development of lasers, it was recognized that noble gas lasers were far too inefficient for any possibility of a directed energy weapon. Three-state ruby lasers were not much better. However, in 1961 Johnson and Nassau demonstrated laser action using trivalent neodymium in calcium tungstate.[3] This was a significant development, as it opened the class of efficient four-state trivalent rare earth laser dopants. Later in 1961, Elias Snitzer succeeded in obtaining laser action from trivalent neodymium (Nd) doped in barium crown glass.[4] This development showed that laser action could be obtained in a material that could be fabricated in large sizes (see Figure 10.1 for a photograph of the very large NOVA glass laser).

In the period following the experiments of Johnson, Nassau, and Snitzer, solid-state laser research exploded into a flurry of experiments demonstrating laser action from a wide variety of di- and trivalent rare earth dopants in an incredible variety of laser hosts. More details on this history can be found in the review papers of the era by Kiss and Pressley[5] and by Young.[6] For a modern list of di- and trivalent rare earth dopants (and their hosts), see Weber.[7] (Laser dopants and their hosts are discussed in Section 10.3.)

Hundreds of important research results were obtained during this prolific period. Of special interest is the demonstration of laser action in Nd-doped yttrium aluminum garnet (YAG) by Joseph Geusic et al. at Bell Laboratories[8] and the demonstration of laser action from the trivalent dopants thulium, holmium, ytterbium, and erbium in YAG by Johnson et al., also at Bell.[9] Other Nd:YAG milestones are the first demonstration of

---

[1]Joan L. Bromberg, *The Laser in America, 1950–1970* (Cambridge, MA: The MIT Press, 1991), p. 102.

[2]Arthur Schalow has said, "I often joked that no matter what you told the press about lasers, it always came out as a 'death ray' or a cure for cancer—or both!" Quoted in Jeff Hecht, *Laser Pioneers*, revised ed. (Boston, MA: Academic Press, 1992), p. 92.

[3]L. F. Johnson and K. Nassau, *Proc. IRE* (Correspondence) 49:1704 (1961); L. F. Johnson, G. D. Boyd, K. Nassau, and R. R. Soden, *Phys. Rev.* 126:1406 (1962); and L. F. Johnson, G. D. Boyd, K. Nassau, and R. R. Soden, *Proc. IRE* (Correspondence) 50:213 (1962).

[4]E. Snitzer, *Phys. Rev. Lett.* 7:444 (1961).

[5]Z. J. Kiss and R. J. Pressley, *Proc. IEEE* 54:1236 (1966).

[6]C. Gilbert Young, *Proc. IEEE* 57:1267 (1969).

[7]Marvin J. Weber, ed, *Handbook of Laser Science and Technology, Vol. I, Lasers and Masers* (Boca Raton, FL: CRC Press, Inc., 1982); and the more recent supplement: Marvin J. Weber, ed, *Handbook of Laser Science and Technology, Sup. I, Lasers* (Boca Raton, FL: CRC Press, Inc., 1991).

[8]J. E. Geusic, H. M. Marcos, and L. G. Van Uitert, *Appl. Phys. Lett.* 4:182 (1964).

[9]L. F. Johnson, J. E. Geusic, and L. G. Van Uitert, *Appl. Phys. Lett.* 7:127 (1965).

Q-switched Nd:YAG operation by Geusic et al.,[10] and the first mode-locking of Nd:YAG by DiDomenico et al.[11] (Q-switching is discussed in Section 6.2 and mode-locking in Section 6.4.)

Another important pattern began to emerge during the solid-state laser development period of the 1960s. Unlike gas laser development, solid-state laser development was driven by the available materials technology. This pattern was certainly apparent during the early days of Nd:YAG laser development. When Geusic had originally identified YAG as a possible laser host, there were no high optical quality YAG crystals available. Geusic worked with LeGrand G. Van Uitert (of the Chemistry Research Department at Bell Laboratories) to develop a Czochralski crystal-growth process specifically for making high optical quality Nd:YAG crystals. This growth process was then handed off to Union Carbide. The laser developments in Nd:YAG precisely paralleled the continuing improvements in the crystal-growth technology.[12]

During this developmental period, Nd:YAG began to emerge as the leading solid-state material for commercial solid-state laser development. YAG is a mechanically robust and high optical quality material with good thermal properties. Neodymium is a four-state laser ion and (when doped into YAG) yields a laser system whose pump bands match well with standard commercial light sources. The combination of the two led to the introduction of the lamp-pumped Nd:YAG laser as a standard commercial laser source (see Section 10.4).

Nd:glass was another laser material to achieve prominence during this period. Glass is a less robust material than YAG and has significantly poorer thermal conductivity. However, the isotropic nature of glass permits higher doping concentrations for the laser ion and glass can be manufactured in very large sizes. Thus, Nd:glass became a favorite laser material for large laser systems. The crown jewels of these efforts are the huge glass lasers built by Lawrence Livermore National Laboratories for inertial confinement fusion (see Figure 10.1). These lasers began with Janus (1974; with an output energy on the order of 80 joules), and proceeded through Cyclops (1975), Argus (1976), Shiva (1977), and Nova (1984); gaining roughly an order of magnitude energy during each development cycle. An overview of the LLNL laser program is given by Emmett et al.[13] and Lawrence Livermore National Laboratory publishes a detailed annual report.

Another important thread in solid-state laser systems is development of semiconductor diode pumping of solid-state lasers (see Section 10.5.5). It was recognized extremely early that the light from semiconductor junctions could efficiently pump solid-state lasers. The first proposal of this concept was by Newman in 1963, and the experimental demonstration consisted of using 880 nm radiation from an LED-like source to excite fluorescence in Nd:$CaWO_4$.[14] Ochs and Pankove in 1964 used an array of LEDs in a transverse geometry to pump a Dy^{2+}:CaF_2 laser.[15] The first semiconductor diode laser pumped solid-state laser

---

[10] J. E. Geusic, M. L. Hensel, and R. G. Smith, *Appl. Phys. Lett.* 6:175 (1965).

[11] M. DiDomenico, J. E. Geusic, H. M. Marcos and R. G. Smith, *Appl. Phys. Lett.* 8:180 (1966).

[12] Joan L. Bromberg, *The Laser in America, 1950–1970* (Cambridge, MA: The MIT Press, 1991), pp. 176–7.

[13] J. L. Emmett, William F. Krupke, and J. I. Davis, *IEEE J. of Quantum Electron.* QE-20:591 (1984).

[14] R. Newman, *J. Appl. Phys.* 34:437 (1963).

[15] S. A. Ochs and J. I. Pankove, *Proc. IEEE* 52:713 (1964).

Figure 10.1 The NOVA laser bay at Lawrence Livermore National Laboratory. Note the technician in the center right for scale. (Courtesy of LLNL, Livermore, CA)

(GaAs laser diodes transversely pumping a U^{3+}:CaF_2 laser) was demonstrated in 1964 by Keyes and Quist.[16]

However, during the early 1960s, semiconductor diode pumping of lasers was unimpressive compared to lamp pumping. Lamp technology was very well developed (primarily driven by photographic applications) and semiconductor laser sources were still in their infancy. Thus, research into semiconductor diode pumping of solid-state lasers took a distinct second place to development of high-power lamp-pumped solid-state lasers.

This situation began to change in the mid-1960s and early 1970s. With higher-power laser diodes available commercially, interest was rekindled in using semiconductor diode lasers to pump solid-state laser hosts. In 1968, Ross demonstrated the first semiconductor diode laser pumped Nd:YAG laser.[17] In Ross's initial experiment, a single GaAs laser diode was used in a transverse geometry. The Nd:YAG rod was operated at room temperature, but the laser diode was cooled to 170K so that the output spectrum of the laser diode would better match the Nd:YAG pump bands.

---

[16] R. J. Keyes and T. M. Quist, *Appl. Phys. Lett.* 4:50 (1964).

[17] M. Ross, *Proc. IEEE* 56:196 (1968).

Transverse pumping was popular in the early experiments because a large number of sources could be grouped around the laser rod. However, transverse pumping is inefficient, since much of the pump light passes through the rod and does not contribute to laser action. In 1973, end-pumping was demonstrated by Rosenkrantz[18] and modeled by Chesler and Singh.[19] In end-pumping the diode laser source is colinear with the laser beam. End-pumping has the advantage of efficiency, since the majority of pump light is absorbed by the laser rod. However, it has a disadvantage in that only a limited number of laser diodes can be packed around the end of a rod. Ross's demonstration of semiconductor diode laser pumping of Nd:YAG and Rozenkrantz's demonstration of end-pumping triggered a number of papers exploring LED and semiconductor diode laser systems.[20] (See Section 10.5.6 for more discussion on transverse and end-pumping with semiconductor diode lasers.)

Significant research in semiconductor diode pumping of solid-state lasers did not appear until the mid-1970s. This was quite late in the development cycle for laser technology. As a consequence, the maturation of semiconductor diode laser pumping leveraged off many well-developed laser technologies. Thus, a number of advancements occurred simultaneously in many parts of the world. Some excellent overview articles are available in this area,[21] and so only the high points are summarized here.

- Q-switching, mode-locking, and injection seeding are all possible in diode laser pumped solid-state lasers. Representative papers include Owyoung et al.,[22] Schmitt et al.,[23] and Alcock et al.[24]
- Doubling and frequency up-conversion are widely used and representative papers include Pollack et al.,[25] Fan et al.,[26] Risk et al.,[27] and Dixon et al.[28]
- More complex diode pumped geometries can be used in order to meet specific needs. Representative papers include Trutna et al.,[29]Kane et al.,[30] and Reed et al.[31]

---

[18]L. J. Rozenkrantz, *J. Appl. Phys.* 43:4603 (1973).

[19]R. B. Chester and S. Singh, *J. Appl. Phys.* 44:5441 (1973).

[20]For a more complete review of this era, see T. Y. Fan and R. L. Byer, *IEEE J. of Quantum Electron.* 24:895 (1988).

[21]See for example, R. L. Byer, *Science* 239:742 (1988); and T. Y. Fan and R. L. Byer, *IEEE J. of Quantum Electron.* 24:895 (1988).

[22]A. Owyoung, G. R. Hadley, P. Esherick, R. L. Schmitt, and L. A. Rahn, *Opt. Lett.* 10:484 (1985).

[23]R. L. Schmitt and L. A. Rahn, *Appl. Opt.* 25:629 (1986).

[24]I. P. Alcock and A. I. Ferguson, *Opt. Comm.* 58:417 (1986).

[25]S. A. Pollack, D. B. Chang, and N. L Moise, *J. Appl. Phys.* 160:4077 (1986).

[26]T. Y. Fan, G. J. Dixon, and R. L. Byer, *Opt. Lett.* 11:204 (1986).

[27]W. P. Risk, J. C. Baumert, G. C. Bjorklund, F. M. Schellenberg, and W. Lenth, *Appl. Phys. Lett.* 52:85 (1988).

[28]G. J. Dixon, Z. M. Zhang, R. S .F. Chang, and N. Djeu, *Opt. Lett.* 13:137 (1988).

[29]W. R. Trutna, D. K. Donald, and M. Nazarathy, *Opt. Lett.* 12:248 (1987).

[30]T. J. Kane, A. C. Nielsson, and R. L. Byer, *Opt. Lett.* 12:175 (1987).

[31]M. K. Reed, W. J. Kozlovsky, and R. L. Byer, *Opt. Lett.* 13:204 (1988).

- Guided wave structures, such as fiber lasers, offer many advantages over the diode pumping of conventional geometry lasers. Fiber lasers are reviewed by Digonnet.[32]
- A class of materials called *stoichiometrics* exist where neodymium is a component of the crystal rather than a dopant. These materials can sustain a higher ion concentration than Nd:YAG and thus offer more efficient absorption of pump light. Stoichiometric materials are reviewed by Danielmeyer et al.[33] and Huber.[34]

10.2 APPLICATIONS

The primary commercial example of a conventional solid-state laser is Nd:YAG. Nd:YAG lases at 1.064 μm and can be frequency doubled (see Chapter 7) to 532 nm, tripled to 355 nm and quadrupled to 266 nm. Continuous wave Nd:YAG lasers are available in power levels up to several hundred watts and pulsed Nd:YAG lasers are available with pulse energies up to a few joules per pulse.

Nd:YAG lasers can also be operated Q-switched (which produces pulses in the tens of nanoseconds) or mode-locked (which produces pulses in the tens to hundreds of picoseconds). (See Section 6.2 for more information on Q-switching and Section 6.4 for more information on mode-locking.)

Q-switched Nd:YAG lasers are frequently used for laser machining processes that require rapid removal of relatively small quantities of material. Examples include trace and link blowing in the electronics industry, laser marking, and laser hole drilling. An important market niche is the use of Q-switched and frequency-doubled Nd:YAG lasers for laser marking of silicon wafers. A developing market niche is the use of Q-switched Nd:YAG lasers for removal of unwanted body hair.[35]

Frequency-doubled cw Nd:YAG lasers compete with argon-ion lasers for the moderate power green laser market. Thus, Nd:YAG lasers compete with argon ion lasers in such applications as printing, display technology, stereolithography, and retinal photocoagulation.

A developing application for Nd:YAG lasers is laser-enhanced bonding.[36] The process uses a laser to drive a polymer adhesive into the material being joined. The process provides a replacement for more traditional sewing, taping, gluing, or ultrasonic welding. Specific applications for the technology include bookbinding, laminating, textiles, injection molding, and carpeting.

---

[32]Michael J. F. Digonnet, *Rare-Earth Doped Fiber Lasers and Amplifiers* (New York: Marcel Dekker, 1993); and *Selected Papers on Rare-Earth Doped Fiber Laser Sources and Amplifiers,* ed M. J. F. Digonnet, SPIE Milestone Series Vol. MS37, 1992.

[33]H. G. Danielmeyer and F. W. Ostermayer, Jr., *J. Appl. Phys.* 43:2911 (1972).

[34]G. Huber, "Miniature Neodymium Lasers," in *Current Topics in Materials Science,* Vol. 4, ed E. Kaldis (Amsterdam: North-Holland, 1980), pp. 1–40.

[35]*Laser Focus World* January: 62–3 (1996).

[36]*Laser Focus World* August: 32 (1995).

Additional medical applications for Nd:YAG lasers include bleaching birthmarks,[37] removing tattoos,[38] and photothermolysis to remove large-diameter leg veins.[39]

Nd:YAG lasers also have a large scientific market. They are commonly used in laser radar applications (Lidar), laser spectroscopy, laser spectrophotometry, and laser metrology applications (such as seafloor mapping[40]) that require moderate power in the green and rugged design.

10.3 LASER MATERIALS

Two major classes of solid-state laser host materials exist: crystalline solid-state hosts (such as Nd:YAG) and isotropic solid-state hosts (principally glass). In these solid-state lasers, a host material with desirable mechanical and thermal properties (such as YAG) is doped with an impurity with desirable laser properties (such as neodymium).

The structural and laser properties of the material are inter-related since the atomic environment around the dopant atom determines the exact nature of the laser transition. As an example, neodymium is used as the dopant atom in both Nd:glass and Nd:YAG lasers. In Nd:YAG lasers, the atom is caged in a crystalline lattice where the immediate environment around each atom is well-ordered and symmetric. The resulting laser transition is relatively narrow (approximately 5 angstroms at 300K) with a wavelength of 1.064 μm. In Nd:glass lasers, the atom is in an amorphous structure where the immediate environment around each atom is poorly ordered and different for each atom. The resulting laser transition is broader than Nd:YAG (approximately 300 angstroms at 300K) with a wavelength ranging from 1.062 μm (silicate glass) to 1.054 μm (phosphate glass).

Many solid-state lasers use trivalent rare earths as the active ions. The rare earths have a partially filled 4f shell, and the various observed transitions occur near this shell. All the trivalent laser transitions are four-state, and the general transitions are given in Figure 10.2. Nd:YAG is a good example of a trivalent rare-earth doped laser material.

An important aspect of the efficient operation of solid-state lasers is effective transfer of the pump energy to the upper laser state. It is possible to co-dope many of the rare earth lasers with transition metal ions. This permits the wide absorption band of the transition metal ion to pump the narrower rare earth laser transition. Chromium-doped Nd:GSGG is a good example of a co-doped laser material.

There are a number of good references on laser host materials. Perhaps the most comprehensive are the detailed summaries given by Weber.[41] Industry and manufacturer

---

[37] *Laser Focus World* May: 67 (1996).

[38] *Laser Focus World* August: 62 (1995).

[39] *Laser Focus World* August: 6 (1995).

[40] *Laser Focus World* November: 53 (1995).

[41] Marvin J. Weber, ed, "Solid State Lasers," *Handbook of Laser Science and Technology, Vol. I, Lasers and Masers*, (Boca Raton, FL: CRC Press, Inc., 1982) Marvin J. Weber, ed, pp. 21–295; and the updated version, "Solid State Lasers," *Handbook of Laser Science and Technology, Supp. I, Lasers* (Boca Raton, FL: CRC Press, Inc., 1991), pp. 3–216.

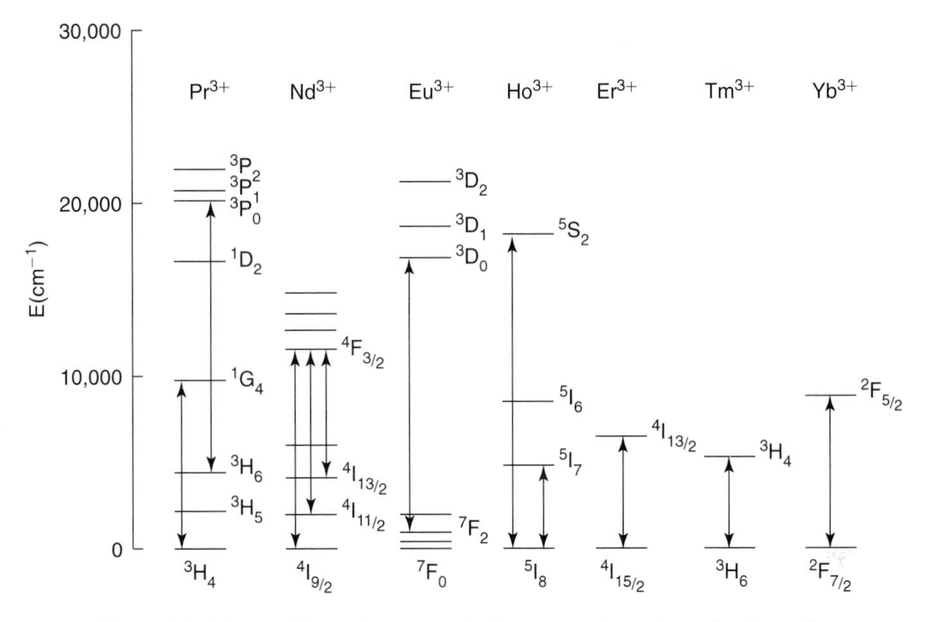

Figure 10.2 Many solid-state lasers use trivalent rare earths as the active ions. The rare earths have a partially filled 4f shell, and the various observed transitions occur near this shell. All the the trivalent laser transitions are four-level. (From Z. J. Kiss and R. J. Pressley, "Crystalline Solid Lasers," *Proc. IEEE* 54:1236 (1966), Figure 2. ©1966 IEEE.)

references are also extremely valuable.[42] In addition, some of the more classic references such as Kiss and Pressley,[43] Nassau,[44] Kaminskii,[45] and Danielmeyer[46] remain very relevant.

10.3.1 Crystalline Laser Hosts

A very large number of crystalline laser hosts have been explored over the years. However, as the laser industry has matured, only a few hosts have survived the transition into the commercial marketplace. These "workhorse" hosts are the garnets, glasses, sapphire, and the lithium fluorides (such as LiSAF).

The most useful laser materials have traditionally been the garnets. The general formula for the garnets is ($C_3A_2D_3O_{12}$). The C ion is in a relatively large dodecahedrally coordinated sited, the A ion is in an octahedrally coordinated site, and the D ion is in a

[42] *The Photonics Design and Applications Handbook, 1996 Book 3*, 42d ed. (Pittsfield, MA: Laurin Publishing Co., 1996).

[43] Z. J. Kiss and R. J. Pressley, *Proc. IEEE* 54:1236 (1966).

[44] K. Nassau, "The Chemistry of Laser Crystals," in *Applied Solid State Science*, Vol. 2. ed R. Wolfe (New York: Academic Press, 1974).

[45] A. A. Kaminskii, *Laser Crystals* (New York: Springer-Verlag, 1980).

[46] H. G. Danielmeyer, "Progress in Nd:YAG Lasers," in *Lasers*, Vol. 4, ed A. K. Levine and A. J. DeMaria, (New York: Marcel Dekker, 1976).

tetrahedrally coordinated site. Important laser materials of the garnet structure include yttrium aluminum garnet (YAG, $Y_3Al_5O_{12}$), yttrium gallium garnet (YGG, $Y_3Ga_5O_{12}$), gadolinium gallium garnet (GGG, $Gd_3Ga_5O_{12}$), lanthanum lutetium garnet (LLG, $La_3Lu_5O_{12}$), yttrium gadolinium garnet ($Y_3Gd_5O_{12}$), yttrium scandium gallium garnet (YSGG, $Y_3Sc_2Ga_3O_{12}$), yttrium scandium aluminum garnet (YSAG, $Y_3Sc_2Al_3O_{12}$), and gadolinium scandium gallium garnet (GSGG, $Gd_3Sc_2Ga_3O_{12}$).

To dope garnets, rare earths can be substituted in the C site and metals in the A site. Garnets have been doped with the trivalent rare earths Pr^{3+}, Nd^{3+}, Eu^{3+}, Ho^{3+}, Er^{3+}, Tm^{3+}, and Yb^{3+}. Garnets are also co-doped with chromium Cr^{3+} and titanium Ti^{3+}.

Sapphire (Al_2O_3) can be doped by impurities that replace the aluminum atom. However, the aluminum site is relatively small, so transition metals instead of rare earths are used as dopants. Examples include ruby (three-state, doped with Cr^{3+}) and titanium-sapphire (four-state, doped with Ti^{3+}). Related to sapphire are the beryl crystals such as alexandrite (four-state, $BeAl_2O_4$, doped with Cr^{3+}) and emerald (four-state, $Be_3Al_2Si_6O_{18}$, doped with Cr^{3+}). Ti:sapphire, alexandrite, and emerald lasers are of special interest because the large linewidth permits tunable laser operation. Tunable solid-state lasers are discussed in Chapter 5.

Other popular laser hosts are the cubic fluoride crystals such as CaF_2, SrF_2, BaF_2. There is a charge compensation problem in these materials if they are doped with rare earths such as Nd^{3+}. Thus, the divalent (Nd^{2+}) rare earths are more commonly used as dopants for the cubic fluoride crystals. This charge compensation problem can also be addressed by fluoride crystals such as lanthanum fluoride LaF_3 and $LiYF_4$ (YLF) or with oxide hosts such as Y_2O_3, Gd_2O_3, and Er_2O_3.

10.3.2 Glass Laser Hosts

Crystalline solid-state lasers are hard, thermally conductive, and stable at high temperatures. However, they are also anisotropic, difficult to fabricate in large sizes (the crystals must be grown), and possess a maximum limit on the quantity of dopant atoms. (In the majority of the crystalline solid-state laser materials, the dopant site is restricted in size. Thus, there is typically a limit to the quantity of doping permitted before the material becomes so strained that laser operation is impaired. This limit is typically 1 to 3 atomic %.)

An alternative approach is to use glass as a host for the dopant atoms. Glass lasers are easy (and economical) to fabricate in any size, uniform in material composition, and possess high gains due to high limits on doping concentration. However, glass lasers have some disadvantages, including poor thermal conductivity and greater fragility than crystals.

In addition, glass hosts and crystalline hosts differ significantly in linewidth. Crystalline hosts have linewidths on the order of a few angstroms while glass hosts have linewidths on the order of many hundreds of angstroms. Thus, glass lasers are often used as tunable lasers or for short pulse generation, while crystalline lasers are used where narrow linewidths are required (such as spectroscopy).

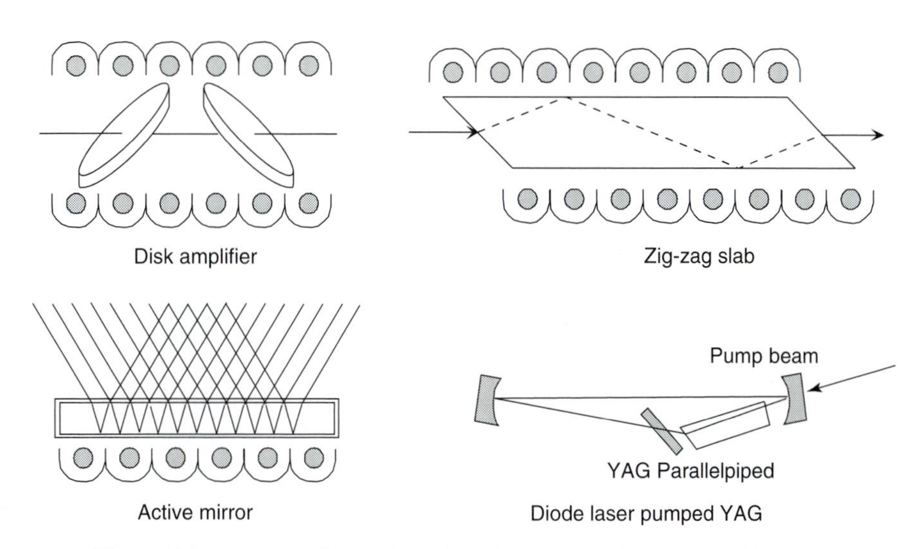

Figure 10.3 A variety of laser shapes have been developed to meet specific design constraints or to solve specific problems.

Crystalline hosts are visualized as having a highly-ordered lattice interspersed with a regular array of ions. Glass hosts are more difficult to visualize, because the mechanism inhibiting crystallization is poorly understood. The currently accepted model suggests that glass is actually composed of a network of energetically favored clusters that form isolated domains within the material. Each cluster is strongly coupled within itself, but weakly coupled to other clusters. This creates a myriad of different spectroscopic properties for the glasses.

10.3.3 The Shape of the Solid-State Laser Material

A rod geometry is the most common shape for a solid-state laser. However, it is not the only shape. A variety of other shapes have been used to meet specific design constraints or to solve specific problems (see Figure 10.3).

In very high-energy lasers (such as Shiva and Nova) the diameter of the beam is extremely large. The large diameter means that spurious modes (amplified spontaneous emission and confined spontaneous emission) are a major problem. In lasers such as these, the laser material is fabricated in the form of large disks where the disk may be 2 to 3 feet in diameter, but only three to four inches thick.

In Nd:YAG and Nd:glass high-average power rod lasers, the heat distribution in the rod is nonuniform. This results in thermal-focusing effects that eventually limit the maximum average power. If the laser material is fabricated as a zig-zag slab, much of this focusing can be eliminated because each bounce of the slab compensates for the focusing of the previous bounce.

In semiconductor diode array pumped lasers, it is often advantageous to fabricate the material in the form of a rectangular block. This increases the efficiency of the coupling from the semiconductor laser array to the laser material.

10.4 THE LASER TRANSITION IN Nd:YAG

Nd:YAG is probably the most widely used solid-state laser material. More details on Nd:YAG as used in lasers can be found in Koechner,[47] Danielmeyer,[48] and from various manufacturers.[49]

YAG (yttrium aluminum garnet) is a clear, hard $\bar{4}3m$ material[50] whose most familiar application is as a diamond replacement in jewelry. YAG is optically isotropic, of good optical quality, and quite thermally conductive (0.14 W/cm-K). YAG is also nonhygroscopic, melts at 1970C, and has a Knoop hardness of 1215 (around 8.2 on the Moh's hardness scale), which makes it one of the most durable of the common laser crystals.

Doping YAG with neodymium results in a blue to violet crystal. Because the material becomes strained at atomic percentages greater than approximately 1.5%, the majority of Nd:YAG crystals are doped at approximately 1%. It is generally accepted that higher doped material (1 to 1.4%) is better for Q-switched laser performance (due to the higher energy storage) while lower doped material (0.5 to 0.8%) is better for cw (steady-state) laser performance where optical beam quality is important.

By far the most popular dopant for YAG is neodymium. The neodymium ion in YAG forms a characteristic band structure (see Figure 10.4).

The laser pump bands for cw operation are principally the ${}^4F_{3/2}$, the ${}^4F_{5/2}$, the ${}^4F_{7/2}$, the ${}^4H_{9/2}$, and the ${}^4S_{3/2}$ bands. The ground state is the ${}^4I_{9/2}$ band. Thus, the pump bands for cw operation form a manifold centered around 7500 angstroms and 8100 angstroms (see Figure 10.5). (Notice that for pulsed operation, the higher-energy pump bands become important. Therefore, the pump bands for pulsed operation also include the manifolds centered at 5300 angstroms and 5800 angstroms.)

The primary laser transition in Nd:YAG at room temperature (the transition with the lowest threshold energy) is the ${}^4F_{3/2}$ band to the ${}^4I_{11/2}$ band. (In Nd:YAG, crystal-field splitting effects divide the principle bands into $J + 1/2$ doubly degenerate bands. The major Nd:YAG 1.064 μm laser transition originates at the R_2 crystal field split component of the ${}^4F_{3/2}$ band and terminates on the Y_3 crystal field split component of the ${}^4I_{11/2}$ band.)

At room temperature, approximately 40% of the ${}^4F_{3/2}$ band is in the R_2 configuration and approximately 60% in R_1 (from the Boltzman's thermal distribution). As the laser lases, the R_2 population is refilled from the R_1 state.

---

[47]Walter Koechner, *Solid State Laser Engineering*, 4th ed. (Berlin: Springer-Verlag, 1996).

[48]H. G. Danielmeyer, "Progress in Nd:YAG Lasers," in *Lasers*, Vol. 4, ed A. K. Levine and A. J. DeMaria (New York: Marcel Dekker, 1976).

[49]Litton-Airtron and Union Carbide are two major manufacturers of Nd:YAG laser crystals.

[50]Crystals are divided into seven major crystal systems defined by their symmetry properties. These classes are triclinic, monoclinic, orthorhombic, tetragonal, cubic, trigonal, and hexagonal. Each of these classes is further divided into point groups characterized by a number such as $\bar{4}3$m or 6. There are two common numbering systems for point groups, the Hermann-Maugin system (used more commonly by vendors) and the Schoenflies system. Boyd, *Nonlinear Optics* (San Diego: Academic Press, 1992) gives a nice overview of point groups in Chapter 1 from a tensor viewpoint (Table 1.5.1 is especially informative). William Leonard and Thomas Martin, in *Electronic Structure and Transport Properties of Materials* (Malabar, FL: Robert Krieger, 1987), provide a good background in Chapter 5 on crystallographic notation.

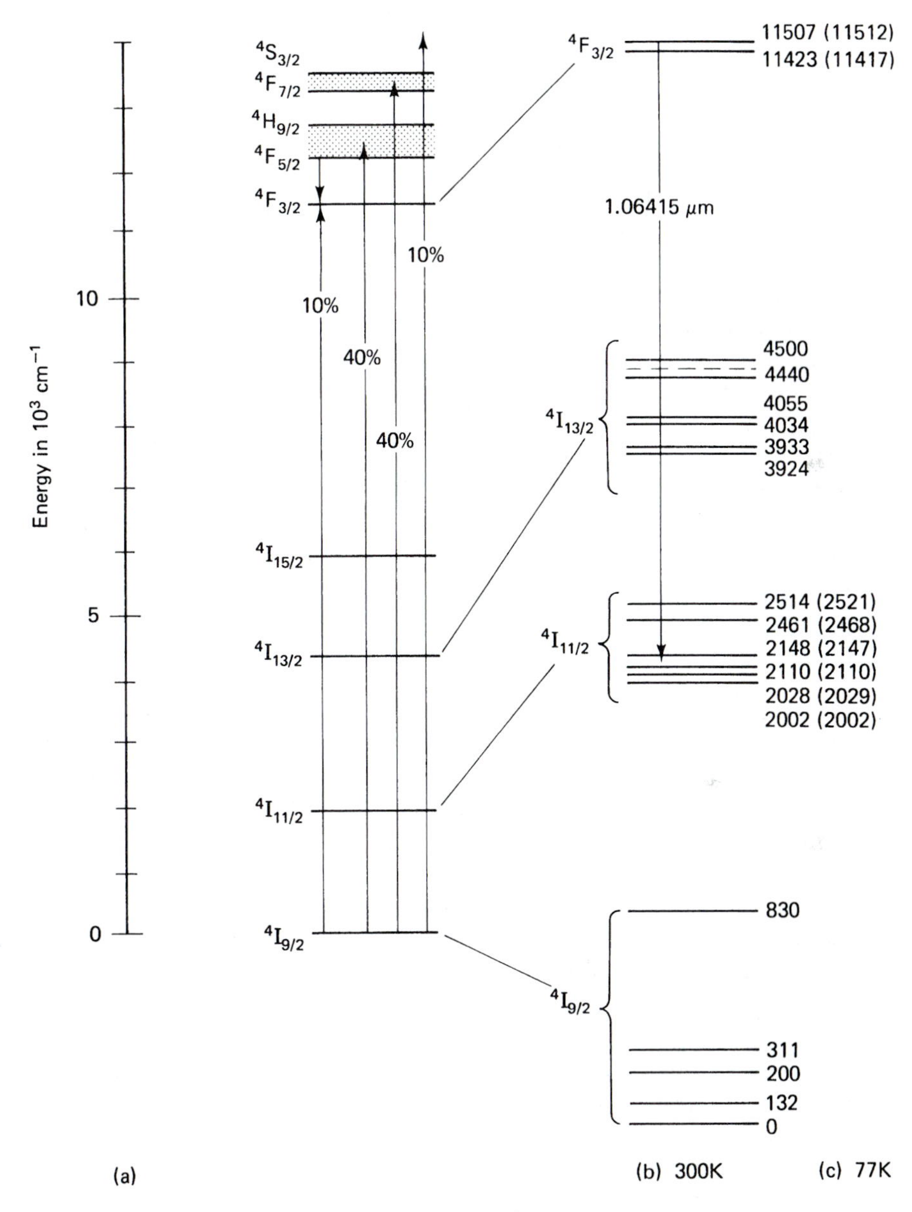

Figure 10.4 The energy band diagram for the Nd^{3+} ion in YAG. From LASER ELECTRONICS 2E. by VERDEYEN, J.T. ©1989, Figure 10.5, p. 317. (Adapted by permission of Prentice-Hall, Inc., Upper Saddle River, NJ)

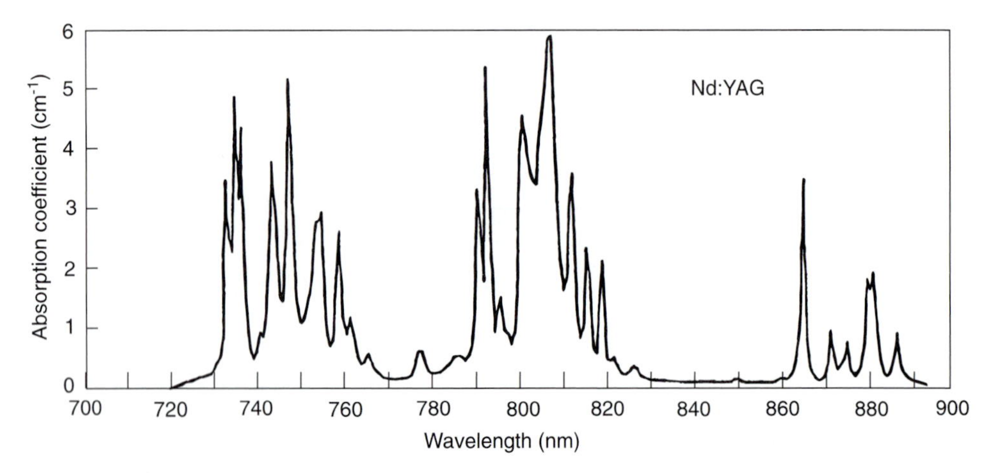

Figure 10.5 The pump bands for cw operation form a manifold centered around 7500 angstroms and 8100 angstroms. The pump bands for pulsed operation also include the manifolds centered at 5300 angstroms and 5800 angstroms. (From *The Photonics Design and Applications Handbook, 1996,* Book 3 of the *Photonics Directory*, 42nd ed. (Pittsfield, MA: Laurin Publishing Co., Inc., 1996), p. H-322. Reprinted with the permission of Laurin Publishing Co., Inc.)

The fluorescence efficiency of Nd:YAG is 99.5%. The probability that an ion excited to the ${}^4F_{3/2}$ state will transition to the ${}^4I_{11/2}$ state is 60%. The majority of the remaining transitions are to the ${}^4I_{9/2}$, ${}^4I_{13/2}$ and ${}^4I_{15/2}$. (${}^4F_{3/2}$ to ${}^4I_{9/2}$ = 25%, ${}^4F_{3/2}$ to ${}^4I_{13/2}$ = 14%, and ${}^4F_{3/2}$ to ${}^4I_{15/2}$ = 1%.)

By inserting a prism, grating, or using a selectively coated mirror, it is possible to lase on any combination of the crystal field split lines of the ${}^4F_{3/2}$ band to the ${}^4I_{11/2}$ band. This offers a range of transitions from 1.05 μm to 1.12 μm. In addition, it is possible to lase from the ${}^4F_{3/2}$ band R_2 state to all the field components of the ${}^4I_{13/2}$ band. This offers a range of transitions from 1.31 μm to 1.35 μm. Finally, if the laser is cooled, it is possible to lase in a near three-state configuration from the ${}^4F_{3/2}$ band R_2 state to some of the field components of the ${}^4I_{9/2}$ band. This offers a range of transitions from .93 μm to .95 μm (and a line at 1.839 μm).

Nd:YAG crystals are typically grown by the Czochralski method (the same method as used for the growth of silicon for semiconductor device substrates) with the $\langle 111 \rangle$ crystal axis in the growth direction. The major factor in the high cost of Nd:YAG is the slow crystal-growth rate (0.5 mm/hr). During the growth of Nd:YAG, crystal facets growing from opposite sides of the crystal form a characteristic core region. This core region is highly strained and nonuniformly doped. Doping densities in the core material may run twice as high as in the surrounding areas. Thus, the core region cannot be used for laser material. Laser material is typically cut from the regions surrounding the core, resulting in potential nonuniformities in dopant distribution. As with many Czochralski processes, the dopant is retained in the melt as the crystal grows. Thus, the dopant density decreases from one end of the boule to the other. This decrease is reflected in the laser rods, which may vary by 10 to 15% in neodymium concentration from one end to the other. The maximum size of YAG rods at this time is on the order of 1 cm in diameter and up to 15 cm long.

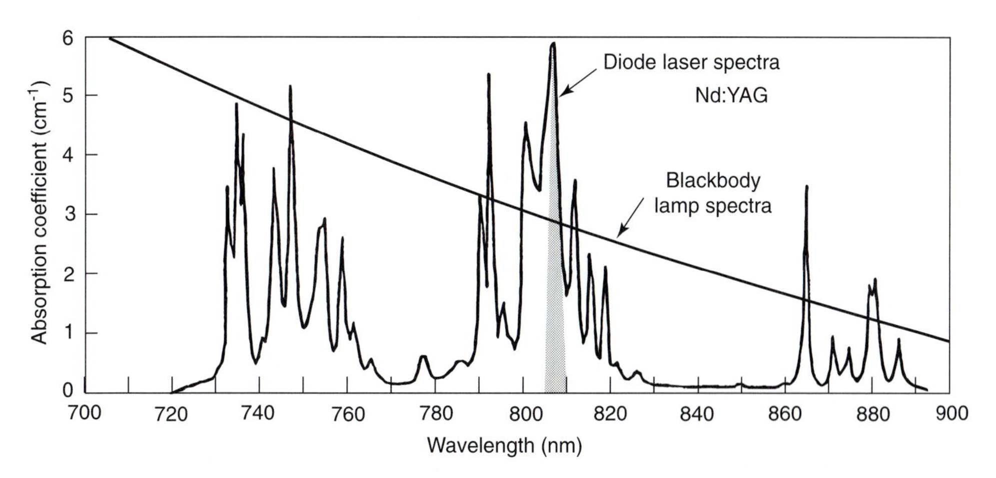

Figure 10.6 The goal in designing optical pumps for solid-state lasers is to match the output spectrum of the optical pump with that of the laser pump bands. Background spectrum from *The Photonics Design and Applications Handbook, 1996 Book 3*, 42d ed. (Pittsfield, MA: Laurin Publishing Co., 1996, p. H-322. Reprinted with the permission of Laurin Publishing Company.)

10.5 PUMP TECHNOLOGY

Most solid-state lasers are pumped with optical sources. The goal in designing optical pumps for solid-state lasers is to match the output spectrum of the optical pump with that of the laser pump bands.

Optical pump sources can be divided into two broad categories. One category is black- and greybody radiators, of which filament lamps are the best example. The other category is pump sources with a line emission spectra, of which semiconductor lasers are the best example.

There is typically a trade-off involved in choosing an optical pump source for a laser (see Figure 10.6). For the highest efficiency, the pump source should have the largest possible overlap between the pump bands of the laser and the output emission of the pump source. However, for the input maximum power, the pump source should have the largest possible amount of power delivered to the pump bands of the laser.

Typically the same pump source cannot meet both conditions. For example, krypton arc lamps are a continuum source with moderately poor overlap with the pump bands of solid-state lasers. However, state-of-the-art krypton arc lamps can deliver up to 20,000 watts of pump power per lamp.[51] In contrast, a semiconductor diode laser can be constructed so that the laser radiation almost precisely coincides with the pump bands of the laser. However, semiconductor lasers (even laser diode arrays) are difficult to construct with greater output power than a few hundred watts.

The design of lamp pumped solid-state lasers is quite different than the design of diode laser pumped solid-state lasers. Therefore, this section will first cover noble gas discharge

[51]For example, see EG&G lamp FX 77C-13.

lamp pumped solid-state lasers using the Spectra-Physics Quanta-Ray GCR-2 Nd:YAG laser as a representative example. Then, diode pumped solid-state lasers will be covered, using the Coherent Radiation DPSS 1064 Nd:YAG laser as a representative example.

10.5.1 Noble Gas Discharge Lamps as Optical Pump Sources for Nd:YAG Lasers

Noble gas discharge lamps are typically used in higher power Nd:YAG lasers. Details on design of noble gas discharge lamps for laser applications can be found in Koechner[52] and from the various lamp manufacturers.[53]

Noble gas discharge lamps are a compromise between blackbody radiators and line sources. They have a significant blackbody component generated by recombination radiation from gas ions capturing electrons into bound states (free-bound) and from Bremsstrahlung radiation. This component is usually characterized by a lamp *effective temperature*, which is the temperature of the blackbody profile that best matches the lamp's continuum radiation. Lamp effective temperatures are typically 4,000 to 5,000K for cw lamps and 5,000 to 15,000 K for pulsed lamps. (For more information on blackbody radiators, see Section 2.2.2.)

Superimposed on the continuum component is a line structure determined by the current density J (amps/cm^2) and the type of noble gas. For low-current densities, the spectrum is dominated by discrete transitions between the bound energy states of the gas atoms and ions (see Figure 10.7). This line structure is typically chosen to match the pumpbands of the laser of interest.

At high-current densities, the spectrum is dominated by recombination and Bremsstrahlung radiation. Thus, the spectrum is basically a continuum or blackbody spectrum. Notice as the current density in the high-current mode is increased, the peak of the continuum curve shifts downward in wavelength. Thus, the higher the current density, the bluer the lamp.

The optical properties of the lamp envelope will also affect the spectral profile. Noble-gas discharge lamps are frequently fabricated from fused silica and will transmit lamp radiation from 200 nm to 4 μm. However, the UV radiation transmitted by fused silica may be undesirable in certain lasers because it damages many coolants and creates color centers in certain laser materials. Therefore, it is possible to obtain lamp tubes that have been doped with metals to increase the UV absorption. However, increasing the absorption also increases the thermal loading of the tube wall and thus decreases the lamp lifetime.

It is also possible to dope the tube wall with a fluorescent material (cerium being a common choice). The fluorescent material will absorb light in the UV and re-emit it at the pumpbands of the laser. However (as with metal-doped fused silica lamps) reductions in lamp lifetime may be experienced due to increased thermal loading in the quartz envelope.

Krypton and xenon are the two noble gases most commonly used in laser lamp design. Xenon is the typical fill material, because it yields a higher radiation density for a given electrical input than the other noble gases. Xenon is also very efficient, as it converts 40 to 60% of the input electrical power into optical energy in the 200 nm to 1 μm region.[54]

---

[52]Walter Koechner, *Solid State Laser Engineering*, 4th ed. (Berlin: Springer-Verlag, 1996).

[53]EG&G and ILC are two major manufacturers of noble gas discharge lamps for laser applications.

[54]J. H. Goncz and P. B. Newell, *J. Opt. Soc. Am.* 56:87 (1966).

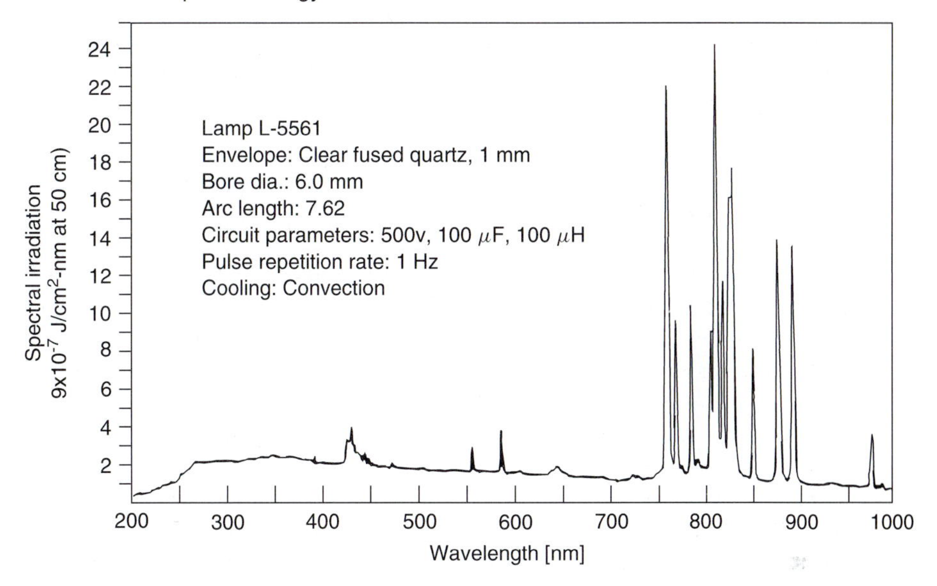

Figure 10.7 At low current densities, the spectrum of a noble gas discharge lamp is dominated by discrete transitions between the bound energy states of the gas atoms and ions. (Reproduced with the permission of ILC Technology, Sunnyvale, CA)

The one exception is for low-power Nd:YAG lasers, where the output spectrum from krypton is a better match to Nd:YAG because the two strongest lines of krypton (7600 angstroms and 8100 angstroms) provide a close good match to the Nd:YAG pump bands. Although krypton is much less efficient (25 to 35%), the close match of the line structure overcomes the loss in efficiency. However, at a power level of around $2 \cdot 10^5$ W/cm^2, the efficiency of xenon in creating a background continuum overcomes the advantageous match between krypton's line continuum and the pump bands.[55] Past $2 \cdot 10^5$ W/cm^2, xenon is the preferred pump source.

Noble gas discharge lamps are typically designed so that the plasma completely fills the lamp. They consist of a linear or helical tube (typically of quartz), two electrodes that are sealed into the tube, and a gas fill. The helical tube clearly has the better optical transfer characteristics. However, thermal design considerations generally make linear lamps a better choice unless the application absolutely requires the highest possible input intensity. Therefore, helical tubes are almost exclusively reserved for solid-state laser materials (such as ruby) that are difficult to pump. Notice that it is extremely difficult to fabricate noble gas discharge lamps in any geometry other than a linear tube or a helix. Thus, all commercial laser systems (including square, disk, and slab geometries) use either linear or helical noble gas discharge lamps.

Details of lamp construction. Typical noble gas discharge lamps are shown in Figures 10.9 and 10.10. Typical wall thicknesses are 1 to 3 mm, typical internal diameters are 3 mm to 20 mm, and lengths range from 2 inches to 4 feet. The gas fill is usually at a

[55] J. R. Oliver and F. S. Barnes, *IEEE J. Quantum Electron.* QE-5:225 (1969); and J. R. Oliver and F. S. Barnes, *Proc. IEEE* 59:638 (1971).

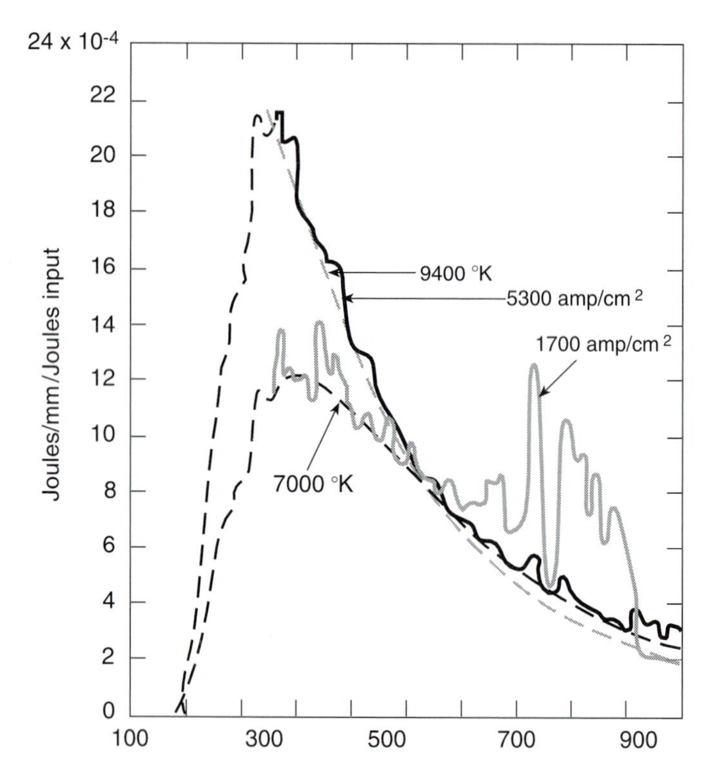

Figure 10.8 At high current densities, the spectrum of noble gas discharge lamps is a continuum or blackbody spectrum dominated by recombination and Bremsstrahlung radiation. (Courtesy of EG&G, Salem, MA)

pressure of 300 to 700 torr. The tube wall is often doped with cerium or erbium to improve the lamp efficiency by absorbing ultraviolet radiation from the lamp and emitting infrared radiation in the pump bands.

The anode and cathode are sealed into the lamp. The anode is typically rounded, while the cathode is pointed for arc stability. The anode is typically fabricated of a material whose work function is higher than that of the cathode. (For example, the anode can be pure tungsten at 4 eV and the cathode can be 2% thoriated tungsten at 2 eV. Alternatively, the anode can be 2% thoriated tungsten at 2 eV and the cathode can be a tungsten matrix at 1.5 eV.) The difference in work functions makes it easier to start the lamps. (However, note that the lamps are polarized and will pass current in only one direction without damage!)

It is interesting to note that in cw lamps, the kinetic energy of electrons incident on the anode is much larger than on the cathode. Thus, the anode heats up much more than the cathode. Rounded anodes also conduct more heat from the plasma, since they contact more of the plasma surface than the pointed cathode.

In constructing the lamps, a great deal of attention must be paid to the electrode seal. Quartz has a relatively low thermal expansion coefficient, while tungsten has a high thermal expansion coefficient. Therefore, some intermediate material must be used between the tungsten and the quartz to avoid lamp breakage due to differing thermal expansion coefficients.

One technique is to use one or several layers of borosilicate glasses of intermediate thermal expansion coefficient to grade the thermal expansion from the electrode to the quartz (see Figure 10.9). The borosilicate layering technique has the advantage of robustness under high power loading. It also permits *baking out* of the lamp and electrodes at high temperatures to aid in long-term gas purity. However, the seal is difficult to fabricate and requires the quartz envelope *necking* down around the electrode to assist with heat transfer during operation.

A second method is to use a copper rod, one end of which is brazed to the electrode and the other end of which is welded to a copper-plated nickel cup. The quartz is then coated with a platinum flash. The seal between the copper-plated nickel cup and the platinum coated quartz is made with a low-temperature indium solder (see Figure 10.10). The soldering technique has some advantages in geometrical simplicity and also provides a cooling path for the electrode through the copper. Soldered lamps can also be filled through the copper neck, which avoids the necessity for a filling nipple. However, the indium solder cannot be operating above its softening point (around 180°C) and so the lamp must be kept cool during operation.

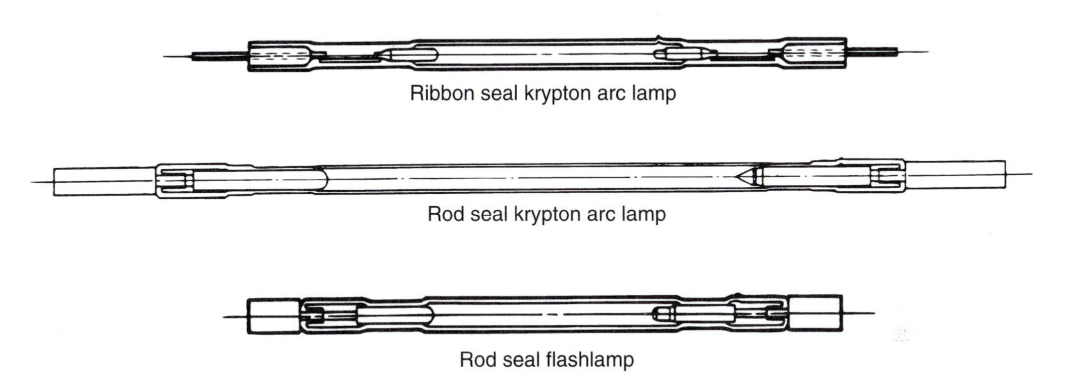

Figure 10.9 One type of lamp uses one or several layers of borosilicate glasses of intermediate thermal expansion coefficient to grade the thermal expansion from the electrode to the quartz. (Reproduced with the permission of ILC Technology, Sunnyvale, CA)

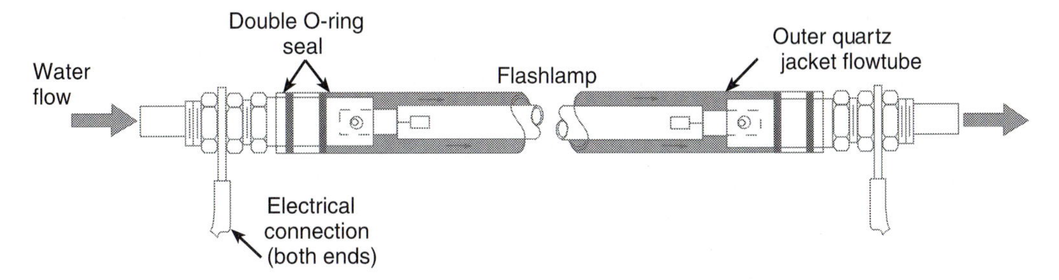

Figure 10.10 Another common lamp type uses a solder connection between the lamp and electrode. (Courtesy of EG&G, Salem, MA)

It is generally true that the higher the fill pressure, the better the conversion efficiency from electrical power to optical power. However, increasing fill pressures make the lamp significantly more difficult to start.

It is interesting to notice that the lamp plasma itself can be opaque to light. The absorption increases with current and with wavelength. Even moderate pumping densities can lead to plasmas that are nearly opaque at the wavelength of the pump bands. In a large diameter noble gas discharge lamp driven at high current densities, the majority of the radiation from the center of the lamp will be reabsorbed before reaching the lamp's surface. Thus, the majority of the emitted radiation comes from a thin sheath near the surface of the plasma. This is the principle effect limiting the diameter of noble gas discharge lamps.

Noble gas discharge lamps fail either catastrophically (they blow up) or gradually (they eventually get so dim or hard to start that they are replaced). Catastrophic failure generally occurs in pulsed lamps and originates from two effects. The first is when the shock wave from the expanding plasma breaks a lamp. The second is when excess heating of the envelope leads to excess thermal stress. The first mechanism is peak power dependent, the second is average power dependent. The lamp lifetime is a strong function of the ratio between the operating energy and the explosion energy (see Figure 10.11) so lamps are usually operated highly underrated.

Gradual failure generally occurs in cw lasers and is typically due to erosion of the electrodes, reduction of fill pressure, and buildup of deposits on the lamp wall. Cathode sputtering of the electrode material on the walls caused by the bombardment of xenon ions onto the negative lamp electrode is the most significant effect. Treating the cathode with oxides and cesium has been found to significantly reduce this sputtering.

In most commercial pulsed lasers, a small filament current (called a simmer current) is run continuously. This reduces the voltage required to start the lamp, as well as cathode sputtering and shot-to-shot jitter. When the full lamp voltage is applied to the lamp, the simmer filament expands with a radial velocity which is proportional to the input energy. The cylindrical shock wave stresses the tube wall axially toward the electrodes. If the stress exceeds the breaking stress of the lamp then the lamp will explode immediately. If the stress is a bit less than the breaking stress, then the lamp will develop microcracks and either crack or explode some time later.

Thermal loading of lamp-pumped Nd:YAG lasers. Consider a typical continuous wave Nd:YAG laser with an output power of 300 watts and an input power of 12 kilowatts. Assuming a quantum efficiency of 50% (low) this means that 600 watts are absorbed in the laser. Thus 11,400 watts are *not* absorbed in the laser. The majority of this power is optical power from the lamps outside the pumpbands of the laser. This excess power will be absorbed by the cavity and by the lamps, thus dramatically increasing the temperature of the laser. Roughly 10 to 20% of the electrical power will be dissipated as heat through the electrodes and 30 to 50% as heat through the envelope. In addition to causing mechanical overheating problems (seals and so on), thermal gradients will cause thermal focusing in the laser rod.

Typically the lamps, the cavity, and the Nd:YAG rod are cooled by water (or by ethylene glycol, ethanol, etc.). The usual pattern is to first take the incoming cold water and confine it to the region of the laser rod with a flow tube. This will remove the heat

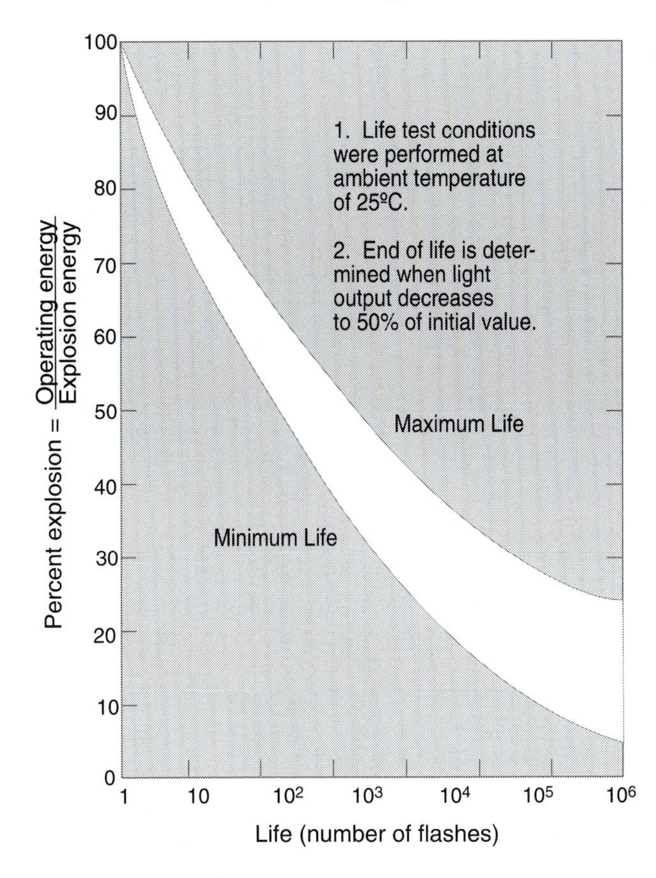

Figure 10.11 Lamp lifetime as a function of the ratio between the operating energy and the explosion energy. (Courtesy of EG&G, Salem, MA)

deposited in the rod that was not converted into laser light. Next, the water is allowed to flow through the major part of the laser cavity to remove the heat deposited in the reflectors and in the cavity walls. Finally, the water can be confined to the region around the lamps with a flow tube. This removes the heat absorbed in the quartz envelope.

Many variations on this theme are possible depending on the total power dissipation in the laser. For example, in extremely high average power lasers, a water cooling loop is provided through the electrodes to avoid destroying them. In very low power lasers, water cooling may only be provided over the lamps. In some extremely low power lasers, it may even be possible to use air cooling.

10.5.2 Power Supplies for Noble Gas Discharge Lamps

Noble gas discharge lamps are typically used in higher power Nd:YAG lasers. Details on design of noble gas discharge power supplies for laser applications can be found in Koechner[56] and from the various laser power supply manufacturers.[57]

---

[56]Walter Koechner, *Solid State Laser Engineering*, 4th ed. (Berlin: Springer-Verlag, 1996).

[57]Universal Voltronics, EMCO High Voltage Company, Kaiser Systems and Laser Drive, Inc., manufacture power supplies for laser applications. Additional manufacturers can be found in *The Photonics Buyers' Guide to*

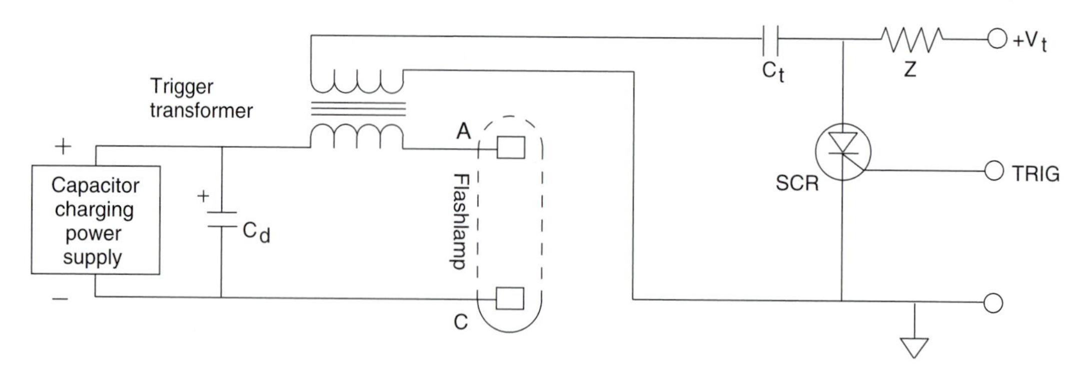

Figure 10.12 Lamps can be triggered by charging a capacitor to approximately 1 kV using a voltage multiplier circuit. Then, the capacitor is discharged through an SCR to trigger a step-up isolation transformer connected to the main drive circuitry of the lamp. (Courtesy of EG&G, Salem, MA)

All noble gas discharge lamps share a common problem. The voltage required to create a plasma in the lamp (typically several kV) is much greater than the voltage required to run the lamp (typically several hundred volts). Therefore, all noble gas discharge lamp circuits have one part of the circuit whose function is to start the lamp and another part of circuit whose function is to run the lamp.

One method commonly seen in commercial systems is to charge a capacitor to approximately 1 kV using a voltage multiplier circuit. Then, the capacitor is discharged through an SCR to trigger a step-up isolation transformer connected to the main drive circuitry of the lamp (see Figure 10.12).

Another method is to charge the lamp to the desired voltage, and then pulse 10 kV onto a wire wrapped around the lamp. This will ionize the plasma and fire the lamp (see Figure 10.13). (As an aside, this is exactly the same way that the flashbulb in a disposable camera works. For an excellent miniature version of a noble gas discharge lamp system, disassemble a cheap disposable camera.)

Power supply circuits for continuous wave lasers. Power supplies for continuous wave lasers are usually straightforward in design (just large). The major requirement is power (voltage regulation is usually not an issue). The only unusual design constraint is the necessity to start the lamp with a high voltage pulse.

A typical configuration is illustrated in Figure 10.14. This configuration is similar to a three-phase bridge rectifier circuit, except SCRs (silicon controlled rectifiers) take the place of diodes. Each phase of the three-phase power supply is transferred through a pair of SCRs. One SCR transfers current on the positive phase of the cycle, the other on the negative phase. By suitable firing of the SCRs, the output voltage can be controlled electronically. An LC filter helps eliminate ripples.

The plasma starting voltage is obtained by discharging a capacitor (not shown) through the primary of the trigger transformer. (It turns out that a slightly higher voltage than the op-

---

Products and Manufacturers, 1996 Book 2 of the *Photonics Directory*, 42d ed. (Pittsfield, MA: Laurin Publishing Co., Inc., 1996).

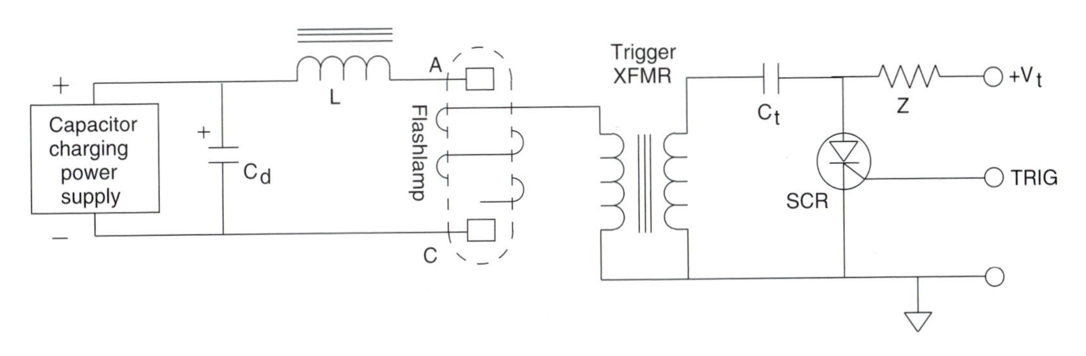

Figure 10.13 Lamps can be triggered by charging the lamp to the desired voltage, and then pulsing a high voltage onto a wire wrapped around the lamp. This will ionize the plasma and fire the lamp. (Courtesy of EG&G, Salem, MA)

erating voltage is required for clean and reliable starting.) Therefore, a small low current high voltage supply is added to charge the filter capacitor immediately prior to firing the lamp.

The major difference between a conventional high current DC switching power supply and a switching power supply designed to drive noble gas discharge lamps is the necessity for an approximately 30 kV high voltage trigger circuit to start the lamps. (This particular circuit uses the isolation transformer trigger method of Figure 10.12.)

Power supply circuits for pulsed lasers. Power supplies for pulsed lasers are much more difficult to construct than power supplies for continuous wave lasers because the lamp plasma needs to be triggered at each pulse. In the traditional method (see

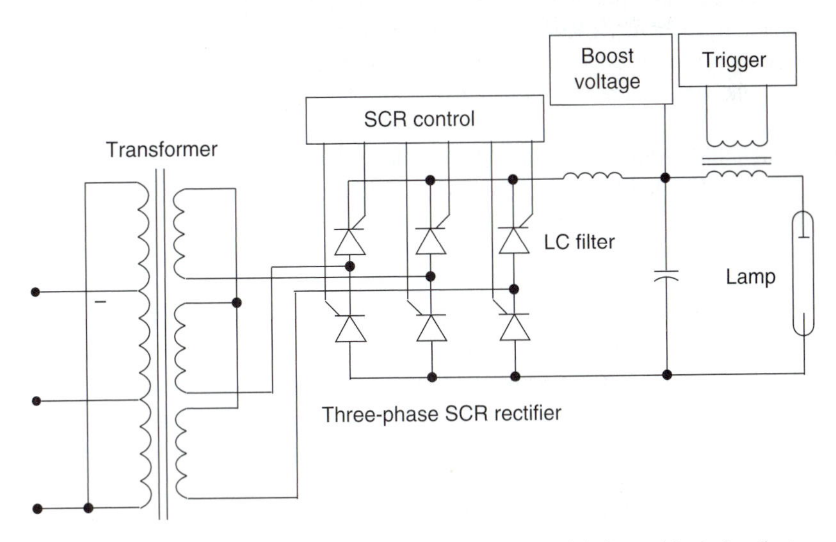

Figure 10.14 Power supplies for cw lasers are usually straightforward in design (just large). The major requirement is power, and voltage regulation is usually not an issue.

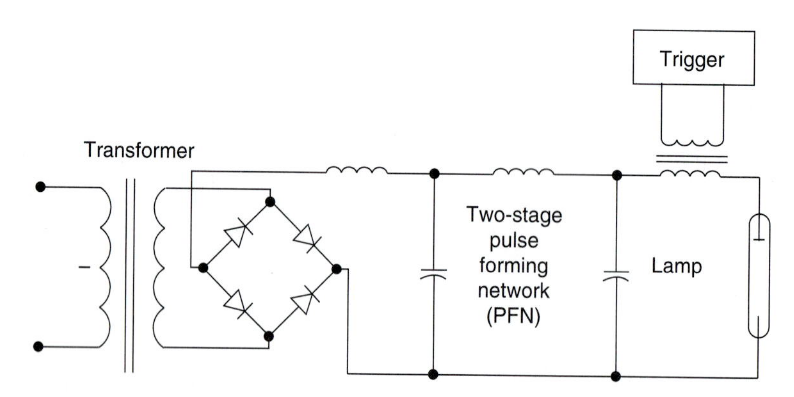

Figure 10.15 In the traditional method for pulsing a lamp, an LC bank of the desired energy capacity is charged up to the desired voltage. The lamp is off and thus there is a large DC voltage across the lamp. Then, a high voltage trigger is applied to the lamp. The lamp itself acts like a switch, and the LC bank is discharged through the lamp.

Figure 10.15), an LC bank of the desired energy capacity is charged up to the desired voltage. The lamp is off and thus there is a large DC voltage across the lamp. Then, a high-voltage trigger is applied to the lamp. (This can either be applied with an isolation transformer or with a wire wrapped around the lamp.) The lamp itself acts like a switch and the LC bank is discharged through the lamp. The length and character of the resulting pulse is formed by a combination of the dynamic impedance of the lamp and the LC network.

In the simmer method, the lamp is started once, and a low current discharge (the simmer) is continually maintained by a small power supply (see Figure 10.16). Simmer circuits are very popular, as they offer increased flashlamp life, reduced coolant degradation, reduced EMI/RFI, and improved efficiency. Some sort of electronic switch (an SCR, ignitron, or krytron) is placed between the simmering lamp and the main LC bank. The LC bank is charged up to the desired voltage by a second HV power supply. At the appropriate time, the electronic switch is triggered and the LC bank is discharged through the simmering lamp. (A diode is incorporated to avoid destroying the simmer supply!) The length and character of the resulting pulse is formed primarily by the LC network.

10.5.3 Pump Cavities for Noble Gas Discharge Lamp-Pumped Lasers

The pump cavity is the structure that efficiently transfers the pump light to the laser material. In lamp-pumped Nd:YAG lasers, the pump cavity is typically a large geometry optical structure. This is necessary in order to simultaneously transfer pump light to the laser material and to permit room for effective cooling of both the laser and the lamps. Additional discussion of pump cavities can be found in Koechner.[58]

---

[58]Walter Koechner, *Solid State Laser Engineering*, 4th ed. (Berlin: Springer-Verlag, 1996).

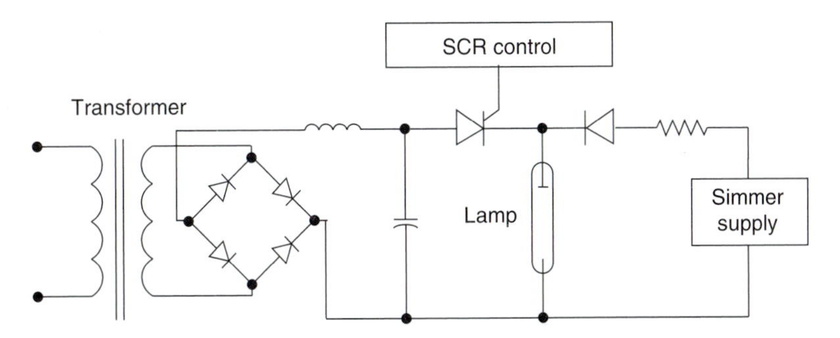

Figure 10.16 In the simmer method for pulsing a lamp, the lamp is started once, and a low current discharge (the simmer) is continually maintained by a small power supply.

Pump cavity materials. The cavity must be coated with a highly reflecting material in order to optimize the power transfer from the lamps to the laser material. There are two major approaches to this design problem. One approach is to coat the cavity with a thin evaporated or electroplated metal. This forms an coating where an image of the lamp is focused on the rod. The other approach is to coat the cavity with a diffuse reflective surface such as MgO or $BaSO_4$. Such surfaces provide a nonimaging transfer of the pump light to the rod and can reduce gain nonuniformities in the rod.

Metal evaporated films. Gold-, silver-, and copper-evaporated films have the highest reflectivity at the pump bands of most solid-state laser material (see Figure 8.4). However, copper and silver oxidize easily. Therefore, virtually all cavities for cw Nd:YAG lasers use gold as the cavity reflector material. (This design issue becomes more interesting for pulsed Nd:YAG lasers. In these lasers, a significant amount of the pumping comes from the higher energy pump bands at 5300 angstroms and 5800 angstroms. Silver is much more reflective than gold at the shorter wavelengths, and an alternative solution is to use silver coated with some protective layer to avoid the oxidation problem.)

Diffuse reflectors. Diffuse reflectors are usually formed of a ceramic compressed powder (such as MgO or $BaSO_4$). This powder is typically squashed between two quartz plates fabricated in the shape of the reflector structure. Since it is quite difficult to obtain quartz fabricated in peculiar shapes, these reflectors are usually limited to closely coupled cavities with a circular reflector shape. These reflectors are commonly used in pulsed Nd:YAG lasers to maximum the reflectivity over the pump bands while minimizing the both the focusing effects and the cost.

Pump cavity geometries. Noble gas discharge lamps are typically cylindrical. Solid-state gain materials are also frequently cylindrical. This permits an efficient optical pump cavity that exploits the properties of a highly reflecting ellipse. In this method, a cylindrical flashlamp is located on one foci of an ellipse and a laser rod of the same diameter as the plasma in the flashlamp is located on the other foci. The elliptical cylinder is capped with two reflective endpieces that create the equivalent of an infinitely long elliptical cylinder. Traditional elliptical cavities are fabricated with a variety of eccentricities (see Figure 10.17). Large eccentricity cavities tend to maximize the direct transfer of lamp radiation to the rod. Small eccentricity cavities tend to be easier to fabricate and to cool.

Multiple ellipses are commonly used in laser cavities (see Figure 10.18). Multiple ellipses permit more lamps, and thus higher output laser powers. Use of multiple lamps also allows temporal inhomogeneities in the lamp discharge (due to random plasma effects) to be averaged over the lamps. Thus, the shot-to-shot jitter in multiple lamp cavities is lower than in single lamp cavities.

Closely coupled geometries are another common variation (see Figure 10.19). In a closely coupled geometry, the lamps and rod are placed as close together as mechanical constraints permit. (The most extreme example of a closely-coupled geometry is the configuration where the lamp and rod are simply wrapped in aluminum or silver foil!) Closely-coupled geometries are generally more efficient than conventional geometries. However, they are more difficult to cool and suffer from both thermal and gain inhomogeneities. Closely-coupled geometries are rarely seen in commercial lasers.

Rods are certainly not the only way that laser materials are fabricated! Other common geometries are the disk amplifier, active mirror, and the zig-zag slab (see Figure 10.3). All of the nonrod geometries suffer from the fact that the most economical way to make discharge lamps is by using cylindrical lamp geometries. Although other lamp geometries have been tried, they have proven to either be unsuccessful or uneconomical.

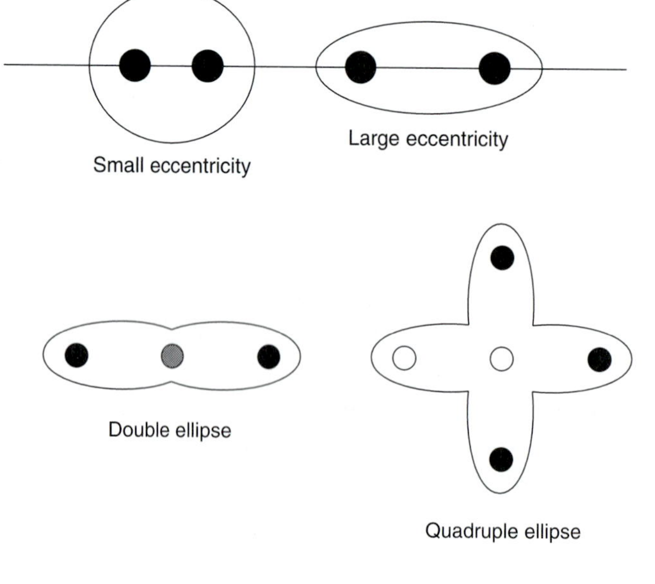

Figure 10.17 Noble gas discharge lamps are typically cylindrical. Solid-state gain materials are also frequently cylindrical. This permits an efficient optical pump cavity where a cylindrical flashlamp is located on one foci of an ellipse and a laser rod of the same diameter as the plasma in the flashlamp is located on the other foci.

Figure 10.18 Multiple ellipses permit more lamps, and thus higher output laser powers. Use of multiple lamps also allows temporal inhomogeneities in the lamp discharge (due to random plasma effects) to be averaged over the lamps.

Figure 10.19 In a closely coupled geometry, the lamps and rod are placed as close together as mechanical constraints permit.

Pump light distribution in rods. Inhomogeneities in pump light eventually translate into distortions in output laser beams. There are three effects important in determining pump light distribution: the geometrical properties of the pump cavity, refractive focusing occurring in the rod and non-uniform absorption of pump light.

Polished rods act like cylindrical lenses in that the pump light is focused into the center of the rod. The competing effect is absorption in the rod. If the rod is a relatively poor absorber, the light intensity in the center of the rod will be higher. If the rod is a good absorber, then the light intensity at the periphery will be higher. A uniform profile of pump light can be created by balancing the focusing against the absorption.

There are a number of interesting options to alter the balance between the focusing effect from the rod polish and the exponential (Beer's law) absorption of pump light. The focusing can be reduced by roughening the surface of the rod (frosting) or the rod can be immersed in a fluid whose index is close to that of the rod (an index matching fluid). The focusing can be increased in high absorption materials by cladding the rod with a high index transparent material (such as sapphire for ruby or glass for Nd:glass).

10.5.4 Spectra-Physics Quanta-Ray GCR Family

The Spectra-Physics Quanta-Ray GCR family consists of noble gas discharge lamp pumped Nd:YAG lasers. They operate pulsed and are available in various combinations of energy and repetition rate ranging from the 10 Hz, 2.2 J/pulse GCR-350 to the 100 Hz, 325 mJ/pulse GCR-190. The GCR-209 is illustrated in Figure 10.20.

The heart of the GCR lasers is a rectangular module containing an Nd:YAG rod and either one or two lamps (see Figure 10.21). The basic module can be used as either an oscillator or amplifier module.

A basic module consists of an Nd:YAG rod mounted in either a single or double elliptical cavity (see Section 10.5.3) and optically pumped by either one (single ellipse) or two (double ellipse) xenon noble gas discharge lamps (see Section 10.5.1). The reflectors possess a gold reflecting layer. All the aluminum parts in the cavity are protected to minimize electrochemical corrosion effects in the head.

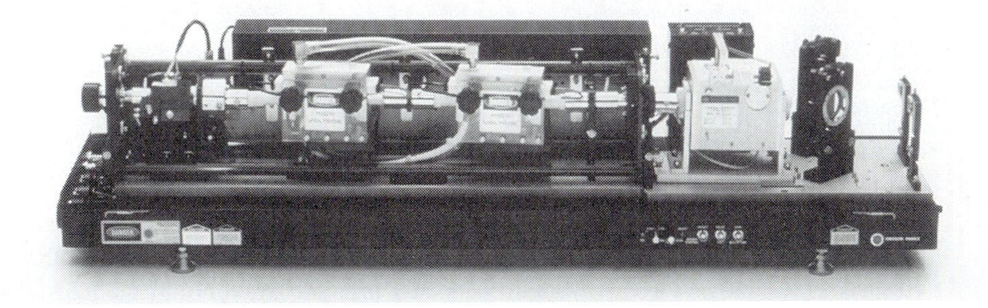

Figure 10.20 The Spectra-Physics Quanta-Ray GCR family consists of noble gas discharge lamp pumped Nd:YAG lasers. They operate pulsed and are available in various combinations of energy and repetition rate ranging from the 10 Hz, 2.2 J/pulse GCR-350 to the 100 Hz, 325 mJ/pulse GCR-190. (Courtesy of Spectra-Physics, Mountain View, CA)

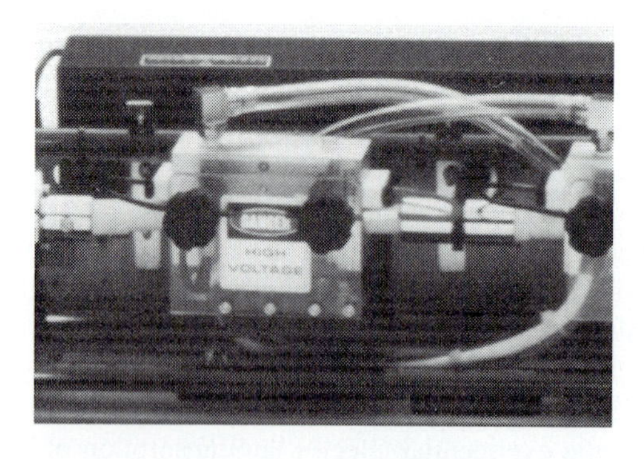

Figure 10.21 The heart of the GCR lasers is a basic rectangular module containing an Nd:YAG rod. One or two modules are used as a starting oscillator and additional modules are used as amplifiers. (Courtesy of Spectra-Physics, Mountain View, CA)

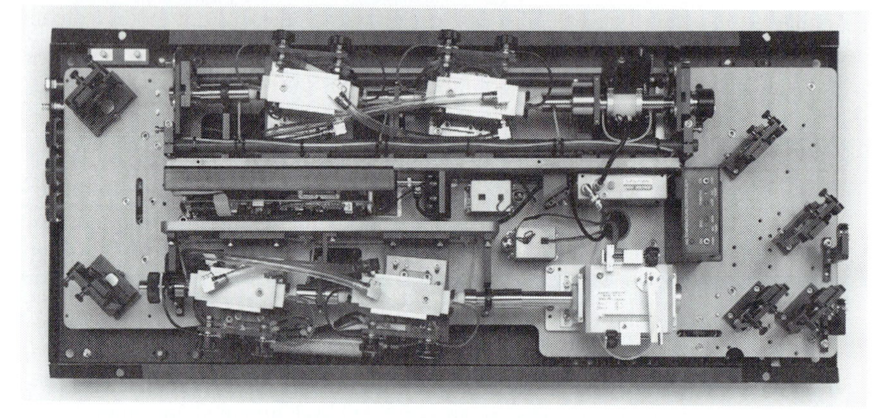

Figure 10.22 A top view of the two-rod oscillator and two-rod amplifier GCR laser with an included harmonic generator. (Courtesy of Spectra-Physics, Mountain View, CA)

A basic design philosophy in the GCR lasers is that there are either two flashlamps or four flashlamps in a system. Thus, a system could consist of an oscillator only (two lamps total), one oscillator and one amplifier (with one or two lamps each, for two or four lamps total), a two-module oscillator (with one or two lamps each, for two or four lamps total), a two-module oscillator and one amplifier (one lamp in each oscillator and two in the amplifier), and a two-module oscillator and two amplifiers (one lamp in each of the four modules). A top view of the two-module oscillator and two amplifier configuration is given in Figure 10.22.

In the three-module configuration, single lamp heads are used as oscillator heads and double lamp heads are used as amplifier heads. Each of the oscillator modules contains one laser rod and produces approximately 500 mJ of energy. Both modules are installed inside an L-shaped resonator fabricated from graphite epoxy composite to minimize thermal expansion. Between the two modules is a *c*-axis rotator that provides thermal birefringence compensation. (Since the modules are functionally identical, switching the *e*- and *o*-axes permits one module to be used to compensate for the other. See Section 8.3 for more information on birefringence in crystals.)

The module is water-cooled with deionized water flowing in a closed cycle recirculating loop. The water temperature is controlled in order to maintain constant thermal lensing in the rod. The water is initially confined to a flow tube and circulates over the Nd:YAG rod. Then the water is returned to the housing and cools the lamps and housing in a flooded cavity configuration. The heat is removed from the cooling water by either a water-water heat exchanger (which requires external process water) or a water-air heat exchanger.

The GCR lasers are evolutionary descendents of the Quanta-Ray DCR lasers. The original DCR (Diffraction Coupled Resonator) lasers used a geometrically unstable cavity similar to that discussed in Section 4.5. In particular, the output mirror incorporated a small reflecting dot and the output mode was a very characteristic donut shape. This cavity configuration was desirable since greater energy could be extracted from the Nd:YAG rod. In addition, the unstable geometry was much less sensitive to misalignment. However, the hard-edged dot of the DCR output mirror created diffraction ripples on the output beam.

In the GCR (Gaussian Coupled Resonator) lasers, the original DCR dot reflector has been replaced by a more complex reflector. The function of this proprietary reflector is to soften the edge of the dot. The resulting output mode can be quite close to the Gaussian mode of a stable resonator. The peak reflectivity in the center of the reflector is approximately 30% and the average reflectivity over the entire output coupler is approximately 5%. Interestingly enough, the pulsed GCR Nd:YAG is such a high gain laser that it can lase with an output coupler on BOTH ends of the laser.

GCR lasers can be operated Q-switched (see Section 6.2), mode-locked (see Section 6.4) and injection seeded. A multilayer film polarizer sets the cavity polarization for the Q-switch. The polarizer also serves as an input for an optional injection seeding laser. (For more information on multilayer films, see Section 8.2.) In addition, an integral harmonic conversion unit provides frequency conversion into the second, third, and fourth frequency harmonics of Nd:YAG (see Chapter 7).

10.5.5 Semiconductor Lasers as Solid-State Laser Pump Sources

Semiconductor diode laser pumping of solid-state lasers was demonstrated soon after the invention of the semiconductor laser.[59,60] However, for a number of years, the high cost of semiconductor lasers (in combination with the large number of laser diodes required!) essentially prohibited the possibility of viable commercial diode laser pumping. That changed in the mid-80s, as the costs of semiconductor lasers dropped dramatically due to widespread commercial semiconductor laser applications.

The semiconductor laser diode offers a number of advantages over flashlamp pumping. Specifically:

Match to the laser pumpbands: The laser diode provides an excellent match to the semiconductor pump bands. Although the conversion efficiency of electricity to light in a laser diode (20 to 50%) is typically less than that of noble gas discharge lamps (60 to 70%), virtually all of the emitted light is useful for pumping the laser.

---

[59]R. J. Keyes and T. M. Quist, *Appl. Phys. Lett.* 4:50 (1964).

[60]M. Ross, *Proc. IEEE* 56:196 (1968).

Reduction in the amount of deposited heat: The very good match between the laser diode emission and the pump band absorption means that little energy is deposited in the laser material as heat. This minimizes thermal stress, lensing, and birefringence effects.

Improved pump source lifetime: Typical noble gas discharge lamps last 10^7 shots in pulsed operation, and 200 to 500 hours in cw operation. Typical diode laser pumps last 10^9 shots in pulsed operation, and 10,000 hours in cw operation.

Reduction in UV damage: Noble gas discharge lamps typically deposit significant UV in the cavity. This degrades materials in the pump cavity as well as components of the coolant. Diode lasers have no significant emission in the UV.

Dramatically simpler power supplies: Noble gas discharge lamps used in cw lasers require high-voltage power supplies. Noble gas discharge lamps used in pulsed lasers require elaborate high-voltage switching and pulseforming networks. Diode laser power supplies and switching circuits are typically cheaper and easier to construct.

This section discusses the use of semiconductor diode lasers in their capacity as pumps for solid-state lasers. For more information on semiconductor diode lasers themselves, see Section 12.3.

Design of semiconductor lasers as diode laser pumps. Semiconductor lasers are available to pump a number of common laser materials. Of the most importance are 808 nm devices from $Ga_{0.91}Al_{0.09}As$ materials intended to pump Nd-doped materials (primarily Nd:YAG). For pumping Yb materials, strained layer InGaAs/GaAs MQWs and superlattices are used with an operating wavelength near 980 nm. For pumping Cr:LiSAF and Cr:LiCAF, AlGaInP lasers in the 640-690 nm region are used. (For more information on specific semiconductor lasers, see Section 12.3.)

The major limitation in using laser diodes for pumping solid-state lasers is the (relatively) small output power of the laser diode. At first glance, the strategy would seem to be to make the diode larger (a wider stripe). However, this is not as effective as it should be because the current tends to break up into localized filaments. These localized filaments are high current density and can lead to localized damage of the laser facets and a reduction in the diode lifetime. Broad area diodes (such as the SDL 2300 series) are typically only available in aperture widths up to 500 μm.[61]

The majority of laser diodes used for pumping solid-state lasers are constructed in the form of arrays of many small laser diodes. Each small diode is relatively narrow (perhaps 5 to 7 μm wide) and the diodes are separated by 10 to 15 μm. (See Figure 10.23).

There are small linear arrays which include 10 to 20 laser diodes. These arrays are typically several hundred microns long and can emit output powers in the range of a watt. There are larger linear arrays (composed of 10 to 20 groups of 10 to 20 elements) that are available in widths of up to 1 cm. These arrays can emit output powers up to 20 W.

The SDL-3400 family of cw linear array laser diodes is an example of this type of larger array (see Figure 10.24). These are typically 20 ten-stripe laser subarrays spaced on 500 μm centers where the spacing is determined by thermal limitations. Typical output

[61]The SDL-2300 is offered with power outputs from 0.5 watt to 4 watts and apertures from 50 μm to 500 μm.

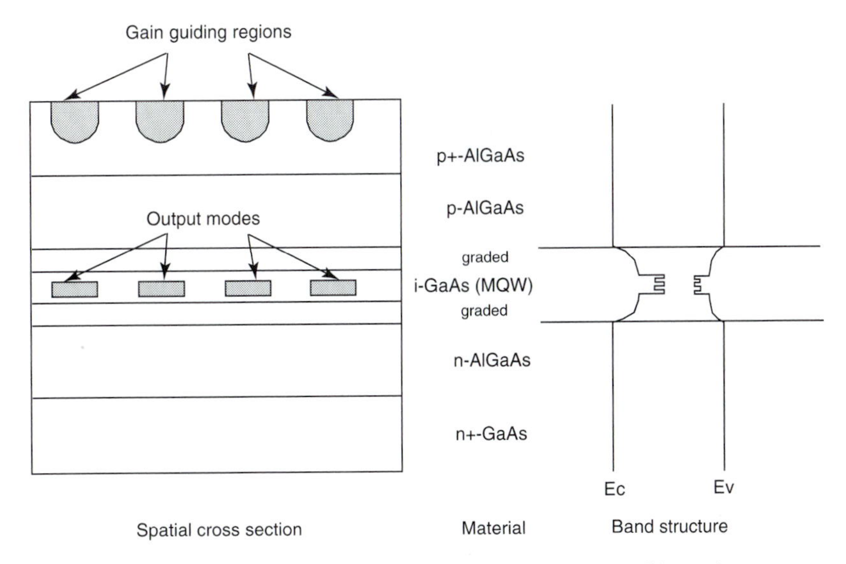

Figure 10.23 The majority of laser diodes used for pumping solid-state lasers are constructed in the form of arrays of many small laser diodes. Each small diode is relatively narrow (perhaps 5 to 7 μm wide) and the diodes are separated by 10 to 15 μm.

Figure 10.24 The SDL-3400 family is an example of a fanily of large linear array laser diodes. (Courtesy of SDL, Inc., San Jose, CA)

power is 20 W, with typical operating current and voltage of 25 A at 1.86 V. The center wavelength ranges from 804 nm to 810 nm and is thermally tuned. The divergence is 15 degrees full angle parallel to the junction and 36 degrees half-angle perpendicular to the junction.[62]

It is possible to stack the larger linear arrays and create a matrix up to 1 cm^2 in dimension. Cooling becomes a major consideration in these larger arrays and pulsed operation is common. The SDL-3200 family of quasi-cw stacked array laser diodes is an example of this type of two-dimensional array (see Figure 10.25). Quasi-cw devices operate in a long pulse width mode for optimum energy transfer into solid-state laser materials with upper state

[62]Information is from the 1996–7 SDL product catalog.

Figure 10.25 The SDL-3200 family is an example of a family of quasi-cw second stacked array laser diodes. (Courtesy of SDL, Inc., San Jose, CA)

lifetimes on the order of 100 to 1000 μs. Low-thermal impedance water-cooled packages allow high average power and high duty factor. Output energy density up to 2.5 kW/cm^2 is available, as is a duty factor of 20%.

A typical quasi-cw stacked array laser diode is the SDL-3251-C6. This diode has an output power of 600 W (qcw), series resistance of 0.018 Ω, and typical operating current and voltage of 115 A at 11.4 V. The center wavelength ranges from 785 nm to 810.5 nm and is thermally tuned (0.3 mm/°C). The divergence is 10 degrees half-angle parallel to the junction and 30 degrees half-angle perpendicular to the junction. Efficiency is 36–46%, the pulse width is 400 μs and the duty cycle is 1.3%. A more dramatic example of a quasi-cw stacked array laser diode is the SDL-3251-HJ. This diode has an output power of 4800 W (qcw), series resistance of 0.144 Ω, and typical operating current and voltage of 115 A at 91.2 V. The center wavelength ranges from 785 nm to 810 nm and is thermally tuned (0.3 nm/°C). The divergence is 10 degrees parallel to the junction and 30 degrees perpendicular to the junction. Efficiency is 36 to 46%, the pulsewidth is 400 μs and the duty cycle is 2%.[63]

Thermal management is a key aspect of the design of laser diode arrays. The primary effect of a change in the diode temperature is a change in the wavelength of the laser light. Additionally, higher diode temperatures also mean higher thermal gradients. Higher thermal gradients generate a wider range of operating wavelengths and thus a wider linewidth for the diode spectrum.

In diode pumped lasers with a short absorption path (such as transversely pumped lasers), it is critical to have a narrow spectral range from the diode to maximize the absorption in the pump bands. However, in optically thicker materials (such as end-pumped lasers), the difference in absorption coefficients at various wavelengths becomes much less important.

Thermoelectric coolers (TE coolers) are usually used for thermal management in smaller units where precise temperature stability is critical. These coolers not only provide a way to remove excess heat efficiently, but are also solid-state devices that can be incorporated in feedback control loops. Thus, it is possible to dynamically control the temperature (and thus the operating wavelength) of TE-cooled lasers. However, for larger arrays, the cost of TE coolers becomes prohibitive. Luckily, larger arrays usually require less accurate thermal management, and liquid cooling methods can be used.

[63]Information is from the 1996–7 SDL product catalog.

Electrical systems for pumping laser diode arrays. Laser diode power supplies are simple in comparison to noble gas discharge lamp power supplies. However, there are some real traps in developing power supply technologies for laser diodes. These include:

> *Device sensitivity:* Laser diodes are delicate (especially compared to noble gas discharge lamps). They do not tolerate reverse biases (even a few volts reverse bias will destroy them) or voltages and currents in excess of their rated limits. Even a few mA more than their nominal operating current can result in dramatically reduced lifetimes. Additionally, minor power supply glitches (such as discharging capacitors through the diode upon shut down of the supply) can result in their destruction.
>
> *Series resistance:* Laser diodes have a very low series resistance. Thus, the internal resistance of the power supply must be minimized for efficient operation.
>
> *Static buildup:* Static buildup across the leads can be sufficient to destroy devices on the shelf. Leads must be shorted or diodes stored in conductive wrappers to prevent this.

However (in spite of these design difficulties) diode laser power supplies for cw lasers clearly possess advantages of size and power over power supplies used for driving noble gas discharge lamps.

The major positive advantages of laser diode pump sources become apparent when designing pulsed laser diode power supplies (see Figure 10.26). In a pulsed diode pumped solid-state laser, the diode array drivers replace the pulse forming network of a flashlamp pumped laser system. The diode-array driver is a low-voltage switching network that supplies a constant current pulse to the diode arrays. The low voltages permit the use of power MOSFETs as switching elements. The electrical energy stored in electrolytic capacitors is transferred to the diode arrays on each laser pulse.

10.5.6 Pump Cavities for Diode Laser Pumped Solid-State Lasers

Diode lasers possess significantly simpler pump cavity geometries than lamp pumped lasers. This is because the light output from diodes is very directional. Thus, the emphasis in constructing optical cavities for diode pumped lasers is not on collecting the light, but rather on matching the mode of the diode laser light to the geometry of the gain material.

Two major geometries appear in diode-pumped solid-state lasers. In the *end-pumped* geometry, the light is directed down the length of the rod (see Figure 10.27). This geometry maximizes the efficiency, because essentially all the light is absorbed. However, the end of the laser is a finite size, and so output powers are typically limited by the ability to focus light from a single diode (or diode array or fiber coupled diode array) down the end of the rod.

In the *side-pumped* geometries, the light is directed into the rod from the side (see Figure 10.28). These geometries are similar to the lamp-pumped geometries in that significant light may transfer through the laser material. There is also an increased risk of inhomogeneities. However, a great deal more light can be delivered to the material.

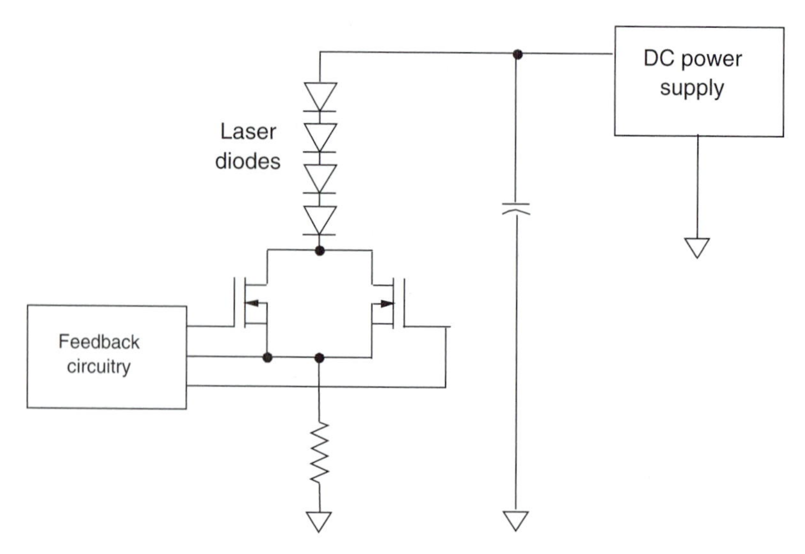

Figure 10.26 Pulsed laser diode power supplies are quite simple. (Compare with Figures 10.15 and 10.16.)

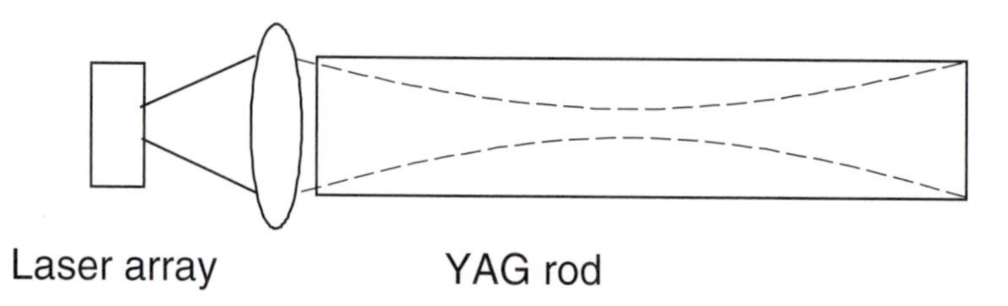

Figure 10.27 In the end-pumped geometry, the light is directed down the length of the rod.

End-pumping laser rods with single diode lasers or diode laser arrays. The key concept in diode-pumping solid-state lasers is to make the output mode of the diode laser precisely match the operating mode (usually $TEM_{0,0}$) of the laser. The simplest possible configuration is to end-pump a solid-state laser rod with a single laser diode. However, laser diodes tend to lase with an output mode possessing a highly elliptical beam due to the difference between the horizontal and vertical dimensions of the lasing region of the diode (see Figure 10.29).

There are two major problems associated with this output mode. The first problem is the large astigmatism. The second is the huge divergence angle (40 degrees is not uncommon for the vertical direction). Typically, a large aperture lens is placed close to the laser in order to capture as much energy as possible. Then, some type of astigmatism compensation is used. Finally, the beam is refocused into the optimal configuration for mode matching into the rod.

Traditional techniques for compensating for the astigmatism include coupling the beam into a fiber or using a GRIN lens. However, both of these techniques intrinsically strip off

Figure 10.28 In the side-pumped (transverse) geometries, the light is directed into the rod from the side.

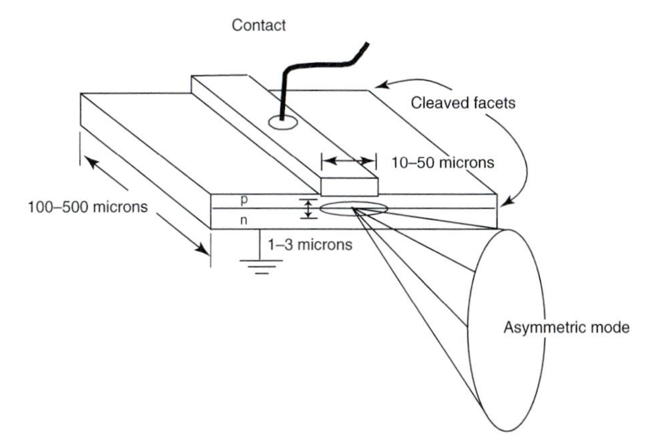

Figure 10.29 Laser diodes tend to lase with an output mode possessing a highly elliptical beam due to the difference between the horizontal and vertical dimensions of the lasing region of the diode. (From LASER ELECTRONICS 2E. by VERDEYEN, J. T., ©1989, Figure 11.1, p. 362. Adapted by permission of Prentice-Hall, Inc., Upper Saddle River, NJ)

part of the beam as the beam mode matches into the optical element. Thus, fiber or GRIN lens coupling techniques are inefficient. More efficient techniques include using aspherical lenses (expensive) or anamorphic prisms[64] (cheaper but more complex to assemble). The anamorphic prism technique is probably the most commonly used in commercial diode pumped lasers because it is highly efficient and relatively low cost (see Figure 10.30).

End-pumping of laser rods with diode laser arrays (rather than single diodes) is also done. However, some care must be taken in these designs, as it is much more difficult to mode-match laser arrays into the relatively small end face of the rod. Therefore, the additional power advantage achieved with the laser diode array may be offset by the mode matching disadvantages of the geometry.

End-pumping of laser rods with fiber-coupled laser arrays is an effective solution to the design problem of mode matching laser arrays into the small end face of the rod. However,

[64] R. Trebino, *Appl. Opt.* 24:1130 (1985).

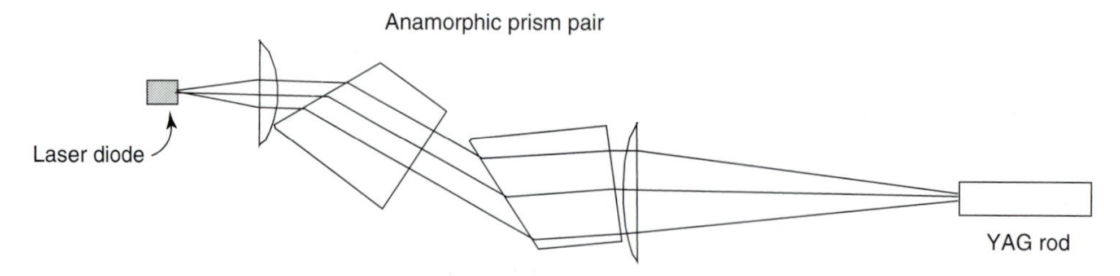

Figure 10.30 Anamorphic prisms are often used for compensating for the astigmatism of a laser diode.

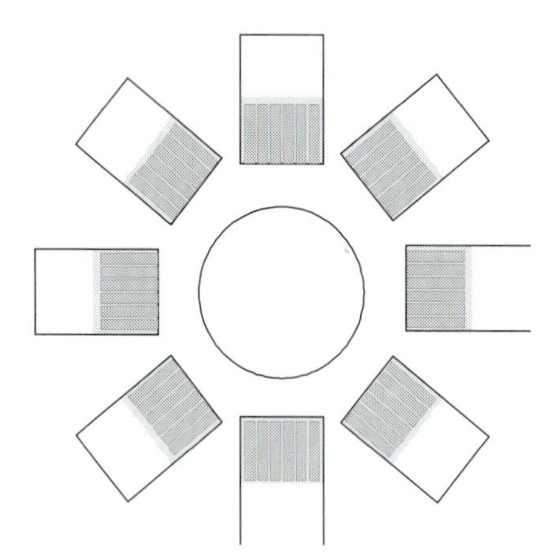

Figure 10.31 Laser diode arrays are frequently arranged around the laser rod. The diode arrays are typically cooled from the back with an impingement cooling method and the rod is cooled by a water jacket.

creating the fiber bundle requires coupling a fiber into the end face of each individual laser diode in the array. This adds significant cost and complexity to the pump system

Side-pumping laser rods with diode laser arrays. Although it is possible to side-pump with single lasers, the geometry does not provide any real advantages. Therefore, laser arrays are almost exclusively used for side-pumping.

In one common technique, laser diode arrays are arranged around the laser rod (see Figure 10.31). The diode arrays are typically cooled from the back with an impingement cooling method. The rod is typically cooled by a water jacket.

In the focusing or reflecting optics techniques, some type of modifying element is used to better focus the light into the rod (see Figure 10.32). This can be a microlens configuration or some type of reflective optical system.

Side-pumping laser slabs with diode laser arrays. Laser arrays are ideally suited for side-pumping laser slabs as they offer an extended source without the disadvantages of noble gas discharge lamps. A key aspect of the design of laser diode-pumped slab lasers is directing the lasing mode through the region of highest gain without distorting the mode due to thermal, stress, and gain variations. As a consequence, most slab geometry lasers use some sort of multiple bounce geometry to compensate for distortions seen unequally by the laser beam (see Figure 10.33).

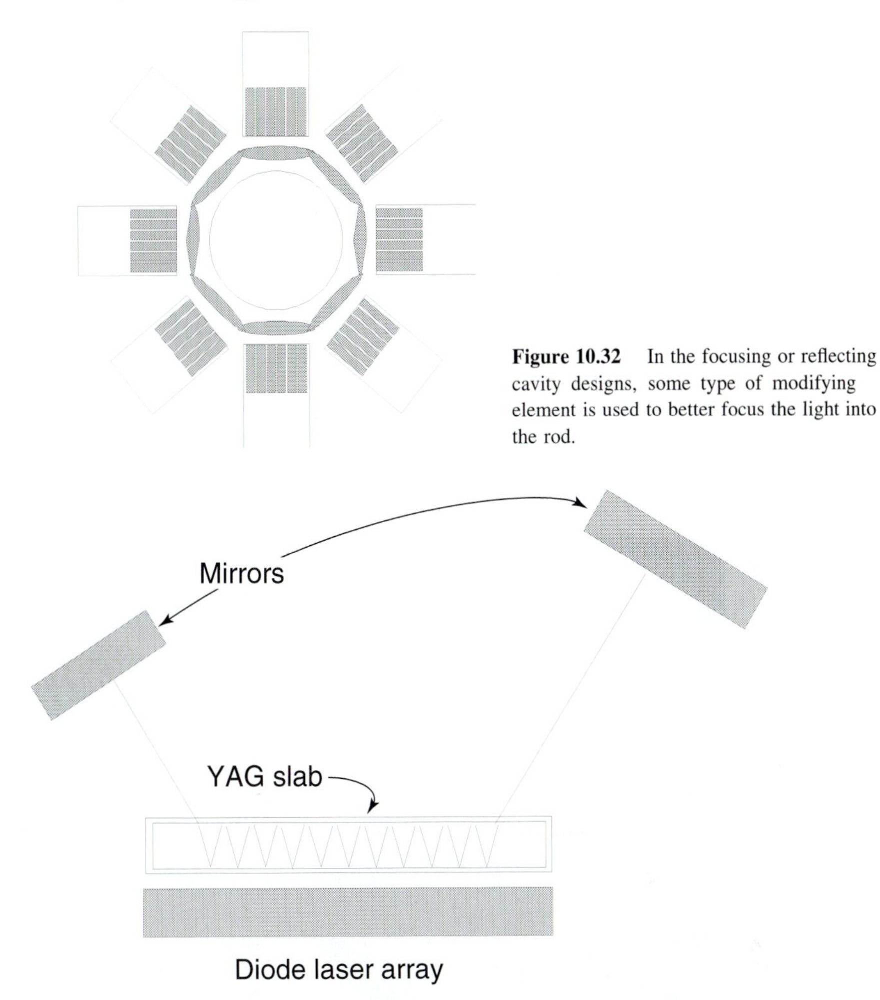

Figure 10.32 In the focusing or reflecting cavity designs, some type of modifying element is used to better focus the light into the rod.

Figure 10.33 A key aspect of the design of laser diode-pumped slab lasers is directing the lasing mode through the region of highest gain without distorting the mode due to thermal, stress, and gain variations. As a consequence, most slab geometry lasers use some sort of multiple bounce geometry to compensate for distortions seen unequally by the laser beam.

10.5.7 Coherent DPSS 1064 Laser Family

The Coherent DPSS 1064 is a diode-pumped cw Nd:YAG laser (see Figure 10.34). It is available in output powers ranging from 100 to 300 mW.

The heart of the DPSS 1064 is a patented ring laser cavity (see Figure 10.35). An infrared beam (emitted by a thermoelectrically cooled AlGaAs semiconductor laser diode) is focused into a block of Nd:YAG laser material. The resonator uses only two mirrors (plus the refraction through the Nd:YAG crystal) to create the ring geometry. The ring is unidirectional and the preferred direction is maintained by using the Faraday effect in

Nd:YAG in combination with a Brewster-cut rotator plate (see Section 11.4.1 for more on ring lasers).

Coherent DPSS lasers can also incorporate an intracavity doubling element and operate at the second harmonic frequency of Nd:YAG (see Chapter 7 for more information on intracavity doubling). This creates a compact cw single longitudinal mode solid-state green laser source that competes very well with small argon-ion lasers.

EXERCISES

Overview of conventional solid-state lasers

10.1 (design) The concept of laser weapons ("death rays") has been a driving force behind much laser research. Do some library research (for example, check recent issues of trade journals such as *Laser Focus World* and *Photonics Spectra*, as well as the popular press such as *Time*

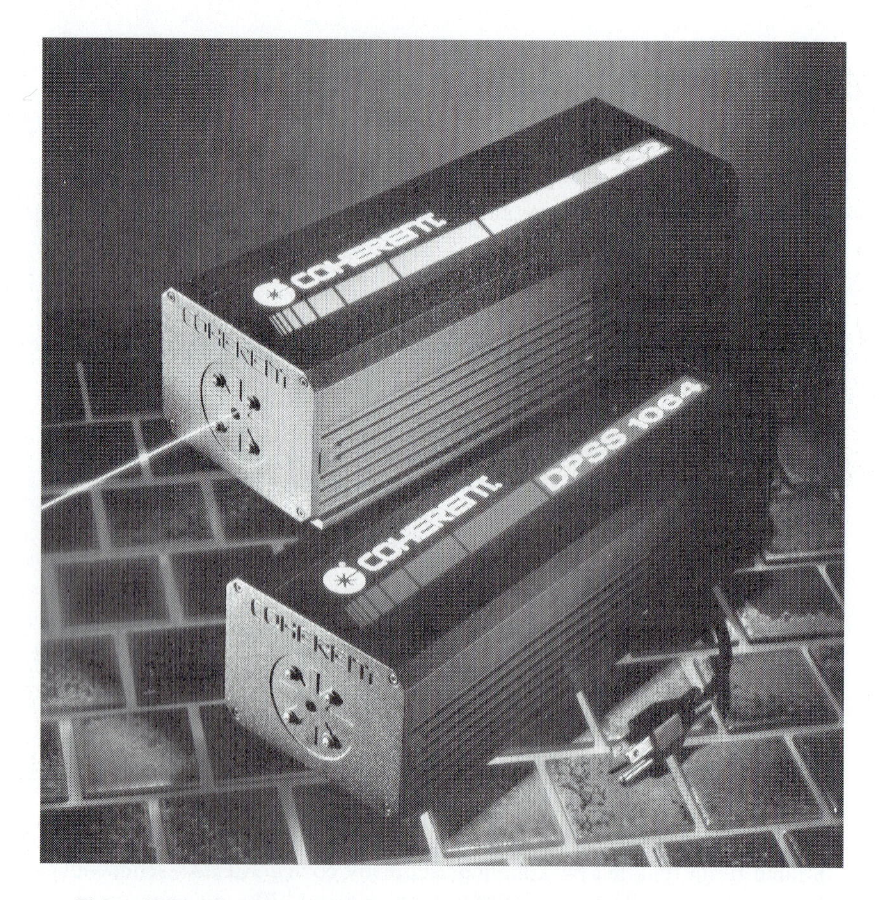

Figure 10.34 The Coherent DPSS 1064 is a diode-pumped cw Nd:YAG laser. It is available in output powers ranging from 100 to 300 mW. (Courtesy of the Coherent, Inc., Laser Group, Santa Clara, CA)

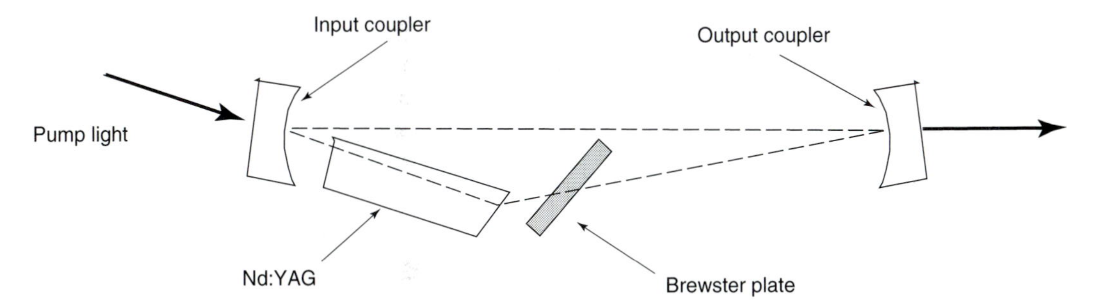

Figure 10.35 The heart of the DPSS 1064 is a patented ring laser cavity. (Courtesy of the Coherent, Inc., Laser Group, Santa Clara, CA)

Magazine and *US News and World Report*) and determine the current state of the art in laser weapons. Consider the advantages and disadvantages of laser weapons. Do you think that laser weaponry is a viable technology? Your answer should include:

(a) a one- to two-paragraph description of the state of the art in laser weapons (including references!),

(b) a concise table (supported by references) detailing the advantages and disadvantages of laser weapons, and

(c) a one- to three-paragraph summary of your opinion (supported by facts, references, and calculations) of the viability of laser weapon technology.

10.2 (design) Consider the question of pumping a laser rod with another laser. There are two possible geometries. The pump laser can be directed down the length of the rod (end-pumped) or it can be directly transversely across the rod (transverse-pumped). Assume a solid-state laser rod 50 mm long and 6 mm in diameter. Assume that the absorption at the pump line is $\alpha = 0.2\ \text{cm}^{-1}$. Assume that the pump laser is a well-behaved $\text{TEM}_{0,0}$ laser with a starting beam waist of 0.5 mm. Design two optical systems for pumping the laser rod. One should be transverse and one should be end-pumped. Compute the light absorbed in the rod for each design. Your answer should include:

(a) A sketch of each of the two designs,

(b) a brief (one to three sentences) description of the logic behind the designs, and

(c) a calculation of the pump light absorbed in the rod for each geometry.

10.3 (design) Diode-pumped fiber lasers are seen by many researchers as a viable low-cost alternative to diode-pumped or conventionally-pumped solid-state lasers. Do some library research (for example, check recent issues of trade journals such as *Laser Focus World* and *Photonics Spectra*, and recent issues of academic journals such as the *IEEE Journal of Quantum Electronics*, *Applied Physics Letters*, and *Optics Letters*) and determine the current state of the art in diode-pumped fiber lasers. Do you think that diode-pumped fiber lasers have a future? Your answer should include:

(a) a one- to two-paragraph description of the state of the art in diode-pumped fiber lasers (including references!), and

(b) a one- to three-paragraph summary of your opinion (supported by facts, references and calculations) of the commercial viability of diode-pumped fiber lasers.

A possible starting point is Michael J. F. Digonnet, *Rare-Earth Doped Fiber Lasers and Amplifiers* (New York: Marcel Dekker, 1993); and *Selected Papers on Rare-Earth Doped Fiber Laser Sources and Amplifiers*, M. J. F. Digonnet (SPIE Milestone Series, Vol. MS37, 1992).

Conventional solid-state laser host materials

10.4 (design) Contact a few vendors and obtain data sheets for Nd:YAG and one or two laser glasses. (You may want to coordinate with your classmates to avoid having 30 people call the same vendor!) Create a table comparing the various optical and mechanical properties of Nd:YAG against the glasses you selected. Which material would you pick for a high energy per pulse laser intended for laser surgery? Why? Your answer should include:

(a) your comparison table, and

(b) a concise (a few sentences) description of which material you would pick for the intended application and why.

(Possible vendors include Litton/Airtron, Union Carbide, and Casix for crystals; Schott, Hoya, and Corning for glasses. Check the *Photonics Buyers Guide to Products and Manufacturers* (Pittsfield, MA: Laurin Publishing Co., Inc.) for additional vendors.)

10.5 The zig-zag slab illustrated in Figure 10.36 represents an interesting engineering problem. The objective is to create a slab with a known number of bounces where the beam is incident on the front face at Brewster's angle, and where the laser beam is not clipped by the edges of the slab.

(a) Develop a method to calculate the length of an ideal laser slab given the number of bounces, the thickness of the slab, the index of refraction of the laser material, and the index of refraction of the material outside the slab.

(b) Calculate the center-to-center length of an ideal slab, ten bounces long, with a thickness of 15 mm, a slab index of refraction of $n = 1.7$, and an external index of refraction of 1.0.

(c) Is it possible to construct a zig-zag slab where the beam enters at Brewster's angle, but the slab does not meet the total internal reflection condition?

10.6 Consider an Nd:YAG laser with a lasing wavelength of 1.064 microns and a linewidth of 4.5 angstroms. Also consider a silicate Nd:glass laser with a lasing wavelength of 1.061 microns and a linewidth of 2.88 nm.

(a) Using a computer if possible, construct a plot which contains both Lorentzian and Gaussian lineshapes for both the Nd:YAG and the Nd:glass lasers (four graphs on one plot). The plot should plot normalized intensity on the y-axis versus wavelength on the x-axis. Normalize to the line with the largest peak gain.

(b) Do you think a silicate Nd:glass laser can be used as an amplifier for an Nd:YAG laser? Use the plot to justify your answer. At one time, LLNL actually built a laser system like this (Shiva). How do you think they got it to work?

The laser transition in Nd:YAG

10.7 Consider the band diagram of Nd:YAG given in Figure 10.4. Notice that it is possible to lase on any combination of lines starting from the $^4F_{3/2}$ (R2) band and ending on the $^4I_{9/2}$ band, $^4I_{11/2}$ band, or the $^4I_{13/2}$ band.

(a) Compute the wavelengths of the possible transitions in Nd:YAG.

(b) Comment on whether you think it is possible to build a laser that would have a selector switch to permit lasing on any of these lines.

Noble gas discharge lamps as pump sources

10.8 Consider the problem of pumping a laser using a blackbody flashlamp. Assume the laser is Nd:YAG with two pump bands, one at 810 nm and one at 750 nm. For the purposes of this calculation, assume the pump bands are Gaussian in profile and have a linewidth of 25

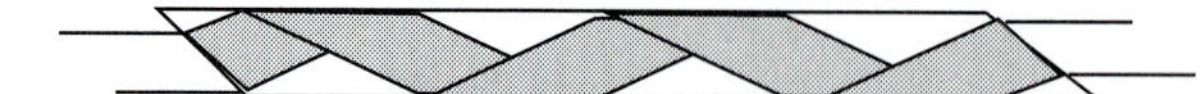

Figure 10.36 A zig-zag slab.

nm each. Assume that pump transition is pumped with a blackbody flashlamp centered at 780 nm.

(a) Graphically or mathematically determine the total percentage of flashlamp energy that overlaps with the pump transition.

(b) Compare your idealized absorption profile with the real absorption profile of Figure 10.5 and comment briefly on the validity of the calculation.

(c) What does this calculation tell you about the efficiency of Nd:YAG?

10.9 (design) Purchase a small disposable camera with a flash (typically $< \$20$). Disassemble the camera and deduce the operation of the flash circuit by inspection. Explain how the disposable camera flash is similar to a noble gas discharge lamp as used for pumping a pulsed laser. Your answer should include:

(a) a circuit diagram of the disposable camera flash, and

(b) a brief (a few paragraphs) description on the similarities and differences between the disposable camera flash and a noble gas discharge lamp as used for pumping a pulsed laser.

10.10 (design) Assume you want to design a single lamp elliptical cavity laser head for an Nd:YAG laser rod 5 mm in diameter and 60 mm long to be operated in a pulsed laser. Develop a simple lamp concept. Contact a few vendors and obtain data sheets and pricing for lamps similar to your lamp concept. (You may want to coordinate with your classmates to avoid having 30 people call the same vendor!) Select your favorite lamp for this application and prepare a brief justification of why you chose it. Your answer should include:

(a) the lamp you selected, its associated data sheets from the vendor, and its price, and

(b) a concise (a few sentences) description of why you chose this lamp.

(Possible vendors include EG&G and ILC. Check the *Photonics Buyers Guide to Products and Manufacturers* [Pittsfield, MA: Laurin Publishing Co., Inc.] for additional vendors.)

10.11 (design) Consider the simmered circuit in Figure 10.37. Assume the lamp impedance to be

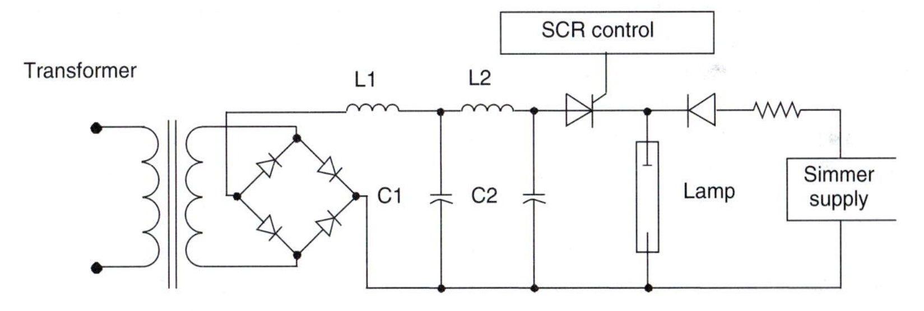

Figure 10.37 A simple simmered circuit for driving noble gas discharge lamps.

negligible. Design a pulse forming network (that is, chose values for the inductors L_1, L_2 and the capacitors C_1, C_2) that will create a roughly square current pulse with an energy of 300 J/pulse, a peak current of 500 A, and a pulse width of 300 μs. Your answer should include:

(a) a sketch of your proposed circuit,

(b) a brief (a few sentences) description of the rationale for your design, and

(c) a SPICE (or equivalent) plot of the output current pulse.

10.12 (design) Consider a laser incorporating a 50 mm long Nd:YAG rod that is 6 mm in diameter. Assume the laser uses a single noble gas discharge lamp that has a plasma that is also 6 mm in diameter and 50 mm long. (Assume the actual lamp outer diameter is 10 mm and the total length, including electrodes, is 100 mm.) Design a pump cavity for this laser. Assume that the absorption in the pump bands is $\alpha = 0.2\ \text{cm}^{-1}$, assume that the cavity will be coated with a material that is $R = 95\%$ reflective at the pump bands, assume that only the first bounce off the walls matters, and (finally) assume that the plasma is opaque to the pump light. Your answer should include:

(a) a sketch of your pump cavity design,
(b) a brief (few sentences) description of the rationale for your design, and
(c) a calculation for the percentage of lamp energy that is actually absorbed in the rod.

10.13 Consider the problem of balancing the focusing and the absorption in a cylindrical laser rod. Assume an Nd:YAG ($n = 1.82$) rod 5 mm in diameter. Assume a collimated ray-like (as opposed to Gaussian) beam 3 mm in diameter is incident on one side of the rod. Assume the absorption is zero.

(a) Where is the focus generated by the cylindrical lens of the rod?
(b) Assume the input beam is 25 watts. Compute the intensity deposited in the rod as a function of distance. In other words, create a plot where the x-axis is distance and the y-axis is watts/cm^2.
(c) Repeat the calculation of (b), but assume a nonzero absorption.
(d) Attempt to determine an optimal absorption which creates a flat intensity profile where focusing is balanced against absorption.

10.14 (design) Obtain data sheets from two major manufacturers of conventional noble gas discharge lamp pumped Nd:YAG lasers. Compare and contrast the laser systems. (You may wish to coordinate with your classmates to avoid having a large number of people contacting the same vendor!) Your answer should include:

(a) a table comparing the various features of the lasers, and
(b) several paragraphs discussing the relative strengths and weaknesses of the laser systems.

Semiconductor lasers as solid-state laser pump sources

10.15 (design) Consider the anamorphic prism pair of Figure 10.30. Assume you have a laser diode with 10 degrees full angle divergence one direction and 30 degrees full angle divergence in the orthogonal direction. Assume you want to couple the diode into a 5 mm diameter Nd:YAG rod. Design an anamorphic prism pair to do this. Your answer should include:

(a) a dimensioned sketch of your anamorphic prism pair design, and
(b) a brief (a few sentences) description of the rationale for your design, and
(c) supporting calculations for the critical aspects of the design.

10.16 (design) Consider again the question of pumping a laser rod with another laser. As was discussed in an earlier question, there are two possible geometries. The pump laser can be directed down the length of the rod (end-pumped) or it can be directly transversely across the rod (transverse-pumped). Assume a solid-state laser rod 50 mm long and 6 mm in diameter. Assume that the absorption at the pump line is $\alpha = 0.2\ \text{cm}^{-1}$. Assume that the pump laser is a laser diode with 10 degrees full angle divergence one direction and 30 degrees full angle divergence in the orthogonal direction. Design two optical systems for pumping the laser rod. One should be transverse and one should be end-pumped. Compute the light absorbed in the rod for each design. Your answer should include:

(a) a sketch of each of the two designs,
(b) a brief (one to three sentences) description of the logic behind the designs, and
(c) a calculation of the pump light absorbed in the rod for each geometry.

10.17 (design) Consider the ring laser cavity of Figure 10.35. Assume a reasonable dimension for the diameter of the Nd:YAG rod, and determine a set of self-consistent dimensions for the picture. In other words, select angles for the rod geometry that minimize loss but still permit the ring geometry. (Don't forget that the beam shouldn't be clipped by any of the elements!) You may assume the cavity is unidirectional clockwise. Your answer should include:

(a) a dimensioned sketch of your ring design,

(b) a brief (few sentences) description of the rationale for your design, and

(c) supporting calculations for the critical aspects of the design.

10.18 (design) Obtain data sheets from two major manufacturers of diode-pumped Nd:YAG lasers. Compare and contrast the laser systems. (You may wish to coordinate with your classmates to avoid having a large number of people contacting the same vendor!) Your answer should include:

(a) a table comparing the various features of the lasers, and

(b) several paragraphs discussing the relative strengths and weaknesses of the laser systems.

11

Transition-Metal Solid-State Lasers

Objectives

- To summarize the sequence of historical events leading to the development of the transition-metal solid-state laser.
- To summarize commercial applications for transition-metal solid-state lasers.
- To compare and contrast the energy band structure and major laser properties for the primary commercial solid-state laser materials (ruby, alexandrite, Ti:sapphire, Nd:YAG).
- To describe the design of ring laser cavities.
- To compare and contrast ring laser cavities with linear cavities.
- To describe the importance of birefringent filters.
- To compute the intensity transmittance for a birefringent filter.
- To describe the construction of a commercial laser pumped continuous wave Ti:sapphire laser.
- To describe the design principles underlying femtosecond pulse laser design. This would include such issues as group velocity dispersion (GVD), self-phase modulation (SPM), femtosecond pulse temporal measurement, colliding pulse mode-locking (CPM), grating pulse compression, solitons, and Kerr-lens mode-locking (KLM).
- To describe the construction of a commercial ultrashort pulse Ti:sapphire laser.

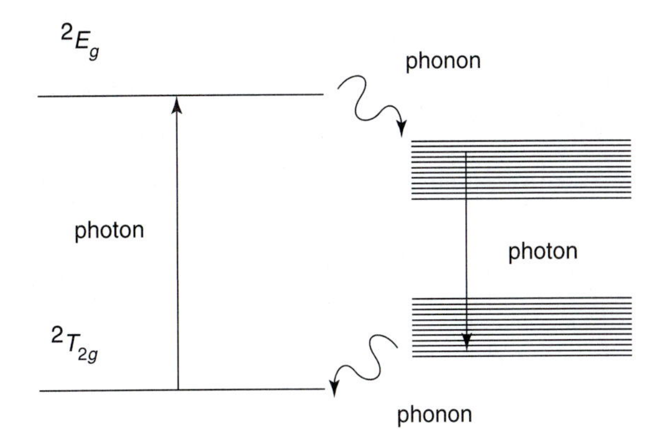

Figure 11.1 The transition-metal solid-state tunable lasers use metals in the fourth row of the periodic table as the active ions. These metals can produce transitions that involve phonons as well as photons (often called vibronic or phonon-terminated transitions). Such transitions can create tunable four-level laser behavior.

11.1 HISTORY

The history of transition-metal solid-state tunable lasers is exceptionally fascinating. For the HeNe, argon-ion and Nd:YAG lasers (even the diode pumped Nd:YAG lasers) the majority of the laser science was in place by the mid-1960s and commercial development proceeded rapidly after that. Transition-metal tunable solid-state lasers are quite different. Transition-metal tunable solid-state lasers are barely mentioned in review papers on tunable laser technology as recently as 1982.[1]

Ti:sapphire lasers (the current stars of the solid-state tunable laser market) were discovered by Moulton in 1982.[2] However, early results with Ti:sapphire were not promising due to difficulties with material growth.[3] It was only after the materials problems were solved that the true potential of the Ti:sapphire laser was realized. As a consequence, much of the laser development (including the remarkable self-mode-locking properties of Ti:sapphire discussed in Section 11.5) has occurred relatively recently.

The transition-metal solid-state tunable lasers use metals in the fourth row of the periodic table as the active ions. The transition-metals have a partially filled 3d shell, and the various observed transitions occur near this shell. 3d electrons interact more strongly with the crystal field than the 4f electrons in conventional solid-state lasers such as Nd:YAG. This can produce transitions that involve phonons as well as photons (often called vibronic or phonon-terminated transitions). Such transitions are rather peculiar, as they can create four-level laser behavior between two level transitions. A schematic of a vibronic transition is illustrated in Figure 11.1.

In a vibronic transition an optical photon is used to make the transition from the ground state to the pump state. Then the electron decays to the upper laser state by releasing a phonon (an acoustical quanta similar to a photon). The laser action occurs between the upper and lower laser states. The lower laser state then decays to the ground state by releasing

[1]B. D. Guenther and R. G. Buser, *IEEE J. of Quantum Electron.* QE-18:1179 (1982).

[2]P. F. Moulton, *Solid State Research Report,* DTIC AD-A124305/4 (1982:3) (Lexington: MIT Lincoln Lab., 1982), pp. 15–21.

[3]P. Lacovara and L. Esterowitz, *IEEE J. of Quantum Electron.* QE-21:1614 (1985).

another phonon. Thus, four-state laser behavior is obtained from a system that is effectively two-state. More importantly, since a wide variety of phonon transitions are possible, the upper and lower laser states consist of large manifolds of states. Therefore, highly tunable laser action is possible.

The first vibronic laser was reported by Johnson et al. at Bell Laboratories in 1963.[4] It was a divalent transition-metal laser using Ni^{2+} in MgF_2. It stimulated some early work by McCumber in the theory of vibronic lasers.[5] However, it was cryogenically cooled and did not excite much commercial interest.

Further efforts by Johnson and his colleagues during the mid to late 1960s resulted in several more cryogenically cooled divalent transition-metal lasers. These included Co^{2+} in MgF_2 and V^{2+} in MgF_2.[6]

A major advancement occurred in 1976 when Morris and Cline[7] observed that alexandrite ($BeAl_2O_4$:Cr^{3+} or chromium doped chrysoberyl, tunable from 700 nm to 818 nm) would lase on a vibronic transition. Walling et al. confirmed these results and demonstrated Q-switching behavior.[8] Alexandrite was particularly interesting at the time of its discovery because it lased at room temperature and increased in output power as the temperature increased.[9]

The successful use of Cr^{3+} in a beryl crystal led to several other interesting vibronic lasers. In particular, in 1982 Shand and Walling,[10] and independently Buchert et al.,[11] showed that emerald ($Be_3Al_2(SiO_3)$:Cr^{3+}, another type of chromium-doped chrysoberyl and tunable from roughly 700 nm to 800 nm) would lase as a vibronic laser at room temperature. Chromium was also found to generate vibronic laser performance in gadolinium scandium gallium garnet (GSGG).[12]

These encouraging results in chromium-doped materials led to a rebirth in tunable solid-state laser research. Ti:sapphire (the crown jewel of modern tunable solid-state lasers)

---

[4]L. F. Johnson, R. E. Dietz, and H. J. Guggenheim, *Phys. Rev. Lett.* 11:318 (1963).

[5]D. E. McCumber, *Phys. Rev.* 134:A299 (1964); D. E. McCumber, *J. Math. Phys.* 5:508 (1964); and D. E. McCumber, *Phys. Rev.* 136:A954 (1964).

[6]L. F. Johnson, R. E. Dietz, and H. J. Guggenheim, *Appl. Phys. Lett.* 5:21 (1964); L. F. Johnson and H. J. Guggenheim, *J. Appl. Phys.* 38:4837 (1967); L. F. Johnson and H. J. Guggenheim, *J. Appl. Phys.* 38:4837 (1967); and L. F. Johnson, H. J. Guggenheim and R. A. Thomas, *Phys. Rev.* 149:179 (1966).

[7]R. C. Morris and C. F. Cline, "Chromium-Doped Beryllium Aluminate Lasers," U.S. Patent #3,997,853, Dec. 14, 1976.

[8]J. C. Walling, H. P. Jenssen, R. C. Morris, E. W. O'Dell, and O. G. Peterson, Annual meeting Opt. Sci. Amer., San Francisco, CA, 1978; J. C. Walling, H. P. Benson, R. C. Morris, E. W. O'Dell, and G. Peterson, *Opt. Lett.* 4:182 (1979); J. C. Walling, O. G. Peterson, H. P. Jenssen, R. C. Morris, and E. W. O'Dell, *IEEE J. Quantum Electron.* QE-16:1302 (1980); and C. L. Sam, J. C. Walling, H. P. Jenssen, R. C. Morris, and E. W. O'Dell, *Proc. Soc. Photo-Opt. Inst. Eng. (SPIE)* 247:130 (1980).

[9]M. L. Shad and H. Jenseen, *IEEE J. of Quantum Electron.* QE-19:480 (1983).

[10]M. Shand and J. Walling, *IEEE J. of Quantum Electron.* QE-18:1829 (1982).

[11]J. Buchert, A. Katz, and R. R. Alfano, *IEEE J. of Quantum Electron.* QE-19:1477 (1983).

[12]E. V. Zharikov, N. N. Il'ichev, S. P. Kaltin, V. V. Laptev, A. A. Malyutin, V. V. Osiko, V. G. Ostroumov, P. P. Pashinin, A. M. Prokhorov, V. A. Smirnov, A. F. Umyskov, and I. A. Shcherbakov, *Sov. J. Quantum Electron.* 13:1274 (1983).

was discovered in 1982 by Moulton at MIT Lincoln Labs.[13] Although sapphire is the oldest laser material (ruby is Cr^{3+} in sapphire) the discovery of the broadly tunable nature of Ti^{3+} in sapphire was quite unexpected. A review report on tunable solid-state lasers published in 1982[14] and a review paper on alexandrite lasers in 1985[15] do not even mention Ti:sapphire.

Part of the delay in Ti:sapphire emerging as a viable commercial tunable solid-state laser was materials-based. Early Ti:sapphire crystals showed an absorption at the lasing wavelengths that was approximately an order of magnitude higher than the absorption in high-quality sapphire. A number of possible defects were proposed[16] and after much investigation the residual absorption in vertical-gradient-freeze (VGF) crystals was shown to be due to quadruply ionized titanium (Ti^{4+}) substituting for the aluminum in the sapphire.[17,18] Growth and annealing methods have significantly reduced this problem in modern commercial Ti:sapphire material.

In spite of its many advantages, Ti:sapphire does suffer from a few disadvantages. In particular, its short upper state lifetime (3.2 μs) makes it quite difficult to pump with a lamp. Although lamp-pumped Ti:sapphire lasers have been built,[19] most commercial Ti:sapphire lasers are pumped with argon-ion or doubled Nd:YAG lasers.

Several other materials have seen some commercial interest as possible lamp pumped laser materials. In particular $LiCaAlF_6:Cr^{3+}$ and $LiSrAlF_6:Cr^{3+}$ have seen some interest as possible tunable commercial laser sources.[20] A number of other chromium-doped materials including Cr:forsterite and Cr:YAG are also showing strong potential.[21]

Transition-metal solid-state tunable lasers are still being actively developed. Barnes[22] and Budgor et al.[23] provide good overview treatments of this developing field. In addition, there are three special issues in IEEE journals on tunable lasers.[24]

---

[13]P. F. Moulton, *Solid State Research Report.* DTIC AD-A124305/4 (1982:3) (MIT Lincoln Lab., Lexington, 1982), pp. 15–21, reported by P. F. Moulton, "Recent Advances in Solid-State Lasers," *Proc. Con. Lasers Electro-opt.*, Anaheim, CA, 1984, paper WA2.

[14]B. D. Guenther and R. G. Buser, *IEEE J. of Quantum Electron.* QE-18:1179 (1982).

[15]J. C. Walling, D. F. Heller, H. Samelson, D. J. Harter, J. A. Pete, and R. C. Morris, *IEEE J. of Quantum Electron.* QE-21:1568 (1985).

[16]P. Lacovara and L. Esterowitz, *IEEE J. of Quantum Electron.* QE-21:1614 (1985).

[17]A. Sanchez, A. J. Strauss, R. L. Aggarwal, and R. E. Fahey, *IEEE J. of Quantum Electron.* 24:995 (1988).

[18]R. Aggarwal, A. Sanchez, M. Stuppi, R. Fahey, A. Strauss, W. Rapoport, and C. Khattak, *IEEE J. of Quantum Electron.* 24:1003 (1988).

[19]P. Lacovara, L. Esterowitz and R. Allen, *Opt. Lett.* 10:273 (1985).

[20]S. A. Payne, L. L. Chase, H. W. Newkirk, L. K. Smith, and W. F. Krupke, *IEEE J. of Quantum Electron.* 24:2243 (1988); and S. A. Payne, L. L. Chase, L. K. Smith, W. L. Kway, and H. W. Newkirk, *J. Appl. Phys.* 66:1051 (1989).

[21]C. Pollock, D. Barber, J. Mass, and S. Markgraf, *IEEE J. of Sel. Topics in Quantum Electron.* 1:62 (1995).

[22]Norman P. Barnes, "Transition Metal Solid State Lasers," in *Tunable Lasers Handbook*, ed F. J. Duarte (San Diego: Academic Press, 1995).

[23]A. Budgor, L. Esterowtiz, and L. G. DeShazer, eds, *Tunable Solid State Lasers II* (Berlin: Springer Verlag, 1986).

[24]*IEEE J. of Quantum Electron.* QE-18 (1982); QE-21 (1985); and *IEEE J. of Sel. Topics in Quantum Electron.* (1995).

11.2 APPLICATIONS

Transition-metal solid-state tunable lasers provide two major features. First, they are tunable over a broad range of visible and near IR wavelengths. Second, they can be used to produce extremely short pulses.

The tunability feature means that these lasers are ideal for spectroscopic applications. This not only includes traditional scientific spectroscopy, but also medical diagnostic spectroscopy. For example, Ti:sapphire lasers have been used to perform an optical version of conventional mammography.[25] There are also potential applications for absorption, Raman, and fluorescence spectroscopy in medical imaging.[26]

Solid-state lasers compete with dye lasers for medical applications requiring both tunability and intensity. Primary among these are cosmetic surgery for port wine birthmarks, telangiectasia, warts, stretch marks, acne scars, removing tattoos, and psoriasis.[27] Tunable solid-state lasers also compete with dye lasers for medical applications such as shattering kidney stones.[28]

In addition, the extremely short pulses possible with tunable solid-state lasers are finding application in micromachining. Femtosecond-pulsed Ti:sapphire lasers can be used for micromachining holes in metal and polymer substrates as well as for ablating photoresist films and cutting traces on semiconductor materials.[29] Ti:sapphire lasers compete with Nd:YAG, diode-pumped Nd:YAG, and excimer lasers for this extremely important market.

11.3 LASER MATERIALS

Ruby, alexandrite, and Ti:sapphire are the major transition-metal solid-state laser materials. Although ruby is not used commercially as a tunable laser, it does have a tunable vibronic transition. Interestingly enough, the band structure of alexandrite is quite similar to ruby; except in alexandrite the vibronic transition is the important one and the narrow line transition is not used. In contrast, Ti:sapphire has crystalline and mechanical properties virtually identical to ruby, but a dramatically different band structure.

A number of publications can provide additional information for the interested reader. Overview treatments are given by Weber,[30] Koechner,[31] and Duarte,[32] while more specific

---

[25] *Laser Focus World,* Feb.: 38 (1996).

[26] *Laser Focus World,* Feb.: 72 (1996).

[27] *Laser Focus World,* May: 66-7 (1996).

[28] *Laser Focus World,* May: 66-7 (1996).

[29] *Laser Focus World,* January: 22 (1996).

[30] Marvin J. Weber, ed, *Handbook of Laser Science and Technology, Vol. I, Lasers and Masers* (Boca Raton, FL: CRC Press, Inc., 1982); and more recently, Marvin J. Weber, ed, *Handbook of Laser Science and Technology, Supplement I, Lasers* (Boca Raton, FL: CRC Press, Inc., 1991).

[31] Walter Koechner, *Solid State Laser Engineering*, 4th ed. (Berlin: Springer-Verlag, 1996).

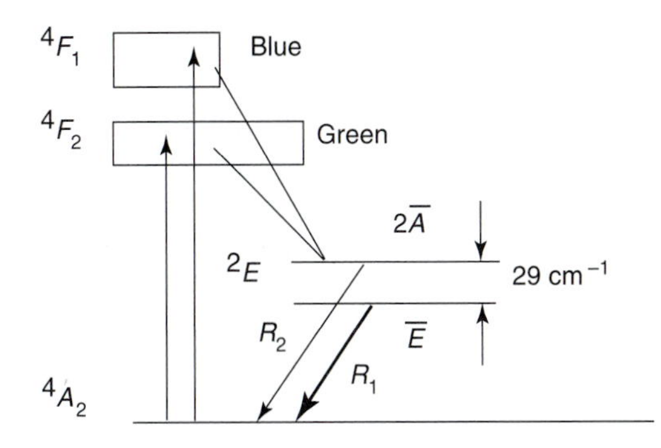

Figure 11.2 The energy band diagram for ruby.

information can be obtained from the wide variety of review papers on alexandrite[33] and Ti:sapphire.[34,35] Manufacturer data sheets and application notes are also very useful.[36]

11.3.1 Ruby—Primary Line at 694.3 nm

Ruby (chromium-doped Al_2O_3) is a red or pink hexagonal crystal whose most familiar application is jewelry. Ruby is an optically uniaxial crystal[37] that is hard (Moh's hardness of 9), of good optical quality, and extremely thermally conductive (0.42 W/cm-K at 300K). Ruby is nonhygroscopic, refractory, and is generally considered the most durable of the common laser crystals (with the possible exception of Ti:sapphire). Ruby crystals are typically grown by the Czochralski method (the same method as used for the growth of silicon). Ruby can be grown at 0, 60, or 90 degrees to the optic axis, and laser material is usually grown at 60 degrees.

Sapphire is doped with Cr^{3+} to obtain ruby. The Cr^{3+} substitutes for the Al^{3+} in the crystal. Typical dopings are 0.05 weight percent of Cr_2O_3. However, excess chromium can distort the crystal structure and concentrations are sometimes reduced to 0.03 weight percent to enhance the optical beam quality.

The energy diagram for ruby is given in Figure 11.2. Ruby is three-state and is the only commercially viable three-state laser system. The laser pump bands are principally the $^4F_1$ and the $^4F_2$ bands. The ground state is the $^4A_2$ band. The two pump bands form manifolds centered around the blue (400 nm) and green (555 nm). The pump bands are

[32] F. J. Duarte ed, *Tunable Lasers Handbook* (San Diego: Academic Press 1995).

[33] J. C. Walling, D. F. Heller, H. Samelson, D. J. Harter, J. A. Pete, and R. C. Morris, *IEEE J. of Quantum Electron.* QE-21:1568 (1985).

[34] A. Sanchez, A. J. Strauss, R. L. Aggarwal, and R. E. Fahey, *IEEE J. of Quantum Electron.* 24:995 (1988).

[35] R. Aggarwal, A. Sanchez, M. Stuppi, R. Fahey, A. Strauss, W. Rapoport, and C. Khattak, *IEEE J. of Quantum Electron.* 24:1003 (1988).

[36] Major crystal suppliers are Union Carbide (ruby, alexandrite and Ti:sapphire) and Litton Airtron (alexandrite).

[37] A uniaxial crystal is one where two of the Cartesian directions have one index of refraction n_o and the third has a different index of refraction n_e. See Section 8.3 for a discussion of uniaxial and biaxial crystals.

each quite wide, with the blue band about 0.05 microns wide and the green band about 0.07 microns wide.

The lifetime in the pump bands is extremely short, with the ions cascading almost immediately to the metastable ^2E states. The upper ^2E state is termed the $2\overline{A}$ state and the lower is termed the $\overline{E}$ state. The $2\overline{A}$ and $\overline{E}$ states are separated by 29 cm^{-1}, which gives a population ratio at thermal equilibrium of 87%. Thus, while fluorescence in ruby occurs from both the $2\overline{A}$ state to the $^4A_2$ (termed the R_2 transition at 692.9 nm) and from the $\overline{E}$ state to the $^4A_2$ (termed the R_1 transition at 694.3 nm), laser action first occurs on the R_1 transition. Once laser action has begun, the rapid relaxation time from the $2\overline{A}$ to the $\overline{E}$ transition prohibits laser action starting on the R_2 line. The only way to start laser action on the R_2 line is to suppress the R_1 line by special dielectric coated mirrors or internal cavity absorbers. (Interesting enough, even though lasing occurs primarily on the R_1 and R_2 lines, sidebands have been observed on the long wavelength side, in particular at 767 nm, attributed to vibronic lasing.)

Since ruby is uniaxial, its absorption coefficient is a very strong function of the polarization direction of the light (see Figure 11.3). This property strongly affects the beam quality. The best optical quality ruby is grown with the crystal axis at 60 degrees to the boule axis. When such a ruby rod is pumped in a diffuse reflecting pump cavity, pump light parallel to the c-axis will be absorbed differently than pump light perpendicular to the c-axis. This will cause the pump distribution (and thus the laser output beam) to be elliptical.

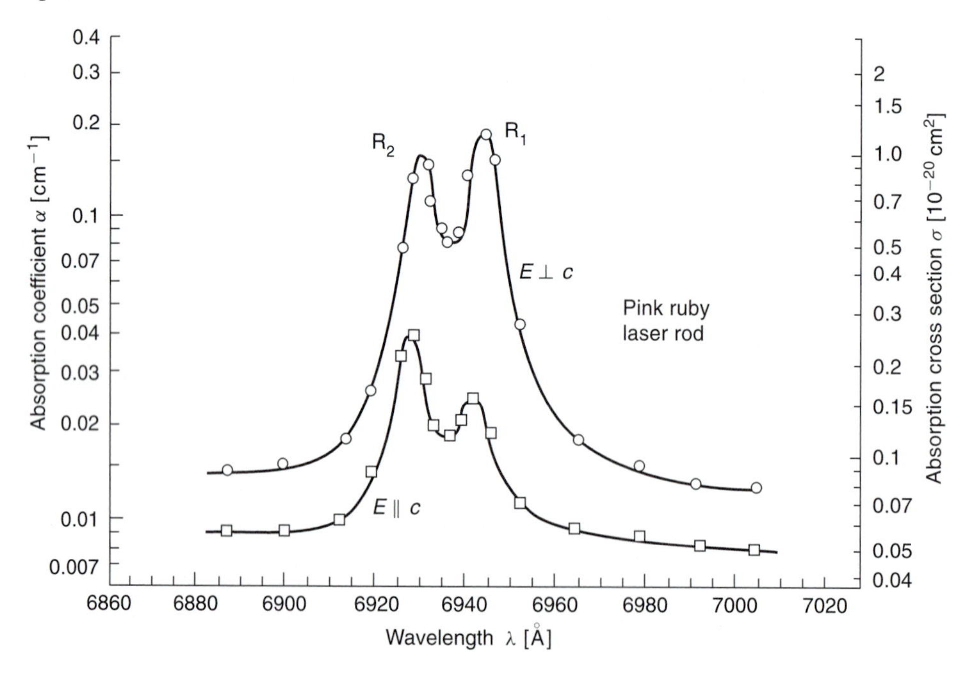

Figure 11.3 Since ruby is uniaxial, its absorption coefficient is a very strong function of the polarization direction of the light. (From D. C. Cronemeyer, *J. Opt. Soc. Am.* 56:1703 (1966). Reprinted with the permission of the Optical Society of America.)

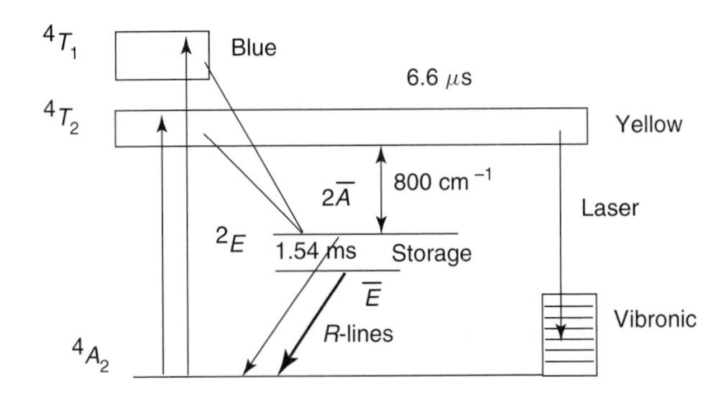

Figure 11.4 The energy band diagram for alexandrite.

11.3.2 Alexandrite—Tunable from 700 nm to 818 nm

Alexandrite ($BeAl_2O_4{:}Cr^{3+}$ or chromium-doped chrysoberyl) is a hard orthorhombic material. Chrysoberyl itself is considered a semiprecious jewelry material and is commonly called oriental topaz. It ranges in color from yellow through green to brown. When chrysoberyl is doped with chromium, the material turns emerald green and displays a secondary red color when viewed in artificial light. (As an aside, one variety of chrysoberyl occurs in a crystal form consisting of parallel arrangements of fibers. When cut as a cabochon, it is called cat's-eye or tiger's-eye.)

Alexandrite is biaxial,[38] hard, of good optical quality, and quite thermally conductive (0.23 W/cm-K as compared with 0.14 W/cm-K for YAG and 0.42 W/cm-K for ruby). Alexandrite is nonhygroscopic, melts at 1870°C, and has a Moh's hardness of 8.5 (which makes it harder and more durable than YAG, but somewhat less than ruby). Additionally, alexandrite has a very high thermal fracture limit (60% of ruby and five times that of YAG).

Doping the yellowish chrysoberyl with chromium results in an emerald green alexandrite crystal. Alexandrite is biaxial and the crystal appears green, red, or blue, depending on the angle and lighting conditions. The principle axes of the indicatrix are aligned with the crystallographic axes.[39] Lasers are usually operated with light parallel to the *b*-axis because the gain for polarization in this direction is roughly ten times that of any other direction.

As with ruby, the Cr^{3+} occupies the aluminum sites in the crystal. However, there are two different aluminum sites in alexandrite. One site has mirror symmetry, the other has inversion symmetry. Most of the chromium substitutes for aluminum in the larger mirror site (about 78%), which (luckily!) is the dominant site for laser action. The doping in alexandrite can be a great deal higher than with ruby. Doping concentrations as high as 0.4 weight percent still yield crystals of good optical quality (although 0.2 to 0.3 weight percent is somewhat more common).

The energy diagram for alexandrite is given in Figure 11.4. Alexandrite can be operated as either a three-state system or as vibronic four-state system (note the similarity to ruby!). The laser pump bands are principally the $^4T_1$(higher) and the $^4T_2$ (lower) bands. The ground state is the $^4A_2$ band. The two pump bands form manifolds centered around

---

[38] A biaxial crystal is one where all three of the Cartesian directions have different indices of refraction. See Section 8.3 for a discussion of uniaxial and biaxial crystals.

[39] See Section 8.3 for more discussion on the indicatrix.

the blue (410 nm) and yellow (590 nm). The pump bands are each quite wide, with widths approximately 1000 angstroms.

In a fashion similar to ruby, there is a metastable ^2E state. As with ruby, laser action can occur on the R lines of the ^2E state and can generate three-state laser behavior at similar wavelengths (680.4 nm). The major difference between the R-state lasing in alexandrite and ruby is that alexandrite possesses a higher threshold and lower efficiency. Thus, alexandrite is not used as a ruby replacement.

The major value of alexandrite is in its four-state tunable laser energetics. When operated as a four-state laser, alexandrite is a vibronic laser. Lasing action occurs between vibronic rather than purely atomic states. Thus, the emission of a photon is also accompanied by the emission of phonons as the state returns to equilibrium. This is what provides the tunability. The laser wavelength is determined by which of the vibronic states is the top laser state. Any energy not carried off by the photon will then be emitted by phonons to restore the system to its ground state.

Laser action occurs by emission from the $^4T_2$ state to the excited vibronic states within the $^4A_2$ band. Subsequent phonon emission returns the $^4A_2$ band to equilibrium. The $^4T_2$ state is much shorter lived (6.6 μs) than the metastable ^2E states (1.54 ms) and is quite close in energy (800 cm^{-1}), so the metastable ^2E states serve as a storage state for the $^4T_2$ state.

In any laser with a very broad pumping curve, there is the risk that the laser photons themselves will be reabsorbed. Luckily, in alexandrite, there is a deep minimum in excited-ion absorption across the primary laser tuning range. Therefore, excited-ion absorption is not an issue over much of the range. However, the long wavelength tuning limit in alexandrite is due to excited-ion absorption. At higher wavelengths than 818 nm, the excited-ion absorption cross-section is larger than the laser cross-section, and laser action is suppressed.

Alexandrite has exceptionally fascinating behavior with temperature. As the temperature increases in most four-state lasers, the lower laser state population increases, and laser action is reduced. However, the laser output from alexandrite increases with temperature up to a temperature of around 200°C, and then decreases abruptly. This is because there are two competing temperature effects in alexandrite. Although the lower state population does increase with temperature, there is also the coupling between the metastable ^2E storage states and the $^4T_2$ state. As the temperature increases, the population in the upper laser state increases, partly counteracting the increase in population in the lower state. Thus, alexandrite possesses an improvement in performance with temperature at wavelengths greater than 730 nm.

Unfortunately, the lifetimes of the upper and storage states somewhat reduce the advantages of higher temperatures. When the temperature is increased, the population in the $^4T_2$ state increases. However, this state has a shorter lifetime. Thus the effective upper state lifetime (the population-ratioed combination of the two upper state states) is reduced. At some temperature, the combined upper state lifetime is shorter than the pump pulse width and energy is lost in fluorescence. This also reduces the advantages of increasing temperature.

One final temperature effect is worth mentioning. At higher temperatures, the peak of the gain curve shifts to longer wavelengths. This is due to increased phonon populations as well as a shift of the R lines and enhancement of the long wavelength vibronic transitions. The net result is a rather dramatic change in laser wavelength with increasing temperature.

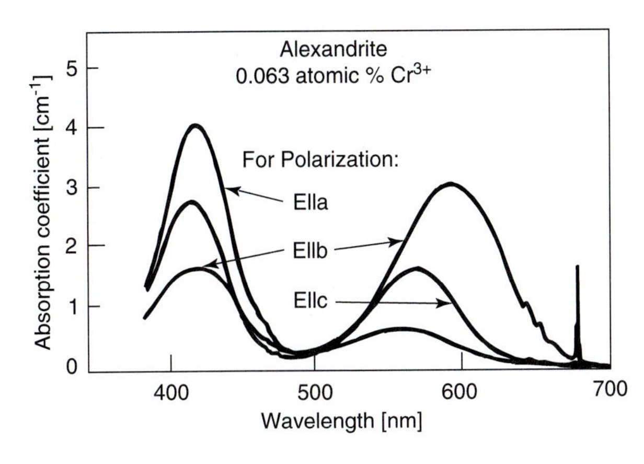

Figure 11.5 Since alexandrite is biaxial, its absorption coefficient is a very strong function of the polarization direction of the light, for all three directions. (Modified from J. C. Walling, H. P. Jenssen, R. C. Morris, E. W. O'Dell, and O. G. Peterson, *Opt. Lett.* 4:182 (1979), Figure 3. Reprinted with the permission of the Optical Society of America).

Since alexandrite is biaxial, its absorption coefficient is a very strong function of the polarization direction of the light, in all three directions! (See Figure 11.5.) This creates some of the same orientation and pumping inhomogeneities observed in ruby.

11.3.3 Ti:sapphire—Tunable from 670 nm to 1090 nm

Ti:sapphire was developed relatively late in laser evolution. However, since the discovery of laser action in Ti:sapphire in 1982, Ti:sapphire has become one of the most widely used solid-state laser materials.

Recall that ruby is chromium-doped Al_2O_3. Ti:sapphire is titanium-doped Al_2O_3. Thus, ruby and Ti:sapphire have many of the same mechanical and optical properties. Ti:sapphire is also an optically uniaxial crystal that is hard (Moh's hardness of 9), of good optical quality, and extremely thermally conductive (0.42 W/cm-K). Ti:sapphire is nonhygroscopic, refractory and is even more durable than ruby due to a slight advantage obtained with the titanium doping.

Sapphire is doped with Ti^{3+} to obtain laser quality Ti:sapphire. The Ti^{3+} substitutes for the Al^{3+} in the crystal. Typical dopings range from 0.03 to 0.15 weight percent of titanium (slightly higher than the chromium in ruby). However, the similarities between ruby and Ti:sapphire do not include the spectroscopy. The energy diagram for Ti:sapphire (Figure 11.6) possesses only two states, but the vibronic nature of the transitions makes it possible to absorb from the bottom of the ground state to the upper vibronic manifold and to lase from the bottom of the upper vibronic manifold down to the vibronic ground states.

This peculiar spectroscopy is driven by the single $(3d)^1$ electron. The $(3d)^1$ state (which would be degenerate in a free transition-metal) is split by the cubic field when the metal is substitutionally doped into the aluminum site in the sapphire. The result is a doubly degenerate excited state ^2E_g and a triply degenerate ground state $^2T_{2g}$. These excited and ground states are further split by the trigonal field and spin orbit coupling. The result is a bottom manifold of states $^2T_{2g}$ and a top manifold ^2E_g. One phonon of energy 172 cm^{-1} couples to the excited state ^2E_g state and two phonons with energies of 220 and 260 cm^{-1} couple to the ground state $^2T_{2g}$. (In comparison, ruby has three d electrons and a rather more conventional energy state diagram!)

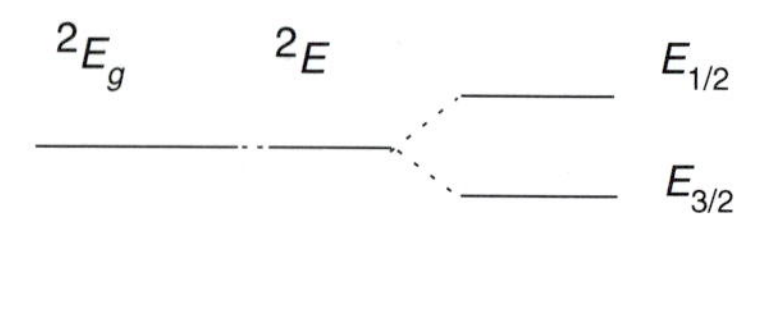

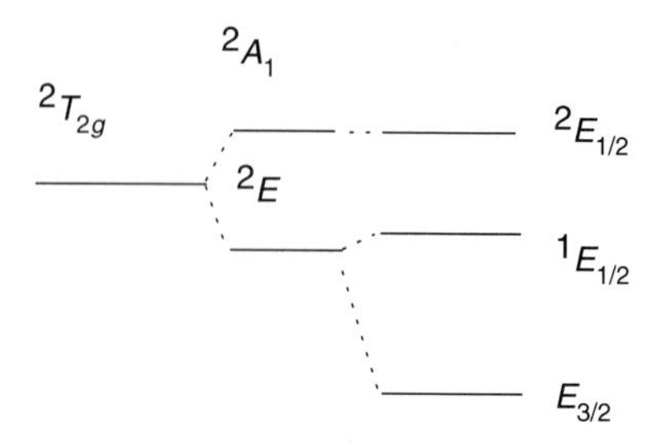

Figure 11.6 The energy band diagram for Ti:sapphire.

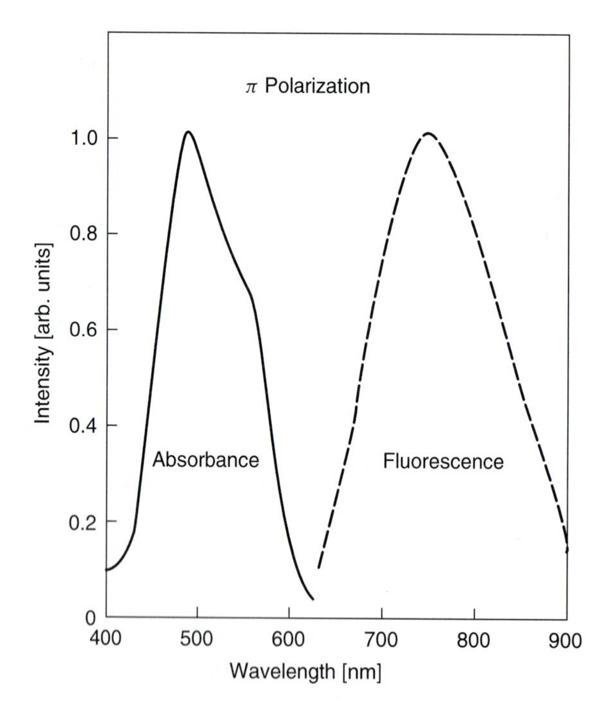

Figure 11.7 Ti:sapphire possesses very large polarization-dependent absorption bands in the blue-green (centered around 490 nm with widths of roughly 150 nm). Similarly, Ti:sapphire possesses very large polarization dependent emission bands in the red/IR (centered around 780 nm with widths of roughly 230 nm). (From P. F. Moulton, *J. Opt. Soc. Am. B* 3:125 (1986). Reprinted with the permission of the Optical Society of America.)

Ti:sapphire possesses very large polarization dependent absorption bands in the blue-green (centered around 490 nm with widths of roughly 150 nm; see Figure 11.7) Absorption of radiation polarized along the optic axis π is nearly twice that of the perpendicular polarization σ. The absorption is generated by vibronic transitions between the ground state $^2T_{2g}$ and the excited state ^2E_g. Similarly, Ti:sapphire possesses very large polarization dependent emission bands in the red/IR (centered around 780 nm with widths of roughly 230 nm). Nearly three times as much light is emitted polarized along the optic axis π as perpendicular to it σ. The emission is generated by transitions between vibronic transitions from the excited state ^2E_g to vibronic transitions to the ground state $^2T_{2g}$.

It is a delightful property of the vibronic states of Ti:sapphire that the absorption and emission bands barely overlap. Thus, a laser photon generated within the emission band is unlikely to stimulate absorption in the absorption band.

Although Ti:sapphire has a reasonable cross-section ($4.1 \cdot 10^{-19}$ cm^2), it also has a relatively short upper state lifetime (3.2 μs). As a consequence, it is a difficult material to pump with flashlamps. However, flashlamp operation has been achieved, with outputs on the order of 3J/pulse at 2% efficiency. Flashlamps used for Ti:sapphire must typically be altered (for example with a dye) to convert the blue-UV light of the lamp into the blue-green absorption profile of the laser. The usual design strategy for successful operation of a flashlamp pumped Ti:sapphire laser is to create a *very* short flashlamp pulse (less than 10 μs) through clever design of the pulse-forming network.

The blue-green absorption profile of Ti:sapphire is beautifully matched to the output of cw argon lasers, cw internally-doubled Nd:YAG lasers (particularly diode-pumped cw YAGs) and pulsed doubled Nd:YAG lasers. Using a Q-switched doubled Nd:YAG laser as the pump gives the additional advantage of permitting gain-switched operation. (Gain switching is similar to Q-switching, except in gain-switching, the laser gain is turned on very quickly, rather than the laser loss being turned off very quickly.)

The phenomenal bandwidth of Ti:sapphire is ideal for mode-locking. Ti:sapphire lasers have been mode-locked with acousto-optic mode-locking, passive mode-locking, injection-seeding, and coupled-cavity mode-locking. In 1991 it was discovered that Ti:sapphire lasers will self-mode-lock. A flurry of advances occurred in this technology, resulting in elegant methods for self-mode-locking Ti:sapphire lasers with temporal pulse lengths of less than 10 fs. This advancement was a major change in laser development, as until that time all ultrashort pulse lasers were dye lasers (see Section 11.5).

11.3.4 Comparison between Major Solid-State Laser Hosts

Consider the solid-state hosts as shown in Table 11.1. Notice that ruby and Ti:sapphire are the mechanically more robust and thermally conductive. Nd:YAG and alexandrite are both four-state lasers with long upper state populations (permitting lamp pumping), while

TABLE 11.1 SOLID-STATE LASER HOSTS

	Ruby	Nd:YAG	Alaexandrite	Ti:sapphire
Hardness	9	8.2	8.5	9.1
Thermal conductivity (W/cm-K)	0.42 (300 K)	0.14 (300 K)	0.23 (300 K)	0.42 (300 K)
Upper-state lifetime	3.0 ms (300 K)	230 ηs (300 K)	260 ηs (300 K)	3.2 ηs (300 K)
σ_{21} (cross-section) (cm^2)	$2.5 \cdot 10^{-20}$	$6.5 \cdot 10^{-19}$	$1 \cdot 10^{-20}$	$4.1 \cdot 10^{-19}$
Linewidth (Å)	5.3 (11 cm^{-1}) (300 K)	4.5 (300 K)	1000	2300

Ti:sapphire has a very short upper state population (generally requiring laser pumping). Nd:YAG and Ti:sapphire have relatively large cross-sections while alexandrite and ruby have smaller cross-sections. Ti:sapphire has the widest linewidth and ruby the narrowest.

11.4 TI:SAPPHIRE LASER DESIGN

Ti:sapphire lasers differ from conventional solid-state lasers in two major ways: First, Ti:sapphire lasers are typically pumped by an argon-ion or Nd:YAG laser. As a consequence, the resonant cavity geometry is often a ring or folded-Z cavity to allow the pump beam to interact with the laser material while avoiding pump light in the output beam. (Ring laser cavities are discussed in Section 11.4.1.) Second, Ti:sapphire lasers have very large tuning curves. Tuning elements such as etalons do not possess sufficient range to tune Ti:sapphire lasers. Instead, birefringent filters are usually employed. (Birefringent filters are discussed in Section 11.4.2).

11.4.1 Ring Lasers

The first ring resonator was demonstrated in 1963 by Tang et al. in ruby.[40] Although the primary applications for early ring resonators were in optical gyroscopes,[41] ring lasers did not see much commercial application as resonator structures until the development of the dye laser.[42,43] Not only did the ring ease some of the geometrical constraints in building dye lasers, the counter-propagating nature of the mode permitted the construction of the first femtosecond pulse lasers (see Section 11.5).

Ring laser geometries are often necessary with Ti:sapphire lasers, because of the geometrical requirements imposed by pump lasers. Some especially elegant designs have been implemented, including diode-pumped monolithic ring resonators where the entire ring is inside the gain material[44].

Basics on ring resonators. The prototypical laser resonator consists of two mirrors facing each other (see Figure 11.8). This configuration creates electric and magnetic field standing waves in the resonator with a period of one-half of an optical wavelength. These standing waves interact with only part of the volume of the laser material, thus creating spatial inhomogeneities in the gain. This effect is termed *spatial hole burning*.

If the laser is running on multiple longitudinal modes (creating many overlapping standing wave regions), or if the laser is a gas laser (where the atoms are in constant motion), this effect is not very significant. However, for single-mode solid-state lasers, spatial hole burning can significantly reduce the output power of the operating laser mode.

---

[40] C. L. Tang, H. Statz, and G. deMars, Jr., *J. Appl. Phys.* 34:2289 (1963).

[41] W. Chow et al., *Rev. Mod. Phys.* 57:61 (1985).

[42] F. P. Schafer and H. Muller, *Opt. Commun.* 2:407 (1971).

[43] J. M. Green, J. P. Hohimer, and F. K. Tittel, *Opt. Commun.* 7:349 (1973).

[44] T. J. Kane and R. L. Byer, *Opt. Lett.* 10:65 (1985).

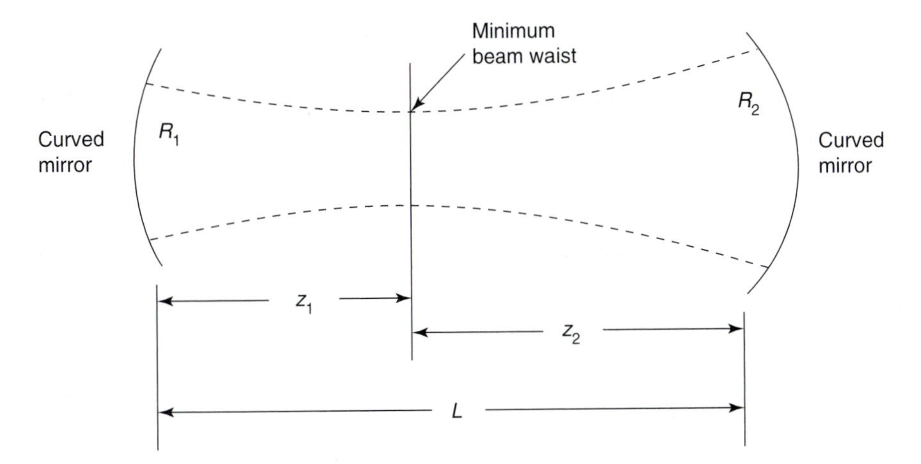

Figure 11.8 The prototypical laser resonator consists of two mirrors facing each other.

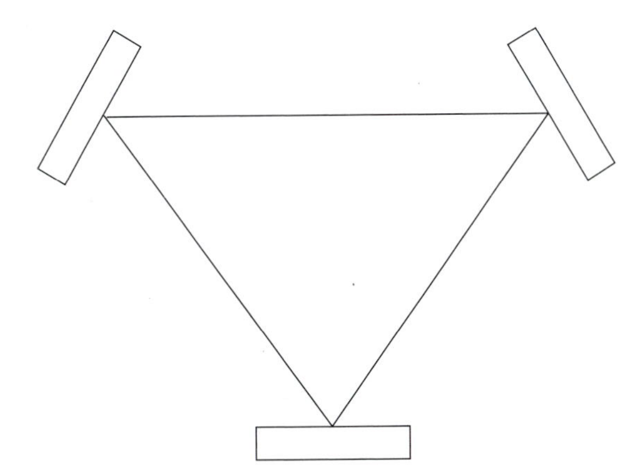

Figure 11.9 A simple ring resonator will have longitudinal modes (the back-and-forth length of a conventional laser is equivalent to the round-trip length of a ring laser) and standing wave patterns (formed by the interaction between the left and right traveling waves in the ring).

Spatial hole burning can be reduced by generating circularly polarized light in the laser material,[45] mechanically moving the material,[46] or by phase modulating the beam entering the material.[47] However, one of the most successful methods for eliminating spatial hole burning is to structure the resonator in a unidirectional ring geometry.

A simple ring resonator (such as in Figure 11.9) is not very different from a conventional resonator. A simple ring resonator will have longitudinal modes (the back-and-forth length of a conventional laser is equivalent to the round-trip length of a ring laser) and standing wave patterns (formed by the interaction between the left and right traveling waves in the ring).

However, ring lasers can also be made to operate in only one direction. Such unidirectional ring lasers no longer possess standing waves and do not create spatial hole

[45] D. A. Draegert, *IEEE J. of Quantum Electron.* QE-8:235 (1972).

[46] H. G. Danielmeyer, *Appl. Phys. Lett.* 16:124 (1970).

[47] H. G. Danielmeyer, *Appl. Phys. Lett.* 17:519 (1970).

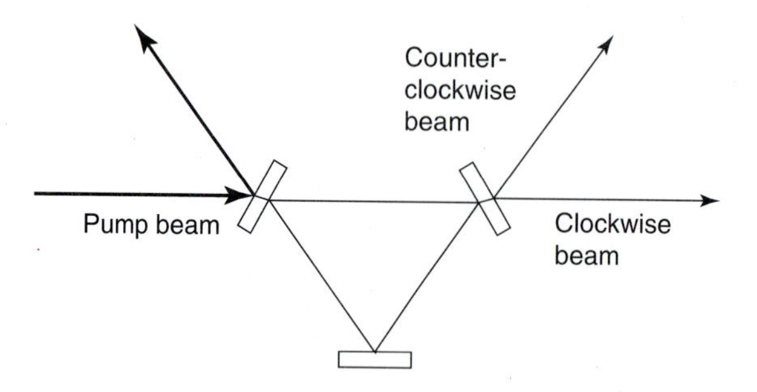

Figure 11.10 One of the advantages of a ring laser is that it is relatively easy to configure the resonator so that the back-reflected beam from a pump laser is not reflected into the resonator of the pump laser.

burning effects. Eliminating spatial hole burning provides several advantages. First, since the laser material is more homogeneous, there is more overall gain available to the longitudinal modes. This increases mode competition between adjacent longitudinal modes, making it possible to pump the laser harder and still retain single longitudinal mode operation. Additionally, since the gain is more homogeneous, more power can be extracted by a single operating mode. The combination of increased pumping range and increased gain means that unidirectional ring lasers can deliver significantly more power than conventional resonators.

Ring lasers possess some additional advantages. If a laser is being used as a pump for a ring laser, it is relatively easy to configure the resonator so that the back-reflected beam is not returned back into the resonator of the pump laser (see Figure 11.10). The optical system formed by a plane parallel mirror of a conventional resonator and the output mirror of the pump laser can act like a Fabry-Perot cavity attached to the laser. The interaction between this external Fabry-Perot and the resonator of the pump laser can significantly alter the frequency and intensity properties of the pump laser. (This is the origin of the oscillation that occurs when aligning a laser cavity with a HeNe laser and backreflecting the alignment beam precisely down the bore of the alignment laser.)

Ring lasers also possess more flexibility in cavity design, particularly for unstable resonators or for lasers that have a number of sensitive intracavity elements. Additionally, unidirectional ring lasers inherently have an ordering to the elements in the cavity. (For example, the gain material might be followed by a doubling crystal and then by a birefringent filter.) Inherent ordering of optical components is important in many applications.

Creating unidirectional oscillation. One method for generating unidirectional oscillation is to create a small external cavity aligned with one of the propagating directions of the ring laser (see Figure 11.11). If the laser is running counter-clockwise (CCW), the traveling wave in the laser does not interact with the external cavity. However, if the laser is running clockwise (CW), the beam is back reflected into the CCW direction and supports the CCW mode.[48]

This method works for certain laser materials. However (even when it works) the method does not completely suppress the light in the unwanted direction. Perhaps more

[48]The terms cw (for continuous wave) and cw (for clockwise) are both commonly used in laser engineering. There is no standard notation to distinguish them, and so the meaning must be deduced from context. Here I will use CW for the latter.

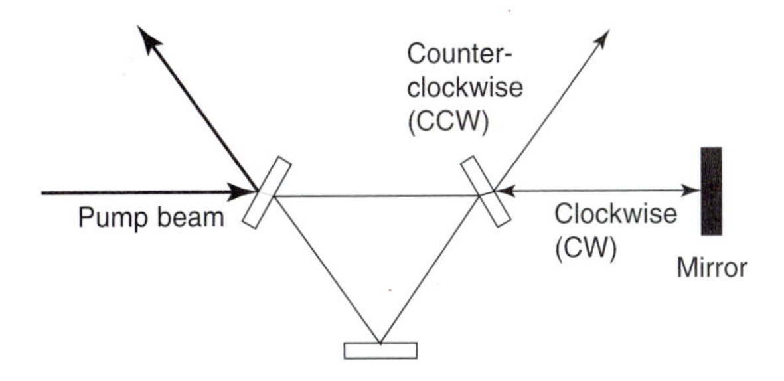

Figure 11.11 One method for generating unidirectional oscillation in a ring laser is to create a small external cavity aligned with one of the propagating directions of the ring laser.

importantly, the method can enhance mode-beating (particularly if there are multiple longitudinal modes running in the ring) and interferometric effects.

A more reliable method for generating unidirectional oscillation in a ring laser is to use an optical diode constructed from a Faraday rotator, a half-wave plate, and a Brewster plate or polarizer (see Figure 11.12).

A Faraday rotator (as used in a ring laser) is a device that changes the angle of linear polarization of light as a function of an applied DC magnetic field. The angle is changed in the *same* direction, independent of the direction of the wave propagation.

To explain this further, consider a vertically polarized wave traveling in the CCW direction and entering a Faraday rotator as shown in Figure 11.13. Assume the Faraday rotator changes the direction of linear polarization from vertical to 45 degrees to the vertical.

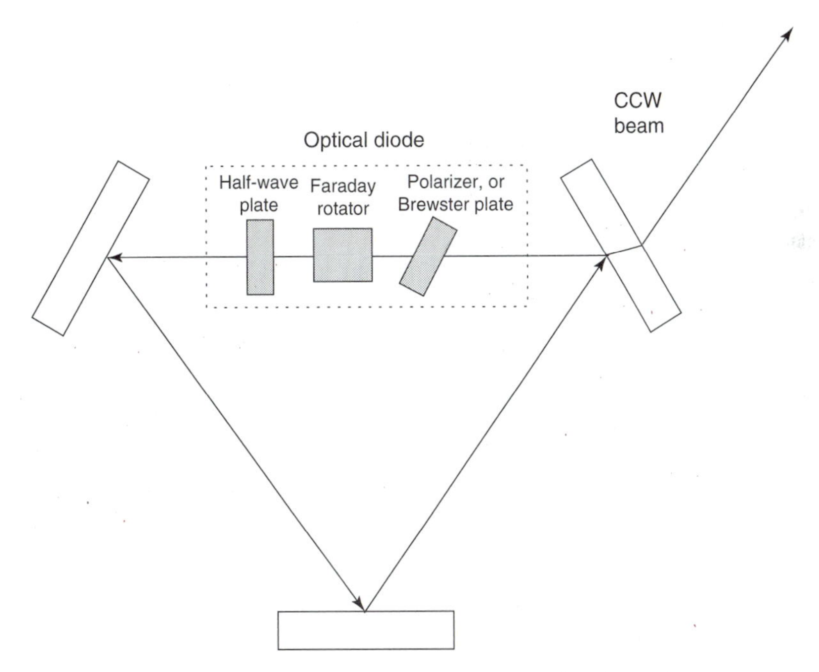

Figure 11.12 Another method for generating unidirectional oscillation in a ring laser is to place an optical diode (constructed from a Faraday rotator, a half-wave plate, and a Brewster plate or polarizer) in the resonator.

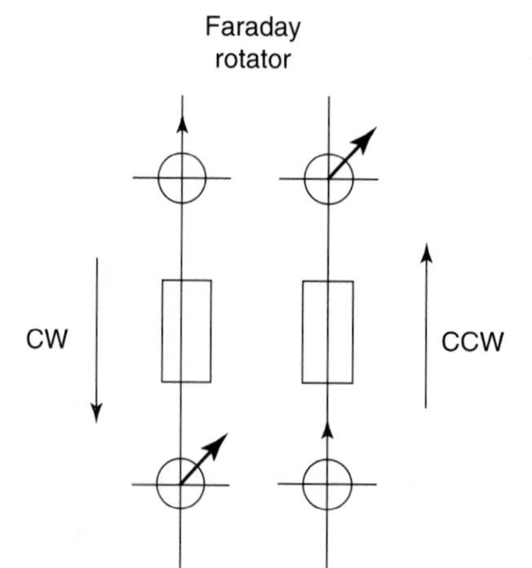

Figure 11.13 Vertically polarized light passing through a Faraday rotator in clockwise (CW) and counter-clockwise (CCW) directions.

Now, assume a vertically polarized wave traveling in the CW direction is incident on the same Faraday rotator. The rotator will also change the direction of linear polarization from vertical to 45 degrees to the vertical. The angle has been changed in the same direction independent of the propagation direction of the wave.

Contrast this behavior with that of a half-wave plate. Consider a vertically polarized wave traveling in the CCW direction and entering a half-wave plate as shown in Figure 11.14. Assume the half-wave plate changes the direction of linear polarization from vertical to 45 degrees to the vertical. Now, assume a vertically polarized wave traveling in the CW direction is incident on the same half-wave plate. The rotator will change the direction of linear polarization from vertical to 315 degrees to the vertical (that is, 45 degrees in the *other direction* from vertical).

The unusual nature of Faraday rotators can be explained in another way. Assume that a Faraday rotator is configured with a mirror so that a beam enters the rotator, bounces off the mirror and enters the rotator again (but traveling the other direction). (See Figure 11.15.) Consider a vertically polarized wave traveling in the CCWdirection and entering the Faraday

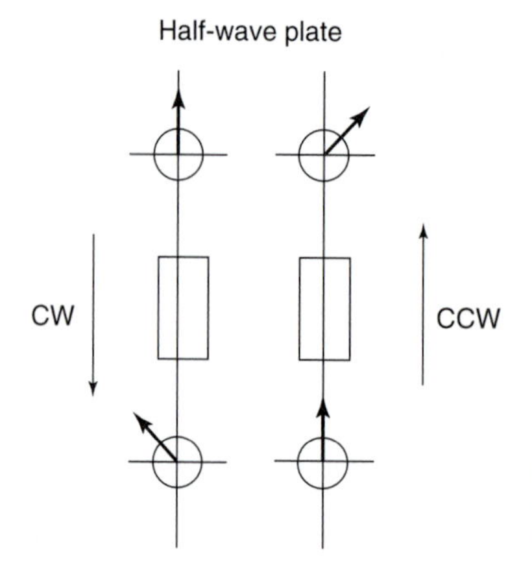

Figure 11.14 Vertically polarized light passing through a half-wave plate in clockwise (CW) and counter-clockwise (CCW) directions.

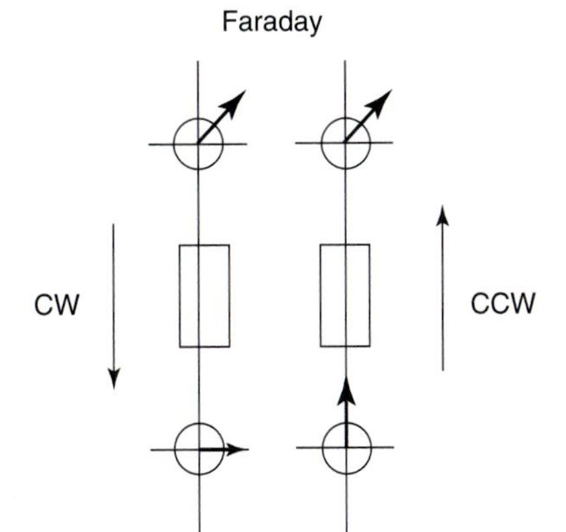

Figure 11.15 Vertically polarized light passing through a Faraday rotator in the CCW direction, being back-reflected, and passing back through the Faraday rotator in the CW direction.

rotator. Assume the rotator changes the direction of linear polarization from vertical to 45 degrees to the vertical. Now, the beam will be directed back through the rotator by the mirror, and will again be altered by 45 degrees in the same direction. The beam is now 90 degrees from its original polarization.

Contrast this with a half-wave plate. Assume that a half-wave plate is configured with a mirror so that a beam enters the half-wave plate, bounces off the mirror, and enters the half-wave plate again (but traveling the other direction). (See Figure 11.16.) Consider a vertically polarized wave traveling in the CCW direction and entering the half-wave plate. Assume the half-wave plate changes the direction of linear polarization from vertical to 45 degrees to the vertical. Now, the beam will be directed back through the half-wave plate by the mirror, and will be altered by 45 degrees in the reverse direction. The beam is at the same polarization as the original!

The contrasting properties of Faraday rotators and half-wave plates allow the creation of optical diodes. As an example, consider the optical system illustrated in Figure 11.17. Here, a Brewster plate (or polarizer) is oriented to pass vertical polarization. The Brewster plate is followed by a Faraday rotator and then by a half-wave plate. Consider the CCW wave first. A vertically polarized wave is first passed with no loss by the Brewster plate.

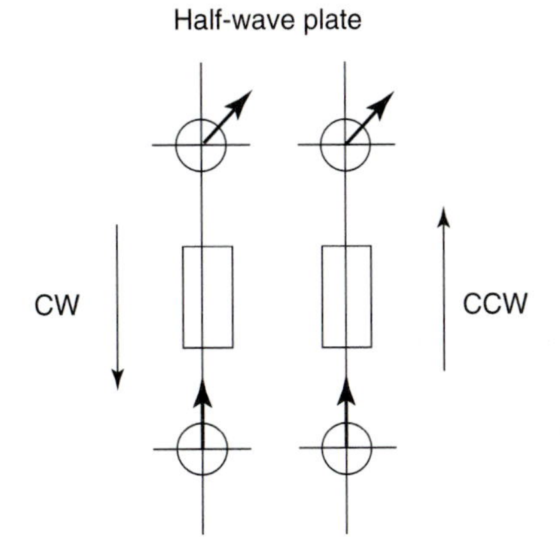

Figure 11.16 Vertically polarized light passing through a half-wave plate in the CCW direction, being back-reflected, and passing back through the half-wave plate in the CW direction.

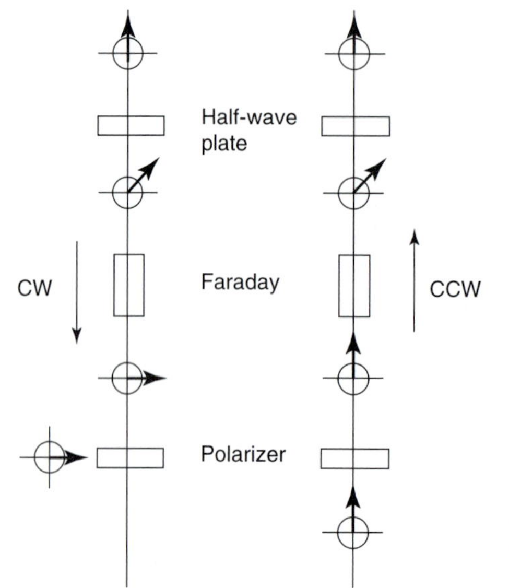

Figure 11.17 The contrasting properties of Faraday rotators and half-wave plates allow the creation of an optical diode. This system passes CCW light unchanged, but attenuates CW light.

Then, the Faraday rotator changes the direction from vertical to 45 degrees to the vertical. The half-wave plate then returns the polarization direction to the vertical. The original wave polarization has been recovered.

Now, consider the CW wave. A vertically polarized wave enters the half-wave plate and the direction is changed from vertical to 45 degrees to the vertical. The Faraday rotator then continues the change to 90 degrees to the vertical. The CW wave (now perpendicular to the CCW) experiences high loss from the Brewster plate (or polarizer) and is attenuated.

11.4.2 Birefringent Filters

A tunable laser (such as Ti:sapphire) requires some type of adjustable optical tuning element in order to run with a narrow linewidth. Spatially dispersive elements, such as prisms or gratings, can be used to provide this tuning. However, it is extremely difficult to design good tuning systems for wide bandwidth tunable lasers using spatially dispersive elements.[49]

An alternative strategy is to use etalons. Recall from Chapter 3 that a conventional etalon makes a very good high-resolution filter. However, a single etalon can only be tuned over its free spectral range. Thus, etalons are well-suited for isolating narrow linewidth features (such as a single argon-ion longitudinal mode) but are less well-suited for broad tuning.

Birefringent filters offer an effective alternative to dispersive elements or etalons. Multiple element birefringent filters offer excellent resolution over a very broad spectral tuning range. The first use of a birefringent filter as a tunable narrow band laser filter was

---

[49] A. L. Bloom, *Opt. Engineering* 11:1 (1972).

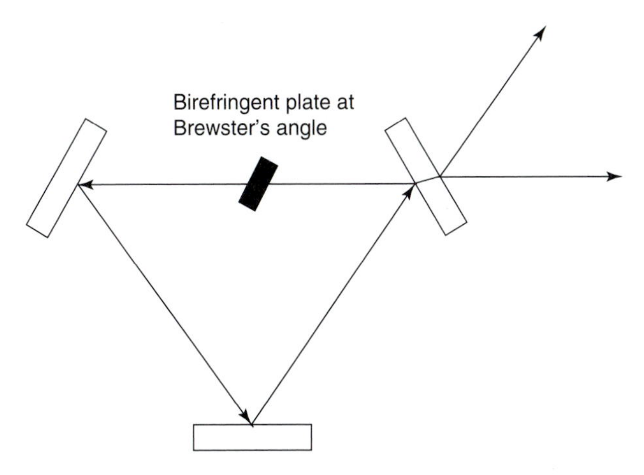

Figure 11.18 A birefringent filter is a plate of birefringent material placed inside a laser cavity at Brewster's angle. The birefringent filter will transmit a set of wavelengths corresponding to an integral number of full wave retardations in the crystal.

in 1972 by Yarborough.[50] Birefringent filters are analyzed by Bloom,[51] Preuss and Gole,[52] and Valle and Moreno.[53] Negus et al. hold a patent on a birefringent filter for use in a tunable pulsed laser cavity.[54]

A birefringent filter is a plate of birefringent material placed inside a laser cavity at Brewster's angle (see Figure 11.18). When the wavelength of the light incident on the filter corresponds to an integral number of full wave retardations in the birefringent element, then the beam emerges unchanged. However, if the wavelength of the incident light is anything other than an integral number of full wave retardations, then its polarization will be altered and it will suffer loss at the Brewster surface. Thus, the birefringent filter will transmit a set of wavelengths corresponding to an integral number of full wave retardations in the crystal (see Figure 11.19).

A birefringent filter can be tuned by rotation around its own axis (the rotational direction that does not change the Brewster angle orientation). This alters the index of refraction seen by the incident beam and thus the wavelength corresponding to an integral number of full wave retardations.

A single birefringent filter possesses a sinusoidal-like transmittance profile (see Figure 11.19) that is usually too broad for use as a tuning element in a Ti:sapphire laser. However, multiple birefringent filters can be used to provide greater selectivity. For example, Figure 11.20 illustrates the transmittance of a three-stage quartz birefringent filter with thicknesses of 0.5 mm, 1 mm, and 7.5 mm.

Birefringent filters can be analyzed with a matrix technique in a similar fashion to multilayer dielectric films.[55] Of special interest is the intensity transmittance for a single

[50] J. M. Yarborough and J. Hobart, post-deadline paper, CLEA meeting (1973).

[51] A. Bloom, *J. Opt. Soc. of Am.* 64:447 (1974).

[52] D. R. Preuss and J. L. Cole, *Appl. Opt.* 19:702 (1980).

[53] P. J. Valle and F. Moreno, *Appl. Opt.* 31:528 (1992).

[54] D. K. Negus et al., "Birefringent Filter for Use in a Tunable Pulsed Laser Cavity," U.S. Patent #5,164,946, Nov. 17, 1992.

[55] D. R. Preuss and J. L. Cole, *Appl. Opt.* 19:702 (1980).

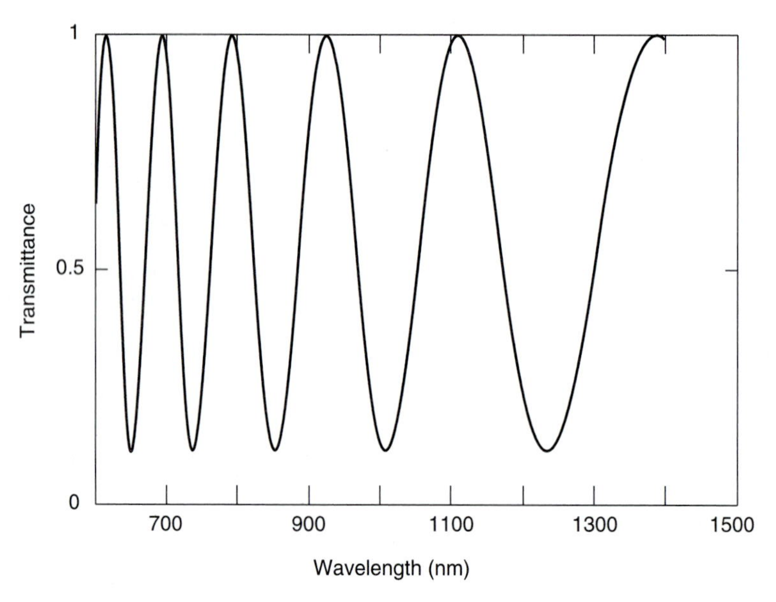

Figure 11.19 A single birefringent filter possesses a sinusoidal-like transmittance profile.

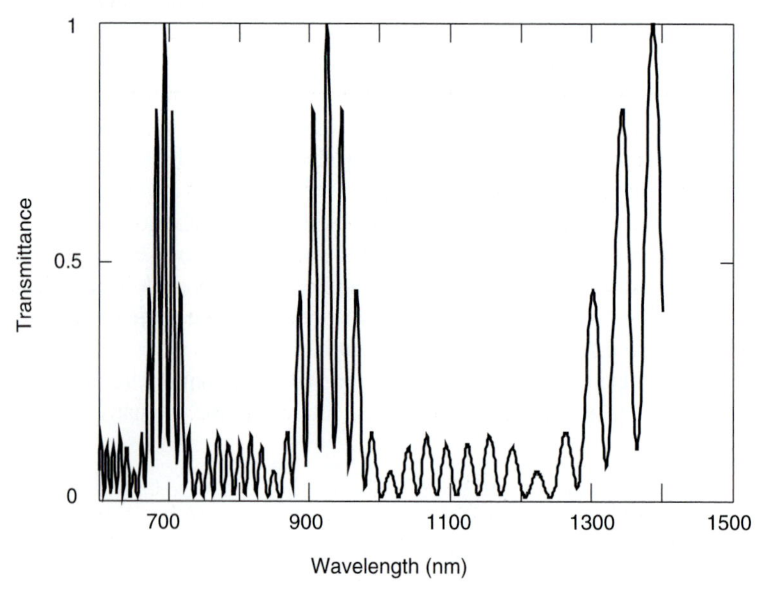

Figure 11.20 Multiple birefringent filters can be used to provide greater selectivity. This illustrates the transmittance of a three-stage birefringent filter with thicknesses of 0.5 mm, 1 mm, and 7.5 mm.

birefringent filter given by

$$I_{\mathrm{TM}} = 1 - \sin^2(2\phi)\frac{n_o^4 - n_o^2\cos^2(\theta)}{\left(n_o^2 - \cos^2(\phi)\cos^2(\theta)\right)^2}.$$

$$\sin^2\left(\frac{\pi d}{\lambda}\frac{n_e\left[1-\cos^2(\theta)\cos^2(\phi)/n_e^2 - \cos^2(\theta)\cos^2(\phi)/n_o^2\right]}{\left[1-\cos^2(\theta)\sin^2(\phi)/n_e^2 - \cos^2(\theta)\cos^2(\phi)/n_o^2\right]^{1/2}} - \frac{\pi d}{\lambda}\frac{n_o}{\left[1-\cos^2(\theta)/n_o^2\right]^{1/2}}\right) \tag{11.1}$$

where $\theta = \pi/2 - \theta_B$ for Brewster angle incidence (where θ_B is Brewster angle), ϕ is the tuning angle (ranges from 0 to 90 degrees), and d is the thickness.[56]

Example 11.1

Calculate the transmittance for a three-stage birefringent filter fabricated from quartz. Assume the smallest element is 0.5 mm thick and that the elements are in the ratio 1:2:15. Assume the filter is in at Brewster's angle and that the rotational angle is 50 degrees. Plot the transmittance from 600 nm to 1400 nm.

Solution. Programs such as Mathcad are well-suited for this type of calculation. Figure 11.20 illustrates the results of a Mathcad calculation for this problem.

11.4.3 Coherent Model 890 and 899 Ti:Sapphire Lasers

The Coherent Model 890 laser is a laser-pumped cw broadband Ti:sapphire laser (see Figure 11.21). The output power depends on the pump and can be as high as 2.5 watts with the SW mirror set and a 15 watt argon-ion pump.

Three sets of standard optics are used to access the Ti:sapphire tuning range. Using these three sets of optics, the wavelength can be tuned from 690 nm to beyond 1100 nm (see Figure 11.22).

In order to accommodate both low- and high-power pump lasers, the Model 890 can be configured for either high- or low-power pumping. The goal of the variable pump scheme is to optimize the overlap between the $\mathrm{TEM}_{0,0}$ resonator mode and the pump beam. This maximizes the conversion efficiency and minimizes thermal lensing in the Ti:sapphire.

The laser wavelength is tuned with a birefringent filter (see Section 11.4.2). Smooth continuous tuning is accomplished by minimizing the competition between the birefringent tuning filter and the natural birefringence of Ti:sapphire. To achieve this, the crystal is mounted such that the polarization vector of the intracavity mode is aligned parallel to the c-axis of the crystal.

The alignment of the polarization and optic axis vectors is achieved via a face-normal rotation adjustment. This alignment prevents gaps or skips in the output tuning curve. The Model 890 resonator is a folded linear resonator (see Figure 11.23). The backbone of the 890 resonator is a 5 cm diameter, solid, stainless-steel bar. The bar has a high thermal mass, which reduces the system sensitivity to changes in the ambient temperature. Mirrors

[56]D. R. Preuss and J. L. Cole, *Appl. Opt.* 19:702 (1980).

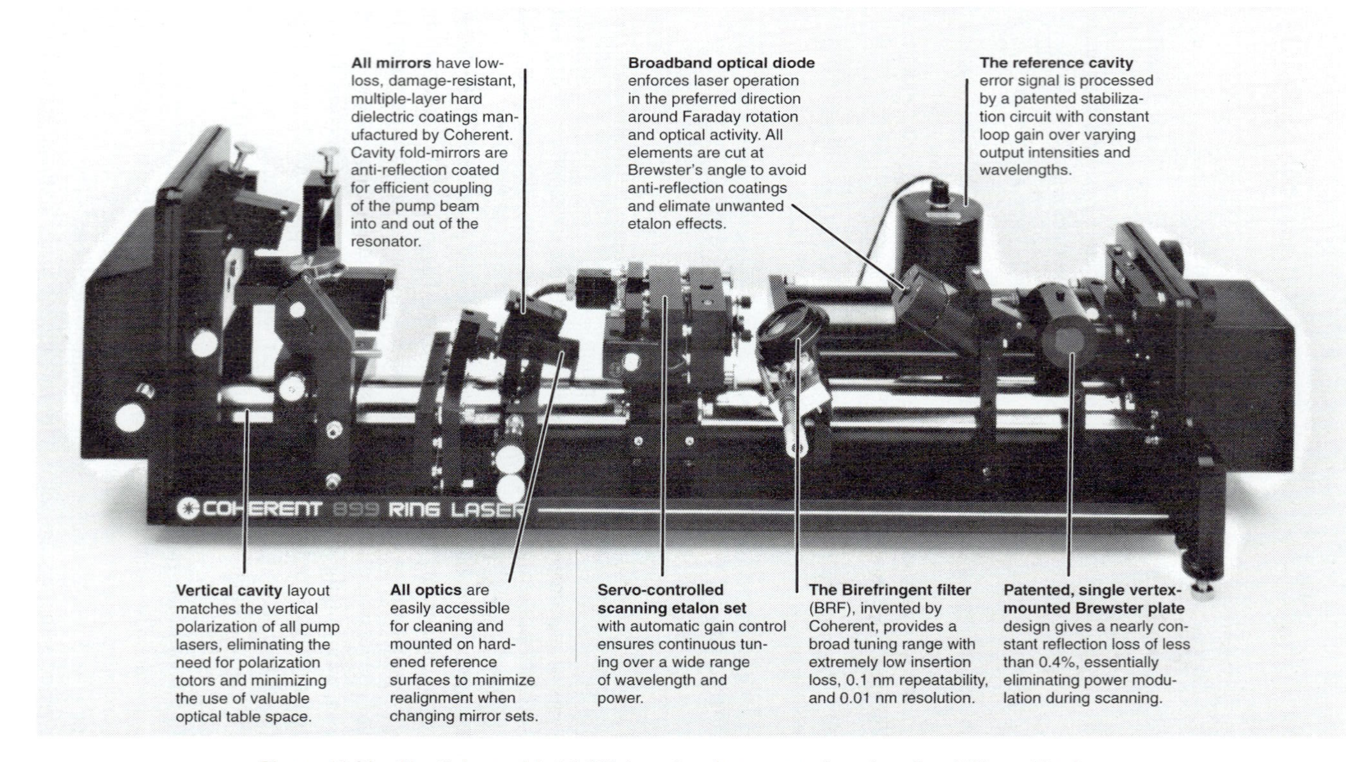

Figure 11.21 The Coherent Model 890 laser is a laser-pumped cw broadband Ti:sapphire laser. The output power depends on the pump and can be as high as 2.5 watts with the SW mirror set and a 15 watt argon-ion pump. (Courtesy of the Coherent, Inc., Laser Group, Santa Clara, CA)

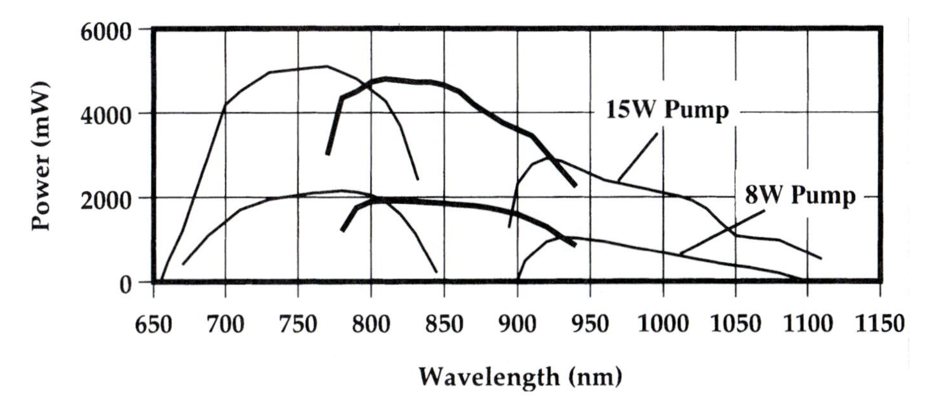

Figure 11.22 The Coherent Model 890 laser is a laser-pumped cw broadband Ti:sapphire laser. Three sets of standard optics are used to access the Ti:sapphire tuning range. Using these three sets of optics, the wavelength can be tuned from 690 nm to beyond 1100 nm. (Courtesy of the Coherent, Inc., Laser Group, Santa Clara, CA)

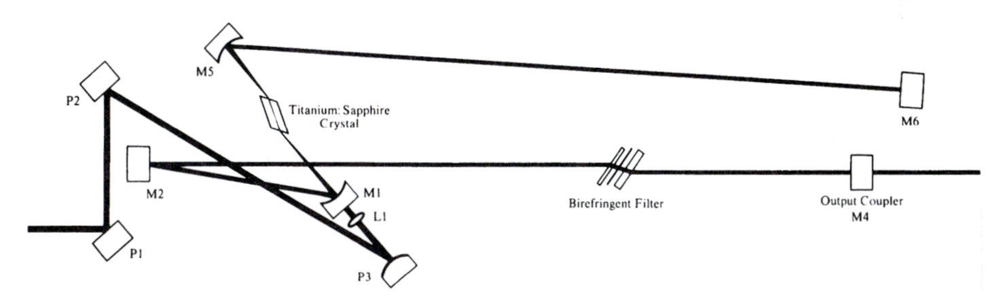

Figure 11.23 The Model 890 resonator is a folded linear resonator. (Courtesy of the Coherent, Inc., Laser Group, Santa Clara, CA)

M1 and M5 are transparent to the pump light.Mirrors M1, M2, M5, and M6 are highly reflective over the Ti:sapphire laser wavelengths. Mirror M4 is the output mirror for the laser. Mirrors P1, P2, and P3 are highly reflecting at the pump wavelengths. Thus, the pump beam enters the cavity along the M1 to M5 axis and the laser action occurs between the M4 and M6 output mirrors. The remaining mirrors are simply to assure that the pump beam can effectively pump the laser material without interacting with the Ti:sapphire laser output beam.

Coherent also sells a more elaborate version of the Model 890, which is the Model 899 Ti:sapphire ring laser. As with the model 890, three sets of standard optics are used to access the Ti:sapphire tuning range. Using these three sets of optics, the wavelength can be tuned from 700 nm to beyond 1000 nm (see Figure 11.24). The output power depends on the pump and can be as high as 3 watts with the SW mirror set and a 20 watt argon-ion pump.

The Model 899 resonator is a folded ring resonator (see Figure 11.25) using an Invar (rather than stainless-steel) backbone. Mirrors M1 and M5 are transparent to the pump light. Mirrors M1, M2, and M5 are highly reflective over the Ti:sapphire laser wavelengths.

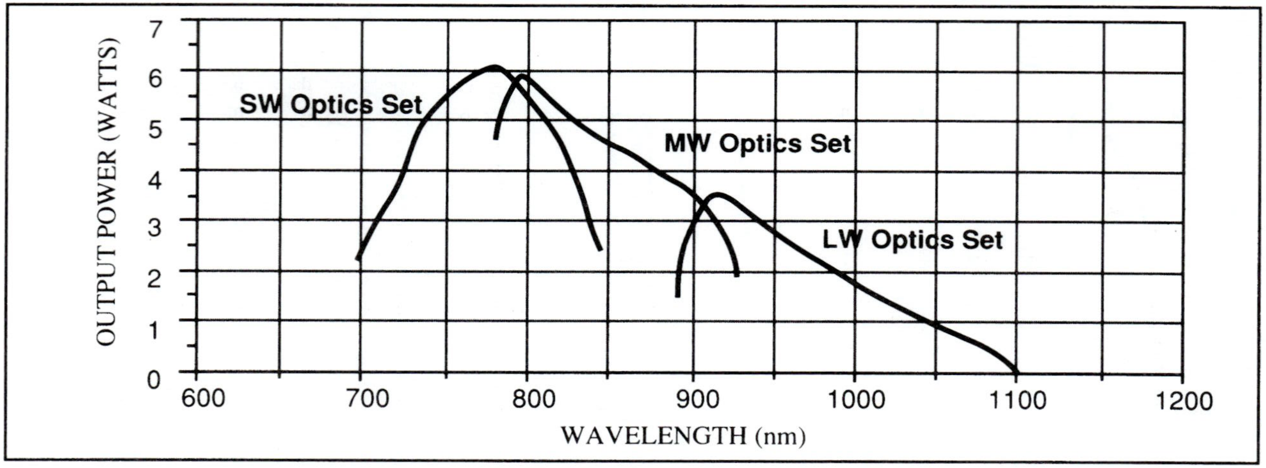

Figure 11.24 In the Model 899 Ti:sapphire ring laser, three sets of standard optics are used to access the Ti:sapphire tuning range. Using these three sets of optics, the wavelength can be tuned from 700 nm to beyond 1000 nm. (Courtesy of the Coherent, Inc., Laser Group, Santa Clara, CA)

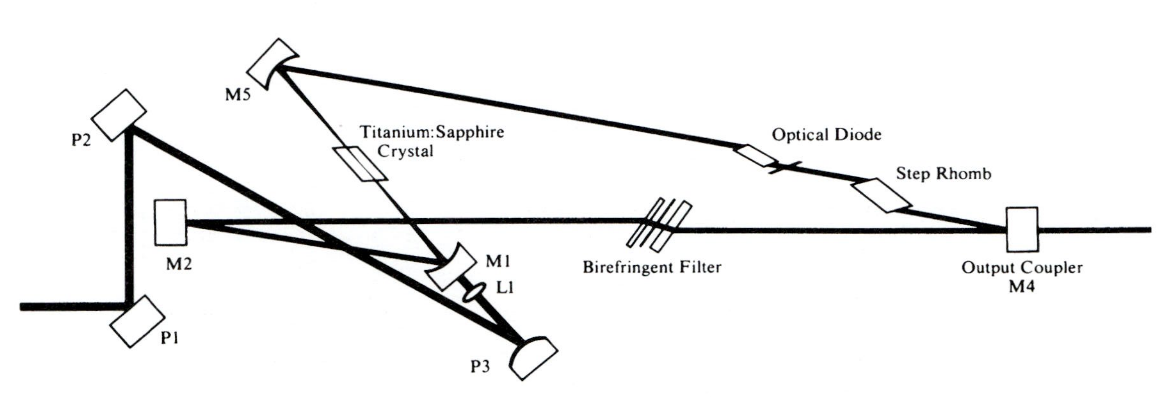

Figure 11.25 The Model 899 resonator is a folded ring resonator. (Courtesy of the Coherent, Inc., Laser Group, Santa Clara, CA)

Mirror M4 is the output mirror for the laser. Mirrors P1, P2, and P3 are highly reflecting at the pump wavelengths. Thus, the pump beam enters the cavity along the M1–M5 axis and the laser action occurs around the M1, M2, M4, and M5 ring. The optical diode and step rhomb are required to assure unidirectional operation around the ring (more on ring lasers in Section 11.4.1).

11.5 FEMTOSECOND PULSE LASER DESIGN

Since the discovery of mode-locking in 1964, there have been three generations of mode-locked lasers.

> *The first-generation lasers* are in the subnanosecond regime, fabricated from solid-state materials such as Nd:YAG and Nd:glass, and using either active loss mechanisms (such as acousto-optic mode lockers) or passive loss mechanisms (such as saturable absorbers). These lasers are ultimately limited in pulsewidth by the relatively narrow spectral bandwidth of the gain material and the inability of the saturable absorbers to track with the pulse narrowing.
>
> *The second-generation lasers* are in the femtosecond regime, using organic dyes as the gain materials, and with passive mode-locking using saturable absorbers in combination with rapid dye gain saturation. These lasers have demonstrated narrow pulse widths, but are difficult to align and maintain.
>
> *The third-generation lasers* are in the low femtosecond regime, using solid-state vibronic materials such as Ti:sapphire, with passive soliton-like pulse shaping formed by balancing self-phase modulation against group velocity dispersion. These lasers also demonstrate narrow pulse widths, but are straightforward to construct.

Ultra-short pulse optical systems are an extremely rich research area. For the reader interested in learning more, there are a number of special issues on ultrafast optics,[57] as well as review papers such as Keller et al.,[58] Krausz et al.,[59] and Spielmann et al.[60]

11.5.1 Dispersion in Femtosecond Lasers

Dispersion is the ability of a material to alter the frequency character of light. There are two main dispersive properties important in the design of femtosecond pulse lasers.

The first dispersive property of interest is group velocity dispersion (GVD). Group velocity dispersion is the tendency of various frequencies of light to propagate at slightly

---

[57] *IEEE J. of Quantum Electronics,* Special Issues on Ultrafast Optics and Electronics: April 1983, January 1986, February 1988, December 1989, and October 1992.

[58] U. Keller, W. Knox, and G. 'tHooft, *IEEE J. of Quantum Electron.* 28:2123 (1992).

[59] F. Krausz, M. Fermann, T. Brabec, P. Curley, M. Hofer, M. Ober, C. Spielmann, E. Wintner, and A. Schmidt, *IEEE J. of Quantum Electron.* 28:2097 (1992).

[60] C. Spielmann, P. Curley, T. Brabec, and F. Krausz, *IEEE J. of Quantum Electron.* 30:1100 (1994).

different speeds in certain materials. In materials with a normal or positive GVD, the longer wavelengths travel faster than the shorter ones, thus red shifting the pulse.

The second dispersive property of interest is self-phase modulation (SPM). SPM is an intensity-dependent phase shift that manifests itself either spatially or temporally. (SPM is often termed the Kerr effect.)

Spatial SPM can also be described as self-focusing. As the intensity in the center of the pulse increases, the index of refraction of the material increases, and the pulse focuses.

Temporal SPM is a time-dependent phase shift that occurs as the pulse sweeps through the dispersive material. The rising intensity on the front edge of the pulse increases the index of refraction. This will delay the individual oscillations, and thus red shift the rising edge. The reverse effect occurs on the trailing edge (blue shifting the trailing edge). Thus, temporal SPM chirps the pulse.

For ultrashort pulse generation, the round-trip time in the resonator for all frequency components of the light must be the same. Otherwise, frequency components with different phase shifts will no longer add coherently and the mode-locking will break down (think of an unmode-locked laser with random phases).

In a normal laser, temporal SPM will cause a red shift of the pulse and the GVD will also cause a red shift of the pulse. Thus, in order to achieve transform-limited pulse widths, it is necessary to incorporate some type of dispersion compensation that blue shifts the pulse.

Prisms are the most common way of introducing GVD compensation. Although the glass in prisms has normal dispersion (red shifts the pulse), the geometry can be arranged so that the blue components transverse the prism path in a shorter length of time than the red components. An example of a prism pair where the GVD is a function of L is given in Figure 11.26.

11.5.2 Nonlinearities Used to Create Femtosecond Pulses

In an unmode-locked laser, all of the longitudinal modes will be running with random phases. These random phases may be synchronized by the addition of a suitable nonlinearity. A laser is referred to as passively or self-mode-locked when this nonlinearity is generated within the laser and does not depend on external influences.

For passive mode-locking to work, the introduced nonlinearity must create an amplitude or phase instability. If an amplitude instability is created, this amplitude instability should provide gain to the most intense initial fluctuation and loss to the others. In addition, this amplitude instability should be shorter than the minimum size of the initial fluctuation.

In lasers where the upper state lifetime is short in comparison to the cavity round-trip time (principally dye lasers) these requirements on the introduced nonlinearity are easily met

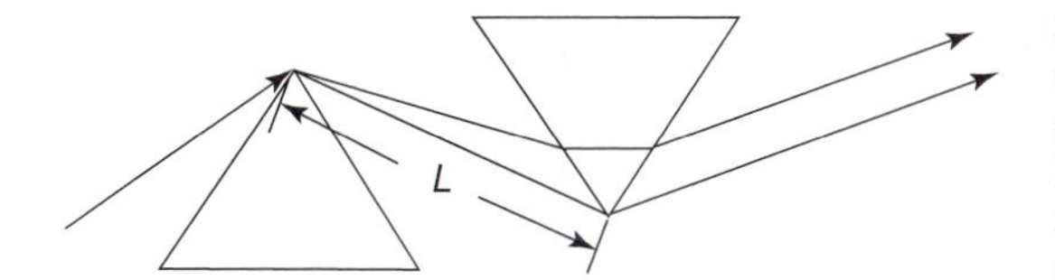

Figure 11.26 Prisms are the most common way of introducing GVD compensation. A prism geometry can be arranged so that the blue components transverse the prism path in a shorter length of time than the red components.

by using a slow saturable absorber. The combined action of the rapid saturable gain (due to the laser) and the slower saturable absorption (due to the dye) creates a gain window that follows the pulse width down to the dephasing time of the dye (usually tens of femtoseconds).

To illustrate this in more detail, visualize a short spontaneous fluctuation. As the fluctuation goes through the saturable absorber, the intensity of the pulse will bleach the absorber and the absorber loss will drop dramatically. As the fluctuation goes through the gain material (with a short upper state lifetime), it will dynamically reduce the available gain for a period following the pulse. Thus, there is a window between the drop in loss due to the dye bleaching and the drop in gain due to dynamic gain saturation. This window favors the more intense pulse (see Figure 11.27).

Now consider the situation with a typical solid-state laser. In solid-state lasers, the upper state lifetime is usually long and the lasers do not experience dynamic gain saturation on time scales compatible with mode-locked pulses. Thus, passive mode-locking techniques in solid-state lasers rely on saturable absorber dyes with extremely short lifetimes. In these systems, only the dynamic bleaching of the dye works to shorten the pulse. These systems are also limited by the finite-state excited lifetime of the saturable absorber dye. Such lifetimes are picoseconds in most organic dyes. Even the most exotic semiconductor saturable absorbers have lifetimes in the hundreds of femtoseconds (and are severely wavelength dependent).

However, it is also possible to mode-lock a laser using spatial or temporal SPM. SPM has the interesting characteristic that it almost instantaneously follows the variation in the optical field intensity. This ultra-fast effect can be transformed into a saturable absorber response by inclusion of compensating elements in the laser cavity.

For example, spatial SPM (self-focusing) can be used to create an intensity-dependent loss. Spatial SPM will introduce an intensity-dependent change in both the location and the width of the resonator mode beam waist. By introduction of a hard or soft aperture, the self-focusing effect can be converted into an intensity-dependent loss. (The process is also referred to as Kerr-lens mode-locking or KLM.)

Kerr-lens mode-locking can be used in many solid-state lasers, either by relying on the intrinsic self-focusing of the laser material or by adding a self-focusing material. In many cases (for example, Ti:sapphire) it is not even necessary to include a true aperture.

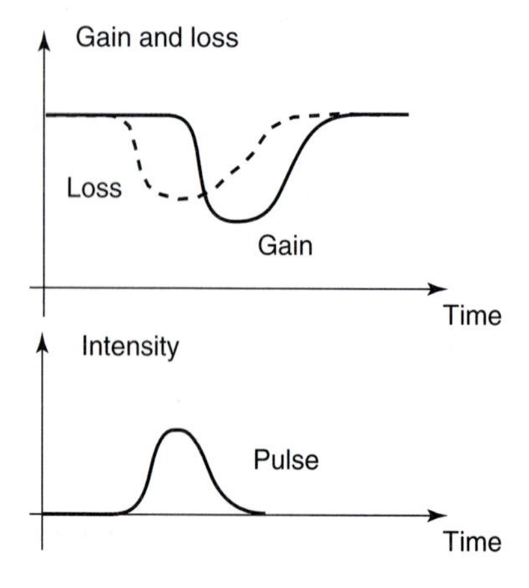

Figure 11.27 In lasers where the upper state lifetime is short in comparison to the cavity round-trip time, there is a window between the drop in loss due to the dye bleaching and the drop in gain due to dynamic gain saturation. This window favors the more intense pulse.

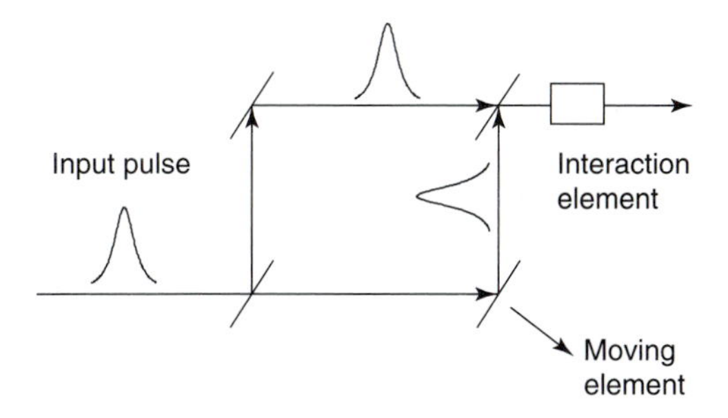

Figure 11.28 The majority of femtosecond pulse measurement techniques depend on the use of two pulses, with one delayed in time with respect to the other.

The higher gain seen by a smaller mode will experience gain saturation and this will serve as a soft aperture in the cavity.

11.5.3 Measuring Femtosecond Pulses

Once laser pulse widths drop below a few nanoseconds, it becomes difficult to measure them with conventional photodetectors and electronics. Thus, a variety of elegant techniques have appeared to measure these short pulses.

The majority of these techniques depend on use of two pulses, with one delayed in time with respect to the other (see Figure 11.28). These two pulses are typically created by a beam splitter and the relative time delay introduced by a moving mirror (such as a mirror on a speaker). The pulses are then allowed to re-interact in some way, and a trace obtained form this interaction. By changing the position of one pulse temporally with respect to the other, an autocorrelation pattern can be obtained (admittedly, at one point per pulse).

There are a number of possible interaction elements. Two of the most common are second harmonic crystals and two-photon phosphors. In the second-harmonic method, two arms of an interferometer are arranged so that the pulses have orthogonal polarizations. A second-harmonic crystal is cut so that harmonic light is only produced when both polarizations are present. The method thus measures the overlap of the pulse with its delayed replica. The two-photon method is somewhat simpler in concept than the second-harmonic method. In the two-photon method, the interaction material is simply a two-photon phosphor. When two photons are simultaneous absorbed in the fluorescent material, a single photon of twice the frequency is emitted and detected by a camera or photodetector.

11.5.4 Colliding Pulse Mode-Locking

Early work in femtosecond dye lasers primarily focused on resonator design. A large number of designs were proposed and tested. However, in 1981, Fork et al. came up with the seven-mirror colliding pulse mode-locking (CPM) ring laser.[61,62] CPM rings are so successful that all femtosecond dye lasers since then use some form of CPM ring geometry.

A typical set-up for a CPM system is shown in Figure 11.29. The laser gain material is a flowing jet of rhodamine 6G dye excited by an argon laser. The saturable absorber is a thin jet (10 μm) of absorber dye.

[61] R. L. Fork, B. I. Greene, and C. V. Shank, *Appl. Phys. Lett.* 38:671 (1981).

[62] R. L. Fork, C. V. Shank, R. Yen, and C. A. Hirlimann, *IEEE J. of Quantum Electron.* QE-19:500 (1983).

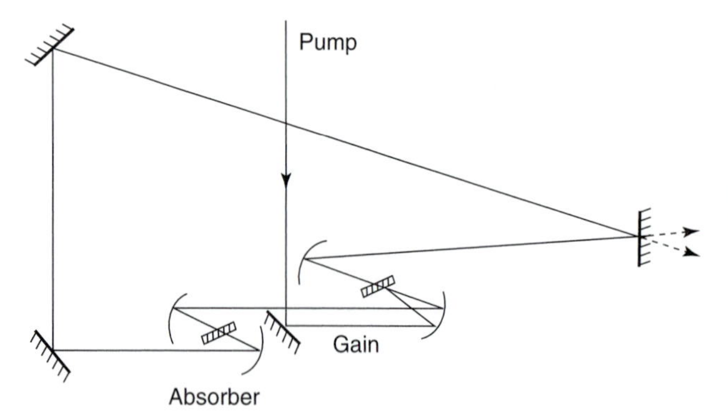

Figure 11.29 In colliding pulse mode-locking, two counter-propagating pulses are synchronized so as to overlap in a saturable absorber dye. (From R. L. Fork, C. V. Shank, R. Yen, and C. A. Hirlimann, "Femtosecond Optical Pulses," *IEEE J. of Quantum Electron.* QE-19:500 (1983). ©1983 IEEE.)

In CPM, two counter-propagating pulses are synchronized so as to overlap in a saturable absorber dye. The interference of the two pulses will create a standing wave pattern in the saturable absorber dye. The minimum loss condition in the dye corresponds to the maximum constructive interference for the laser pulses.

The system is self-synchronizing because the lowest loss condition always occurs when the two counter-propagating pulses overlap in the saturable absorber. Other conditions are more lossy and rapidly lose out to the lowest loss situation.

Competition between pulses for gain is minimal because the gain recovery time is swift in comparison to the round-trip cavity time. The interval between pulse arrival times in the gain media can be made large and equal for both pulses by separating the gain and absorber jets by one-quarter of the round-trip path length. (The idea here is for the pulses to overlap in the saturable absorber dye jet, but not in the gain media.)

A figure of merit for saturable absorber dye mode-locking is the ratio of the optical field intensity which saturates the absorber, to the optical field intensity which saturates the gain material. The higher this ratio, the more stable the laser. CPM rings offer a minimum of a factor of two improvement over conventional resonators (the factor of two is because two pulses saturate the absorber, while only one pulse saturates the gain material). An additional improvement results from the nonlinearity of the saturable absorber dye. Finally, the increased stability of CPM rings permits removal of dispersive cavity elements which may broaden the pulse. The net result is that CPM rings reduce the pulse width by a factor of 4 to 10 over conventional resonators.

11.5.5 Grating Pulse Compression

Grating pulse compression is a classical technique for reducing the temporal length of pulses by using self-phase modulation (see Figure 11.30).

An input pulse is directed through a dispersive material which broadens the spectrum of the pulse through self-phase modulation. (Single-mode fiber is the usual dispersive material because the fiber yields spatially uniform frequency broadening which is almost entirely due to self-phase modulation.) The resulting pulse (of the same temporal width, but with a factor of 3 to 5 broader spectrum) is then reflected between two gratings. The grating pair introduces a phase shift, which is a quadratic function of frequency and is opposite in

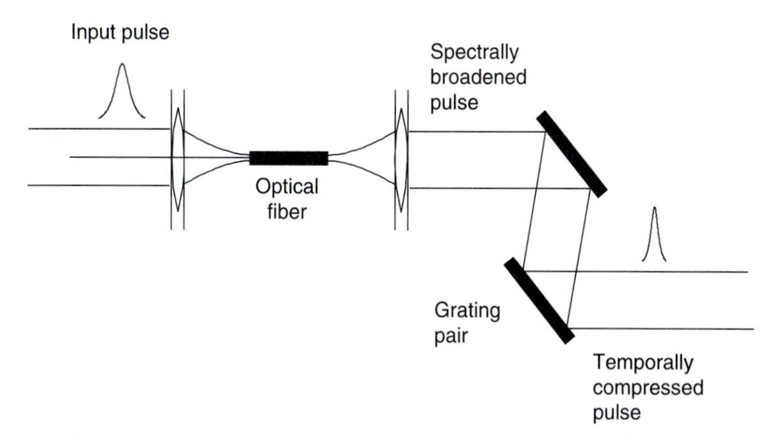

Figure 11.30 Grating pulse compression uses self-phase modulation in a fiber to broaden the spectrum of the pulse, followed by a grating pair to temporally compress the pulse.

sign to that of the dispersive medium. The end result is a temporally shorter pulse with a broader spectrum.

11.5.6 Solitons

Once the CPM ring laser geometries became well-established, research began to focus on dispersive properties of the resonator. In 1984, Martinez et al. proposed the idea that balancing group velocity dispersion (GVD) against self-phase modulation (SPM) might provide soliton-like pulse shaping and further reduce the temporal pulse width.[63] However, Martinez et al. observed that this effect was unlike to occur in conventional CPM ring lasers because of a competition between a negative contribution to the intracavity SPM, which arises from the time-dependent saturation of the absorber dye, and the positive SPM, which arises from the fast Kerr effect in the dye solvent. This causes an intracavity total SPM that is small and negative. Their suggestion was that one of these contributing sources would need to be reduced for the soliton behavior to be observed.[64]

This concept of balancing the GVD against the SPM to form solitons was first demonstrated by Valdmanis et al. in 1985 in a six-mirror CPM ring cavity with four prisms used as dispersion tuning elements[65,66] (see Figure 11.31). In this configuration, four fused-silica prisms were used with the beam running through the apex of each prism. The dispersion was tuned by changing the length L. With excess negative GVD, the laser maintained stable operation, but the pulse width gradually increased. For excess positive GVD, the laser was either not stable or ran with a stable mode but with long pulse widths. Valdmanis et al. hypothesized that the fast Kerr effect in the dye solvent was compensating for the negative

---

[63] O. E. Martinez, R. L. Fork, and J. P. Gordon, *Opt. Lett.* 9:156 (1984).

[64] O. E. Martinez, R. L. Fork, and J. P. Gordon, *J. Opt. Soc B,* 2:753 (1985).

[65] J. A. Valdmanis, R. L. Fork, and J. P. Gordon, *Opt. Lett.* 10:131 (1985).

[66] J. A. Valdmanis and R. L. Fork, *IEEE J. of Quantum Electron.* QE-22:112 (1986).

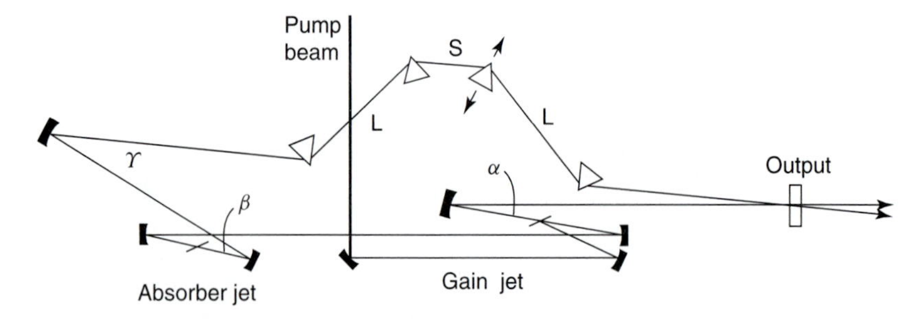

Figure 11.31 The concept of balancing the group velocity dispersion against the self-phase modulation to form solitons was first demonstrated by Valdmanis et al. in 1985 in a six-mirror CPM ring cavity with four prisms used as dispersion tuning elements. (From J. A. Valdmanis and R. L. Fork, "Design Considerations for a Femtosecond Pulse Laser Balancing Self Phase Modulation, Group Velocity Dispersion, Saturable Absorption, and Saturable Gain," *IEEE J. of Quantum Electron.* QE-22:112 (1986). ©1986 IEEE.)

introduced GVD by the prisms and thus creating the soliton formation. In essence, the Kerr effect generated a pulse with a positive chirp (a positive linear chirp has a phase greater than zero and can be considered qualitatively to have the red light coming first and the blue light coming second), which required the addition of a negative GVD to generate pulse compression.

11.5.7 Kerr-Lens Mode-Locking (KLM) in Ti:Sapphire

Prior to 1991, a variety of mode-locking techniques had been applied to Ti:sapphire. These included synchronous pumping, acousto-optic modulation, passive mode-locking, injection seeding, and additive pulse mode-locking. The only common feature was that the techniques were difficult and not very successful (the shortest pulse width was approximately 300 femtoseconds using injection seeding).

However, in 1991, Spence et al. demonstrated a new mode-locking technique in Ti:sapphire.[67] This technique was the first demonstration of Kerr-lens mode-locking and had the elegant feature that the critical nonlinearity was produced by the Ti:sapphire crystal itself (rather than by an additional optical element).

The cavity naturally mode-locked down to 100 fs and was astonishingly simple (containing only two prisms for dispersion compensation; see Figure 11.32). The simplicity of the set-up meant that his results were not well accepted until the same effect was confirmed in Ti:sapphire and observed in a variety of other lasers including Cr:LiSAF, Cr:LiCAF, Nd:glass, Nd:YAG, Nd:YLF, and Cr:forsterite (for a more complete review of Kerr lens mode-locking, see Krausz[68]).

The demonstration of self-mode-locking by Spence, coupled with the recognition of the importance of dispersion compensation in femtosecond pulse lasers, sparked a flurry of new activity in the early 1990s. Cavity designs simplified dramatically and research effort

[67] D. E. Spence, P. N. Kean, and W. Sibbett, *Opt. Lett.* 16:42 (1991).

[68] F. Krausz, M. Fermann, T. Brabec, P. Curley, M. Hofer, M. Ober, C. Spielmann, E. Wintner, and A. Schmidt, *IEEE J. of Quantum Electron.* 28:2097 (1992).

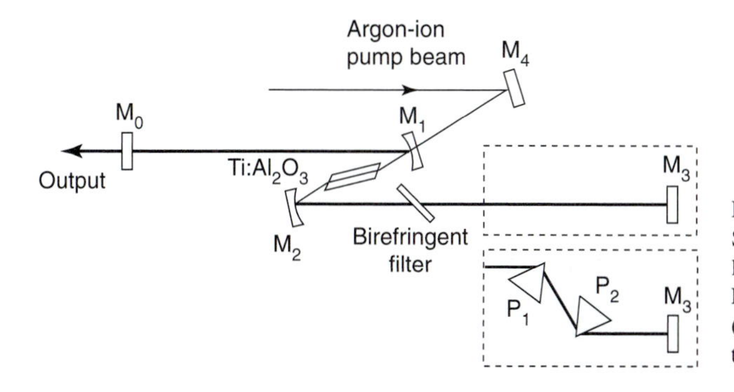

Figure 11.32 In 1991, Spence, Kean, and Sibbett first demonstrated Kerr-lens mode-locking in Ti:sapphire. (From D. E. Spence, P. N. Kean, and W. Sibbett, *Opt. Lett.* 16:42 (1991). Reprinted with the permission of the Optical Society of America.)

moved toward more sophisticated methods of dispersion compensation. The understanding of subtleties in material dispersion quickly led to Ti:sapphire lasers producing pulses in the 10 fs range without external pulse compression.[69, 70] The extraordinary mechanical simplicity of modern femtosecond pulse Ti:sapphire laser systems is stimulating commercial interest.[71]

11.5.8 Coherent Mira Femtosecond Lasers

The Coherent Mira lasers are ultrashort pulse Ti:sapphire lasers. The Mira 900-F is a femtosecond pulse laser with output pulse widths from 100 to 200 fs. The laser is tunable from less than 700 nm to more than 1000 nm by changing mirror sets. The Mira 900-P is very similar in overall design to the 900-F, except the optical cavity has been optimized for operation near 1 to 2 ps (see Figure 11.33).

The optical cavity for the femtosecond Mira 900-F is illustrated in Figure 11.34. The pump beam enters at M4, and mirrors M4 and M5 are transparent to the pump light. Mirrors M2, M3, M4, M5, M6, and M7 are highly reflective over the Ti:sapphire wavelengths. Mirror M1 is the output mirror for the laser. The cavity can also be run as a cw cavity by simply moving prism P1 and using the mirror combination M1, M2, M3, M4, M5, M8, and M9.

The Mira 900-F is typically pumped by an 8 to 15 W argon-ion laser. Average output power ranges from 270 mW (LW mirror set pumped by an 8 watt argon-ion laser) to 1100 mW (SW mirror set pumped by a 14 watt argon-ion laser). The repetition rate is 76 MHz with a beam diameter of 0.7 to 0.8 mm and a divergence of 1.5 to 1.7 mrad.

The Mira 900-F uses Kerr-lens mode-locking (see Section 11.5.7), where the nonlinear element is the Ti:sapphire crystal. A hard aperture is provided, rather than relying on the soft aperture of the nonlinear focusing. GVD compensation (see Section 11.5) is accomplished using a pair of prisms P1 and P2. A birefringent filter (see Section 11.4.2) is incorporated for broadband tuning.

Femtosecond Ti:sapphire lasers using GVD compensation do not start mode-locking spontaneously. Some transient change in cavity length is typically required to initiate the

[69] M. Asaki, C. Huang, D. Garvey, J. Zhou, H. Kapteyn, and M. Murnane, *Opt. Lett.* 18:977 (1993).

[70] I. Christov, M. Murnane, H. Kapteyn, J. Zhou, and C. Huang, *Opt. Lett.* 19:1465.

[71] W. Knox, "Practical Lasers Will Spawn Various Ultrafast Applications," *Laser Focus World* June: 135–41 (1996).

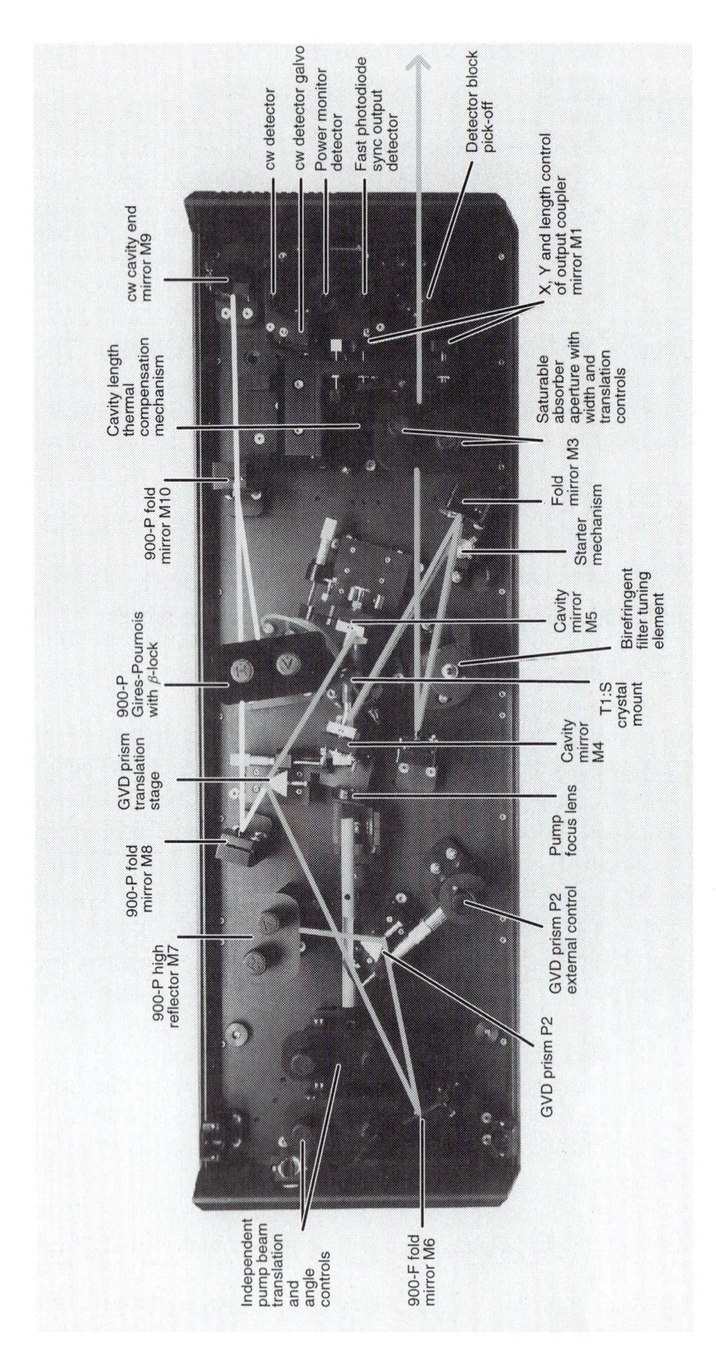

Figure 11.33 The Mira 900-P is a picosecond pulse Ti:sapphire laser. (Courtesy of the Coherent, Inc., Laser Group, Santa Clara, CA)

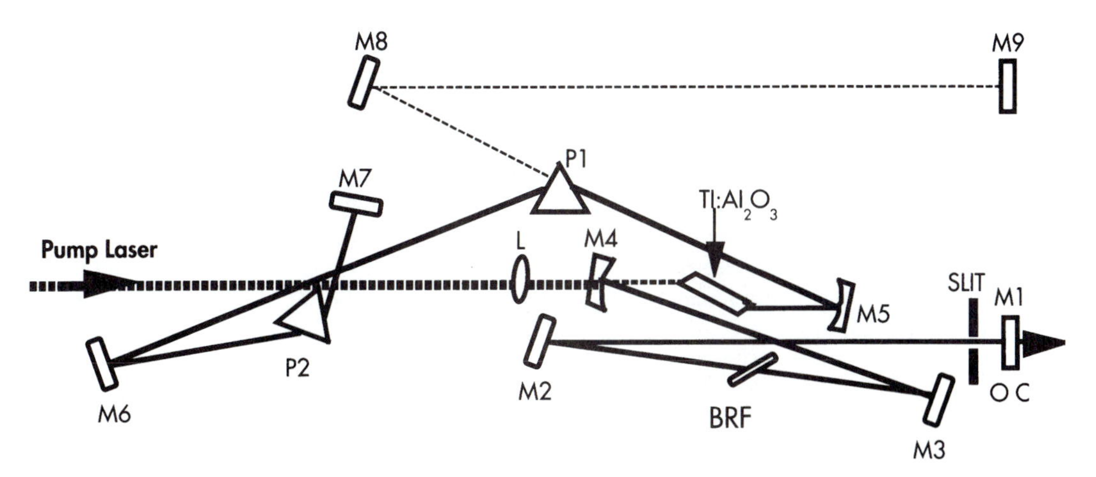

Figure 11.34 A schematic of the Mira femtosecond laser cavity. (Courtesy of the Coherent, Inc., Laser Group, Santa Clara, CA)

mode-locking process. (Ti:sapphire researchers typically start the lasers mode-locking by tapping on a mirror!) The Mira laser incorporates a novel transient-length variation technique that creates a sufficiently short pulse to start the mode-locking process.

The optical cavity for the picosecond Mira 900-P is illustrated in Figure 11.35. The pump beam enters at M4, and mirrors M4 and M5 are transparent to the pump light. Mirrors M2, M3, M4, M5, M8, and M10 are highly reflective over the Ti:sapphire wavelengths. Mirror M1 is the output mirror for the laser. The cavity can also be run as a cw cavity by simply moving prism P1 and mirror M10 and using the mirror combination M1, M2, M3, M4, M5, M8, and M9.

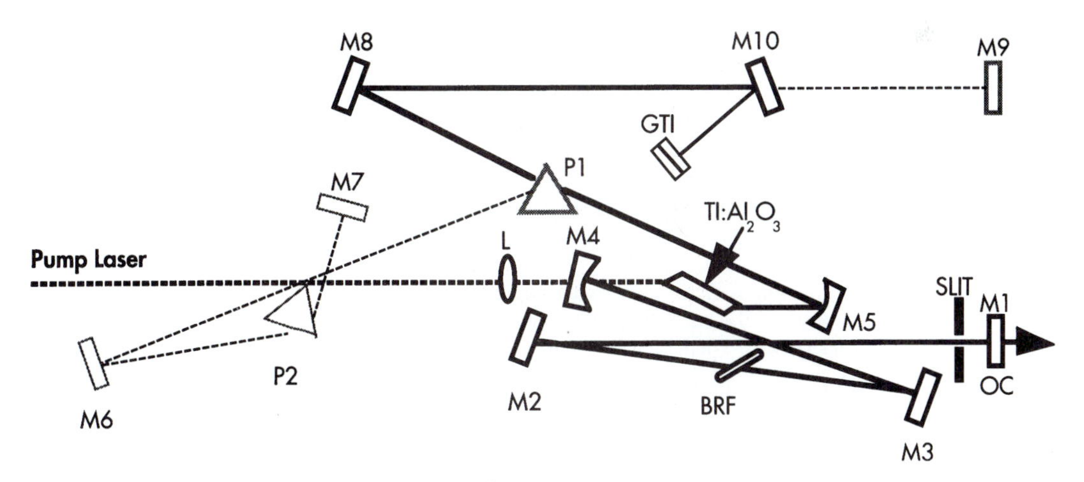

Figure 11.35 A schematic of the Mira picosecond laser cavity. (Courtesy of the Coherent, Inc., Laser Group, Santa Clara, CA)

The Mira 900-P is typically pumped by an 8 to 15 W argon-ion laser. Average output power ranges from 270 mW (LW mirror set pumped by an 8 watt argon-ion laser) to 1300 mW (SW mirror set pumped by a 14 watt argon-ion laser). The repetition rate is 76 MHz with a beam diameter of 0.7 to 0.8 mm and a divergence of 1.5 to 1.7 mrad.

The Mira 900-P also uses Kerr-lens mode-locking (see Section 11.5.7), where the nonlinear element is the Ti:sapphire crystal. Again, a hard aperture is provided rather than relying on the soft aperture of the nonlinear focusing alone. However (unlike the Mira 900-F) GVD compensation (see Section 11.5) is accomplished by removing prism P1 and using a Gires-Tournois interferometer.[72] A birefringent filter (see Section 11.4.2) is incorporated for broadband tuning.

EXERCISES

Overview of transition-metal solid-state tunable lasers

11.1 (design) Obtain a copy of B. D. Guenther and R. G. Buser, *IEEE Journal of Quantum Electronics* QE:18:1179 (1982). Read it and compare the predictions of 1982 with the actual development of tunable solid-state lasers between 1982 and today. Your answer should include:
- **(a)** a brief summary (one or two paragraphs) on how it was perceived in 1982 that tunable lasers would evolve,
- **(b)** a brief summary (one or two paragraphs) on the actual evolution, and
- **(c)** a table comparing the predicted development with the actual development.

11.2 (design) Obtain a copy of C. Pollock, D. Barber, J. Mass, and S. Markgraf, *IEEE Journal of Selected Topics in Quantum Electronics* 1:62 (1995). Read it and summarize the predictions for the future in the development of solid-state lasers. Your answer should be in the form of:
- **(a)** a few sentences describing the current perception on solid-state laser development, and
- **(b)** a table or list summarizing future developments.

Transition-metal solid-state tunable laser materials

11.3 (design) Obtain manufacturers' data sheets for ruby, Nd:YAG, alexandrite, and Ti:sapphire. (You may wish to coordinate with your classmates to avoid having a large number of people contacting the same vendor.) Construct a table comparing the optical and physical properties of these materials. Compare and contrast the materials. Your answer should include:
- **(a)** a summary table, and
- **(b)** several paragraphs discussing the relative advantages and disadvantages of the materials.

Ring lasers

11.4 (design) In using a HeNe laser as an alignment laser, it is quite common to observe a large oscillation when the back reflection from a flat optical component is redirected back into the alignment HeNe. Explain what is going on.

11.5 (design) Consider an optical system composed of a polarizer, a Faraday rotator, and a half-wave plate. Further assume that the Faraday rotator has been built so that it changes the linear polarization by 20 degrees. (For example, for a vertically polarized input CCW wave, the direction of the linear polarization emerging from the Faraday rotator is 20 degrees from the

---

[72] Anthony E. Siegman, *Lasers* (Mill Valley, CA: University Science Books, 1986), pp. 348–49.

normal.) Using diagrams similar to Figure 11.17, explain the effect that this element would have if installed in a typical ring laser.

11.6 (design) Creating unidirectional oscillation is a very important design issue in ring lasers. The text mentions using an external cavity or a Faraday rotator to create unidirectional oscillation. Propose at least one other method for achieving unidirectional oscillation in a ring laser. Your answer should include:
(a) a sketch of your idea,
(b) a reference if it is not an original idea, and
(c) a few sentences discussing the idea.

Birefringent crystals

11.7 Assume that you have two identical polarizers oriented with their polarization axes at 90 degrees to each other. Assume you have an incident unpolarized white light source of an intensity of 100 watts/cm^2. Assume both polarizers have Δ_0=0.5 and Δ_{90}=5.5 over the range of the white light source. How much light is transmitted through the polarizers? Now, add a third identical polarizer in-between the two crossed polarizers, oriented at 45 degrees to each original polarizer. How much light is now transmitted through the three polarizers? How much light would be transmitted if the three polarizers were perfect? Explain.

11.8 (design) Use foam core and pins to construct a model of the indicatrix for a negative uniaxial crystal. (This can be done by cutting two foam-core ellipses with n_e and no major axes, and one foam-core circle with an n_o radius. Then, one of the ellipses is cut in half, and one is cut in quarters. The structure is then reassembled as an indicatrix using the pins. The model can be further enhanced by adding a k-vector and the associated ellipse using a different color of foam core.)

11.9 (design) Repeat the previous problem, but build a positive uniaxial indicatrix.

11.10 (design) Repeat the previous problem, but build an indicatrix for a biaxial crystal. Biaxial crystals have *two* optical axes. Locate them on the indicatrix once it is built.

11.11 Consider the general problem of propagation through a positive or negative uniaxial crystal. Determine an equation that allows the calculation of $n_o(\theta, \phi)$ and $n_e(\theta, \phi)$ for an arbitrary θ and ϕ given n_o and n_e of the original crystal.

11.12 Consider the following uniaxial crystals.

Crystal	n_o	n_e
Quartz	1.5443	1.5534
Rutile (TiO_2)	2.616	2.903
Calcite	1.6584	1.4864

Assume that you are propagating a beam at 35 degrees to the optic axis (as measured from the normal) and midway between the x- and y-axes (i.e., at 45 degrees to each axis). Determine the observed $n_o(\theta, \phi)$ and the $n_e(\theta, \phi)$ for all three systems.

Wave plates

11.13 Assume that you are on an interview for a job in optics and the interviewer hands you two identically appearing optical elements. She tells you that they are a quarter-wave plate and a polarizer. She asks you to figure out which is which. Describe what method you would use to distinguish between the elements using only items present in a conventional interviewer's office.

11.14 Consider a wave plate fabricated from calcite ($n_o = 1.6584$, $n_e = 1.4864$). How thick must the wave plate be in order to operate as a quarter-wave plate with a $\lambda/2$ phase shift? Assume the wave plate is to be used with a 632.8 nm laser.

Birefringent filters

11.15 Create a program that predicts the transmission of a birefringent filter. Use this program to calculate the transmittance of a three-stage filter fabricated from quartz. Assume the smallest element is 0.6 mm thick and that the ratio is 1:2:15. Assume the filter is installed at Brewster's angle and that the rotational angle is 50 degrees.

11.16 The calculated transmission function of a birefringent filter really does not illustrate the true value of the filters when installed inside a laser cavity. Combine the birefringent filter analysis of Section 11.4.2. with the pumping constraints discussed in Section 5.6 to predict the output intensity (as a function of wavelength) for a Ti:sapphire laser with a birefringent filter installed inside the laser. Assume the Ti:sapphire laser is the Coherent model 890, discussed in Section 11.4.3. Assume birefringent filter is the one in Exercise 11.15. Your answer should include:

(a) a brief (a few sentences) analysis of how you attacked the problem,
(b) a listing of your assumptions,
(c) a concise summary of your final equations, and
(d) a plot of intensity (y-axis) versus wavelength (x-axis) for the laser.

The Coherent model 890 and 899 Ti:sapphire lasers

11.17 (design) Obtain data sheets from two major manufacturers of cw Ti:sapphire lasers. Compare and contrast the laser systems. (You may wish to coordinate with your classmates to avoid having a large number of people contacting the same vendor.) Your answer should include:

(a) a table comparing the various features of the lasers, and
(b) several paragraphs discussing the relative strengths and weaknesses of the laser systems.

The design of femtosecond lasers

11.18 (design) Obtain copies of O. E. Martinez, R. L. Fork and J. P. Gordon, *Opt. Lett.* 9:156 (1984); J. A. Valdmanis, R. L. Fork, and J. P. Gordon, *Opt. Lett.* 10:131 (1985); and D. E. Spence, P. N. Kean, and W. Sibbert, *Opt. Lett.* 16:42 (1991). Each of these papers is representative of the state of the art at the time the paper was written. Read the papers, and develop a summary of the development of the state of the art in group velocity dispersion (GVD) compensation between 1984 and 1991. Your answer should include a brief (three to five paragraphs) summary of the development of the state of the state of the art in GVD compensation between 1984 and 1991.

11.19 (design) Measuring the width of femtosecond pulses is a real challenge! One type of indirect measurement technique is discussed in the text. Propose two other possible ways for measuring the pulse width of femtosecond pulses. Your answer should include

(a) two simple sketches of your proposed schemes for measuring the pulse width of femtosecond pulses, and
(b) a few sentences on each approach (with calculations as appropriate) describing the advantages and disadvantages of the approach.

11.20 (design) Do some library research (for example, check recent issues of trade journals such as *Laser Focus World* and *Photonics Spectra*, and recent issues of academic journals such as the *IEEE Journal of Quantum Electronics*, *Applied Physics Letters*, and *Optics Letters*) and determine the current state of the art in femtosecond-pulse lasers. What do you think is the shortest possible pulse length? Your answer should include:

(a) a one- to two-paragraph description of the state of the art in femtosecond pulse lasers (including references!), and

(b) a one- to three-paragraph summary of your opinion (supported by facts, references, and calculations) of the shortest possible pulse length from a femtosecond-pulse laser.

The Coherent Mira femtosecond lasers

11.21 (design) Obtain data sheets from two major manufacturers of femtosecond-pulse Ti:sapphire lasers. Compare and contrast the laser systems. (You may wish to coordinate with your classmates to avoid having a large number of people contacting the same vendor.) Your answer should include:

(a) a table comparing the various features of the lasers, and

(b) several paragraphs discussing the relative strengths and weaknesses of the laser systems.

12

Other Major Commercial Lasers

Objectives

Carbon dioxide lasers

- To summarize the generic characteristics of the CO_2 laser.
- To describe the various energy states of the CO_2 laser and to summarize how these states interact with each other.
- To summarize the sequence of historical events leading to the development of the CO_2 laser.
- To describe the major characteristics of waveguide versus free space CO_2 lasers.
- To describe the construction of a commercial waveguide CO_2 laser.

Excimer lasers

- To summarize the generic characteristics of the excimer laser.
- To describe the various energy states of the excimer laser and to summarize how these states interact with each other.
- To summarize the sequence of historical events leading to the development of the excimer laser.
- To describe the general design principles underlying excimer lasers. These include preionization, corona discharge circuitry, and main discharge circuitry.
- To describe the construction of a commercial excimer laser.

Semiconductor diode lasers

- To summarize the sequence of historical events leading to the development of the semiconductor laser.
- To describe the energy band structure of the semiconductor diode laser.

- To summarize the process of pumping the semiconductor diode laser with a PN-junction.
- To describe the process of creating a semiconductor laser cavity by cleaving the semiconductor material.
- To describe the similarities and differences between homostructure and heterostructure semiconductor diode lasers.
- To describe the importance of vertical and horizontal confinement in designing semiconductor laser structures.
- To describe the major vertical and horizontal confinement structures.
- To describe the general physical principles governing the design of quantum wells, with special emphasis on the importance of the width of the quantum well in determining the optical properties of quantum well laser diodes.

12.1 THE DESIGN OF CARBON DIOXIDE LASERS

CO_2 lasers operate over a series of vibrational and rotational bands in the regions 9.4 and 10.6 μm. They are both high-average-power and high-efficiency laser systems. Commercially available cw CO_2 lasers range in power from 6 watts to 10,000 watts, and custom lasers are available at even higher powers. Small (2 to 3 feet long) CO_2 lasers can produce hundreds of watts of average power at an efficiency of 10%. Larger CO_2 lasers can produce many kilowatts of cw power. CO_2 lasers are widely used in such diverse commercial applications as marking of electronic components, wafers, and chips; marking on anodized aluminum; trophy engraving; acrylic sign making; rapid prototyping of 3D models; cutting of ceramics, textiles, and metals; carpet, sawblade, and sail cutting; drilling; thin film deposition; and wire stripping (see Figure 12.1). They find application in the medical field for laser surgery, and in research for spectroscopy and remote sensing. Military applications include imaging, mapping, and range-finding. They have also been used in inertial confinement fusion as an alternative to large glass lasers.

CO_2 is a laser material totally unlike the materials discussed so far in this text. Conventional lasers lase off of electronic transitions between various atomic states. CO_2 lasers lase off molecular transitions between the various vibrational and rotational states of CO_2. Among other things, this means that CO_2 lasers have a longer wavelength and higher efficiency than most conventional lasers. Additional information on CO_2 lasers can be found in Cheo,[1] Duley,[2] Tyte,[3] and Witteman.[4] Additional information on high peak power and gas dynamic CO_2 lasers can be found in Anderson,[5] Beaulieu,[6] and Losev.[7]

---

[1] Peter K. Cheo, *Handbook of Molecular Lasers* (New York: Marcel-Dekker Inc., 1987).

[2] W. W. Duley, *CO_2 Lasers: Effects and Applications,* (New York: Academic Press, 1976).

[3] D. C. Tyte, *Advances in Optical Electronics, Vol. 1,* ed D. W. Goodwin, (New York: Academic Press, 1970), pp. 129–198.

[4] W. J. Witteman, *The CO_2 laser* (Berlin: Springer-Verlag, 1987).

[5] John Anderson, *Gasdynamic Lasers: An Introduction* (New York: Academic Press, 1976).

[6] J. A. Beaulieu, *Proc. IEEE* 59:667 (1971).

[7] S. A. Losev, *Gasdynamic Laser* (Berlin: Springer-Verlag, 1981).

Figure 12.1 Carbon dioxide laser cutting system. (Courtesy of Synrad, Mukilteo, WA, and Summagraphics)

12.1.1 Introduction to CO_2 Laser States

Consider the CO_2 molecule as depicted in Figure 12.2. There are three normal modes of vibration possible in this molecule: the symmetric stretch mode, the bending mode, and the asymmetric stretch mode. The states are labeled by a notation (p_1, p_2, p_3) where the subscripts refer to the various normal modes and where p is an integer corresponding to the number of quanta in the mode. Thus (001) is the state with one quanta in the asymmetric stretch mode and (200) is the state with two quanta in the symmetric stretch mode. The total vibrational energy of the CO_2 molecule is expressed as

$$E(\nu_1, \nu_2, \nu_3) = h\nu_1\left(p_1 + \frac{1}{2}\right) + h\nu_2\left(p_2 + \frac{1}{2}\right) + h\nu_3\left(p_3 + \frac{1}{2}\right) \tag{12.1}$$

where ν_1, ν_2, ν_3 represent the frequencies of the particular modes.[8]

[8]Amnon Yariv, *Optical Electronics*, 4th ed. (Philadelphia, PA: Saunders College Publishing, 1991), p. 242.

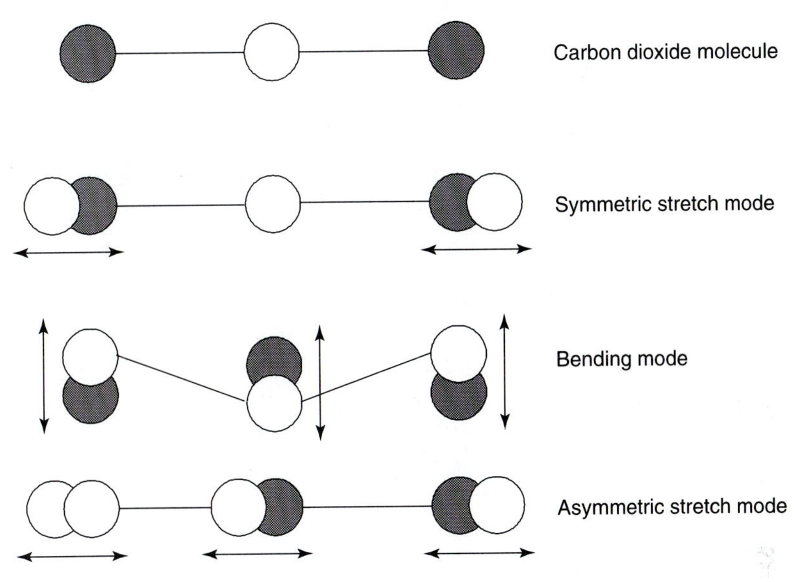

Figure 12.2 Normal modes of the carbon dioxide molecule.

The CO_2 molecules can also rotate, resulting in a series of closely spaced states characterized by the rotational quantum number J. The rotational energies of a given vibrational state i relative to the $J = 0$ level are given as

$$E_{i,J} = hc_o B_i J(J+1) - hc_o D J^2 (J+1)^2 \tag{12.2}$$

where B_i and D are constants.[9]

The principal laser transitions in CO_2 are the (001) to (100) 10.6 μm transitions and the (001) to (020) 9.4 μm transitions (see Figure 12.3). Each of the levels (001), (100), and (020) consists of a series of rotational states. Transitions in CO_2 occur between states where $J_{\text{odd}} \rightarrow (J+1)_{\text{even}}$ (termed the P-branch) and $J_{\text{odd}} \rightarrow (J-1)_{\text{even}}$ (termed the R-branch). (See Figure 12.4.)

If no wavelength discrimination is provided in the cavity, the P-branch of the (001) to (100) 10.6 μm transition will dominate. However, if wavelength selection is provided (by a grating, for example), it is possible to lase on any of the allowed P- or R-branch transitions. Notice, however, that since both the (001)$\rightarrow$(100) and the (001)$\rightarrow$(020) transitions share the same upper laser level, then the (001)$\rightarrow$(100) transition must be suppressed for the (001)$\rightarrow$(100) transition to lase.

The majority of CO_2 lasers contain a mixture of three gases (CO_2, N_2, and He) in a roughly 0.8:1:7 ratio.[10] The CO_2 is the laser gain material. The N_2 has only one excited mode (the symmetrical stretch mode) and the energy of the (1) N_2 vibration nicely aligns with the (001) upper state of the CO_2 molecule (see Figure 12.3). Since the N_2 vibrational states are metastable (very long lifetimes) the energy in the (1) N_2 transition (plus a little kinetic energy) can be transferred to a CO_2 molecule as a means of populating the (001)

---

[9] Amnon Yariv, *Quantum Electronics*, 2d ed. (New York: John Wiley and Sons, 1975), p. 213, Appendix 3.

[10] W. W. Duley, *CO_2 Lasers: Effects and Applications* (New York: Academic Press, 1976), p. 16.

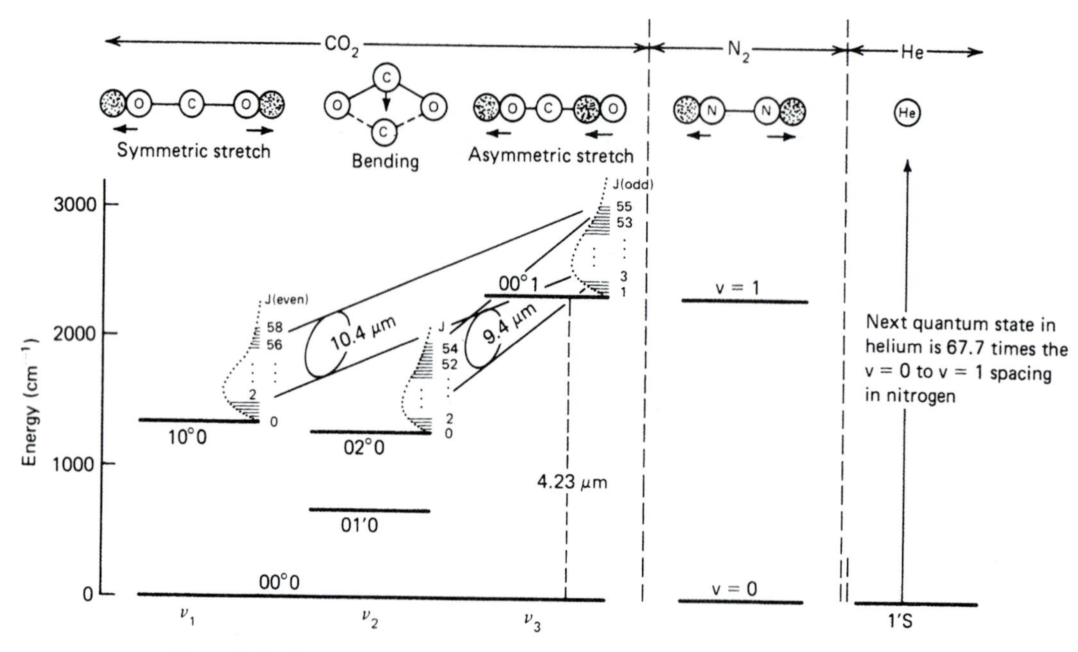

Figure 12.3 Laser states of the carbon dioxide molecule. (From LASER ELECTRONICS 2E. by VERDEYEN, J.T. ©1989, Figure 10.14, p. 336. Adapted by permission of Prentice-Hall, Inc., Upper Saddle River, NJ.)

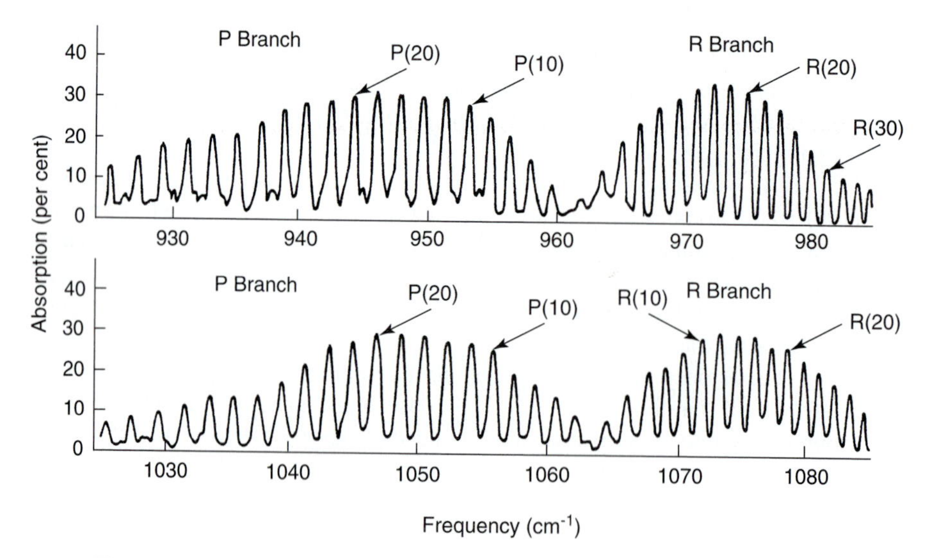

Figure 12.4 Absorption spectrum of the carbon dioxide molecule. (From E. F. Barker and A. Adel, *Phys. Rev.* 44:185 (1933))

upper CO_2 level (notice that the N_2–CO_2 energy transfer is very similar to the He–Ne energy transfer in HeNe lasers; see Section 9.1.3). The helium in the gas mixture provides cooling by means of thermal transfer to the walls (helium is a very thermally conductive gas). Helium also plays a role in optimizing the kinetic energy of the N_2 molecules for maximum energy transfer between the N_2 and CO_2.

Because of the metastable N_2 and the match between the (1) N_2 level and the (001) CO_2 level, the conversion efficiency between input electrical power to power in the upper laser state is 50 to 70%. Since the quantum efficiency is roughly 45%, this means that CO_2 lasers can operate at extremely high efficiencies (10 to 35%).

12.1.2 The Evolution of CO_2 Lasers

The first demonstration of laser action from CO_2 was reported by Patel in 1964.[11,12,13] The concept of using N_2 to transfer vibrational energy from the electrical discharge to the CO_2 was recognized by Legay and Legay-Sommaire in the same year[14] and the idea of incorporating helium for cooling was first proposed by Patel a year later.[15] During this period of rapid development on the CO_2 laser, Patel and other researchers were able to improve Patel's original 1 mW output to roughly 100 watts.[16,17,18]

The first CO_2 lasers were constructed from long tubes of glass where the desired laser mixture flowed through the glass tube (see Figure 12.5). Electrodes in the gas generated a plasma arc to excite the N_2 molecules into their symmetrical stretch mode. Although the very first demonstration of laser action from CO_2 used RF excitation, systems soon converted to DC excitation for increased power.[19]

The original glass tube CO_2 lasers operated at low pressures with the electrical discharge running longitudinally down the cavity. As a consequence, operating pressures were low due to the necessity to create and maintain a plasma over a long distance. However, in 1970, Beaulieu[20] first reported operation of an atmospheric pressure CO_2 laser by exciting the discharge transversely to the cavity (see Figure 12.6). These Transverse Excited Atmospheric (TEA) lasers offered higher gains and greater output powers than longitudinally excited lasers.

---

[11] C. K. N. Patel, *Phys. Rev. Lett.* 12:588 (1964).

[12] C. K. N. Patel, *Phys. Rev. Lett.* 13: 617 (1964).

[13] C. K. N. Patel, *Phys. Rev.* 136:A1187 (1964).

[14] F. Legay and N. Legay-Sommaire, *C. R. Acad. Sci.* 259B:99 (1964).

[15] C. K. N. Patel, P. K. Tien, and J. H. McFee, *Appl. Phys. Lett.* 7:290 (1965).

[16] C. K. N. Patel, *Phys. Rev.* 136:A1187 (1964).

[17] N. Legay-Sommaire, L. Henry, and F. Legay, *C.R. Acad. Sci.* 260B:3339 (1965).

[18] C. K. N. Patel, P. K. Tien, and J. H. McFee, *Appl. Phys. Lett.* 7:290 (1965).

[19] C. K. N. Patel, *Appl. Phys. Lett.* 7:15 (1965).

[20] A. J. Beaulieu, *Appl. Phys. Lett.* 16:504 (1970).

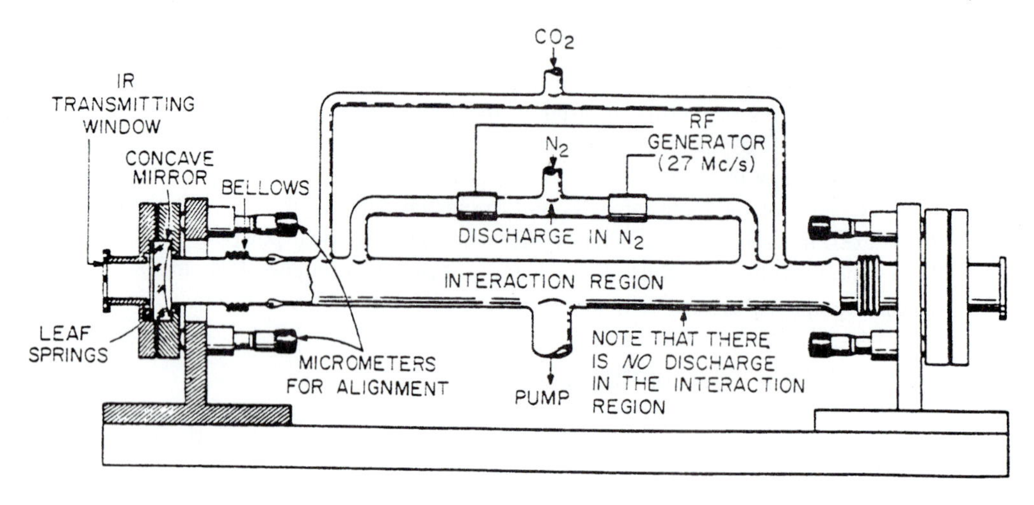

Figure 12.5 Early carbon dioxide laser construction. (From C. K. N. Patel, *Phys. Rev. Lett.* 13: 617 (1964). Reprinted with the permission of the author.)

The CO_2 laser Q-switches exceptionally well and Q-switched operation was reported in 1966 by a number of researchers including Flynn,[21,22] Kovacs,[23] Bridges,[24] and Patel.[25] However, the narrow bandwidth of CO_2 (approximately 50 MHz), means that physically long lasers are required to effectively demonstrate mode-locking. In spite of this difficulty, the first mode-locking of a conventional CO_2 laser was reported in 1968 by Caddes,[26] and Wood and Schwartz.[27] High-peak power can also be obtained from CO_2 lasers by pulsing or gain switching the lasers.[28] TEA lasers are especially well-suited for production of high-peak power CO_2 laser pulses.[29]

In a conventional CO_2 laser, the output power will increase as the gas flow is increased. This increased power is thought to be due to enhanced cooling and more effective removal of dissociation products such as CO and O_2 from the CO_2 discharge.[30] However, in many applications, it is not possible to support the peripheral equipment for handling flowing gases and a sealed laser configuration is required. In a sealed laser, the lack of gas flow means that some mechanism must be provided to regenerate the dissociated gas products

---

[21] G. W. Flynn, M. A. Kovacs, C. K. Rhodes, and A. Javan, *Appl. Phys. Lett.* 8:63 (1966).

[22] G. W. Flynn, L. O. Hocker, A. Javan, M. A. Kovacs, and C. K. Rhodes, *IEEE J. Quan. Elec.* QE-2:378 (1966).

[23] G. W. Flynn, L. O. Hocker, A. Javan, M. A. Kovacs, and C. K. Rhodes, *IEEE J. Quan. Elec.* QE-2:378 (1966).

[24] T. J. Bridges, *Appl. Phys. Lett.* 9:174 (1966).

[25] C. K. N. Patel, *Phys. Rev. Lett.* 16:613 (1966).

[26] D. E. Caddes, L. M. Osterink, and R. Targ, *Appl. Phys. Lett.* 12:74 (1968).

[27] O. R. Wood and S. E. Schwartz, *Appl. Phys. Lett.* 12:263 (1968).

[28] A. E. Hill, *Appl. Phys. Lett.* 12:324 (1968).

[29] W. W. Duley, *CO_2 Lasers: Effects and Applications* (New York: Academic Press, 1976), Chapter 2.

[30] Tyte, D. C., in *Advances in Optical Electronics,* Vol 1, ed D.W. Goodwin (New York: Academic Press, 1970), pp. 167–168.

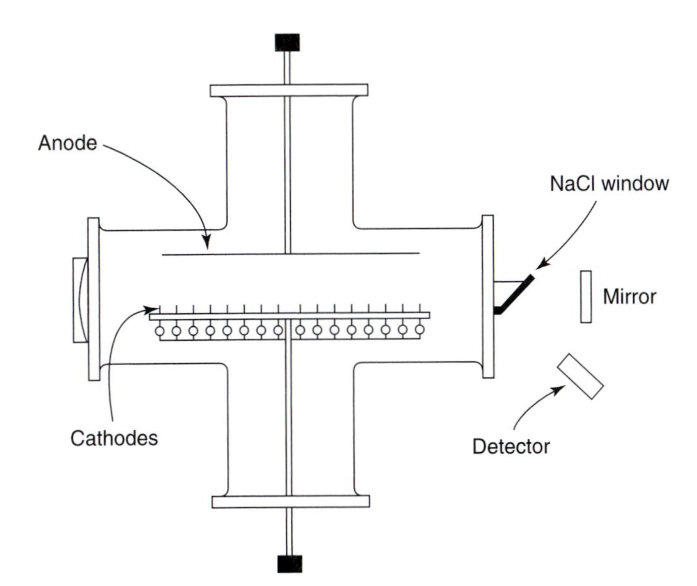

Figure 12.6 A schematic of an early Transverse Excited Atmospheric (TEA) laser. (Reprinted with permission from A. J. Beaulieu, *Appl. Phys. Lett.* 16:504 (1970). ©1970 American Institute of Physics.)

(particularly the oxygen species) back into CO_2. If these products are permitted to react with the tube walls, the chemical equilibrium of the plasma is disturbed and additional dissociation products are formed. Various regeneration methods include adding additional gases, periodically heating the tube, or incorporating catalyst alloys on the electrodes. Sealed lasers demonstrating such regeneration methods were first developed by Wittman in 1965[31] and further developed by Wittman[32] and Carbone.[33]

The initial use of flowing gases to improve the output performance of CO_2 lasers led to the development of another fascinating way to pump CO_2. The basic idea is to begin with a hot equilibrium gas mixture and then to expand the mixture through a supersonic nozzle. This lowers the temperature and pressure of the gas mixture in a time short compared to the upper state lifetime. When this occurs, the upper laser level cannot track with the temperature and pressure changes and so remains at its initial values. In contrast, the lower level population drops dramatically. The result is a population inversion that extends some distance downstream of the supersonic nozzle (see Figure 12.7). Lasers using this type of pumping are called gas dynamic lasers and were first suggested by Konyukhov and Prokhorov[34] in 1966 and demonstrated by Gerry[35] and Konyukhov.[36]

The most spectacular forms of gas dynamic lasers are those run using jet or rocket engines as the pump source. The basic idea is to create a laser gas mixture by burning some type of fuel that generates the CO_2. The fuel source is often ignited with a methanol burner,

---

[31] W. J. Witteman, *Phys. Lett.* 18:125 (1965).

[32] W. J. Witteman, *IEEE J. Quan. Electron.* QE-5:92 (1969).

[33] R. J. Carbone, *IEEE J. Quan. Electron.* QE-5:48 (1969).

[34] V. K. Konyukhov and A. M. Prokhorov, *JETP Lett.* 3:286 (1966).

[35] E. T. Gerry, *IEEE Spectrum* 7:51 (1970).

[36] V. K. Konyukhov, I. V. Matrosov, A. M. Prokhorov, D. T. Shalunov, and N. N. Shirokov, *JETP Lett.* 12:321 (1970).

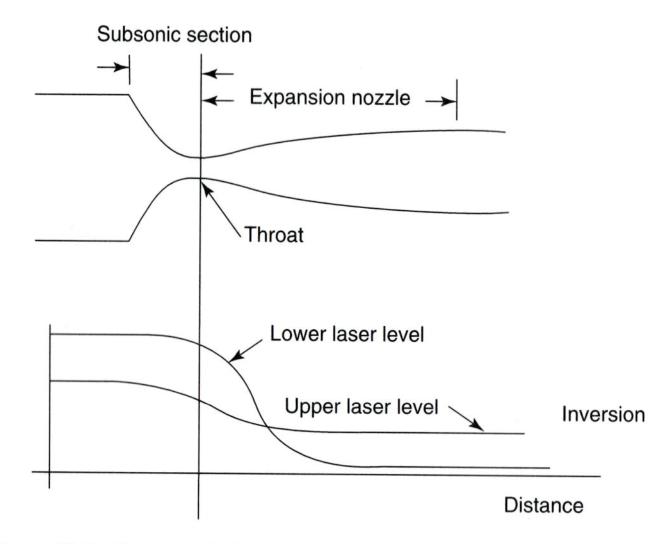

Figure 12.7 Gas dynamic lasers operate by creating a population inversion via gas expansion through a nozzle. (From E. T. Gerry, "Gasdynamic Lasers," *IEEE Spectrum* 7:51–58 (1970). ©1970 IEEE.)

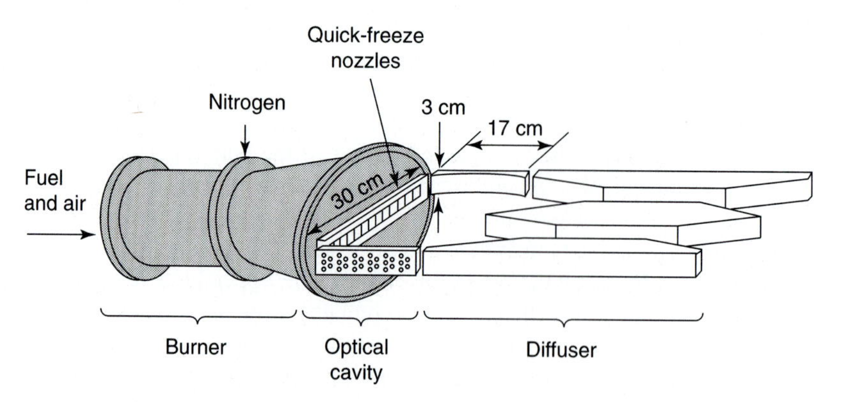

Figure 12.8 The most spectacular forms of gas dynamic lasers are those run using jet or rocket engines as the pump source. (From E. T. Gerry, *IEEE Spectrum* 7:51 (1970). ©1970 IEEE.)

which also injects water into the mixture. (The water is used to decrease the lifetime of the lower laser state.) Extra nitrogen is added to improve the excitation of the CO_2. The resulting mixture is then compressed by the engine and allowed to expand out through a series of supersonic nozzles. The optical cavity is then located sideways across the expansion chamber (see Figure 12.8).[37]

[37]E. T. Gerry, *IEEE Spectrum* 7:51 (1970).

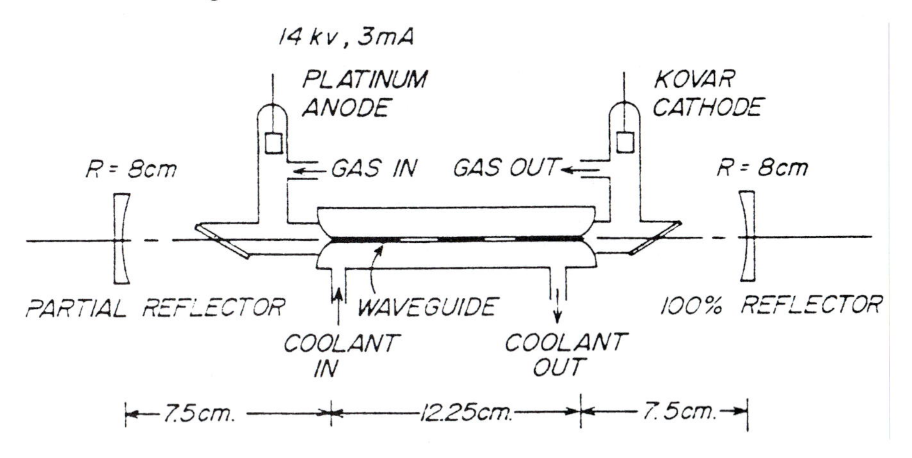

Figure 12.9 The construction of an early waveguide carbon dioxide laser. (Reprinted with permission from T. J. Bridges, E. G. Burkhardt, and P. W. Smith, *Appl. Phys. Lett.* 20:403 (1972). ©1972 American Institute of Physics.)

12.1.3 Waveguide CO_2 Lasers

One very good method for improving CO_2 laser performance is to decrease the bore size of the laser. This increases the number of gas collisions with the bore and significantly enhances the cooling rate (see Figure 12.9). If the electrodes are located transversely (rather than longitudinally) in the laser cavity, then the possibility also exists of using the electrodes themselves as an optical waveguide, thus permitting an even smaller bore size. The use of such a waveguide allows increased gas pressure with the attendant advantages of improved gain and larger linewidth. Operation in a waveguide mode also offers some additional advantages in alignment stability. The concept of a waveguide CO_2 laser was first proposed in 1964 by Marcatili and Schmeltzer[38] and later demonstrated by Steffen and Kneubuhl[39] and Smith.[40] Transverse-excited waveguide lasers are disclosed by Smith in U.S. Patent #3,815,047.[41]

Waveguide lasers use a small bore to confine the laser beam. The bore is itself an optical element, composed of two or four optically reflecting walls. Conventional mirrors are placed on either end of the cavity, but (unlike a conventional free space laser) these mirrors do not define a Gaussian beam in the cavity. Instead, the laser establishes various stable modes inside the bore (not unlike the modes in a laser fiber or a zig-zag slab laser). It is also possible to control the mode formation by introducing artifacts inside the bore that force the development of stable reflecting points.[42]

---

[38] E. A. J. Marcatili and R. A. Schmeltzer, *Bell Sys. Tech. J.* 43:1783 (1964).

[39] H. Steffen and F. K. Kneubuhl, *Phys. Lett.* 27A:612 (1968).

[40] P. W. Smith, *Appl. Phys. Lett.* 19:132 (1971).

[41] Peter W. Smith, "Transversely Excited Waveguide Gas Laser," U.S. Patent #3,815,047, June 4, 1974.

[42] Peter Laakmann, "Sealed Off RF-Excited Gas Lasers and a Method for Their Manufacture," U.S. Patent #5,065,405, November 12, 1991.

Waveguide lasers are typically differentiated from free space lasers by a number called the Fresnel number. This is defined as

$$F = \frac{a^2}{L\lambda_o} \tag{12.3}$$

where a is the beam radius (for a cylindrical laser) or 1/2 the beam width (for a square laser), L is the length, and λ_o the free space wavelength. A laser with a Fresnel number of less than 0.5 is a true waveguide laser. A laser with a Fresnel number of greater than about 10 is a true free space laser. Lasers with Fresnel numbers around 1 are intermediate lasers that have some of the features of both classes.[43]

Waveguide lasers are typically smaller, lighter, easier to align, and cheaper than their glass tube ancestors. They also have significantly lower operating voltages, as the gas discharge must only be sustained transversely across the bore (a few millimeters) rather than longitudinally along the tube (many centimeters).

12.1.4 A Typical Modern CO_2 Industrial Laser

The remainder of this chapter will focus on a family of sealed low-power CO_2 lasers representative of modern commercial lasers used for industrial laser machining applications such as cutting and marking (see Figure 12.10). The specific units under discussion are the very popular series 48 sealed CO_2 lasers manufactured by Synrad in Mukilteo, Washington, U.S.A. These lasers represent an excellent example of modular design and are available in three power levels (10W, 25W, and 50W). (See Figure 12.11.)

CO_2 lasers can be operated with flowing gases or in a sealed configuration. Sealed lasers (such as the series 48 lasers) have obvious advantages in the industrial workplace as they do not require complex gas handling systems. Sealed lasers are only commercially available up to approximately 250 watts. The cross-over point between sealed laser technology and flowing gases technology is roughly 1 kW, and driven primarily by size and manufacturing constraints.[44]

Water-cooling is another critical issue in CO_2 laser design. Although CO_2 lasers are exceptionally efficient, 10% efficiency still means that 90% of the input power ends up somewhere else, usually as heat in the chassis. If the laser gets too hot, then the lower state population increases, and the laser performance drops. With good heat sink design, sealed CO_2 lasers can be operated in air-cooled mode up to approximately 25 watts. Past that power level, water-cooling is typically required.[45]

Design and manufacture of the series 48 module. The basic series 48 module is described in U.S. Patent #5,065,405 (Peter Laakmann, "Sealed Off RF-Excited Gas Lasers and a Method for Their Manufacture," November 12, 1991) and the technology is discussed in U.S. Patent #4,805,182 (Peter Laakmann, "RF-Excited All-Metal Gas laser," February 14,

[43] Peter Laakmann, "Sealed Off RF-Excited Gas Lasers and a Method for Their Manufacture," U.S. Patent #5,065,405, November 12, 1991.

[44] Peter Laakmann, "Using Low Power CO_2 Lasers in Industrial Applications," Synrad Application Note.

[45] Peter Laakmann, "Using Low Power CO_2 Lasers in Industrial Applications," Synrad Application Note.

Figure 12.10 Typical products marked by a carbon dioxide laser. (Courtesy of Synrad)

1989). The key points of the design and manufacturing are described below and additional details may be found in the patents.

The basic series 48 module consists of two extruded aluminum electrodes and two extruded aluminum ground plane strips (see Figure 12.12). The inner surfaces of the electrodes and ground strips are optically reflective at 10.6 μm. (The electrodes are typically anodized with a 5 μm hard anodization to improve discharge stability and RF breakdown characteristics.[46]) The top and bottom electrodes are identical and measure approximately 1 cm by 2 cm by 40 cm long. The left and right ground plane strips are also identical and measure approximately 2 cm by 4 cm by 40 cm long. To reduce costs, the overall shape of the electrodes and ground planes is predefined by the extrusion process and only minor post-extrusion machining operations need to be performed.

The inner surfaces of the electrodes and the ground strips define the optical cavity of the laser. The bore of this cavity measures roughly 5 mm square, which gives the overall

[46]Y. F. Zhang, S. R. Byron, P. Laakmann, and W. B. Bridges, *Cleo '94*, 1994; *Tech. Digest Series*, Vol. 8, 94CH3463-7, pp. 358–9.

Figure 12.11 The Synrad series 48 sealed carbon dioxide lasers. (Courtesy of Synrad)

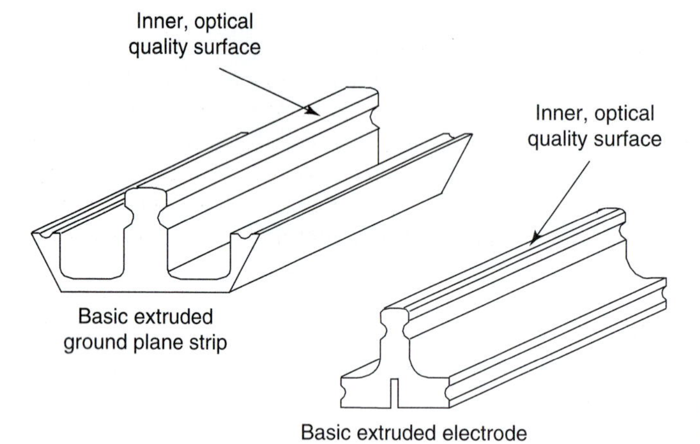

Figure 12.12 The Synrad electrodes and ground plane strips.

laser system a Fresnel number of approximately 1.5 and a diameter to length ratio of approximately 0.015. Thus, the laser operates in the intermediate regime between full waveguide operation and full free space operation. In this intermediate regime, only a fraction of the power of the optical wave interacts with the walls of the cavity. This fraction is high enough to obtain a good optical fill factor, but not so high as to create large losses due to absorption in the optical cavity structure.

If the optical surfaces of the electrodes and the ground strips are perfectly flat and aligned, then spurious reflections from the surfaces may cause aberrations in the output mode pattern. By deliberately introducing artifacts into the cavity geometry, these aberrations may be reduced or eliminated. Two types of artifacts have been found to be successful in reducing mode aberrations. The first artifact consists of slightly tapering the walls of the bore. In this method, the bore is made slightly larger near the end mirror than at the

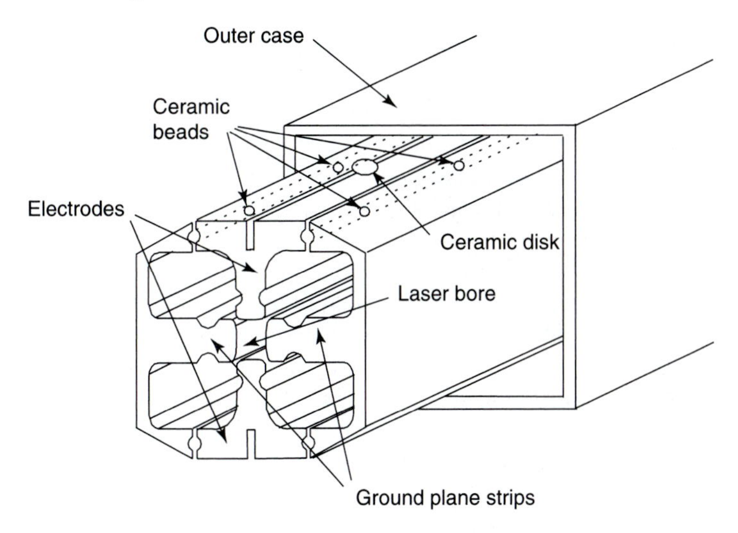

Figure 12.13 The Synrad series 48 cross-section.

front mirror. The taper angle is quite small, typically less than a milliradian. The second artifact consists of introducing small, sharp bends in the optical surfaces. The bends can be in one electrode and its adjacent ground strip, or in two opposing electrodes (or ground strips). If the bend is introduced into the electrode and its adjacent ground strip, then it is a bend on the order of 5 to 10 milliradians. If the bend is introduced into opposing electrodes (or ground strips), then it is on the order of 1 milliradian. These bends prevent reflections off all four walls from adding in phase to produce competing modes and parasitic oscillations.

The two electrodes and two ground planes are assembled in a clean room. The electrodes and ground plane strips are slipped into an outer case as shown in Figure 12.13. Small ceramic beads are used to isolate the electrodes from the ground plane strips. Larger ceramic disks are used to isolate the electrodes from the outer case. Electrical feedthroughs (which also serve as gas fill ports) are provided in the outer case.

A common cause of short lifetimes in CO_2 lasers is the accumulation of water vapor. Water vapor can migrate through the o-ring mirror seals, or be formed by hydrogen combining with dissociated oxygen. An effective way to minimize contamination due to water vapor is to introduce a getter into the cavity. A molecular sieve getter for removal of water vapor is typically inserted into the cavity during the initial assembly. Following the initial assembly, aluminum mirror mounts are welded to both ends of the tube. The laser module is then subjected to a high vacuum bake-out process. This process removes water as well as the majority of volatile contaminants such as hydrocarbons and hydrogen.

The bake-out process is followed by a passivation process where the laser module is exposed to an oxygen-helium plasma. The passivation process produces numerous oxygen species that react with the exposed aluminum parts and create aluminum oxides. (Al_2O_3, sapphire, is one of the many aluminum oxides formed by this process.) These oxides serve to prevent oxygen loss in the finished laser and extend the operating lifetime. The passivation process also serves to clean the bore of the laser and prevent contamination of the mirrors during operation.

The passivation is performed by generating an oxygen plasma within the laser bore. To create all the oxides necessary for effective passivation, the plasma is operated at several different temperatures, and at peak excitation powers that exceed those the laser will experience in operation.

After the passivation is complete, the laser mirrors are added to the module. Although nearly perfect mirror alignment is required for effective operation of a true waveguide laser, the best mode quality and output power performance of an intermediate regime laser is often obtained when the end mirrors are aligned slightly off-axis. The back mirror is a 100% dielectric coating on silicon (3 meter concave for the 10 watt series 48 laser). The front mirror is flat with a partially reflective dielectric coating on ZnSe (95% reflectivity for the 10 watt series 48 laser). The mirrors are mounted against an o-ring seal and three small screws provide alignment adjustment. (Although the o-ring seal at first glance seems like a potentially unsatisfactory element, Synrad experience has shown them to work quite well. The leak rate is approximately 1 cm^3 of helium per year, not the limiting factor in the laser lifetime. The o-ring mounts do not fail catastrophically, and provide a simple and straightforward means to make alignment adjustments.[47])

Once the mirrors have been installed, the laser is again subjected to a short vacuum bake-out. This bake removes water vapor introduced by the mirror mounting operation. The bake is relatively short (120°C for 8 hours for the 10 watt series 48 laser) to avoid damaging the passivation layer.

Once the bake-out is completed, the laser is carefully realigned and permanently filled with the operating gas mixture. The operating mixture consists of roughly 7% xenon, 10% CO_2, 13% N_2, and 67% He.[48] (The xenon is added to lower the overall electron temperature and improve the cross-section of the (1) transition in N_2).[49] The basic laser module is now complete.

The finished laser module can be mounted in one of several different housings. The most common is to mount the laser in a simple housing incorporating an integral heat exchanger. The laser tube fits in the bottom of the housing, and the RF electronics to drive the electrodes mounts in the top (see Figure 12.14).

Expanding the basic module. The basic 10 watt laser is designed to effectively scale to higher powers. Two electrodes, two ground strips, and an RF driver define a 10 watt module. Four electrodes (of the same length as the 10 watt electrodes), two ground strips (twice as long as the 10 watt ground strips), and two RF modules define the 25 watt module.

The two 25 watt modules are combined together in a simple but effective way to form the 50 watt module. The standard 25 watt module lases with its polarization vector

[47]P. Laakmann, *Lasers and Optronics* March: 35–41 (1989)

[48]P. Laakmann, "Using Low Power CO_2 Lasers in Industrial Applications," Synrad Application Note.

[49]W. J. Witteman, *The CO_2 Laser* (Berlin: Springer-Verlag, 1987), pp. 74–75.

Figure 12.14 Synrad CO_2 series 48 internal modules and housings. (Courtesy of Synrad)

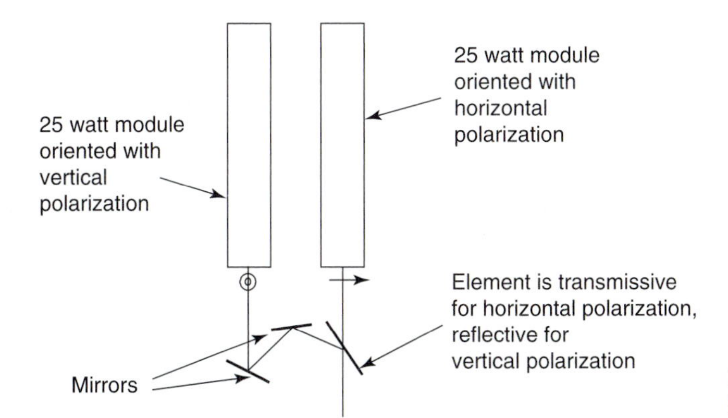

Figure 12.15 Combining the 25 watt beams together for the 50 watt unit.

parallel to the ground planes.[50] If one 25 watt module is installed with its polarization vector vertical, and the second installed with its polarization vector horizontal, then the orthogonally polarized beams can be combined at the output using a polarizing beam splitter (see Figure 12.15).

The same modular scaling strategies are applied to the RF drive electronics. The single 10 watt module uses one RF driver board. The 25 watt module uses two identical driver boards, and the 50 watt module uses four identical driver boards.

[50] Y. F. Zhang, S. R. Byron, P. Laakmann, and W. B. Bridges, *Cleo '94*, 1994; *Tech. Digest Series,* Vol. 8, 94CH3463-7, pp. 358–9.

Design of the RF driver board. Although the very early CO_2 lasers frequently used a radio frequency (RF) discharge,[51,52,53,54,55] this was considered by most researchers to be an inefficient and inconvenient excitation source.[56] It was not until 1977, when Katherine Laakmann began re-examining the use of RF-excitation for waveguide lasers,[57] that the potential value of RF-excitation for small sealed waveguide lasers became apparent.

Conventional CO_2 lasers use a longitudinal high voltage DC current to create the plasma. In a DC longitudinally excited laser, the output power is difficult to control electrically. Changing the voltage across the electrodes changes the character of the plasma. This, in turn, alters the plasma impedance and the ability of the plasma to excite the CO_2 molecules. Thus, a rather small change in the exciting voltage can result in large changes in the laser performance. Furthermore, the length and stability characteristics of the longitudinal excitation make rapid changes in the laser performance difficult.

In the late seventies, Laakmann Electro-optics (started by Peter and Katherine Laakmann) developed the transverse RF-excited CO_2 laser. The major advantage of transverse RF-excited lasers is that the output power can be electrically controlled over a wide range, at rates of up to 20 kHz. (Figure 12.16 illustrates the control of the output power by alteration of the duty cycle for 5 kHz and 20 kHz operation from a 65 watt Synrad CO_2 laser.) Other advantages of RF-excitation include significantly lower voltages (hundreds compared to thousands of volts) and the possibility of longer laser lifetimes due to reductions in gas-dissociation-related damage.

RF-excitation does have its costs. In particular, transistor controlled RF-excited devices have roughly half the wallplug efficiency of similar DC-excited units.[58] Additionally, the power supplies required for RF-excitation are typically more expensive than those in an equivalent DC-excited unit.

The excitation frequency of the RF discharge ranges from a value of approximately $v/2a$ to $50v/a$, where a is the width of the laser bore, and v is the drift velocity of electrons in the laser gas. The drift velocity ranges from approximately $5 \cdot 10^6$ to $1 \cdot 10^7$ cm/sec in a typical CO_2 laser gas mixture.[59] Thus, for a typical transversely excited RF discharge laser system, appropriate excitation frequencies lie in the UHF-VHF region.

In order to avoid radio interference with other services, certain ISM (Industrial, Scientific, Military) frequencies have been set aside by the FCC for industrial uses requiring

---

[51] C. K. N. Patel, *Phys. Rev.* 136:A1187 (1964).

[52] P. Barchewitz, L. Dorbec, R. Farrenq, A. Truffert, and P. Vautier, *C. R. Hebd. Seanc. Acad. Sci.* 260:3581 (1965).

[53] P. Barchewitz, L. Dorbec, A. Truffert, and P. Vautier, *C. R. Hebd. Seanc. Acad. Sci.* 260:5491 (1965).

[54] R. Farrenq, C. Meyer, C. Rossetti, L. Dorbec, and P. Barchewitz, *C. R. Hebd. Seanc. Acad. Sci.* 261:2617 (1965).

[55] C. Rossetti, R. Farrenq, and P. Barchewitz, *J. Chim. Phys.* 64:93 (1967).

[56] Tyte, D. C., in *Advances in Optical Electronics*, Vol. 1, ed D. W. Goodwin (New York: Academic Press, 1970), p. 172.

[57] Katherine Laakmann, "Waveguide Gas Laser with High Frequency Transverse Discharge Excitation," U.S. Patent #4,169,251, September 25, 1979.

[58] P. Laakmann, *Lasers and Optronics* (1989), pp. 35–41.

[59] Katherine Laakmann, "Waveguide Gas Laser with High Frequency Transverse Discharge Excitation," U.S. Patent #4,169,251, September 25, 1979.

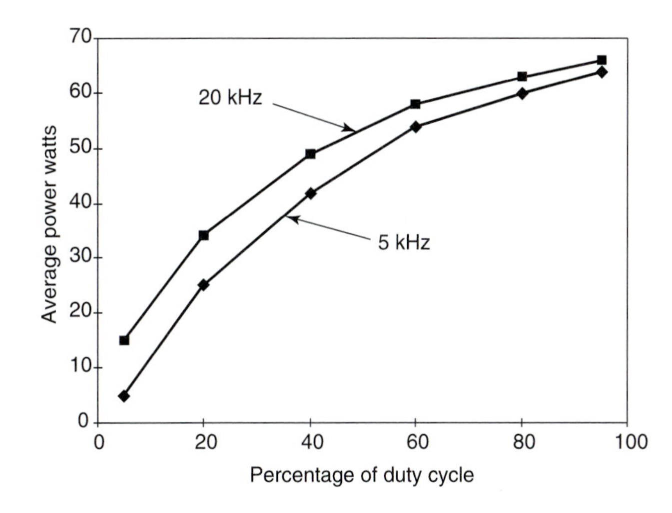

Figure 12.16 Output power versus duty cycle for the 57G Synrad laser. (Courtesy of Synrad)

large amounts of RF power. These ISM frequencies are used for larger RF CO_2 lasers and are 27.12 MHz and 40.68 MHz. Smaller lasers may operate outside of the ISM frequency limits if the shielding is sufficiently good so that RF leakage is below FCC minimums.[60]

Supplying power to the plasma of an RF-excited transversely pumped CO_2 waveguide laser requires careful design. The RF voltage source typically has a very low impedance, while the laser tube may have an impedance of hundreds to thousands of ohms at its operating frequency. A laser tube with a square bore 4.8 mm in width and 37 cm long, with a laser gas pressure of 60 torr; will have an impedance of approximately 200 ohms when operating with an output of approximately 15 watts.[61] The impedance increases as the power decreases, and is several thousand ohms if the plasma is not ignited.

An impedance matching network typical of the Synrad series 48 lasers is described in U.S. Patent #5,008,894, "Drive System for RF-Excited Gas Lasers," by Peter Laakmann. The basic concept is illustrated in Figure 12.17. An RF power supply output stage is connected through a transmission line to the pair of electrodes and ground strips forming the waveguide laser. The collector of the upper transistor in the push-pull output stage of the RF power supply is connected to the top electrode of the laser through the core of a 1/4 wavelength (at the laser operating frequency) transmission line. The lower transistor in the push-pull output stage is connected through the cladding of the transmission line to ground via a blocking capacitor. A coil connects the top electrode to the bottom electrode of the waveguide laser. (This coil serves to neutralize the capacitive reactance and to generate bi-phase excitation of the plasma.[62]) A transformer is used to match the impedance of the final stage to the preceding circuitry in the RF source.

In a small RF-excited CO_2 laser (such as the Synrad 10-50 watt series 48 lasers) the entire power supply and impedance matching networks can be integrated into a single

[60] P. Laakmann, "Using Low Power CO_2 Lasers in Industrial Applications," Synrad Application Note.

[61] P. Laakmann, "Drive System for RF-Excited Gas Lasers," U.S. Patent #5,008,894, April 16, 1991.

[62] P. Laakmann, "Electrically Self-Oscillating RF-Excited Gas Laser," U.S. Patent #4,837,772, June 6, 1989.

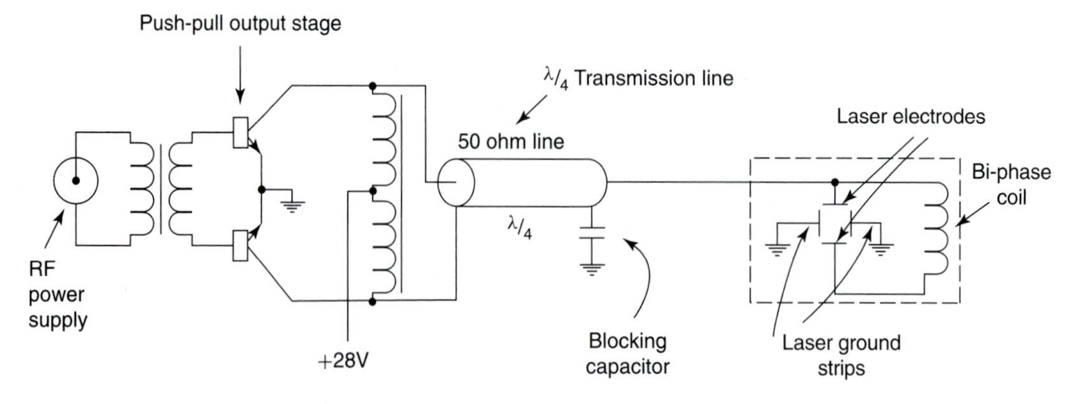

Figure 12.17 An impedance matching network typical of the Synrad series 48 lasers. (From Peter Laakmann, "Drive System for RF-Excited Gas Lasers," U.S. Patent #5,008,894.)

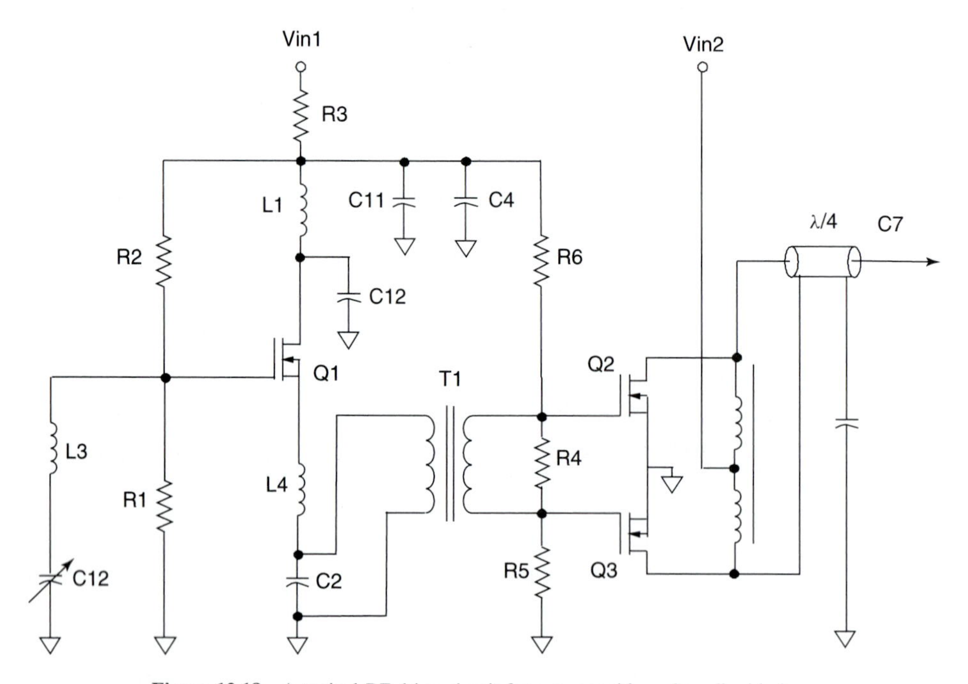

Figure 12.18 A typical RF drive circuit for a waveguide carbon dioxide laser.

unit as illustrated in Figure 12.18. The input stage of the power supply is a relatively conventional RF oscillator circuit. The output stage is a push-pull stage and is integrated into the impedance matching network as described in the previous paragraph and in U.S. Patent #5,008,894.[63]

[63]P. Laakmann, "Drive System for RF-Excited Gas Lasers," U.S. Patent #5,008,894, April 16, 1991.

12.1.5 Optical Components and Detectors for CO_2 Lasers

Window, mirrors and lenses. The long (10.6 μm) wavelength and high average power of CO_2 lasers lead to some difficulties in construction of optical systems. Conventional glass optics cannot be used at CO_2 wavelengths due to high optical absorption and low thermal conductivity. Table 12.1 summarizes the important optical properties of materials typically used for transmissive optics in CO_2 lasers. The alkali halides (such as KBr) tend to have very low absorption, but are hygroscopic and require special coatings to survive in moist environments. The semiconductors (such as ZnSe and GaAs) are much more robust, but with correspondingly higher absorption coefficients. Unfortunately, both germanium (Ge) and silicon (Si) suffer from a thermal runaway problem. Above a certain critical laser intensity, as the sample heats, the absorption increases, and the sample heats more. This feedback mechanism rapidly results in destruction of the sample.

TABLE 12.1 PROPERTIES OF MATERIALS USED IN CO_2 LASERS

Material	Thermal expansion ($\times 10^6$ / °C)	Thermal conductivity ($\times 10^{-2}$ W / cm °C)	α (at 10.6 μm cm^{-1})
NaCl	38.9	6.5	$1.34 \cdot 10^{-3}$
KBr	43	4.8	$5 \cdot 10^{-5}$
ZnSe	7.7	13	$6 \cdot 10^{-3}$
GaAs	5.7	37	$1.2 \cdot 10^{-2}$
Si	4.2	120	2.5
Ge	6	59	$3.6 \cdot 10^{-2}$

W. W. Duley, *CO_2 Lasers: Effects and Applications* (New York: Academic Press, 1976), p. 104, Table 3.4.

Zinc selenide (ZnSe), a well-behaved semiconductor with a relatively high thermal conductivity and low absorption, is the typical choice for CO_2 transmissive optics. ZnSe does not suffer from thermal runaway and has the major advantage of being transmissive from 600 nm to 11 μm. (This allows low power HeNe lasers to be used for alignment of CO_2 systems.) ZnSe is available commercially as in the form of windows, dielectric-coated partially-transmitting mirrors, and lenses.

The very high average power typical of CO_2 lasers means that CO_2 reflecting optics must be thermally conductive as well as poor optical absorbers at 10.6 μm. Dielectric- or metal-coated mirrors with single-crystal silicon substrates are the mirrors of choice for lower-power CO_2 lasers. Higher power lasers use metal coatings (typically silver or gold) on molybdenum or copper substrates. Molybdenum (with a thermal conductivity of 1.33 W/cm °C) offers roughly equivalent thermal conductivity as silicon, but is a much more durable refractory material. Molybdenum may also be used without a surface metal coating of gold or silver. (Uncoated molybdenum mirrors typically have reflectances on the order of 98%.) Copper (with a thermal conductivity of 3.9 W/cm °C) is much more thermally conductive than either silicon or molybdenum, but significantly less rugged. Copper is also difficult to polish and exceptionally easy to scratch during cleaning.

The long wavelength of CO_2 lasers does offer certain advantages during fabrication of optics. Diffraction and focusing effects of scratches are roughly proportional to $(1/\lambda)^4$.

Thus, a relatively poor finish by visible optics standards may be perfectly suitable for CO_2 laser optics. Additionally, the longer wavelength permits fabrication of optics using such methods as diamond turning. If properly fabricated, a diamond turned CO_2 reflector may not need to be polished after fabrication.

CO_2 laser detectors. The long wavelength and high average power of CO_2 lasers also lead to difficulties in optical detection and laser power measurement. Several detection technologies are commonly employed with CO_2 lasers to overcome these problems. These technologies include thermopiles, bolometers, pyroelectric crystals, and long wavelength semiconductor photovoltaic and photoconductive detectors. (For more details on thermopiles, bolometers, and pyroelectric crystals, see Section 8.4.)

The majority of photovoltaic and photoconductive semiconductor devices are fabricated from silicon and have a roughly 400 nm to 1.1 μm response range. A simple example of a photodetector of this type is a PIN photodiode. However, photovoltaic or photoconductive devices do exist with responses in the 10.6 μm range. To move the detector into the 10.6 μm range requires switching to more exotic semiconductors with narrower bandgaps than silicon. Examples include HgCdTe, CdTe, and PbSnTe. However, 10.6 μm infrared photovoltaic or photoconductive detectors do suffer from a special problem. Semiconductor energy states that absorb in the 10.6 μm range are separated by approximately 0.1 eV. Thus, thermal population of the energy states from sources other than the laser beam is quite likely. Therefore, the detectors are often cooled using liquid helium, liquid nitrogen, or thermoelectric coolers.

10.6 μm radiation is far outside of the human visual response. Thus, a number of techniques have been developed to enable humans to locate the beam and make qualitative measurements. The simplest include such things as absorption in foamed polystyrene, wood, or metal. More complex methods include the use of color changes in liquid crystals and visible fluorescence from various materials. One of the most elegant and effective methods is the thermal-quenching of visible luminescence from a ZnCdS phosphor excited continuously with UV light. The usual configuration is to use a UV light to excite the back of a small screen coated with the phosphor. The phosphor glows a brilliant yellow until it interacts with a CO_2 beam. The CO_2 beam quenches the fluorescence and leaves dark areas on the screen.

12.2 THE DESIGN OF EXCIMER LASERS

Excimer[64] lasers operate on the radiative transition between the excited state of a molecule and its ground state. Commercial excimer lasers typically use molecules formed from the combination of heavy noble gases (such as Xe, Kr, and Ar) and halogens (such as F, Cl, Br, and I). Common excimer laser combinations include ArCl (175 nm), ArF (193 nm), KrCl (222 nm), KrF (249 nm), XeBr (282 nm), XeCl (308 nm), and XeF (351 nm). Excimer lasers operate pulsed and are available in energies ranging from several millijoules to handfuls of

[64]In the most correct usage, the word *excimer* is limited to homopolar molecules such as Xe_2. Heteropolar molecules such as XeF are more correctly termed *exciplex* molecules. However, since it is quite difficult to pronounce exciplex laser, the term excimer has grown to mean both classes of lasers.

joules per pulse. Repetition rates range from a few kHz to hundreds of Hz and output average powers are available up to 500 watts.

Excimer lasers are prized for their ability to efficiently produce coherent UV and deep UV radiation. They are used in such diverse commercial applications as laser hole drilling, laser chemical vapor deposition, laser photochemistry, creation of soft x-ray plasmas, semiconductor wafer cleaning, laser machining, laser ablative sputtering, excimer laser surface annealing, deep UV lithography, and laser planarization. They find application in the medical field for coronary angioplasty and photorefractive keratectomy. Research applications include spectroscopy, laser photochemistry, laser doping of semiconductors, and remote sensing.

Excimer lasers, like CO_2 lasers, are very different from the other lasers discussed in this text. Conventional lasers lase off electronic transitions between various atomic states. Excimer lasers lase off the transition between a molecular excited state and a molecular ground state. This means that excimer lasers tend to have a shorter wavelength and higher efficiency than most conventional lasers. Additional information on excimer lasers can be found in Rhodes,[65] Laude,[66] Elliot,[67] and Weber.[68]

12.2.1 Introduction to Excimer Laser States

Consider a gas mixture of a rare gas A (such a krypton) and 0.1 to 0.3% of a halogen gas X (such as fluorine). Assume that the mixture is pumped by an intense electron beam, which forms A^+ and X^- ions.[69] The ions then recombine to form the excited $(A^+X^-)^*$ state as:[70]

$$A^+ + X^- + M \rightarrow (A^+X^-)^* + M \tag{12.4}$$

where A is the rare gas ion, X is the halogen ion and M is any third body (usually another rare gas ion) to assure momentum conservation.

The $(A^+X^-)^*$ molecules are the excited molecules and form the upper state population (with a lifetime on the order of 5 to 15 ns). The AX molecules form the lower state population. However, the lifetime of the AX molecule is extremely short (on the order of tens of femtoseconds). Thus, even though the upper state lifetime is short by the standard of most lasers (tens of nanoseconds as compared to tens or hundreds of microseconds), it is still three orders of magnitude greater than the lifetime of the ground state. (For excimer lasers to operate correctly, it is necessary for the ground state molecule to dissociate rapidly. Systems for which the molecule dissociates slowly do not make successful lasers because the laser bottlenecks in the ground state.)

[65] C. K. Rhodes, ed, *Excimer Lasers* (Berlin: Springer-Verlag, 1984), particularly Chapter 4, pp. 87–138.

[66] Lucien Laude, *Excimer Lasers* (Netherlands: Kluwer Academic Publishers, 1994).

[67] David Elliot, *Ultraviolet Laser Technology and Applications* (San Diego: Academic Press, 1995).

[68] Marvin J. Weber, ed, *Handbook of Laser Science and Technology, Vol. II, Gas Lasers* (Boca Raton, FL: CRC Press, Inc., 1982), particularly Section 3, pp. 273–491; and Marvin J. Weber, ed, *Handbook of Laser Science and Technology, Supp. I, Lasers* (Boca Raton, FL: CRC Press, Inc., 1991), particularly Section 3.3.1, pp. 341–387.

[69] The A^+ and X^- ions are A and X atoms which have lost or gained one electron respectively.

[70] The excited $(A^+X^-)^*$ is a molecule with the same number of electrons as AX, but with one of the electrons in a higher energy state.

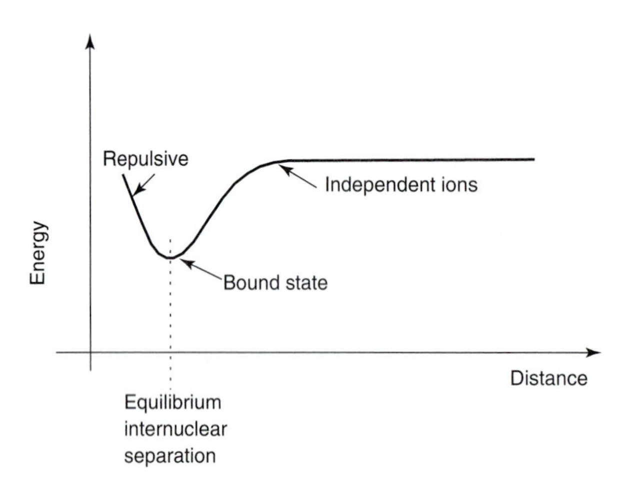

Figure 12.19 The formation of a molecule is driven by a balance between the coulombic attraction of the ions and the repulsive potential that keeps two ions from occupying the same space.

The physics of excimer laser operation is inherently connected to the basic processes of molecule formation. Recall that the formation of a molecule is driven by a balance between the coulombic attraction of the ions and the repulsive potential that keeps two ions from occupying the same space. Thus, the formation of a molecule is a function of the distance between the two ions (see Figure 12.19). If the two ions are widely separated, there is no molecule. If the two ions are trying to occupy the same space, there is no molecule. The balance between the coulombic attraction and the repulsive potential creates a *well* or *pocket* in the potential energy curve. When the two ions are at a distance (the equilibrium internuclear separation) corresponding to this energy well they are *bound* and form the molecule.

The situation in a rare-gas halogen laser is somewhat more complicated than that illustrated in Figure 12.19. This is because of the simultaneous existence of atoms, ions, molecules, and their various excited states. A simplified potential energy diagram of a rare gas halogen excimer laser is illustrated in Figure 12.20. Notice that the upper state manifold has two kinds of potential energy curves. There are covalent curves corresponding to the bonding of an excited atom with another atom ($A^* + X$ or $A + X^*$). There are also ionic curves corresponding to the bonding of two ions (A^+X^-). The ionic curves actually form a family of curves, because there are a variety of electron configurations possible in the $(A^+X^-)^*$ excited state.

Practical excimer laser systems are those for which the ionic $(A^+X^-)^*$ states are the lowest states of the upper manifold. These are systems where the crossings between the ionic and covalent state energies occur for larger internuclear separations than the equilibrium distance. In these systems the dynamics of the reactions lead inevitably to molecules residing in the bound metastable $(A^+X^-)^*$ state. From the $(A^+X^-)^*$ state, the only available downward energy transition is by radiation. There are no competing alternative paths.

The lower state manifold is similar to the upper state manifold, but somewhat less complex. Since the lower state does not have ions, the only potential energy curves are those associated with the various electronic states of the covalent AX. The lower state may (or may not) have a potential well indicating the existence of a bound molecule.

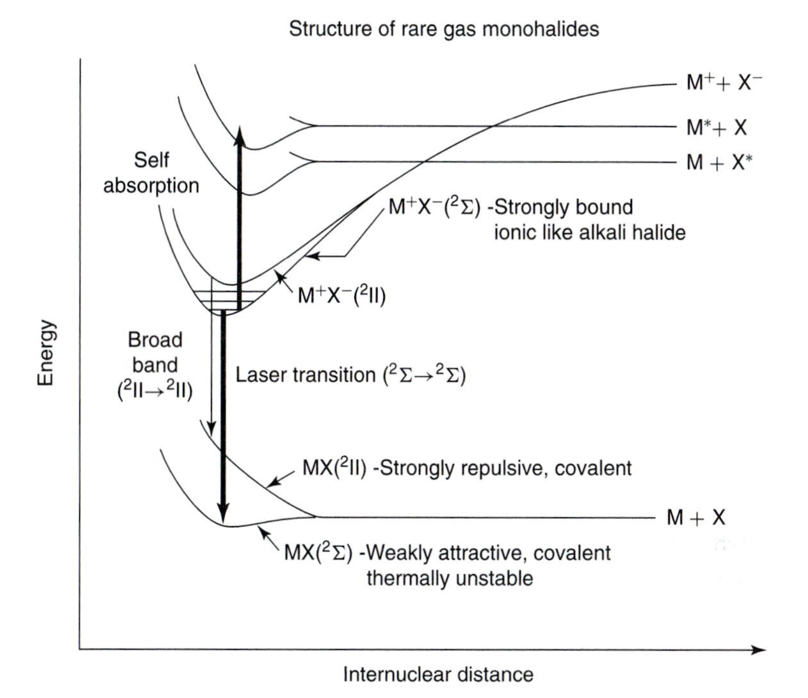

Figure 12.20 A simplified potential energy diagram of a rare-gas halogen laser. (From C. K. Rhodes, ed, *Excimer Lasers* (Berlin: Springer-Verlag, 1984), Figure 4.1, p. 88. ©1984 Springer-Verlag.)

The emission spectrum of KrF illustrates the radiative characteristics of a typical rare gas-halogen transition (see Figure 12.21). The largest peak corresponds to the transition from the bottom of the potential well of the $(A^+X^-)^*$ excited state to the bottom of the potential well of the AX excited state.

Commercially interesting rare gas halide lasers are pumped by electron beam excitation or by electrical discharge. In electron beam excitation, an energetic electron beam is created in vacuum and directed through a thin metal or polymer membrane into the gas mixture. High energy electrons scattering off atoms in the gas create ions, excited atoms, and secondary electrons. The secondary electrons, in turn, create more ions, excited atoms, and electrons. The ions and excited atoms react to form the excited molecules. Under electron beam excitation, ions are produced in preference to excited states (by about a 3:1 ratio) and the primary reaction is

$$A^+ + X^- + M \rightarrow (A^+X^-)^* + M \tag{12.5}$$

where A is the rare gas ion, X is the halogen ion, and M is any third body (usually another rare gas ion), to assure momentum conservation.

Pulsed electrical discharges are more commonly used for commercial lasers because they offer the potential for higher pumping efficiency and higher average power. In an electrical discharge, the low energy electrons drift along in the electric field and increase in energy. Eventually, they reach the energy threshold for excitation of the first excited state

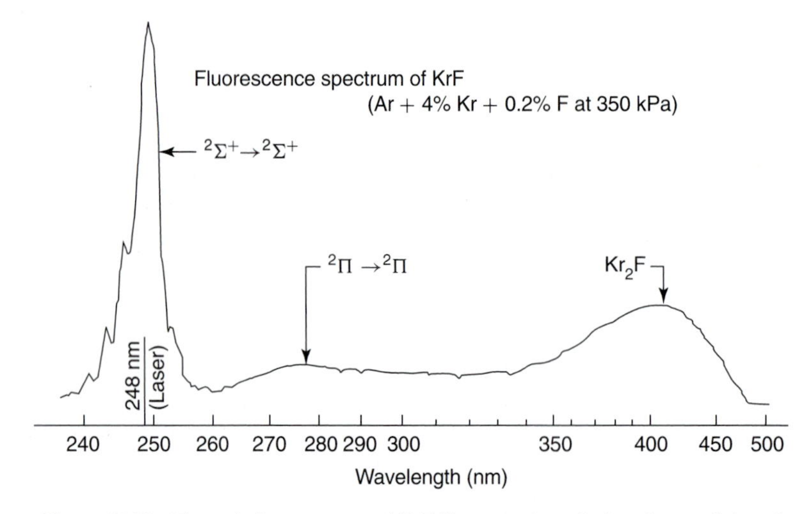

Figure 12.21 The emission spectrum of KrF illustrates the radiative characteristics of a typical rare gas-halogen transition. (From C. K. Rhodes, ed, *Excimer Lasers* (Berlin: Springer-Verlag, 1984), Figure 4.2, p. 91. ©1984 Springer-Verlag.)

of the rare gas atoms. Then, they collide inelastically with the rare gas atom, leaving the gas atom in an excited state and returning the electron to low energy. Under these conditions, a very fast and efficient reaction termed a *harpoon* collision is involved.[71]

$$A^* + MX \rightarrow (A^+X^-)^* + M \tag{12.6}$$

12.2.2 The Evolution of Excimers

The first excimer was demonstrated in 1970 by Basov et al. at the Lebedev Physics Institute.[72] This laser was an Xe_2 excimer that lased at 176 nm and was excited by a relativistic electron beam directed into liquid xenon. In 1972, Koehler et al. at Lawrence Livermore National Laboratories demonstrated an Xe_2 laser using electron beam pumping of gaseous xenon.[73] Laser action from XeBr at 282 nm was first reported in 1975 by Searles and Hart at the Navel Research Laboratory.[74] Searles and Hart measured the optical gain of the XeBr to be 4% per pass over 15 cm of active length. The experiment was originally run at 1 to 4% Br_2, but this was reduced to 0.1 to 1% Br_2 when the Br_2 was discovered to quench the XeBr excited state.

A few weeks later, Ewing and Brau reported laser action from XeCl at 308 nm and KrF at 249 nm.[75] An independent observation of laser action in KrF was also reported by Tisone

[71]D. R. Herschbach, *Adv. Chem. Phys.* 10:319 (1966); pp. 367–379 in particular.

[72]N. G. Basov, V. A. Danilychev, Yu. M. Popov, and D. D. Khodkevich, *JETP Lett.* 12:329 (1970).

[73]H. A. Koehler, M. A. Ferderber, D. L. Redhead, and P. J. Ebert, *Appl. Phys. Lett.* 21:198 (1972).

[74]S. K. Searles and G. A. Hart, *Appl. Phys. Lett.* 27:243 (1975).

[75]J. J. Ewing and C. A. Brau, *Appl. Phys. Lett.* 27:350 (1975).

et al.[76] and Mangano and Jacob.[77] In the Ewing and Brau experiments, laser action was obtained from a mixture of Ar, Xe, and Cl_2 at 89.9:10:0.1 and from a mixture of Ar, Kr, and F_2 at 98.9:1.0:0.1. The output energy was approximately 50 μJ from the XeCl laser and approximately 4 mJ from the KrF laser. Tisone et al. used a mixture of 3000 torr of Ar, 150 torr of Kr, and 6 torr of F_2 and were able to obtain 5.6 J by using 5 cm diameter cavity mirrors. In the Mangano and Jacob experiments, laser action was obtained from a mixture of Ar, Kr, and F_2 at 97.9:2.0:0.1. The maximum output energy in the Mangano and Jacob experiments was 6 mJ. All three experiments used a pulsed electron gun to excite the gas mixture.

Reports of laser action from XeF soon followed, first discovered by Brau and Ewing,[78] but first published by Ault et al.[79] Brau and Ewing used a mixture of Ar, Xe, and F_2 at 99.6:0.3:0.1 where the mixture was excited with a cold cathode electron gun. Ault et al. used a mixture of Ar, Xe, and NF_3 at 250:25:1. The use of NF_3 in the Ault experiments was to both minimize the corrosive effects of F_2 and to avoid quenching due to self-absorption. The measured output energy from the Ault experiments was 5 mJ per pulse, the peak power was 500 kW, and the pulse width was 10 ns.

All of the early excimer and exciplex lasers were electron beam pumped. However, the efficiency advantages of using electrical discharge pumping were recognized very early. Since excimer lasers require very high voltages, longitudinal discharge excitation of excimers was never successfully attempted. Instead, excimer technology leveraged off the existing developments in transversely excited CO_2 lasers. A major advancement occurred in 1975, when Burnham and Djeu at the Navel research laboratories modified a TEA CO_2 laser and demonstrated excimer laser action with an electrical discharge.[80]

12.2.3 General Design Background

Gas flow. In order to operate an electrically discharge excited excimer laser, it is necessary to provide an electric field that exceeds the DC breakdown voltage of the gas mixture by a factor of 2 to 3. This typically requires voltages of 10 to 15 kV/cm. Furthermore, this pulse must be applied in a time short compared to the electronic avalanche time of 20 to 30 ns. This implies a rise time of several kV/nanosecond. Finally, there must be sufficient electrons in the circuit to supply the growing plasma. This requires a very low impedance electrical excitation circuit.

The discharge pulse of an excimer laser creates a number of ionized and excited species. Many of these species are quite long-lived. If another discharge pulse is attempted before these secondary species have decayed, the resulting electrical discharge will be unstable. Therefore, in most commercial excimers, the gas flows across the electrodes. In general, the best results are achieved when the flow rate is high enough to flush the dis-

[76] G. C. Tisone, A. K. Hays, and J. M. Hoffman, *Opt. Commun.* 15:188 (1975). (The Ewing paper was received on June 17, 1975, and the Tisone paper on July 7, 1975.)

[77] J. A. Mangano and J. H. Jacob, *Appl. Phys. Lett.* 27:495 (1975). (The Ewing paper (footnote #75) was received on June 17, 1975, and the Mangano paper on July 7, 1975.)

[78] J. J. Ewing and C. A. Brau, *Appl. Phys. Lett.* 27:350 (1975); and C. A. Brau and J. J. Ewing, *Appl. Phys. Lett.* 27:435 (1975).

[79] E. R. Ault, R. S. Bradford, and M. L. Bhaumik, *Appl. Phys. Lett.* 27:413 (1975).

[80] Jeff Hecht, *Laser Pioneers*, revised ed. (Boston, MA: Academic Press, 1992), p. 46.

charge volume between the electrodes 2 to 3 times before the next pulse. In the case of high power lasers with pulse repetition rates of 500 Hz to 2 kHz, this gas replacement means that velocities of 50 to 150 m/sec are required. Thus, aerodynamics becomes a critical part of excimer laser design.

Preionization. Good electrical discharge uniformity is essential for the successful operation of a pulsed excimer laser. In order to provide a uniform high voltage discharge in high pressure gases, the gas must be preionized immediately prior to the application of the main electrical discharge pulse. Preionization means that a uniform ion or electron cloud is generated in the gas in the discharge region. This cloud serves to *seed* the discharge and prevent the formation of arcs or streamers in the discharge. Once the initial cloud (approximately 10^7 electrons/cm^3) is created, the main discharge will proceed by avalanche ionization to create densities near 10^{15} electrons/cm^3.

Preionization is a very challenging problem because a large number of electrons must be created quickly. Preionization can be accomplished in a variety of ways.[81,82] The two most common are UV light and x-rays. UV light can be generated by spark gaps or by corona discharges. However, UV light does have the limitation that the ionizing range is restricted to a few cm.[83] X-rays can be created in x-ray tubes by collisions of energetic electrons with a high Z material (such as tungsten). X-rays have a much deeper penetration depth, but are more difficult to create. In addition, the use of x-rays in a commercial system requires special licensing and approval.

Multiple spark gaps are commonly used to create UV light for preionization.[84] Early excimer laser systems used a linear sparkboard where each spark gap was separately discharged to create a traveling spark wave down the length of the cavity. Later designs used multiple parallel pins, where the pins were fired simultaneously. Spark gaps provide excellent preionization, but at the cost of some structural complexity. More importantly, spark gaps erode and tend to contaminate the laser gas and optical components of the resonator. Spark gaps are much less frequently used today, as they have been replaced by corona or x-ray preionization methods.

Corona discharges tend to provide lower electron concentrations, but are structurally simpler and do not contaminate the gas. Corona discharges have the additional advantage that they can be easily integrated with the main electrode structure. Thus corona discharges are most commonly used with commercial excimers (such as XeCl) that do not require high electron preionization concentrations.

Another alternative is to use x-ray preionization.[85] Although x-ray preionization was not used in the early development of excimer lasers (due to the difficulty and expense of making short x-ray pulses) advances in x-ray generation technology for lasers have made x-ray preionization a very viable alternative to corona technology. This is especially true for the more exotic excimers which may require higher preionization electron densities.

---

[81] A. J. Palmer, *Appl. Phys. Lett.* 25:136 (1974).

[82] S. C. Lin, *J. Appl. Phys.* 51:210 (1980).

[83] K. Midorikawa, M. Obara, and T. Fujiokam, *IEEE J. of Quantum Electron.* QE-20:198 (1984).

[84] K. Miyazaki et al., *Rev. Sci. Instrum.* 52:201 (1985).

[85] S. Sumida, M. Obara, and T. Fujiokam, *Appl. Phys. Lett.* 33:913 (1978).

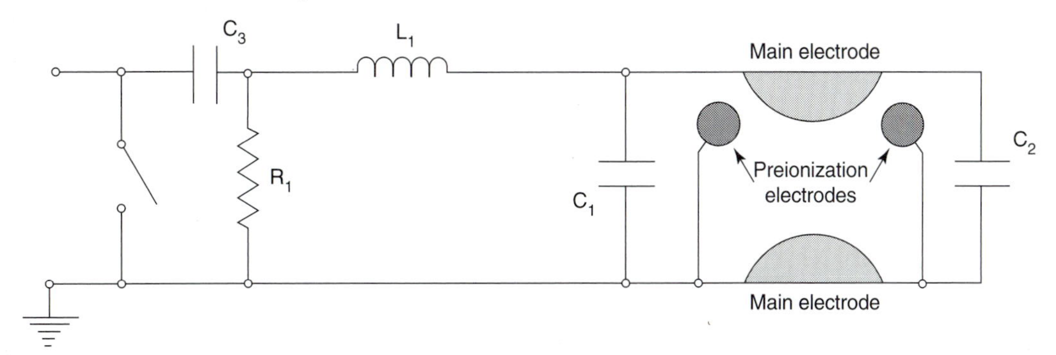

Figure 12.22 A typical corona discharge circuit. The two main electrodes are oriented transversely. The two preionization electrodes are also oriented transversely, and are parallel to the main electrodes. (From E. Müller-Horsche, "Apparatus for Preionizing a Pulsed Gas Laser," U.S. Patent #5,247,531, Sept. 21, 1993.)

Corona discharge circuitry. The corona discharge method is one of the most common commercial preionization techniques.[86,87] In the corona discharge method, UV light is generated by a gas discharge between a metal and a dielectric. The UV light then creates a weak ionization in the discharge region of the electrodes. One of the major advantages of corona preionization is that the dielectric prevents formation of spark channels to the preionization electrodes. Thus, the contamination effects of the sparks are eliminated.

A typical corona discharge circuit is shown in Figure 12.22.[88] The two main electrodes are oriented transversely. The two preionization electrodes are also oriented transversely, and parallel to the main electrodes. The preionization electrodes consist of a conductor with a surrounding dielectric (for example, a quartz sleeve). The preionization electrodes are placed near one of the main electrodes and are connected to the potential of the other electrode. A high-voltage source charges the storage capacitor, and the pulse is triggered by a thyratron (or equivalent high-voltage switching element). The preionization potential is thus driven by the main discharge pulse.

Another option is shown in Figure 12.23.[89] The geometry is similar to Figure 12.22. However, with this system, the preionization potential is driven by a separate set of circuitry from the main discharge pulse. This permits more subtle tuning of the timing between the two events, but at the cost of additional circuitry.

Still another option is shown in Figure 12.24.[90] Here, the clever use of inductors and capacitors generates a very rapid starting spike across the preionization electrodes, which is followed by the main discharge pulse. Suitable choice of inductor and capacitor values permit this circuit to be tuned for optimum output power.

---

[86] G. J. Ernst and A. G. Boer, *Opt. Commun.* 27:105 (1978).

[87] U. Hasson and H. M. Bergmann, *Rev. Sci. Instrum.* 50:59 (1979).

[88] E. Müller-Horsche, "Apparatus for Preionizing a Pulsed Gas Laser," U.S. Patent #5,247,531, Sept. 21, 1993.

[89] E. Müller-Horsche, "Apparatus for Preionizing a Pulsed Gas Laser," U.S. Patent #5,247,531, Sept. 21, 1993.

[90] E. Müller-Horsche, "Apparatus for Preionizing a Pulsed Gas Laser," U.S. Patent #5,247,531, Sept. 21, 1993.

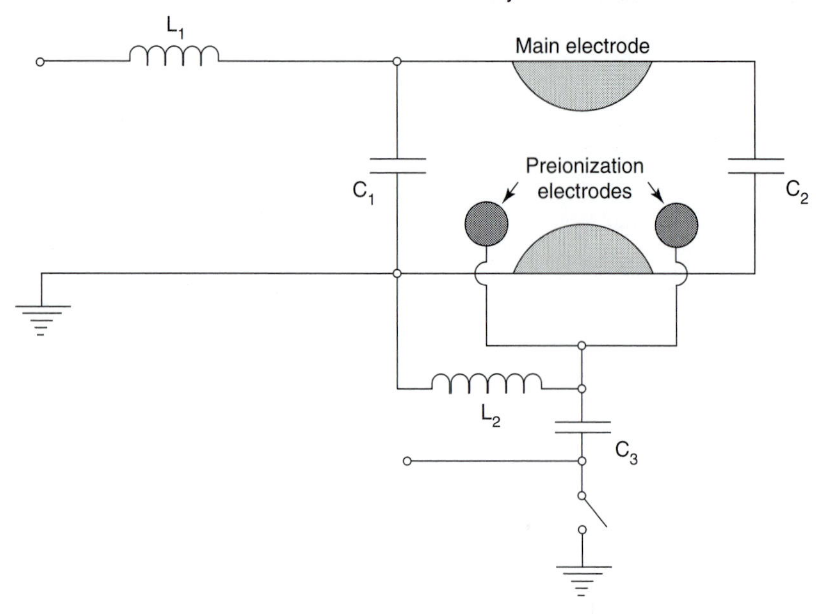

Figure 12.23 Another type of corona discharge circuit. In this circuit, the preionization potential is driven by a separate set of circuitry from the main discharge pulse. This permits more subtle tuning of the timing between the two events, but at the cost of additional circuitry. (From E. Müller-Horsche, "Apparatus for Preionizing a Pulsed Gas Laser," U.S. Patent #5,247,531, Sept. 21, 1993.)

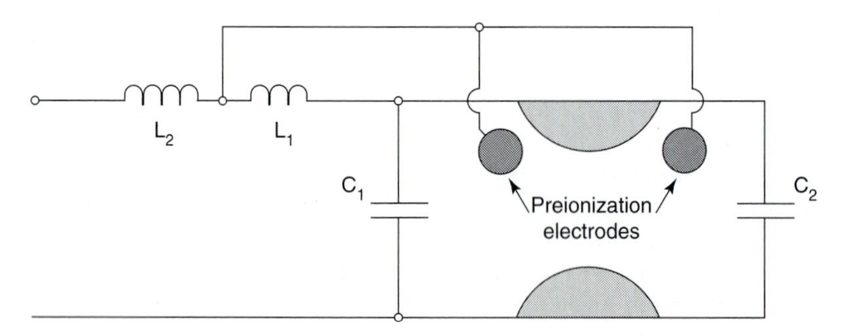

Figure 12.24 In this circuit, the clever use of inductors and capacitors generates a very rapid starting spike across the preionization electrodes, which is followed by the main discharge pulse. (From E. Müller-Horsche, "Apparatus for Preionizing a Pulsed Gas Laser," U.S. Patent #5,247,531, Sept. 21, 1993.)

Main discharge circuitry. Once the preionization stage is complete, the main discharge pulse must be generated. The electrical discharge requirements on discharge excited pulsed excimer lasers are quite stringent. The rise time is short (10^{11} A/sec) and the voltage and power are high (50 kV and 5 kW). Furthermore, the gas (after breakdown) has a very low impedance.

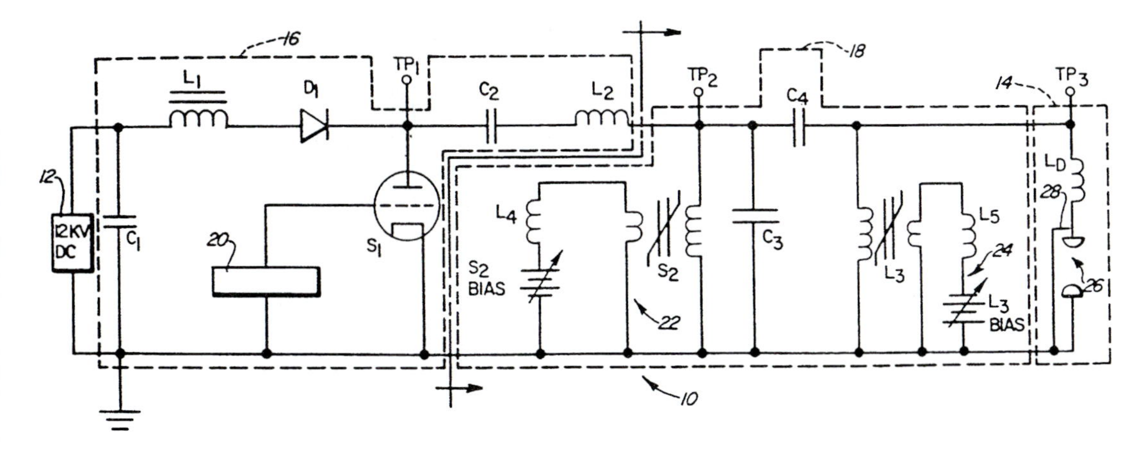

Figure 12.25 A typical discharge circuit for an excimer laser. This circuit provides an electrical interface between a high-voltage, high-impedance power source, and a relatively low-impedance laser load. (From T. S. Fahlen and B. Mass, "Electrical Excitation Circuit for Gas Lasers," U.S. Patent #4,549,091, Oct. 22, 1985.)

An example of a typical discharge circuit is shown in Figure 12.25.[91] The electrical circuit provides an electrical interface between a high-voltage, high-impedance power source and a relatively low-impedance laser load. The circuit includes an electrical excitation circuit 10, a charging circuit 16, a pulse forming network 18, and a laser load 14.

The charging circuit includes a power source capacitor C_1, a charging capacitor C_2 series-connected with an inductor L_2, a charging choke L_1 and an isolating diode D_1. The choke L_1 and charging diode D_1 isolate the power source 12 from the pulse forming network 18. (The capacitor C_1 is significantly larger than the capacitor C_2.) A thyratron S_1 is also included in the charging circuit. The control electrode of the thyratron is connected to a pulse generator.

The pulse-forming network (PFN) includes a saturable inductor switch S_2, a bias power source for S_2, and a choke L_4 connected in series with the bias winding of the saturable inductor switch. The PFN also includes a PFN capacitor C_3 shunted across the saturable inductor switch S_2 and a second PFN capacitor C_4 connected between TP_2 and TP_3. Additionally, the PFN includes a magnetic diode charging inductor L_3, a bias power source for L_3 and a choke L_5 connected in series with the bias winding of the magnetic diode charging inductor. (The choke L_5 provides isolation of the bias power source for L_3 from high-voltage pulses produced by transformer action on the bias winding of the magnetic diode-charging inductor L_3.) The laser load 14 is connected between TP_3 and common. The inductance L_D represents the distributed inductance of the electrode structure of the laser load 14. A preionization circuit 28 is also included.

The electrical excitation circuit performs four relatively separate operations: a slow resonant charge of the charging capacitor C_2, a medium-speed charge of the pulse forming

[91] T. S. Fahlen and B. Mass, "Electrical Excitation Circuit for Gas Lasers," U.S. Patent #4,549,091, Oct. 22, 1985.

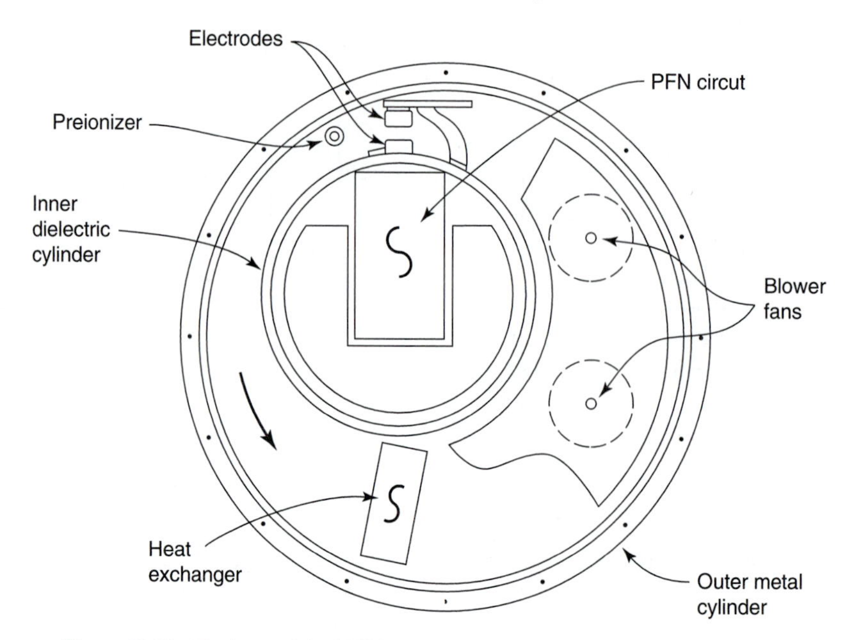

Figure 12.26 The heart of the 5300 is a pressure vessel consisting of two eccentrically mounted cylinders. (From D. J. Clark and T. S. Fahlen, "Gas Transport Laser System," U.S. Patent #4,611,327, Sept. 9, 1986.)

network 18, an inversion of the voltage on half of the pulse forming network, and finally the laser discharge.

12.2.4 A Typical Modern Excimer Laser

The remainder of this section will focus on an industrial excimer laser intended for industrial materials processing applications. The specific unit under discussion is the XMR 5300 laser manufactured by XMR, Inc., in Fremont, CA. This laser is representative of a high-end industrial materials processing excimer laser.

The XMR 5300 is a XeCl laser operating at 308 nm and producing 200 watts at 300 Hz for an energy of approximately 660 mJ/pulse. The repetition rate is selectable from 1 to 300 Hz and the pulse width is 40 to 50 ns. The output beam incorporates a beam homogenizer and is 1.5 cm by 3.3 cm with less than a 5 mrad divergence. The laser operates at a pressure of 5.76 atmospheres with a buffer gas of neon and uses an HCl precursor to generate the chlorine. The design of this laser is discussed in D. J. Clark and T. S. Fahlen, "Gas Transport Laser System," (U.S. Patent #4,611,327), and in T. S. Fahlen, "Gas Discharge Laser Having a Buffer Gas of Neon," (U.S. Patent #4,393,505), and is only summarized here.[92,93]

The heart of the 5300 is a pressure vessel consisting of two eccentrically mounted cylinders (see Figure 12.26). The outer cylinder is metal and constitutes the outer wall of the pressure vessel. The inner cylinder is a dielectric and contains the PFN for the electrodes.

[92]D. J. Clark and T. S. Fahlen, "Gas Transport Laser System," U.S. Patent #4,611,327, Sept. 9, 1986.

[93]T. S. Fahlen, "Gas Discharge Laser Having a Buffer Gas of Neon," U.S. Patent #4,393,505, July 12, 1983.

The gas flow is between the inner cylinder and the outer cylinder and is driven by a set of blower fans. A gas-to-liquid heat exchanger is located near the bottom of the pressure vessel.

The electrodes are placed on the top of the inner dielectric cylinder and the laser is excited transversely. Two pairs of preionizing corona electrodes are located slightly to both sides of the main electrodes. The main discharge plasma is created between the two electrodes near the top of the pressure vessel. Thus, the optical cavity is located parallel to these electrodes.

Physically, the 5300 pressure vessel consists of a nickel-plated stainless-steel cylinder that is 100 cm long and 60 cm in diameter. The vessel contains approximately 200 liters of laser gas. The stainless-steel cylinder has a welded flange with an o-ring groove on each end. End plates are bolted to the welded flanges. Although the system operates at a pressure of roughly 70 psig, the heavy stainless construction has been pressure tested up to 180 psig.

The laser system is designed for easy repair and maintenance. All of the components inside the pressurized vessel are readily accessible. A removable cantilevered arm hinged to the flange holds the end plate so that it can be swung away from the flange. Once removed, the end plate is supported on a wheeled cradle. The inner dielectric cylinder, electrodes, PFN, blowers, and heat exchangers are attached to the end plate and are removed with it. This configuration provides ready access to the sub-assemblies for servicing. Each of the sub-assemblies is a module and can be replaced independently of the others.

The electrodes are two solid-rail electrodes, 79 cm long and 2.5 cm wide, spaced by 2.5 cm. The material of both electrodes is solid nickel. The electrodes are mounted to the outside of the internal dielectric cylinder. Each of the electrodes is bolted to eighteen feedthroughs penetrating the inner dielectric cylinder. Each feedthrough is o-ring-sealed as it enters the inner cylinder. However, the electrodes can be adjusted or removed without replacing the o-rings.

Five to 10 kW of heat is generated in the laser gas from the electrical discharge, the corona preionization, and the blower fans. The gas to liquid heat exchanger maintains the gas temperature at 90°F with a 65°F input coolant temperature at a flow of 3 to 5 gallons per minute. The heat exchanger is a finned tube construction, nickel-plated to minimize damage from the gas mixture. In order to minimize feedthroughs, the heat exchanger is welded to a flange, which is sealed to the end plate with an o-ring.

Figure 12.27 is a perspective figure illustrating the entire system. The outer metal cylinder defines the shape of the entire system. The inner dielectric cylinder can be seen toward the back of the drawing (somewhat hidden by the fans). The window for the optical cavity can be seen near the front top of the pressure vessel. The large triangular housings on each side of the pressure vessel contain the drive assembly for the blower fans.

The double cylinder structure offers a number of advantages. To begin with, it places the pulse-forming network inside of a large steel structure. This serves as an ideal *Faraday cage* and minimizes the laser generating damaging electromagnetic interference from the extremely high voltage and current pulses. The Faraday cage construction similarly reduces the possibility of accident triggering of the PFN by external sources. Locating the PFN close to the electrodes minimizes the overall inductance of the electrical system. Finally, locating the PFN within the pressure vessel minimizes the number of feedthroughs.

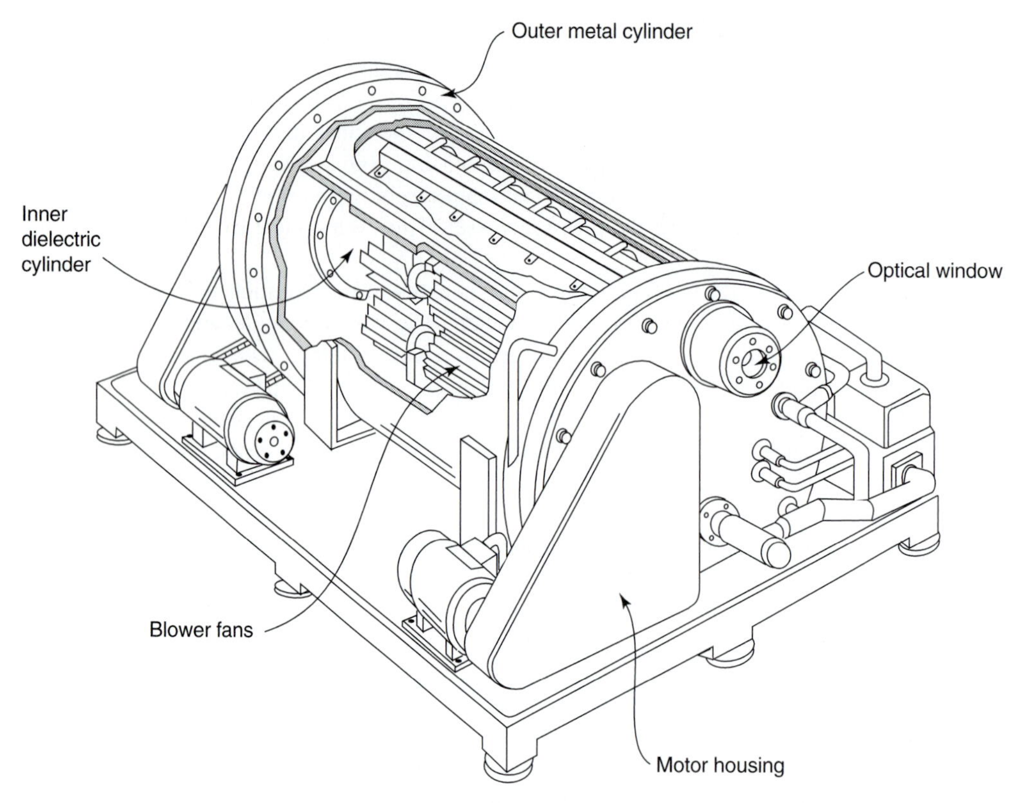

Figure 12.27 A perspective figure illustrating the laser chamber of the XMR 5300. The outer metal cylinder defines the shape of the entire system. The inner dielectric cylinder can be seen toward the back of the drawing (somewhat hidden by the fans). (From D. J. Clark and T. S. Fahlen, "Gas Transport Laser System," U.S. Patent #4,611,327, Sept. 9, 1986.)

The eccentric placement of the cylinders is also a critical part of the design. Notice that the electrodes are located at the narrowest part of the structure. This increases the velocity of the gas over the electrodes and enhances the uniformity of the gas flow. The flow velocity is maximized in the double-cylinder geometry because there are no sharp bends, constrictions, or changes in direction. Additionally, the cylindrical geometry minimizes the surface area exposed to the gas mixture.

One of the major challenges in constructing excimer laser systems is the corrosion caused by the halogen ions. Early excimers were notorious for blowing seals and destroying electrical components. Thus, excimer laser design must include a number of features to simultaneously reduce leakage and enhance the laser lifetime. The 5300 incorporates magnetic couplings for the blowers in order to provide a hermetic seal with no mechanical contact. Within the pressurized vessel, bearings are lubricated with fluorinated lubricants. The blower and coupling bearings inside the vessel are enclosed in a housing, which is pressurized with a small flow of halogen-free gas. In addition, materials used inside the pressure vessel have been selected carefully. The majority of the metals are nickel-plated. Insulating materials are quartz or ceramic.

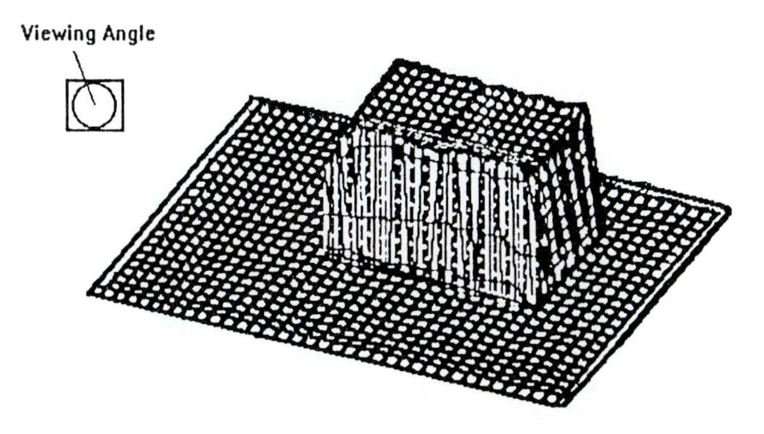

Figure 12.28 Materials processing applications (such as laser planarization or laser annealing) frequently require a "top hat" beam profile with adjustable x and y dimensions. (Courtesy of XMR, Fremont, CA)

The pressure vessel has an input and exhaust port for gas replenishment. Although the 5300 can run for approximately 10^7 pulses without replenishment, eventually parasitic chemical reactions will deplete the halogen and the system will need to be recharged.

The preionization is supplied by a corona wire. The corona wire is a quartz insulated conductor. The discharge volume created by the corona wire is approximately $70 \times 2.5 \times 2$ cm^3.

It is usually necessary to provide a buffer gas in an excimer laser mixture in order to initially support the discharge. Many excimers use helium, because it is chemically inert, inexpensive, has a high ionization potential, and forms stable low pressure discharges. It is also light and has a high specific heat for cooling. However, the 5300 uses neon as a buffer gas. Neon offers many advantages over helium in a system with corona wire preionization. Use of neon as a buffer gas increases both the output power and efficiency of the laser. This enhanced performance appears to be related to increased stability of the plasma.

The electrodes are pulsed at 20 to 50 kV. The rise time of the electrical pulse must be shorter than the upper state lifetime (5 to 15 ns). All of the circuit elements in the PFN are preferably insulated by a bath of transformer oil. The insulating oil is cooled by circulation through a heat exchanger. Electrical connections from the pulse forming network to the external circuitry are made with continuous cables sealed by compression o-rings.

The laser resonator uses a conventional high-reflecting rear mirror and partially transparent front mirror. The laser runs in a pseudowaveguide mode in the vertical direction and in a free-space mode in the horizontal. This creates a beam profile that is Gaussian only in the horizontal direction.

12.2.5 Laser Beam Homogenizers

Materials processing applications (such as laser planarization or laser annealing) frequently require a "top hat" beam profile (see Figure 12.28) with adjustable x and y dimensions.

Since most materials processing lasers lase on some combination of Gaussian modes, an optical element (called a homogenizer) is necessary to convert the Gaussian beam into the "top hat" profile. A large number of techniques have been used to homogenize laser beams. One class of techniques consists of using lenses, white cells, or waveguides to divide the beam into small parts and then recombine the parts into the desired shape.[94,95] A second class of techniques uses prisms or mirrors to fold the beam onto itself.[96]

XMR uses a homogenizer of the beam dividing type. In their homogenizer,[97] two sets of arrays of crossed cylindrical lenses (four groups of cylindrical lenses in total) are used to divide the incoming beam into an array of "beamlets" (see Figure 12.29). A focusing lens then recombines the beamlets into a "top hat" distribution on the exposure plane. In the XMR commercial unit, arrays 3 and 4 (and the focusing lens) are fixed. Arrays 1 and 2 are free to move. The separation of array 1 from array 3 determines the x dimension of the homogenized beam, while the separation of array 2 from array 4 determines the y dimension. Notice that the homogenized "top hat" distribution is always formed at the same location from the homogenizer, independent of the location of the arrays.

12.2.6 Application Highlight

Excimer lasers are employed in a large number of laser materials processing applications. The short wavelength, high energy, and high average power of excimer lasers present unique materials processing benefits.

For example, the polysilicon thin film transistor (poly-Si TFT) is considered to be the next technological step in the development of active-matrix liquid crystal displays.[98] Polysilicon has a number of advantages over amorphous silicon. As one example, polysilicon possesses more than 100 times the carrier mobility of amorphous silicon, thus enabling CMOS peripheral display drivers to be integrated right into the display. Displays with built-in drivers are more reliable than conventional displays, because electrical connections are implemented in metallization layers on the substrate itself instead of with thousands of TAB-bonded interconnects. Another benefit of eliminating TAB-bonding is that screen pitch is no longer limited by interconnection pitch, thus opening the door to super-high-resolution displays suitable for imaging and graphics applications. In addition, high mobility poly-Si TFT transistors offer fast charging speeds to compensate for RC time delays in larger displays.

However, commercializing the use of poly-Si TFT devices for active-matrix liquid-crystal displays has been difficult. The customary approach to creating poly-Si TFT devices is to use a quartz substrate with a layer of deposited amorphous silicon and slowly anneal (700°C for 24 hours) the amorphous silicon into a polysilicon thin film. However, the high cost of quartz substrates and the long process times required for furnace annealing have made

---

[94] Geary, *Optical Engineering* 27:972 (1988).

[95] Iwasaki et al., *Applied Optics* 29:1736 (1990).

[96] Bruno and Liu, *Lasers and Applications* (1987), p. 91.

[97] T. S. Fahlen, S. B. Hutchison, and T. McNulty, "Optical Beam Integration System," U.S. Patent #4,733,944, March 29, 1988.

[98] D. Zankowsky, *Laser Focus World* (1994).

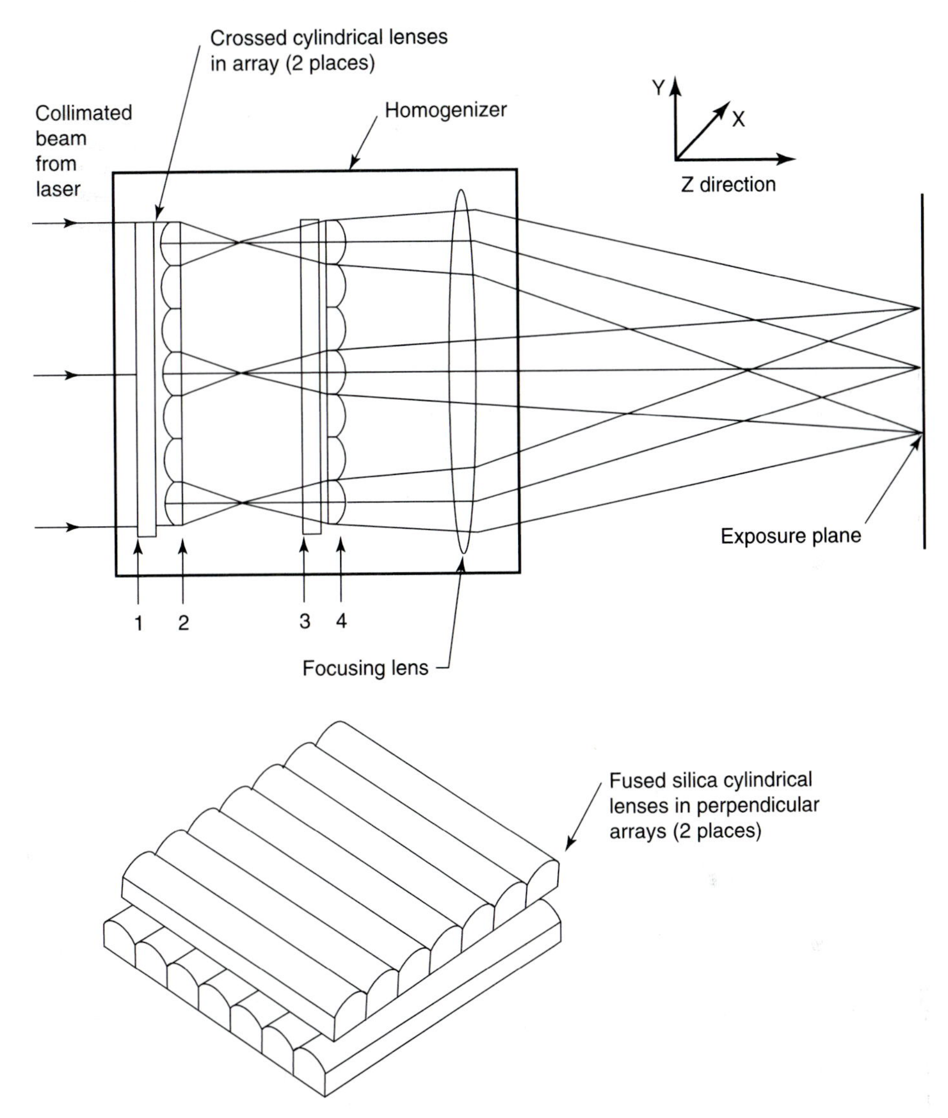

Figure 12.29 In XMR's homogenizer, two arrays of crossed cylindrical lenses (four groups of cylindrical lenses in total) are used to divide the incoming beam into an array of "beamlets." A focusing lens then recombines the beamlets into a "top hat" distribution on the exposure plane. (Courtesy of XMR, Fremont, CA)

this process impractical for high-volume production of large displays. Unfortunately, lower-priced glass substrates cannot be used, because the high temperatures used in conventional annealing processes also melt the glass substrate.

Laser annealing offers an alternative to furnace annealing for creating economical poly-Si TFT devices (see Figure 12.30). In laser annealing, the laser only heats the surface of the material, thus allowing materials other than expensive quartz to be used as substrates.

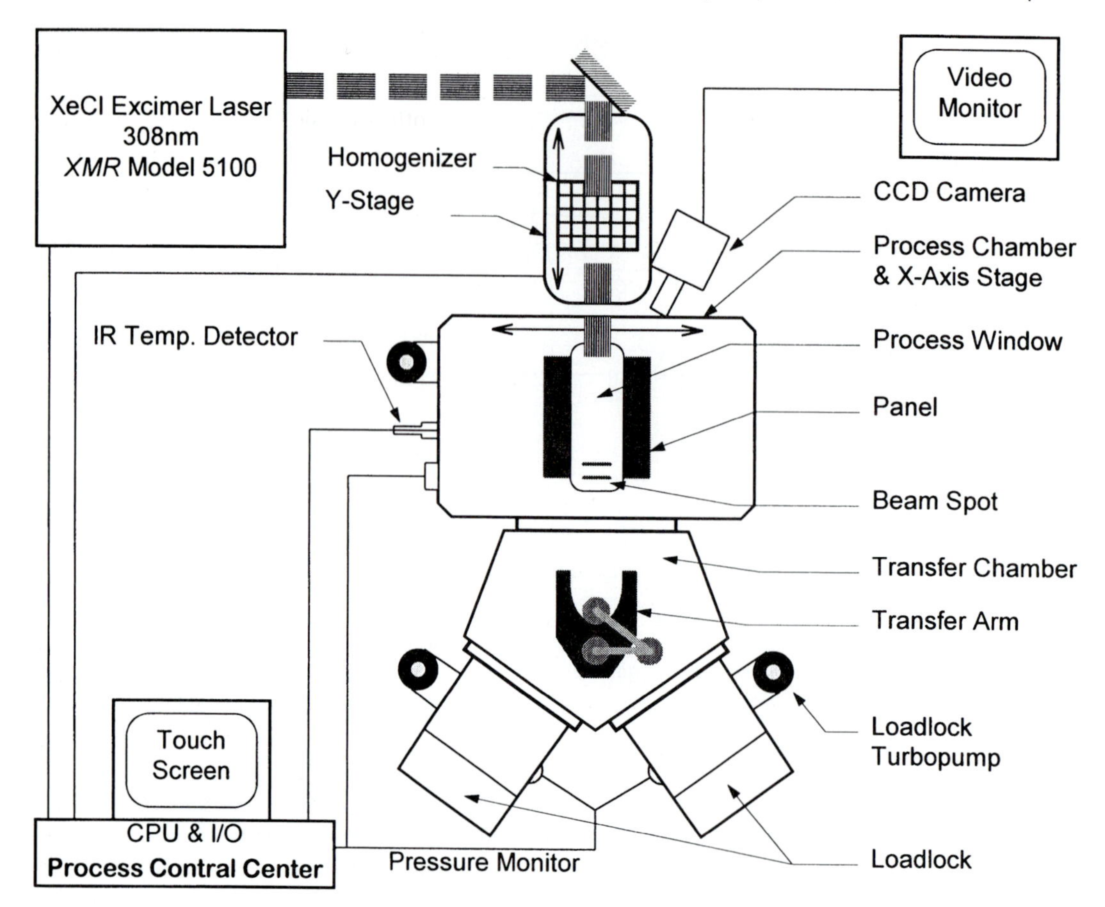

Figure 12.30 Schematic of an excimer laser annealing system for manufacturing poly-Si TFT devices. (Courtesy of XMR, Fremont, CA)

Both argon and excimer lasers have been used to anneal amorphous silicon to polysilicon. However, excimer lasers have a significant advantage over argon lasers in this application. This is because excimers are high power (200+ watt), short pulse (20- to 50-ns) lasers. As a consequence, they only locally heat the surface of the film. The melt region extends just 50 to 100 nm below the surface. Excimer laser annealing is the only current annealing technology that can reliably transform amorphous silicon into polysilicon while the substrate remains at room temperature.

The basic process for creating polysilicon thin films by laser annealing is quite straightforward. Before processing, the substrate is covered with an amorphous silicon thin film. The excimer then scans the glass substrate with a spatially uniform pattern of overlapping rectangular pulses. The UV energy is strongly absorbed by the surface layer of amorphous silicon, which rapidly melts and recrystallizes into polysilicon. The instantaneous local temperature in the surface layer is high enough to achieve excellent crystallinity. However, the average temperature of the glass substrate remains substantially below the damage threshold and is unaffected by the laser crystallization process.

Since laser annealing is a scanning process, it offers considerable present and future process flexibility. The manufacturer has the option of selectively annealing designated areas or moving to larger or smaller substrates. In addition, laser annealing provides the high throughputs required for high-volume active-matrix liquid crystal display manufacturing. The high peak energy of the excimer laser, combined with a 300 Hz pulse repetition rate, enables 360 × 450 mm substrates to be processed in as little as three minutes.

12.3 OVERVIEW OF SEMICONDUCTOR DIODE LASERS

12.3.1 History of Semiconductor Diode Lasers

There are often periods in science when major discoveries are made simultaneously by a number of different researchers. The early history of semiconductor diode lasers follows this pattern.

Consider the research environment for lasers in mid-1962. The existing lasers were lasers with long path lengths and large external resonators. These lasers lased on narrow transitions between well-defined discrete states. Semiconductors (with a high free carrier absorption, wide bandwidth optical transition, and relatively small size) were assumed to have inefficient radiative recombination and were considered an improbable material for laser action.

However, in July 1962 Keyes and Quist presented a paper, "Recombination Radiation Emitted by Gallium Arsenide Diodes," at the Solid State Device Research Conference (SSDRC) in Durham, NH.[99,100] This paper described intense luminescence with a quantum efficiency of approximately 85% from GaAs junctions at 77°K. These results were so startling that they energized a number of research groups to consider the question of semiconductor lasers. These included research groups at General Electric, IBM, and MIT Lincoln Laboratories. (Notice that all of these groups achieved semiconductor laser operation in the latter half of 1962!)

The General Electric group used n-type GaAs wafers with a zinc diffusion to form a degenerate PN-junction. (The importance of the degenerate junction was pointed out by Bernard and Duraffourg in 1961.[101]) The resulting wafers were cut into strips approximately 0.5 mm wide and cemented to plates so that the edges could be lapped and polished. These polished edges formed the Fabry-Perot resonator structure. The resulting diodes were cooled to 77°K and operated in pulsed mode. The paper announcing the first semiconductor laser operation was the General Electric paper submitted in September 1962 and appearing in the November 1, 1962, edition of *Physical Review Letters*.[102]

---

[99]R. J. Keyes and T. M. Quist, "Recombination Radiation Emitted by Gallium Arsenide Diodes," presented at the *Solid State Device Res. Conf.*, 1962.

[100]The material in these papers was later published in R. J. Keyes and T. M. Quist, *Proc. IRE* 50:1822 (1962).

[101]M. G. A. Bernard and G. Duraffourg, *Phys. Stat. Sol.* 1:699 (1961).

[102]R. N. Hall, G. E. Fenner, J. D. Kingsley, T. J. Soltys, and R. O. Carlson, *Phys. Rev. Lett.* 9:366 (1962).

The IBM group also used n-type GaAs wafers with a zinc diffusion to form a degenerate PN-junction. The IBM group was struggling with the concept of how to make a resonator, and so decided to attempt to demonstrate stimulated emission without a resonator. The IBM *Applied Physics Letter* reporting the coherent stimulated emission from a GaAs PN-junction was submitted in October 1962, appearing within days of the publication of the General Electric *Physical Review Letter*.[103]

The MIT Lincoln Laboratories group also used n-type GaAs wafers with a zinc diffusion. They elected to use mechanical polishing to form the resonator structure and achieved laser action within a few weeks of the other groups. Their first paper was submitted in November 1962, and their results were reported in a pair of experimental and theoretical papers.[104, 105] The MIT Lincoln Laboratories group was also noteworthy for being the first group to address the question of applications for the new lasers by demonstrating TV transmission from a rooftop, and then from a nearby mountain top.[106]

Very shortly after the first demonstration of laser action in a degenerate GaAs PN-junction, a number of interesting and important advancements occurred. Just a few weeks after the initial announcement of GaAs diodes by the General Electric group,[107] Holonyak and Bevacqua demonstrated the first visible (710 nm) laser diode in $GaAs_{1-x}P$, using polished facets[108] (the Holonyak and Bevacqua paper was submitted in October 1962). The first cw operation (at 2°K) was observed by Howard et al.,[109] and the first pulsed room-temperature operation by Burns et al.[110]

The issue of polishing versus cleaving to create the small Fabry-Perot cavities was a critical one in early semiconductor laser development. Although cleaving is now considered to be somewhat easier than polishing, the first three lasers to be demonstrated all incorporated polished Fabry-Perot cavities. Holoyak originally attempted cleaving, but returned to polishing after difficulties in cleaving $GaAs_{1-x}P$. The first reported use of cleaving is by Bond et al. in 1963.[111] Dill and Rutz patented the cleaving concept from a disclosure filed on October 20, 1962.[112]

Laser research then began to diverge into a number of separate areas. The idea of using heterostructures in semiconductor diode lasers was a very powerful idea and was

---

[103] M. I. Nathan, W. P. Dumke, G. Burns, R. H. Dill, Jr., and G. Lasher, *Appl. Phys. Lett.* 1:62 (1962).

[104] T. M. Quist, R. H. Rediker, R. J. Keyes, W. E. Krag, B. Lax, A. L. McWhorter, and H. J. Zeiger, *Appl. Phys. Lett.* 1:91 (1962).

[105] A. L. McWhorter, H. J. Zeiger, and B. Lax, *J. Appl. Phys.* 34:235 (1963).

[106] R. H. Rediker, R. J. Keyes, T. M. Quist, M. J. Hudson, C. R. Grant, and R. G. Burgess, "Gallium Arsenide Diode Sends Television by Infrared Beam," *Electron.* 35: 44–45 (1962); and R. J. Keyes, T. M. Quist, R. H. Rediker, M. J. Hudson, C. R. Grant, and J. W. Meyer, "Modulated Infrared Diode Spans 30 Miles," *Electron.* 36: 38–39 (1963).

[107] R. N. Hall, G. E. Fenner, J. D. Kingsley, T. J. Soltys, and R. O. Carlson, *Phys. Rev. Lett.* 9:366 (1962).

[108] N. Holonyak, Jr., and S. F. Bevacqua, *Appl. Phys. Lett.* 1:82 (1962).

[109] W. E. Howard, F. F. Fang, F. H. Dill, Jr., and M. I. Nathan, *IBM J. Res. Develop.* 7:74 (1962).

[110] G. Burns and M. I. Nathan, *IBM J. Res. Develop.* 7:68 (1962).

[111] W. L. Bond, B. G. Cohen, R. C. C. Leite, and A. Yariv, *Appl. Phys. Lett.* 2:57 (1963).

[112] F. H. Dill, Jr., and R. F. Rutz, "Method of Fabrication of Crystalline Shapes," U.S. Patent #3,247,576, filed October 30, 1962; issued April 16, 1966.

first proposed by Kroemer in 1963.[113] However, the heterostructure concept required the development of appropriate materials-processing technologies, such as liquid phase epitaxy (LPE), metal-organic chemical vapor deposition (MOCVD), and molecular beam epitaxy (MBE). Therefore, the original proposal of Kroemer went relatively unnoticed.

In 1963, Nelson first demonstrated the liquid phase epitaxy growth of GaAs on GaAs.[114] In 1967, Woodall et al. demonstrated the growth of the heterostructure $Al_xGa_{1-x}As$ on GaAs.[115] The first LPE pulsed room temperature heterostructure lasers soon followed in 1969[116, 117, 118] and then cw room temperature heterostructure lasers.[119, 120] Quantum confinement was the next major advancement in heterostructures. In 1978, Dupuis et al. first demonstrated a room-temperature quantum well laser.[121] Strained quantum well lasers soon followed.[122, 123, 124]

Distributed feedback lasers (DFB lasers) were another major track in laser development. DFB lasers incorporate an intrinsic grating to force single longitudinal mode operation. Kogelnik and Shank first developed the experimental and theoretical ideas behind DFB lasers.[125] Nakamura et al. reported the DFB semiconductor laser oscillation by photopumping GaAs.[126] Scifres et al.[127] reported the first AlGaAs/GaAs single heterostructure DFB junction lasers. Casey et al.[128] and Nakamura et al.[129] used an AlGaAs/GaAs separately confined heterostructure to achieve the first DFB cw oscillation at room temperature. Room-temperature cw operation of a DFB laser at the fiber-optical communication wavelength of 1.5 μm in InGaAsP/InP was first reported by Utaka et al.[130]

Distributed Bragg reflector lasers (DBR lasers) are a variation on DFB lasers where the reflectors are outside (rather than inside) the active region. Lasing occurs between the

---

[113] H. Kroemer, *Proc. IEEE* 51:1782 (1963).

[114] H. Nelson, *RCA Rev.* 24:603 (1963).

[115] J. M. Woodall, H. Rupprecht, and G. D. Pettit, *Solid-State Device Conf.*, 1967. Abstracts reported in *IEEE Trans. Electron. Devices* ED-14:630 (1967).

[116] H. Kressel and H. Nelson, *RCA Rev.* 30:106 (1969).

[117] I. Hayashi, M. B. Panish, and P. W. Foy, *IEEE J. Quantum Electron.* QE-5:211 (1969).

[118] Zh. I. Alferov, V. M. Andreev, E. L. Portnoi, and M. K. Trukan, *Fiz. Tekh. Poluprovodn.* 3:1328 (1969) [*Sov. Phys. Semicond.* 3:1107 (1970)].

[119] I. Hayashi, M. B. Panish, P. W. Foy, and S. Sumuski, *Appl. Phys. Lett.* 17:109 (1970).

[120] Zh. I. Alferov, V. M. Andreev, D. Z. Garbuzov, Yu. V. Zhilyaev, E. P. Morozov, E. L. Portnoi, and V. G. Trofim, *Fiz. Tekh. Poluprovodn.* 4:1826 (1970). [*Sov. Phys. Semicond.* 4:1573 (1971)].

[121] D. Dupuis, R. D. Dapkus, N. Holonyak, Jr., E. A. Rezek, and R. Chin, *Appl. Phys. Lett.* 32:295 (1978).

[122] E. Yablonovitch and E. O. Kane, *J. Lightwave Technol.* LT-4:504 (1986); 6:1292 (1988).

[123] D. P. Bour, D. B. Gilbert, L. Elbaum, and M. G. Harvey, *Appl. Phys. Lett.* 53:2371 (1988).

[124] P. J. A. Thijs and T. van Dongen, *Electron. Lett.* 25:1735 (1989).

[125] H. Kogelnik and C. V. Shank, *Appl. Phys. Lett.* 18:152 (1971); and H. Kogelnik and C. V. Shank, *Appl. Phys. Lett.* 43:2327 (1972).

[126] M. Nakamura, A. Yariv, H. W. Yen, and S. Somekh, *Appl. Phys. Lett.* 22:515 (1973).

[127] D. R. Scifres, R. D. Burnham, and W. Streifer, *Appl. Phys. Lett.* 25:203 (1974).

[128] H. C. Casey, Jr., S. Somekh, and M. Ilegems, *Appl. Phys. Lett.* 27:142 (1975).

[129] M. Nakamura, A. Aiki, J. Umeda, and A. Yariv, *Appl. Phys. Lett.* 27:403 (1975).

[130] K. Utaka, S. Akiba, K. Sakai, and Y. Matsushima, *Electron. Lett.* 17:961 (1981).

grating mirrors, or between one grating and a conventional facet. The first DBR laser was an AlGaAs/GaAs laser demonstrated by Reinhart et al.[131]

Vertical-cavity surface-emitting lasers (VCSEL) were later developments in semiconductor laser technology.[132] Soda et al. demonstrated the first VCSEL.[133] This was a double heterostructure InGaAsP device operating at 1.3 μm. It used metal mirrors and lased at 77°K. In 1982, Burnham, Scifres and Streifer filed a patent on various designs for VCSEL devices.[134] Epitaxial mirrors for VCSEL devices were first demonstrated in 1983.[135] In 1989, the first cw room temperature VCSEL device was demonstrated by Jewell et al.[136] This device was an quantum well active region sandwiched between n- and p-doped semiconductor Bragg reflectors.

Historical details of this era from the perspective of the major participants are available in a series of review papers in volume QE-23 of the *IEEE Journal of Quantum Electronics*.[137] Additional historical details can be found in the reviews by Hall,[138] Rediker,[139] and in the books by Kressel and Butler,[140] and by Casey and Panish.[141] IEEE has also published a summary of seminal papers on semiconductor diode lasers.[142]

12.3.2 The Basics of the Semiconductor Diode Laser

Energy band structure. The semiconductor laser can be modeled as a two-level laser system. The upper laser state is the conduction band, and the lower laser state is the valence band. The laser wavelength is emitted at the bandgap of the semiconductor.

In order for the semiconductor to have sufficient gain to operate as a laser, it is usually necessary for the electron transition from the conduction band to the valence band to be a direct radiative transition. Gallium arsenide (GaAs) is an example of a direct semiconductor. Silicon is an example of an indirect semiconductor. The distinction between a direct and indirect semiconductor material is often explained using the E (energy) versus k (momentum)

---

[131]F. K. Reinhart, R. A. Logan, and C. V. Shank, *Appl. Phys. Lett.* 27:45 (1975).

[132]J. Jewell, *Scientific American* November: 86–94 (1991).

[133]H. Soda, K. Iga, C. Yitahara, and Y. Suematsu, *Jpn. J. Appl. Phys.* 18:2329 (1979).

[134]R. Burnham, D. R. Scifres, and W. Streifer, U.S. Patent #4,309,670, Jan. 1982.

[135]A. Chailertvanitkul, S. Uchiyama, Y. Kotaki, Y. Kokubun, and K. Iga, *Annual Meet. Jpn. Sec. Appl. Phys.* (1983).

[136]J. L. Jewell, A. Scherer, S. L. McCall, Y. H. Lee, S. Walker, J. P. Harbison, and L. T. Florez, *Electron. Lett.* 25:1123 (1989).

[137]R. N. Hall, *IEEE J. of Quantum Electron.* QE-23:674 (1987); M. I. Nathan, *IEEE J. of Quantum Electron.* QE-23:679 (1987); N. Holonyak, *IEEE J. of Quantum Electron.* QE-23:684 (1987), and R. H. Rediker, *IEEE J. of Quantum Electron.* QE-23:692 (1987).

[138]R. N. Hall, "Injection lasers," *IEEE Trans. Electron Dev.* ED-23:700 (1976).

[139]R. H. Rediker, I. Melngailis, and A. Mooradian, "Lasers, Their Development and Applications at M.I.T. Lincoln Laboratory," *IEEE J. Quantum Electron.* QE-20:602 (1984).

[140]H. Kressel and J. K. Butler, *Semiconductor Lasers and Heterojunction LEDs* (New York: Academic Press, 1977).

[141]H. C. Casey, Jr., and M. B. Panish, *Heterostructure Lasers, Parts A and B* (New York: Academic Press, 1978).

[142]W. Streifer and M. Ettenberg, eds, *Semiconductor Diode Lasers* (New York: IEEE Press, 1991).

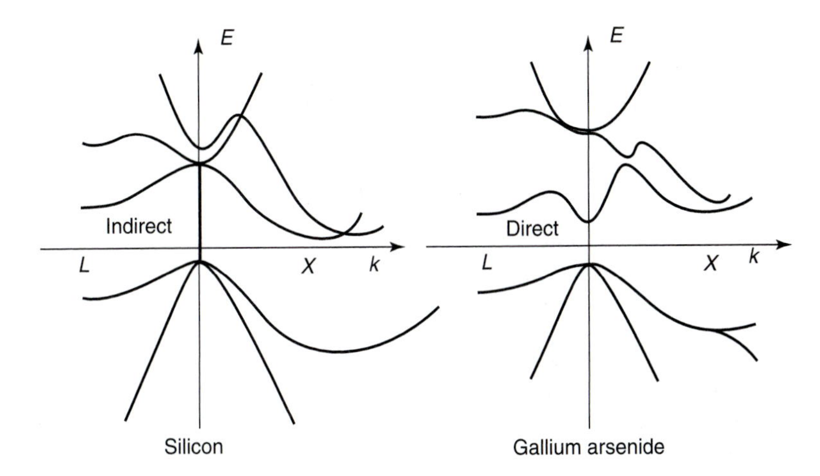

Figure 12.31 The distinction between a direct and indirect semiconductor material is often explained using the E (energy) versus k (momentum) diagram for the semiconductor. If the minima of the conduction band and the maxima of the valence band both occur at $k = 0$, then the semiconductor is direct. If the minima and maxima do not overlap at all, or overlap at $k \neq 0$, then the semiconductor is indirect.

diagram for the semiconductor. If the minima of the conduction band and the maxima of the valence band both occur at $k = 0$, then the semiconductor is direct. If the minima and maxima do not overlap at all or overlap at $k \neq 0$, then the semiconductor is indirect. Examples of E versus k diagrams for indirect and direct materials are given in Figure 12.31.

Pumping the semiconductor diode laser. In order for the semiconductor to lase, it is necessary to create a population inversion between the valence and the conduction bands. Such a population inversion can be created by external pumping (lasers, electron beams, or flashlamps) or by internally pumping (with a PN-junction).

The simplest way to create a population inversion is to pump the semiconductor with another light source. This technique is termed photopumping and is commonly used to test new semiconductor laser materials in order to determine their suitability for laser operation. One way to photopump a semiconductor laser is to use a conventional gas laser source (such as a HeNe or an argon-ion) as a pump source. This has the advantage of simplicity, but tends to overheat the laser material. For this reason, photopumped laser slices are generally thinned and mounted in thermally conductive holders. Another way to photopump is to use a conventional pulsed laser source (such as a Nd:YAG or glass laser). This not only solves the thermal problems, but also permits testing of recombination rates in the material. A very clever way to photopump a diode laser material is with another diode laser. The sample is thinned and mounted directly on the output window of the diode laser. Although the actual diode laser power is relatively small, the coupling efficiency between the diode laser and the photopumped laser is excellent.

However, the majority of commercial semiconductor lasers are electrically pumped using a PN-junction. Under conditions of high current injection in a PN-junction, a re-

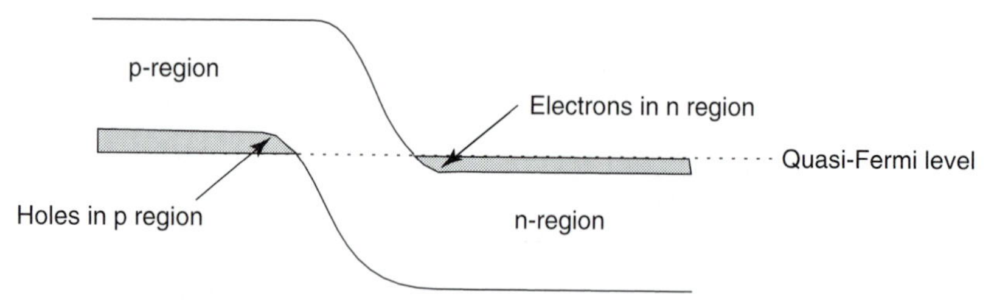

Figure 12.32 In equilibrium, the quasi-Fermi levels of the degenerate p- and n-junction laser material align.

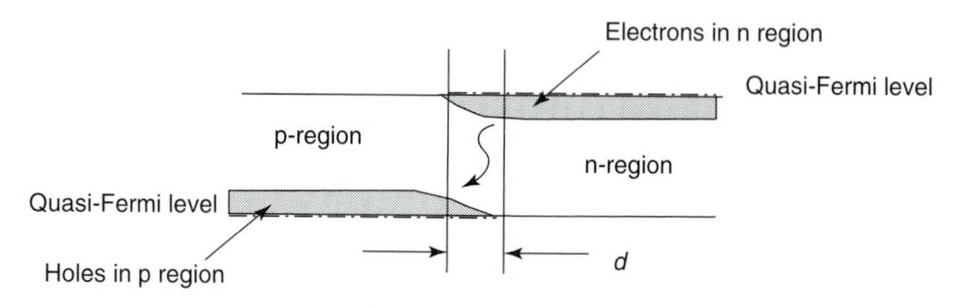

Figure 12.33 The quasi-Fermi levels will misalign by the value of the applied voltage. Under the influence of the forward bias, the holes will drift to the n-region and the electrons will drift into the p-region. The holes and electrons are now spatially coincident and hole-electron recombination occurs.

gion near the depletion region will contain an inverted population of electrons and holes. Appropriate alignment of this region with the cleaved facets will result in laser operation.

Consider the energy band diagram for a simple PN-junction intended for use in a semiconductor photon source. Initially, one side of the junction is heavily doped n and the other side is heavily doped p. In general, both materials are doped degenerately (meaning that the quasi-Fermi level is above the bandedge). In equilibrium the quasi-Fermi levels align as shown in Figure 12.32.

When a positive voltage is applied to the p-region and a negative voltage is applied to the n-region, then the diode is forward biased. The quasi-Fermi levels will misalign by the value of the applied voltage. Under the influence of the forward bias, the holes will drift to the n-region and the electrons will drift into the p-region. The holes and electrons are now spatially coincident and hole-electron recombination occurs (see Figure 12.33).

Notice that in the nonequilibrium situation, the quasi-Fermi levels for the electrons and holes have separated. The spatial distance between the quasi-Fermi levels is essentially the depth d of the active region. In III-V semiconductors (the most common semiconductors for commercial laser diodes), the electrons are much more mobile than the holes. Therefore, the depth of the active region is principally determined by the mobility of the electrons. This distance is approximately given by

$$d = \sqrt{D_n \tau_n} = \sqrt{\frac{D_n n_p}{dn_p/dt}}$$

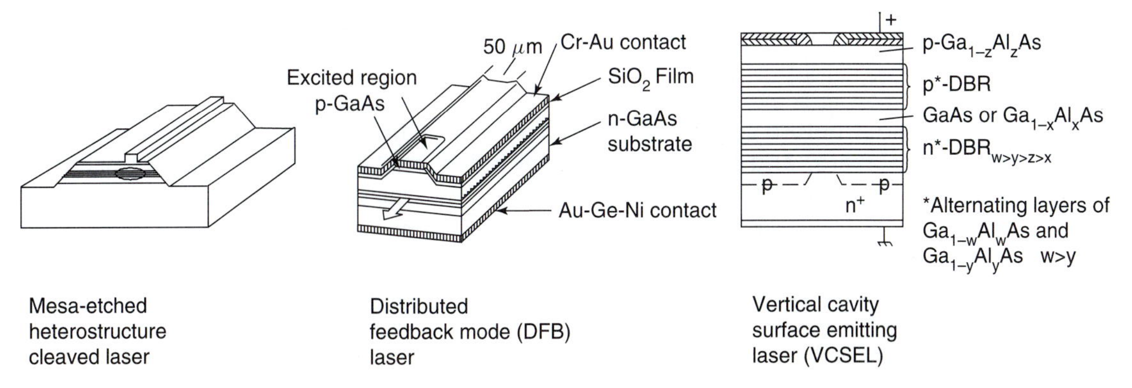

Figure 12.34 A few of the possible semiconductor laser resonator structures: DFB is from K. Aiki, M. Nakamura, and J. Umeda, "Schematic Structure of a GaAlAs/GaAS Distributed Feedback Laser," *IEEE J. of Quantum Electron.* QE-12:597 (1976), ©1976 IEEE; VCSEL is from R. Burnham, D.R. Scifres, and W. Streifer, U.S. Patent #4,309,670 (1982).

where n_p is the electron concentration in the p-region and D_n is the diffusion constant for electrons. Typical numbers for d are on the order of 1 μm. (Again, notice that this is on the order of the same size as the wavelength of the electromagnetic mode being amplified.)

Creating the semiconductor diode laser resonator. There are an astonishing number of possible forms for the semiconductor laser resonator (see Figure 12.34).

In the simplest form, a semiconductor laser consists of small rectangular slab of semiconductor material with two polished or cleaved facets to act as resonator mirrors. The other facets are destroyed in some way (etched, ground, sawn, ion implanted, etc.) in order to avoid spurious laser modes.

The typical horizontal cleaved laser construction can be modified in a number of ways. For example, the facets can be etched using wet or dry processing techniques. External mirrors, gratings, or combinations of mirrors and gratings can be used in place of cleaved facets. External elements can be incorporated off-chip or integrated into the chip.

Semiconductor lasers can be fabricated to lase vertically. With this type of construction, reflecting layers are fabricated on the substrate and the laser operates perpendicularly to the substrate. Reflecting layers can be made from metal, dielectric films, and semiconductor multilayers. Such lasers are called VCSEL.

It is also possible to incorporate gratings directly into the laser design. This results in a laser with a single longitudinal mode forced by distributed feedback from the internal grating. Such lasers are termed DFB lasers. In DBR lasers, a variation on DFB lasers, lasing occurs between the grating mirrors, or between one grating and a conventional facet.

Light propagation in the semiconductor diode laser. In a laser diode, the photons must be modeled as collective entities traveling in a confined fashion down a waveguide. This gives a different character to the issue of modeling light output from a laser diode as compared to a conventional laser. In general, the far-field pattern from a laser diode is determined by the Fourier integral of the electromagnetic field propagating in the waveguide. The electromagnetic field within the waveguide can be calculated using the eigenvalue equations for wave propagation in a slab waveguide (for example, see Casey and Panish[143]).

Homostructure semiconductor diode lasers, heterostructure semiconductor diode lasers, and the importance of lattice matching. Early semiconductor lasers were constructed from n- and p-type layers of the same semiconductor material. Such lasers are called homostructure lasers. Homostructure lasers have the advantage of structural simplicity.

However, semiconductor lasers can also be constructed from n- and p- type layers of different semiconductor materials. Such lasers are called heterostructure lasers. The energy band structure and index of refraction profiles of heterostructure lasers can be tailored to meet a given application.

The majority of commercial heterostructure semiconductor lasers are fabricated from semiconductor materials in columns III and V of the periodic table. [Column III is boron (B), aluminum (Al), gallium (Ga), indium (In), and thallium (Tl); column V is nitrogen (N), phosphorus (P), arsenic (As), antimony (Sb), and bismuth (Bi).] Common laser materials include virtually all combinations of Al, Ga, and In, with P and As. Some work has been done with B, Al, and Ga, with N; as well as Al, Ga, and In with Sb. Very little has been done with Tl and Bi.

Heterostructure lasers require layering these different materials. This is a very complex problem, as the materials have different physical properties. Perhaps the most important of these properties is the lattice spacing. If the materials do not have the same lattice spacing, then dislocations can appear in the semiconductor laser. In addition to the structural difficulties imposed by dislocations, dislocations can also be highly detrimental to semiconductor laser operation as they can serve as a nonradiative sink for carriers.

A very useful diagram for visualizing lattice match in heterostructure lasers is the energy versus lattice constant diagram. (An example of such a diagram for III-V materials is given in Figure 12.35). Notice that only the AlAs-GaAs system is lattice-matched across the entire compositional range. This is one major reason for the widespread use of AlGaAs/GaAs heterostructures on GaAs substrates for semiconductor laser diodes.

12.3.3 Confinement in the Semiconductor Diode Laser

Since the electromagnetic mode in semiconductor lasers is on the order of the size of the laser device, then horizontal and vertical confinement are important issues in semiconductor laser design. These issues usually do not arise in conventional laser design, as conventional lasers are typically operating in a propagation mode where the wavelength is much smaller than the resonator dimensions.

Vertical confinement. Typically vertical confinement is provided by creating material combinations with index of refraction profiles that confine the optical wave. Some of these combinations offer the additional advantage of confining the carriers as well.

---

[143] H. C. Casey, Jr., and M. B. Panish, *Heterostructure Lasers, Parts A and B* (New York: Academic Press, 1978), Chapter 2.

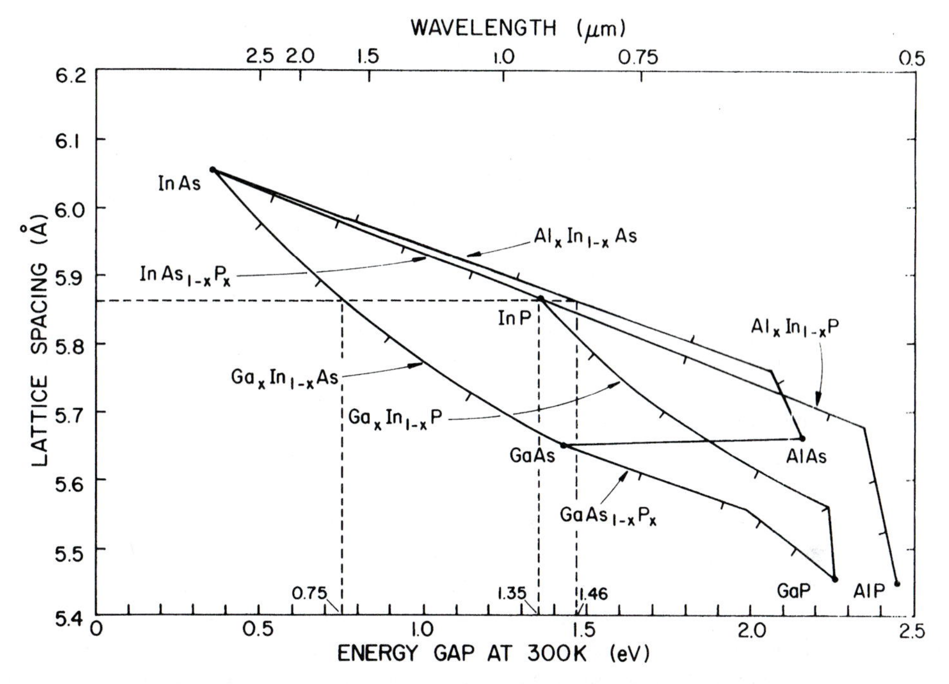

Figure 12.35 The energy versus lattice constant diagram is very useful for visualizing lattice match in heterostructure lasers. (From E. C. H. Parker, *Physics of Molecular Beam Epitaxy* (New York: Plenum Press, 1985), p. 277, Figure 2. Copyright Plenum Press.)

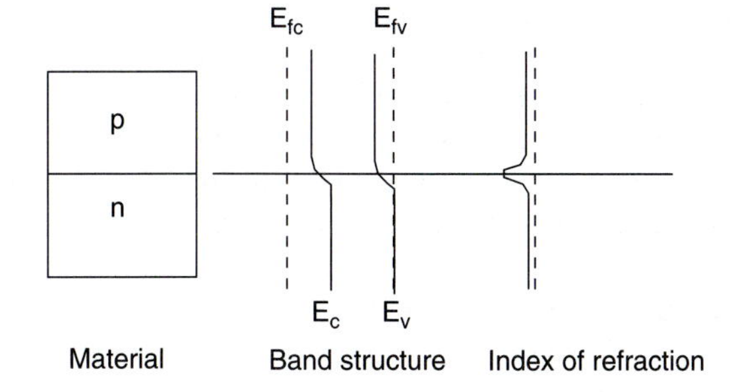

Figure 12.36 A typical homostructure laser consists of one material type with degenerately doped p and n regions. There is only a slight refractive index change at the n-junction to assist in vertical optical confinement of the laser beam.

Homostructure lasers were the very first type of semiconductor laser structure demonstrated. A typical homostructure laser (see Figure 12.36) consists of one material type (for example, GaAs) with degenerately doped p and n regions. There is only a slight refractive index change at the n-junction to assist in vertical optical confinement of the laser beam. Homostructure lasers offer the advantages of simplicity, but the disadvantages of poor confinement.

Single heterostructure lasers were the first type of semiconductor heterostructure laser developed. A typical single heterostructure GaAs laser (see Figure 12.37) is composed of a homojunction GaAs laser diode followed by a p-type AlGaAs layer (with a larger bandgap and lower index of refraction). This geometry creates an index of refraction bump in the p-type GaAs layer. The index of refraction bump provides some vertical confinement for the laser action occurring in the p-type GaAs. Single heterostructure lasers provide significantly more vertical confinement than homostructure lasers, but require more complex heterostructure growth processes.

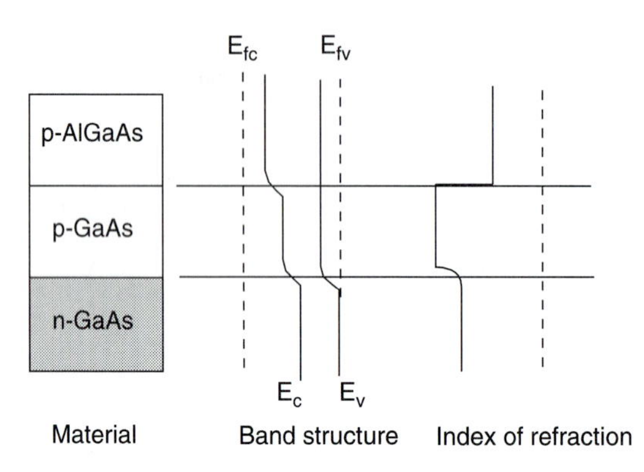

Figure 12.37 A typical single heterostructure laser is composed of a homojunction laser diode followed by a p-type layer with a larger bandgap and lower index of refraction. This geometry creates an index of refraction bump in the p-type GaAs layer. The index of refraction bump provides some vertical confinement for the laser action.

Double heterostructure (DH) lasers have become the standard heterostructure laser type. A typical DH GaAs laser (see Figure 12.38) is composed of a p-type GaAs layer sandwiched between n-type and p-type AlGaAs layers. This geometry creates a large index of refraction bump in the p-type GaAs layer. The index of refraction bump provides vertical confinement for the laser action occurring in the p-type GaAs. DH lasers offer a very nice compromise between the advantages of an index guiding vertical confinement layer, and the disadvantages of heterostructure laser growth.

GRINSCH (graded-index separate confinement heterostructure) lasers offer both optical and electrical confinement in an effort to achieve lower thresholds and narrower beam patterns. A GRINSCH laser is basically a DH laser, but one where the AlGaAs layer has been graded down to meet the GaAs layer (see Figure 12.39). This provides both electron confinement (from the energetic well created by the graded bandgap) and optical confinement (from the index of refraction bump between the AlGaAs and GaAs materials).

GRINSCH lasers frequently find application as quantum well lasers, because the graded bandgap provides a good "funnel" to confine the electrons to the the quantum well regions (see Figure 12.40). MQW-GRINSCH lasers are very popular geometries as they combine many of the best features of DH lasers and quantum wells.

Horizontal confinement. For many semiconductor laser diode applications it is necessary to confine the carriers both in the horizontal and vertical directions. The most popular method for horizontal confinement is to restrict the size of the gain region, a method called *gain confinement*. There are a large number of gain confinement techniques, but the most popular tend to fall into the categories of contact stripe, mesa etch, ion-implantation methods, and regrowth techniques (such as buried heterostructures).

In the contact stripe method, the gain confinement is provided by restricting the spatial extent of the metal contact (see Figure 12.41). This method offers processing simplicity, but provides relatively poor confinement due to electron transport underneath the stripe.

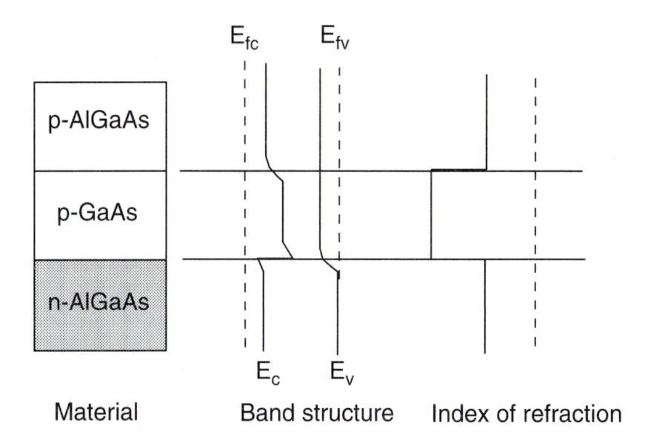

Figure 12.38 A typical double heterostructure laser is composed of a p-type lower bandgap layer sandwiched between n-type and p-type higher bandgap layers. This geometry creates a large index of refraction bump in the p-type layer. The index of refraction bump provides vertical confinement for the laser action occuring in the p-type layer.

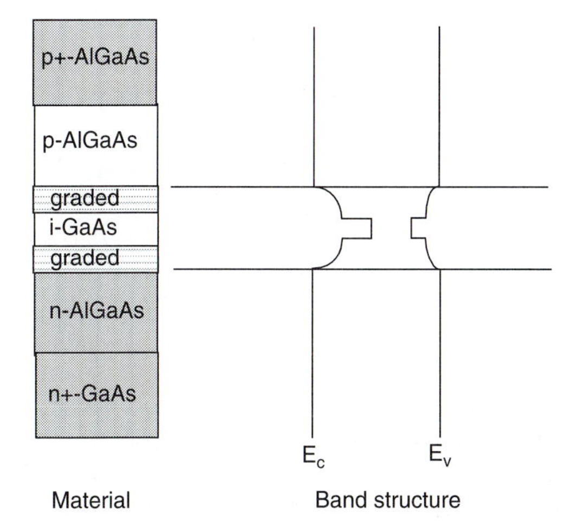

Figure 12.39 A GRINSCH laser is basically a DH laser, but one where the higher bandgap layer has been graded down to meet the lower bandgap layer. This provides both electron confinement (from the energetic well created by the graded bandgap) and optical confinement (from the index of refraction bump between the high and low index materials).

In the mesa etch method, the gain confinement is provided by etching one or several layers of the structure (see Figure 12.42). This method is more difficult to process (as it is both highly nonplanar and requires a sensitive etch stop), but it offers significantly better confinement than the contact stripe method.

In the ion-implantation method, ions are implanted in regions other than the region of interest (see Figure 12.43). This method offers process simplicity and provides a good quality output beam.

In the buried heterostructure method, the gain region is surrounded in both the vertical and horizontal directions by a waveguide of material (see Figure 12.44). Thus, the buried heterostructure process is a two-part method. First, an initial set of layers is grown in an epitaxial system. Then, the wafer is removed from the epitaxial growth system, processed, and returned to the system for new layers to be grown on the material. This process provides outstanding confinement, as the laser is essentially operating in a square waveguide. However, the two-part process cycle is exceptionally complex.

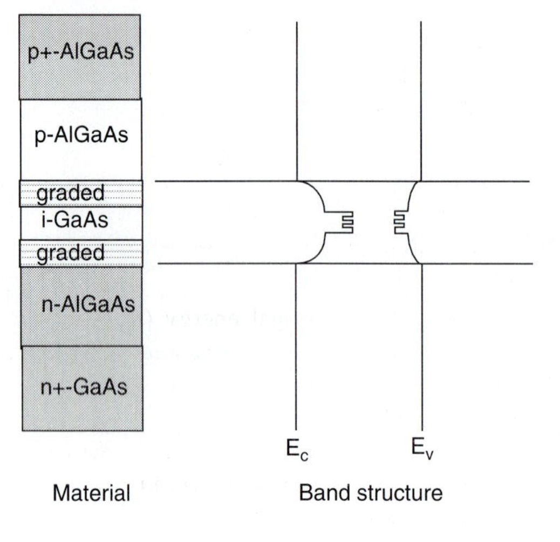

Figure 12.40 GRINSCH lasers frequently find application as quantum well lasers, because the graded bandgap provides a good "funnel" to confine the electrons to the quantum well regions.

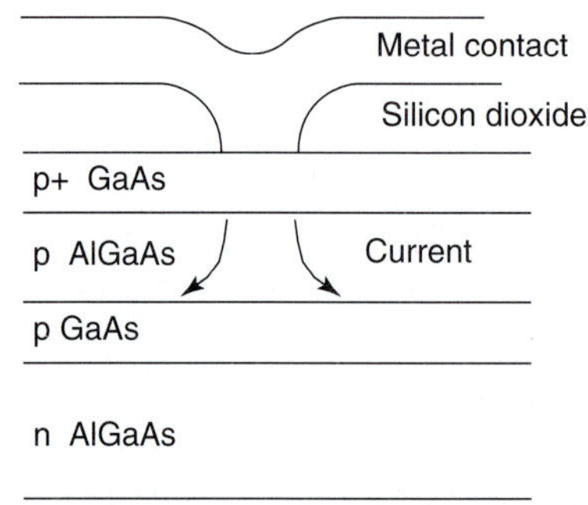

Figure 12.41 In the contact stripe method, the gain confinement is provided by restricting the spatial extent of the metal contact.

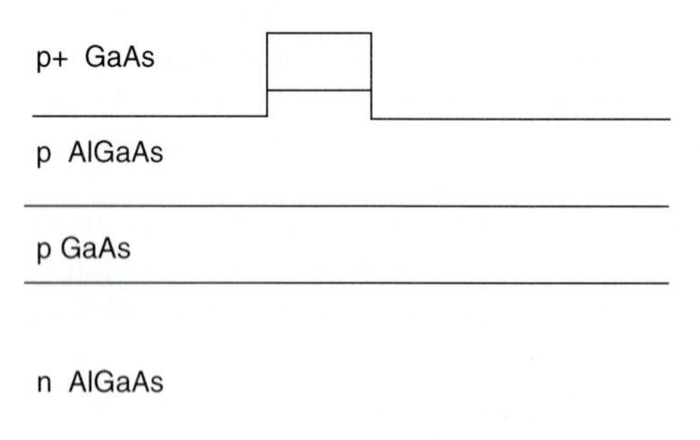

Figure 12.42 In the mesa etch method, the gain confinement is provided by etching one or several layers of the structure.

12.3.4 The Quantum Well Semiconductor Diode Laser

A quantum well is a heterostructure composed of two materials of differing bandgap. The two materials are fabricated in a sandwich with the higher bandgap material forming the outside (or barrier) and the lower bandgap material forming the inside (or well). Under these conditions, energy wells are created in both the conduction and valence bands of the heterostructure (see Figure 12.45).

Discrete quantum states (similar to states in a hydrogen atom) will form in the conduction and valence wells (see Figure 12.46). As with a hydrogen atom, electrons transitioning between these states will produce photons possessing the difference energy between the states.

However, unlike a hydrogen atom, the energy of these quantum well states is a function of the width of the quantum wells. The narrower the quantum well, the larger the resulting transitional energy (to the eventual limit that the transition energy between the first valence and first conduction band state matches the energy gap of the barrier material). The wider the quantum well, the smaller the resulting transitional energy (to the eventual limit that the transition energy between the first valence and first conduction band state matches the energy gap of the well material).

In engineering practice, quantum wells provide the means to manufacture semiconductor lasers at any wavelength between that of the barrier material and the well material. In addition, quantum well lasers provide lower lasing thresholds and higher differential quantum efficiencies than equivalent DH lasers.

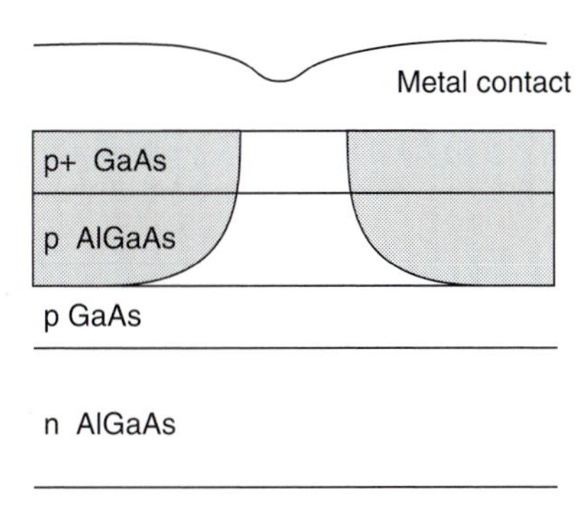

Figure 12.43 In the ion-implantation method, ions are implanted in regions other than the region of interest.

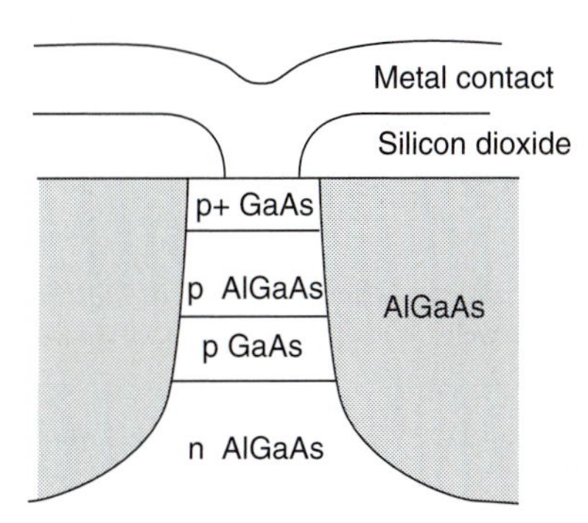

Figure 12.44 In the buried heterostructure method, the gain region is surrounded in both the vertical and horizontal directions by a waveguide of material.

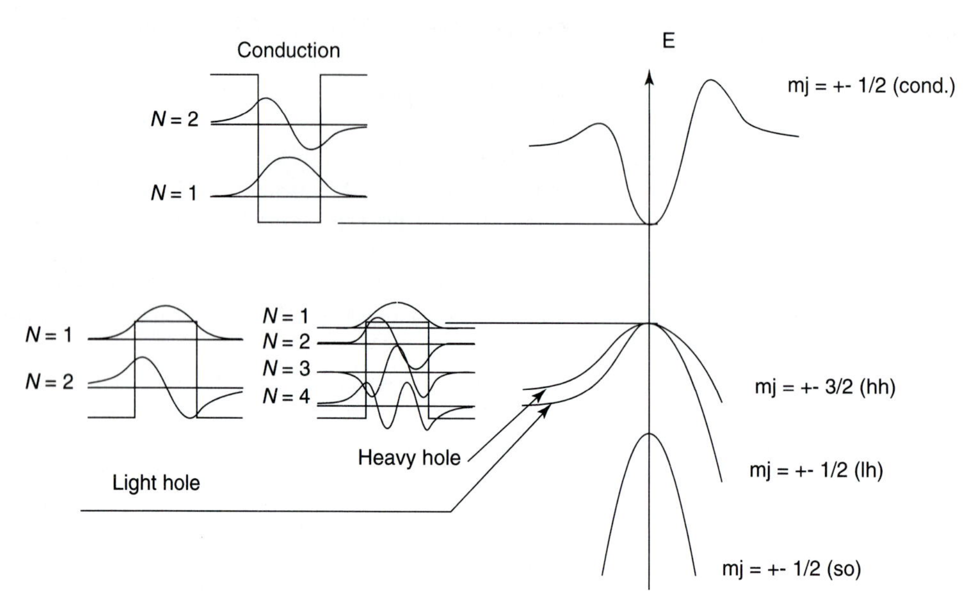

Figure 12.45 A quantum well is a heterostructure composed of two materials of differing bandgap. Under these conditions, energy wells are created in both the conduction and valence bands of the heterostructure.

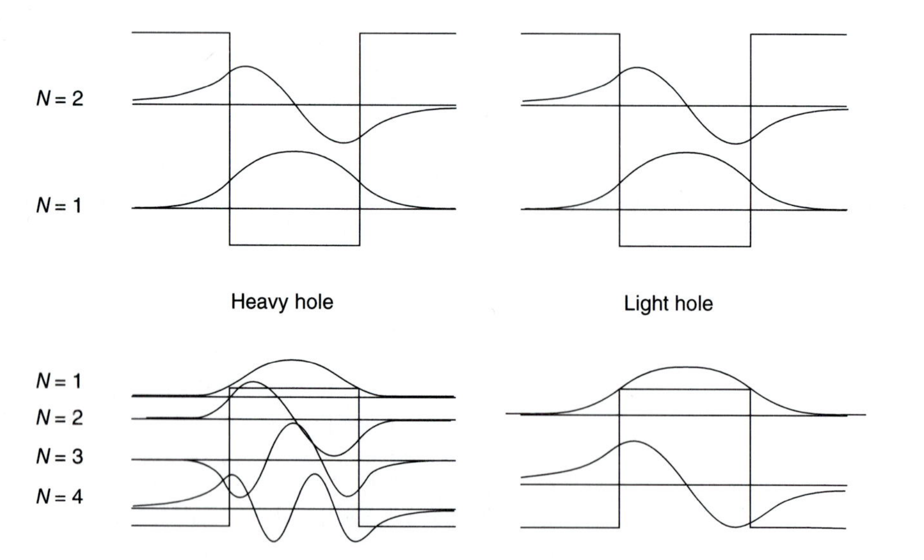

Figure 12.46 Discrete quantum states (similar to states in a hydrogen atom) will form in the conduction and valence wells of a quantum well material.

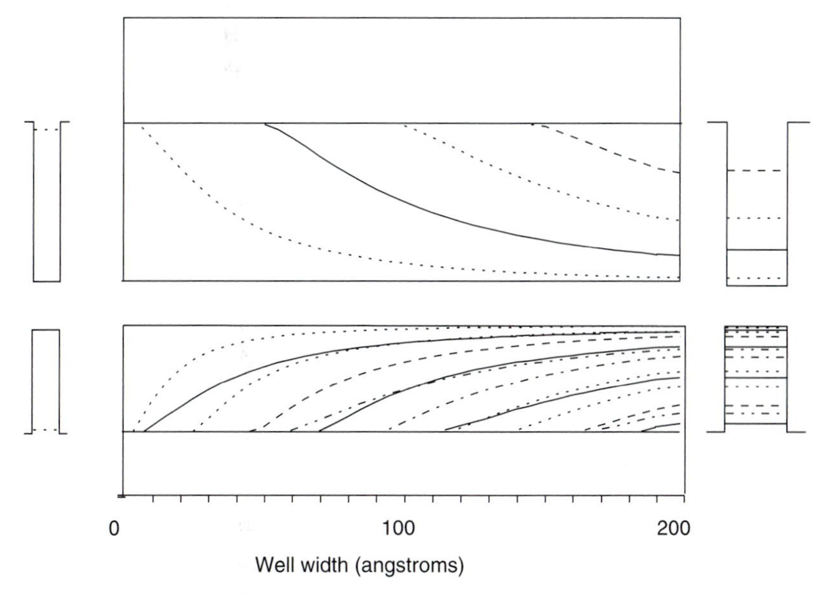

Figure 12.47 The narrower the quantum well, the larger the resulting transitional energy (to the eventual limit that the transition energy between the first valence and first conduction band state matches the energy gap of the barrier material). The wider the quantum well, the smaller the resulting transitional energy (to the eventual limit that the transition energy between the first valence and first conduction band state matches the energy gap of the well material).

12.3.5 Application Highlight: The CD Player

Introduction to the CD audio player. The CD audio player (and its close relative, the CD ROM) are probably the most widely known applications of laser diodes.

A CD disk contains a long string of pits written helically on tracks on the disk. Each pit is approximately 0.5 μm wide and 0.83 μm to 3.56 μm long. Each track is separated from the next track by 1.6 μm. The CD disk is actually read from the bottom. Thus, from the viewpoint of the laser beam reading the disk, the "pit" in the CD is actually a "bump." Interestingly enough, the *edges* of each pit (rather than the pit itself) correspond to binary ones. The signal has been encoded to ensure that there are no adjacent ones. (See Figure 12.48.)

The polycarbonate disk itself is part of the optical system for reading the pits (see Figure 12.49). The index of refraction of air is 1.0 while the index of refraction of the polycarbonate is 1.55. Laser light incident on the polycarbonate surface will be refracted at a greater angle into the surface. Thus, the original incident spot of around 800 μm (entering the polycarbonate) will be focused down to about 1.7 μm (at the metal surface). This helps to minimize the effect of dust and scratches on the surface.

The laser used for the CD player is typically an AlGaAs laser diode with a wavelength (in air) of 780 nm. The wavelength inside the polycarbonate is a factor of $n = 1.55$ smaller (about 500 nm). The pit/bump is carefully fabricated so that it is a quarter-wavelength high (notice a wavelength *inside* the polycarbonate). Light striking the land travels 1/4 + 1/4 =

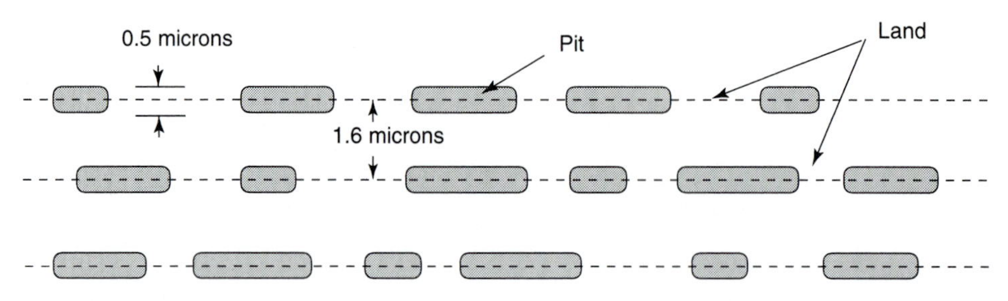

Figure 12.48 A CD disk contains a long string of pits written helically on tracks on the disk. Each pit is approximately 0.5 microns wide and 0.83 microns to 3.56 microns long.

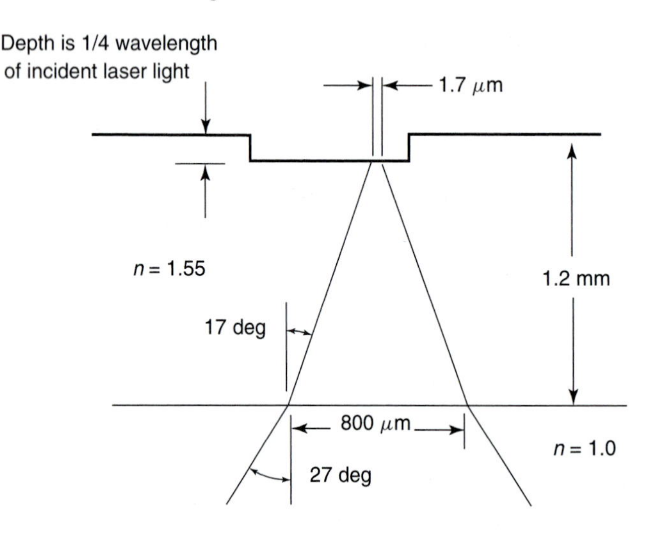

Figure 12.49 The polycarbonate disk is part of the optical system for reading the pits. The original incident spot of around 800 microns (entering the polycarbonate) will be focused down to about 1.7 microns (at the metal surface). This helps to minimize the effect of dust and scratches on the surface.

1/2 of a wavelength further than light striking the top of the pit. The light reflected from the land is then delayed by 1/2 wavelength, and so is exactly out-of-phase with the light reflected from the pit. Thus, these two waves will interfere destructively.

The spacing between pits is also carefully selected. Recall that the image of a beam passing through a round aperture will form a characteristic pattern called an Airy disk. The FWHM center of the Airy disk pattern is a spot about 1.7 μm wide and falls neatly on top of the pit track. The nulls in the Airy pattern are carefully situated to fall on the neighboring pit tracks. This minimizes crosstalk from neighboring pits.

The three-beam optical train. The most common optical train in modern CD players is the three-beam optical train (see Figure 12.50).

The basic operation of the optical train relies on the polarization properties of light. Light is emitted by the laser diode and enters a diffraction grating. The grating converts the light into a central peak plus side peaks. (The main central peak and two side peaks are important in the tracking mechanism.) The three beams go through a polarizing beam splitter. This only transmits polarizations parallel to the page. The emerging light (now polarized parallel to the page) is collimated. The collimated light goes through a quarter-

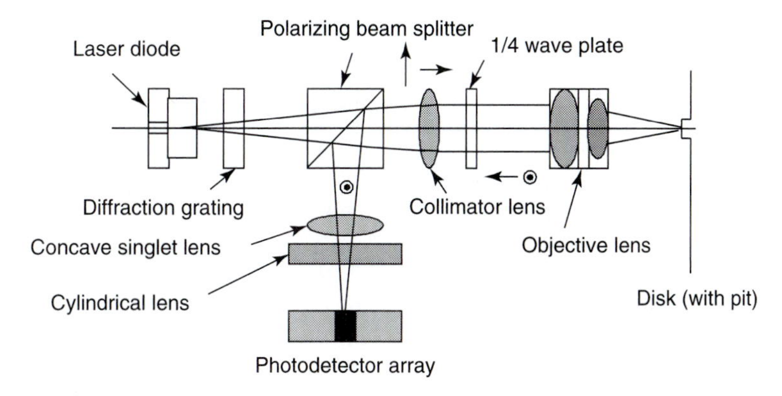

Figure 12.50 The most common optical train in modern CD players is the three-beam optical train.

wave plate and is converted into circularly polarized light. The circularly polarized light is then focused down onto the disk. If the light strikes land, it is reflected back into the objective lens; if the light strikes the pit, it is not reflected. The light then passes through the quarter wave plate again and emerges polarized perpendicular to the original beam (in other words, the light polarization is now vertical with respect to the paper). When the vertically polarized light hits the polarizing beam splitter this time, it will be reflected (not transmitted as before). Thus, it will reflect though the focusing lens and then the cylindrical lens and be imaged on the photodetector array.

Three-beam autofocus. The separation between the laser and the compact disk is critical for the correct operation of the CD player. A clever astigmatic autofocus mechanism is used to maintain this distance. This autofocus mechanism incorporates a cylindrical lens immediately in front of the photodetector array (see Figure 12.51).

If the objective lens is closer to the compact disk than the focal length of the object lens, then the cylindrical lens creates an elliptical image on the photodetector array. If the objective lens is further away from the compact disk than the focal length of the object lens, then the cylindrical lens again creates an elliptical image on the photodetector array. However, this elliptical image is perpendicular to first image. (Of course, if the disk is right at the focal length of the objective lens, then the cylindrical lens does not affect the image and it is perfectly circular.)

So, if the disk is too far away, quadrants D and B will get more light quadrants A and C (see Figure 12.52). Similarly, if the disk is too close, quadrants A and C will get more light than D and B. If things are just right, then all quadrants will get the same amount of light. So, it is possible to build a simple circuit that will maintain the object lens at just the right distance from the disk.

Three-beam tracking. Maintaining the laser beam on the track is also critical for the correct operation of the CD player. The three-beam optical train uses three separate beams to maintain the tracking. These beams are created by the diffraction grating (see Figure 12.53). When the laser beam goes through the diffraction grating, it is split up into a

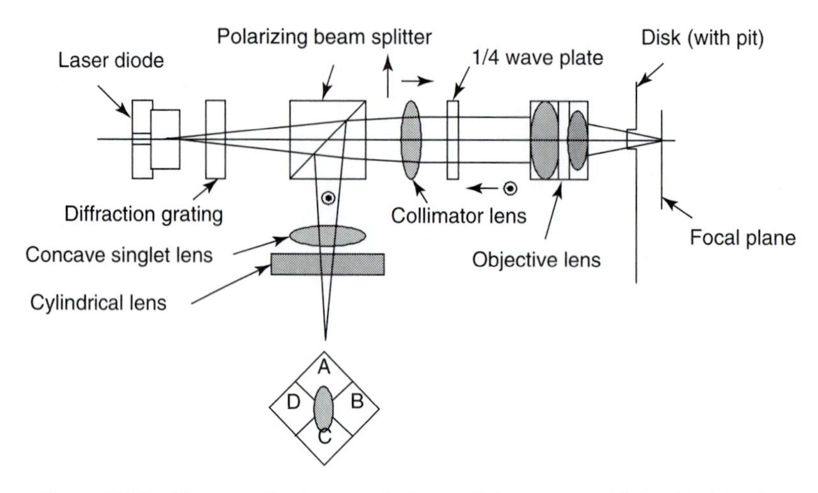

Figure 12.51 The separation between the laser and the compact disk is critical for the correct operation of the CD player. A clever astigmatic autofocus mechanism is used to maintain this distance.

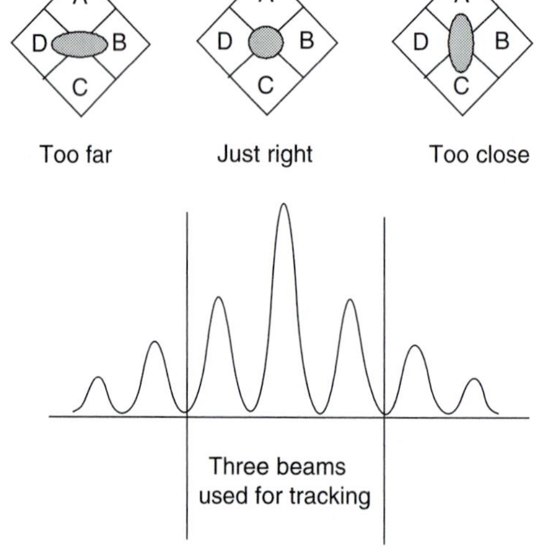

Figure 12.52 If the objective lens is closer to the compact disk than the focal length of the object lens, then the cylindrical lens creates an elliptical image on the photodetector array.

Figure 12.53 The three-beam optical train uses three separate beams to keep the laser beam on the track. These beams are created by the diffraction grating.

central bright beam plus a number of side beams. The central beam and one beam on each side are used by the CD for the tracking system.

To appreciate the tracking mechanism, consider a segment of the CD player containing several tracks (see Figure 12.54). If the optical head is on track, then the primary beam will be centered on a track (with pits and bumps) and the two secondary beams will be centered on land. (The three spots are deliberately offset approximately 20 microns with respect to each other to avoid crosstalk.)

Two additional detectors are placed alongside the main quadrant detector in order to pick up these subsidiary beams. If the three beams are on track, then the two subsidiary photodetectors have equal amounts of light and will be quite bright because they are only

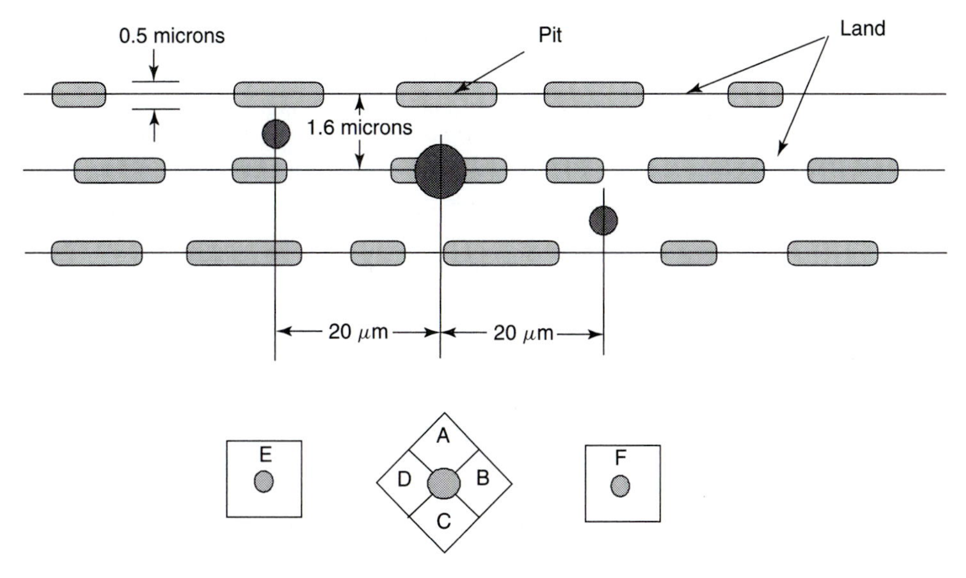

Figure 12.54 If the optical head is on track, then the primary beam will be centered on a track (with pits and bumps) and the two secondary beams will be centered on land.

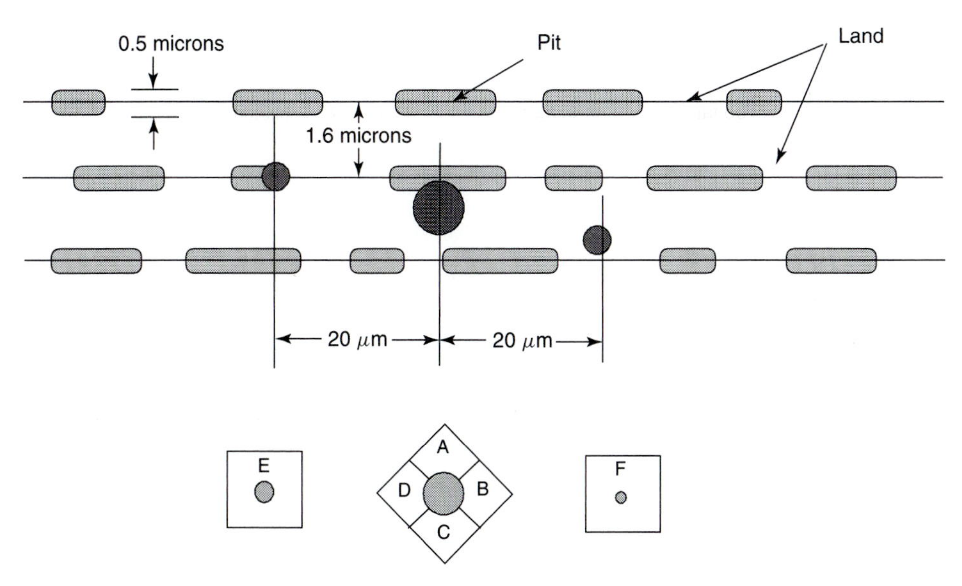

Figure 12.55 If the optical head is off track, then the center spot gets more light energy (because there are fewer pits off track) and the side detectors will be misbalanced.

tracking on land. The central beam will be reduced in brightness because it is tracking on both land and pits (see Figure 12.54). However, if the optical head is off track, then the center spot gets more light (because there are fewer pits off track) and the side detectors will be misbalanced (see Figure 12.55).

Additional optical storage applications using lasers. The CD audio player is just one of many optical storage applications using lasers. The CD recordable (CD-R) is another technology advancement in the optical storage area. The CD-R uses a laser beam to interact with a nonlinear dye and "write" pits in the CD-R disk. CD-R disks are "write once" devices, but offer exceptional performance for archival applications. (This textbook was delivered to the publisher on a CD-R disk.)

Magneto-optical storage is another fascinating laser diode application. Magneto-optical systems use a laser beam to change the orientation of a magnetic material. The process requires heating the magnetic material to near its Curie temperature and using the focused field of the laser to localize the change. The resulting magnetic spots are read optically using either the Kerr or the Faraday effect.

Another optical storage application that is just achieving success in the commercial marketplace is the DVD or digital video disk player. A combination of CD laser technology and clever digital algorithms has resulted in a more robust alternative to the video tape machine.

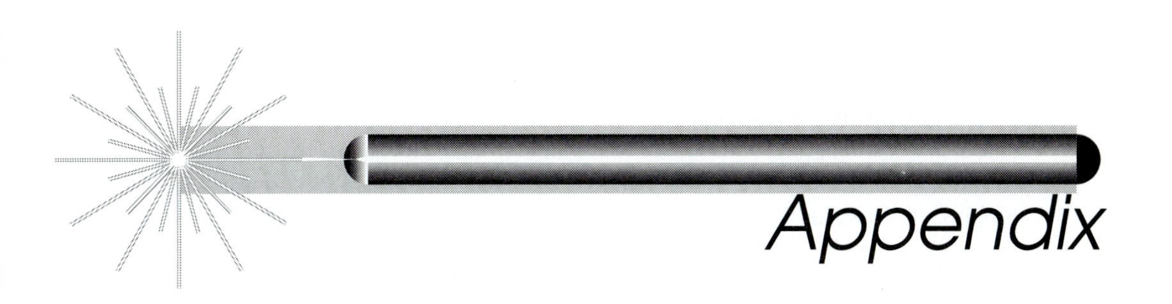

Appendix

A.1 LASER SAFETY

The most important point to remember about lasers is that *they are more dangerous than they look.* You can be killed, blinded, or poisoned. There are very few people working in the laser field who have not had a colleague injured or killed while using a laser. In the author's own personal experience, one colleague was killed by a laser (electrocuted on a copper-vapor) and two colleagues were permanently blinded in one eye (retina detachment, in both cases by a pulsed dye laser). It happens—and it could happen to you!

There are three major ways that lasers can seriously injure or kill a person. These are electrocution, eye damage, and chemical poisoning.

A.1.1 Electrocution

Many lasers operate with voltages on the order of hundreds of volts and currents on the order of tens of amperes. The primary hazard with these lasers is touching an energized laser component and then being unable to release your hand. Eventually your heart goes into fibrillation and you die. Although it is appalling to contemplate, you are no doubt aware of your impending death during the entire process.

Some lasers (such as longitudinally excited CO_2 lasers) operate with extremely high voltages (in the tens of kilovolts) and moderate currents (tens of amperes). Touching an energized laser component can result in a high-voltage arc passing through your hand or body. Such high-voltage arcs can burn the skin and sear internal arteries. An especially frightening scenario occurs when the current passes through the hand, across the chest, and out the opposite foot. Arterial damage to the heart under these conditions can result in death, two to three days following the accident.

Pulsed lasers pose their own unique electrical hazards. Large capacitors are used in these lasers in order to form the pulse that drives the flashlamp. Even in small pulsed lasers, these capacitors may be carrying several hundred joules of energy. Any capacitor capable of carrying more than about 50 J possesses a potential lethal shock hazard. Unshorted capacitors also have an annoying habit of charging on their own from atmospheric electrical disturbances. Plus, charged capacitors have a tendency to arc to neighboring components (or to neighboring mechanical structures) with minor changes in atmospheric conditions. Finally, capacitors occasionally explode, releasing shrapnel as well as potentially toxic liquids or vapors. (It should be no surprise that in large pulsed laser facilities the capacitors are confined to a locked cage a short distance from the laser system.)

Under ideal circumstances, a laser should be designed so that there is no electrical hazard associated with operating the laser. Unfortunately, this is not always the case. Except for specific commercial applications where the laser is sealed up in a box (and only accessed by a trained repairman), it is quite common to be working with a laser that is not optimally designed for electrical safety. It may be an older-model laser, a prototype, or a laser that has been significantly modified for a specific application. *Never assume that a laser is electrically safe—no matter how fancy the box.*

From the author's own personal experience:

> I recall working on a pulsed Nd:glass laser whose power supply worked beautifully during the summer. However, when the humidity rose after the first fall rainstorm, it would begin to arc somewhere in its huge capacitor bank. One day, a colleague and I decided to find the arc. After a great deal of careful hunting, we finally found the arc and fixed the problem. However, in the process of tracking down the arc, we disconnected the crowbar circuit that discharged the capacitors when the laser power switch was turned off. It was late at night when we finally solved the problem, and we forgot to reconnect the crowbar circuit before we went home.
>
> The next day, I fired up the laser as usual and then had some problems with one of the flashlamps. I disconnected the wires from the flashlamp and (as was my usual habit) tapped them on the metal optical table to discharge any small amounts of residual charge. BANG! The wires popped and the tips welded themselves to the table. Some 1000 J had just discharged into my optical table! Nervously, I found my insulated capacitor-discharge cables, dismounted the next set of wires and carefully tapped them on the cable clips. BANG! This went on for six sets of wires. Needless to say, I was quite shaky when I reached the end. Eventually I realized what we had done and reconnected up the discharge circuit.

Working with lasers in a college environment poses its own unique hazards. Students are usually young and inexperienced, and may be quite casual about their own mortality. Frequently stated safety rules and procedures may be ignored. It may be necessary to instigate surprisingly strong safety-enforcement procedures in order to protect students from themselves. Again, from the author's own experience:

> One year, two students had elected to work on a DC longitudinally pumped CO_2 laser for a class project. The rules for working on this laser were particularly well-defined. A number of horror stories had been related to the students about the potential injuries and the possiblity of death resulting from high-voltage injuries. The students were told not to work alone, to wear laser glasses, and to take extreme care when aligning the laser with the cover off.

> The night before the projects were due, one of the students was tired and went home. He then "became restless" and came back into school. He went back into the laser lab alone at 11:00 in the evening and began to align the laser. Because the power cord on the power meter wouldn't reach far enough and the student was unwilling to look for an extension cord, the student had his back to the laser and was adjusting the knobs on the rear mirror it behind his back. His hand slipped about 6 inches and contacted the high-voltage electrode. A 20 kV arc passed through his thumb and out his middle finger.
>
> The student then went to the hospital emergency room. The doctors wanted to keep him under observation, but the student refused and told the doctors he just wanted to go home and sleep. After he had promised to go home, the doctors released him. The student went back to the lab (now about 3:00 in the morning) and continued to run the laser (tired, alone, and without glasses) until 5:30 in the morning.

The following are a general set of rules for working with lasers that may present an electrical hazard.

1. IN ANY LASER LABORATORY—BEFORE MOVING YOUR HANDS, ALWAYS THINK WHERE THEY ARE GOING!
2. When working in (or even visiting) someone else's laser lab, DON'T ASSUME that the laboratory is safe. There may be exposed electrical connections that are very dangerous. ALWAYS ASK BEFORE TOUCHING ANY COMPONENT, FOR ANY REASON.
3. Remember that interlocks may be defeated. This is especially true if the system is under repair or is a research or prototype system. DO NOT ASSUME that a laser system will turn off safely if you do something wrong.
4. **Never** work alone when you are directly operating a high-power laser. Your partner is not only there to turn off the power if you get electrocuted, but also to double-check you and keep you from getting injured in the first place!
5. If you are working in the laboratory and your team partner gets across an electrical line, remember the following emergency responses:
   - **Kill the power!**
   - **Get the individual off the line!** Use some sort of nonconducting object to push or pull them off. However, **DON'T** just simply grab them. You may end up dying too.
   - **Call 911 immediately. (Remember that in Centrex-based phone systems this may mean calling 9-911!)**
   - **Administer first aid**. Knowledge of cardiopulmonary resuscitation (CPR) is essential for individuals routinely working with high-power electrical equipment.
6. Many lasers are water-cooled and many have inadequate or poorly designed plumbing systems. Thus, there is often the potential for mixing water and electricity. Locate and fix water leaks as soon as they are detected.
7. Always remember to turn off the power and DISCHARGE ANY CAPACITORS when working on a laser. If you work much with charged capacitors, it is worth building

up a special set of cables with an insulated handle specifically devoted to discharging capacitors.

8. In general, you should avoid touching any component while the laser is running. Of course, some components require adjustment while the laser is running! So, try to use only one hand when adjusting components. Before reaching into the laser and grabbing the component, first touch it with the back of your hand—if you are shocked, the muscular contraction will throw your hand away from the component. For lasers running in the multi-kV voltage range, use of special high-voltage gloves may be appropriate.

A.1.2 Eye Damage

There are two principal types of eye damage caused by lasers. Lasers with operating wavelengths in the region of approximately 400 nm to 1.4 microns are transmitted by the cornea and absorbed in the retina. Lasers with wavelengths outside this region are absorbed in the cornea, lens, and vitreous humor.

It is important to recognize that a tremendous eye-damage hazard exists with wavelengths from 400 nm to 1.4 microns. In this wavelength region (often called the *ocular focus region*), the lens serves to focus the laser beam on the retina. Thus, the actual laser power density entering the eye can be increased by some 10^5 by the time the light gets to the retina.

Notice that this focusing effect occurs both within the visible (400 nm to 700 nm) AND in the near infrared region (700 nm to 1.4 microns). THUS IT IS POSSIBLE TO ACQUIRE SEVERE RETINAL DAMAGE FROM A LASER BEAM THAT YOU CANNOT SEE! Although this is certainly important when working with Nd:glass or Nd:YAG lasers (1.05 to 1.06 microns), it is especially important to remember when working with near IR semiconductor laser diodes. Semiconductor diode lasers are so incredibly innocuous-looking that they tend to be treated very casually. However, a 15 mW 870 nm semiconductor laser diode has more than enough power to permanently damage your retina.

Although retinal damage is often more severe than corneal damage (primarily due to the focusing effect of the lens) corneal damage cannot be ignored. This is particularly true in the far IR region, where the lasers are often significantly more powerful. Thus, the decreasing sensitivity of the eye may be partially offset by the increasing power of the available lasers.

It it important to recognize that eye damage may not result from laser light coming directly from the laser, but may originate in laser light coming from secondary light paths in the laser lab. For example, the laser light may reflect off the mirror above a sink, off faucets, or off shiny mechanical parts. Thus, the laser lab must be inspected carefully to prevent hazardous secondary reflections. For extremely high-power lasers, even diffuse reflections may be capable of causing damage.

There are a number of personal case histories from individuals who have sustained eye damage from laser beams. In general, individuals are injured because they have removed their protective laser glasses for some reason. In some cases, the glasses were removed because they were "uncomfortable." However, in most cases the individuals operating the laser were adjusting some piece of equipment in the laser path and need to "see" the laser beam (dye lasers and doubled Nd:YAG lasers are notorious for this). In many cases, the users had turned down the laser to a *perceived* safe value in order to make the adjustment. However, permanent eye damage can occur from extremely low-power laser beams. From the author's own personal experience:

> I recall when one of my colleagues was injured by a pulsed dye laser. He had shoved his laser glasses up on his forehead in order to see the laser beam while he was aligning it. In the process, he received a retinal burn in one eye from a glint off an optical component. For the first few days, he only noticed a small blind spot and was not terribly concerned. Then, the blood started to pool behind the retina and the retina began to detach. After a few weeks, the retina was almost totally detached and he only retained a small amount of peripheral vision.
>
> However, the story does not end here. A few months later, he began to suffer twitches and muscular distortion in his face, related to an unconscious attempt to compensate for the original injury. I lost track of him after this—but I expect that the quality of the rest of his life was dramatically reduced due to a few seconds of inattention.

A similar case history is detailed in Winburn's book *Practical Laser Safety*[1]:

> On January 22, 1982, I spent several hours aligning a low-power frequency-doubled Nd:YAG beam through a dye laser set-up. In order to see the 532 nm pump beam propagation, I was not wearing goggles. I had also removed a beam block intended to absorb a Brewster's angle reflection, to observe end pumping of an amplifier cell. The green power was increased to determine the extent of dye lasing without replacing the beam block. I did not put on goggles. While placing a power meter at the dye laser output I leaned over the uncovered amplifier and caught a reflection in my right eye. Because I was in continuous motion looking at the meter and not the beam, I doubt that more than one 10 to 15 nanosecond pulse of ~20 microjoules was focused onto the fovea. While I do remember seeing a green flash—there was no pain. I was not immediately aware of any significant eye damage. It wasn't until I shut the lasers off and returned to my desk to record the day's activity that I realized I had a blind spot comparable to a camera flash, but only in my right eye.

The recommended eye protection for all people who work with lasers is a pair of goggles that are highly absorbing in the spectral region of the laser. Now, this is rather simple for any UV or IR laser, as humans cannot see in these spectral ranges. However, it becomes much more difficult for visible lasers (especially dye lasers) because the glasses that protect the user may also reduce the user's ability to see in the visible spectrum. (This is part of the reason that many laser injuries occur when using tunable dye lasers.) Green argon lasers have the additional problem that many argon laser glasses prohibit the user from using a conventional green phosphor oscilloscope. (Luckily, it is possible to purchase oscilloscopes that possess an amber phosphor and can be seen through most argon laser glasses.)

---

[1]D. C. Winburn, *Practical Laser Safety*, 2d ed. (New York: Marcel Dekker, Inc., 1990), p. 75.

Laser safety goggles are characterized by a minimum safe optical density D_λ, defined as[2]

$$D_\lambda = \log_{10}\left(\frac{H_p}{\text{MPE}}\right) \tag{A.1}$$

where H_p is the power density (or energy density) of the incident laser beam and MPE is the maximum permissible eye exposure (same units as H_p).

The calculation of MPE values for a particular optical system is a detailed operation that is beyond the scope of this text (consult ANSI Z136.1 for details on this calculation). However a few representative values are given in Tables A.1 and A.2 (condensed from Table A3-a,b and A4 of ANSI Z136.1[3])

TABLE A.1 OCULAR MPE VALUES FOR REPRESENTATIVE CW LASER SYSTEMS

Laser type	Wavelength (μm)	MPE (watt/cm^2)	Exposure duration (sec)
He:Cd	0.4416	$2.5 \cdot 10^{-3}$	0.25
Argon	0.4880, 0.5145	10^{-6}	$> 10^4$
HeNe	0.632	$2.5 \cdot 10^{-3}$	0.25
HeNe	0.632	$1 \cdot 10^{-3}$	10
HeNe	0.632	$17 \cdot 10^{-6}$	$> 10^4$
Krypton	0.647	$2.5 \cdot 10^{-3}$	0.25
Krypton	0.647	$1 \cdot 10^{-3}$	10
Krypton	0.647	$28 \cdot 10^{-6}$	$> 10^4$
InGaAlP	0.670	$2.5 \cdot 10^{-3}$	0.25
GaAs	0.905	$0.8 \cdot 10^{-3}$	> 1000
Nd:YAG	1.064	$1.6 \cdot 10^{-3}$	> 1000
InGaAsP	1.310	$12.8 \cdot 10^{-3}$	> 1000
InGaAsP	1.55	0.1	> 10
CO_2	10.6	0.1	> 10

The most important thing to notice about these charts is that the maximum permissible exposure numbers **are really very small!** The second most important thing to notice is that pulsed lasers are a good deal more dangerous (lower MPE) than continuous wave (cw) lasers.

Example A.1

Consider a Nd:YAG laser that has a pulse length of 100 ns per pulse and is operating Q-switched. What is the MPE for this laser?

Solution. Table A.2 shows that a Q-switched Nd:YAG laser with 100 ns per pulse has an MPE of about 5 μJ/cm^2. This is *not* a large energy density!

A.1.3 Chemical Hazards

Many of the chemicals used in dye lasers are highly poisonous. Furthermore, DMSO is often used as a solvent for laser dyes, providing a rapid path for transfer of the laser dye through

[2] ANSI Z136.1-1993, p. 23.

[3] ANSI Z136.1-1993, pp. 71–2.

TABLE A.2 OCULAR MPE VALUES FOR REPRESENTATIVE PULSED LASER SYSTEMS

Laser type	Wavelength (μm)	Pulse length (sec)	MPE (J/cm^2)
ArF	0.193	$2 \cdot 10^{-8}$	$3 \cdot 10^{-3}$
KrF	0.248	$2 \cdot 10^{-8}$	$3 \cdot 10^{-3}$
XeCl	0.308	$2 \cdot 10^{-8}$	$6.7 \cdot 10^{-3}$
XeF	0.351	$2 \cdot 10^{-8}$	$6.7 \cdot 10^{-3}$
Ruby (free-running)	0.6943	$1 \cdot 10^{-3}$	$1 \cdot 10^{-5}$
Ruby (Q-switched)	0.6943	$5\text{–}100 \cdot 10^{-9}$	$5 \cdot 10^{-7}$
Rhodamine 6G	0.500–0.700	$5\text{–}18 \cdot 10^{-6}$	$5 \cdot 10^{-7}$
Nd:YAG (free-running)	1.064	$1 \cdot 10^{-3}$	$5 \cdot 10^{-5}$
Nd:YAG (Q-switched)	1.064	$5\text{–}100 \cdot 10^{-9}$	$5 \cdot 10^{-6}$
CO_2	10.6	$1 \cdot 10^{-3}$	$100 \cdot 10^{-3}$

the skin. Laser dyes should be treated with caution and appropriate safety equipment such as gloves, aprons, and goggles should be used when handling dyes. Well-behaved solvents such as water or ethylene glycol should be used whenever possible. Leftover dye should be treated as hazardous waste.

Many of the crystals used in nonlinear optical research are also poisonous and hygroscopic (water-loving). Handling these crystals with bare hands can result in poisoning. All nonlinear optical crystals should be handled with gloves. Again, broken or damaged optical crystals should be treated as hazardous waste.

There are numerous other chemical hazards associated with particular laser types. For example, excimer lasers tend to release reactive chemicals such as fluorine or chlorine; free-electron lasers have a radiation hazard; chemical lasers have hazards appropriate to the type of chemical; and so on. In all cases, the chemicals associated with the laser should be treated with techniques appropriate to the particular system.

The chemicals used to clean optics should also be treated with caution. For many years, laser optics were cleaned using first acetone, then methanol. However, methanol is a cumulative toxin. Thus, ethanol should be substituted whenever possible. Similar problems exist with most other traditional cleaning techniques.

There has been an excellent trend in recent years for laboratories to be more responsible about chemical use. As part of this responsibility, it is worth reevaluating chemical use in the laser laboratory every few years. In many cases, a chemical that was once thought to be essential has been replaced by a more benign one or even by an alternative nonchemical process. As just one example, harsh chemicals are often used in the stripping of optical fibers. However, an increasing number of optical fibers can be stripped mechanically. Before setting up to strip a fiber using methyl chloride or sodium hydroxide, it is worth checking to see if the fiber can be stripped mechanically.

A.1.4 Other Hazards

Skin damage. Many lasers have the ability to damage the skin. This is particularly true of excimer and CO_2 lasers. A similar system to the MPE limits described in Section A.1.3 exists for skin damage with lasers; however, protection is somewhat less difficult. Flame-retardant long-sleeved shirts and gloves provide adequate protection for most cases. A few representative values for skin MPE are given in Tables A.3 and A.4 (condensed from Table A3-a,b and A4 of ANSI Z136.1[4]).

TABLE A.3 SKIN MPE VALUES FOR REPRESENTATIVE CW LASER SYSTEMS

Laser type	Wavelength (μm)	MPE (watt/cm^2)	exposure duration (sec)
He:Cd	0.441	0.2	> 10
Argon	0.4880, 0.5145	0.2	> 10
HeNe	0.632	0.2	> 10
Krypton	0.647	0.2	> 10
GaAs	0.905	0.5	> 10
Nd:YAG	1.064	1	> 10
CO_2	10.6	0.1	> 10

TABLE A.4 SKIN MPE VALUES FOR REPRESENTATIVE PULSED LASER SYSTEMS

Laser type	Wavelength (μm)	MPE (J/cm^2)	Exposure duration (sec)
ArF	0.193	$3 \cdot 10^{-3}$	$2 \cdot 10^{-8}$
KrF	0.248	$3 \cdot 10^{-3}$	$2 \cdot 10^{-8}$
XeCl	0.308	$6.7 \cdot 10^{-3}$	$2 \cdot 10^{-8}$
XeF	0.351	$6.7 \cdot 10^{-3}$	$2 \cdot 10^{-8}$
Ruby (free)	0.6943	0.2	$1 \cdot 10^{-3}$
Ruby (Q-switch)	0.6943	0.02	$5\text{–}100 \cdot 10^{-9}$
Rhodamine 6G	0.500–0.700	0.03–0.07	$5\text{–}18 \cdot 10^{-6}$
Nd:YAG (free)	1.064	1	$1 \cdot 10^{-3}$
Nd:YAG (Q)	1.064	0.1	$5\text{–}100 \cdot 10^{-9}$
CO_2	10.6	$10 \cdot 10^{-3}$	$1 \cdot 10^{-3}$

Mechanical and physical hazards. Individuals working in laser laboratories often suffer injury from physical hazards simply because laser laboratories are frequently lit at a very low level. Although it is a good practice to attempt to keep the laser laboratory well-lit (for example, by using phase-sensitive detection techniques), there are often circumstances under which the lights will need to be dimmed or turned off. To minimize chances of mechanical injury, observe the following precautions.

1. Avoid placing objects on the tops of cabinets or shelves where they can fall on an unwary laboratory user. If objects *must* be stored high in the air, try to restrict such storage to light objects such as hoses, plastic sheeting, and so on.

[4] ANSI Z136.1-1993, pp. 71–2.

2. Avoid stacking objects on shelves or placing objects near the edge of a shelf where they could be accidently knocked off in the dark.
3. Avoid routing extension cords on the floor. Not only is this an electrical and fire hazard, but it is very easy to trip over extension cords in a darkened room. Route electrical service overhead in wire trays or conduits.
4. Pay attention to the location of water hoses, compressor units, filters, and so on. Equipment that is relatively easy to avoid in a lighted room may become a serious hazard if the room is dark.
5. Before turning off the lights for an experiment, make certain that:
   - Everyone who must move around the room during the experiment knows the path that they need to follow. If the experiment permits, these individuals should have a small light source (penlight flashlight or equivalent).
   - Everyone who is staying put (for example, is taking data) has a comfortable place to sit that is not near any potential hazards.
   - All chemicals are properly stored in an acids or flammables cabinet.
   - All compressed gas bottles are racked up and strapped in, and all possible adjustments to compressed gas pressure have been made.
   - At least one individual has a clear path to the room lights in the event of an unforeseen emergency.

Getting shot or being jailed. In the past, I had not included these as laser hazards. However, during the summer of 1991, several undergraduate students at our university were charged with malicious mischief because they were shining HeNe laser beams at people and houses. In the two cases on record, the individuals who were the recipients of the laser beams thought that the beams were targeting sights for terrorist weapons and contacted the police.

Although this was a unique situation, one perhaps brought about by the tensions in the Persian Gulf at the time, there is increasing use of visible red laser beams (particularly diode lasers) as targeting sights for firearms. Thus, extreme care should be taken when using visible lasers in environments other than the laser laboratory.

Fibers. Optical fibers have crept into laser laboratories as components in many experiments. It is important to recognize that small pieces of optical fiber (particularly single-mode fiber) can be highly dangerous. A typical injury is caused by running your fingers over the edge of a table and accidentally jamming the fiber into your hand like a giant splinter. Since the fiber is small and transparent, it is quite difficult to grab the end with a pair of tweezers and pull it out. In the simplest cases, you can use a microscope and remove the fiber pieces by flexing your hand and using the pain to locate them. In more difficult cases, a doctor will have to use a flensing knife to remove the skin layer by layer (without any anesthetic) until the fiber is located.

To avoid this painful experience, the proper procedure is to take any small pieces of fiber and put them in a clearly labeled jar or box. The jar should be disposed of in your organization's glass disposal area.

A.2 SIGNIFICANT FIGURES

In laser engineering, problems often arise in the casual use of significant figures. For example, spectroscopic data for the wavelength of laser transitions can be measured accurately to at least six significant figures (such as $\lambda = 8265.62$ angstroms). However, transition energies are often given only to two or three significant figures (such as $E_2 = 2.5$ eV). Thus, the conversion between transition energies and wavelengths can cause a serious loss of information. The problem is further complicated by the common use of one and two significant figure approximations for such fundamental constants as h and c_o.

The use of precision and significant figures has been well-defined in sciences such as chemistry and biology. However, the established definitions from the natural sciences are not commonly used in laser engineering. In an attempt to strike a middle ground on this issue, this text has generally followed these rules on significant figures:

1. Constants such as h and c_o are assumed to take the current most accurately known values (typically greater than eight significant figures). To indicate this, such constants will not be replaced by their numerical values in the sample calculations.
2. The number of significant figures in a result are assumed to be equal to the number of significant figures of the least precise number in the calculation.
3. Sample values using textbook values (such as 1.5 eV) are written with a single trailing zero (i.e. 1.50 eV) and are assumed to have sufficient significant figures so that some other number in the calculation determines the significant figures in the result. (This is to avoid writing 1.500000 eV in a sample calculation.)
4. Calculations that contain all textbook values are generally carried out to six significant figures, unless such calculations contain trailing zeros. Trailing zeros are truncated as in Rule #3 above.

A.3 THE ELECTROMAGNETIC WAVE EQUATION

The electromagnetic wave is unusual in that it consists of two in-phase and mutually perpendicular components, one for the electric field $\vec{\mathcal{E}}$ and one for the magnetic field $\vec{\mathcal{B}}$. In free space, the electromagnetic wave is a transverse wave; that is, the direction of the wave motion is perpendicular to the direction of travel. However, in real materials, the presence of lossy elements and free charge may alter the transverse nature of the electromagnetic wave. The equations governing the behavior of the electromagnetic wave can be derived from Maxwell's equations using the following procedure.

A.3.1 Maxwell's Equations

The macroscopic vector form of Maxwell's equations is

$$\vec{\nabla} \cdot \vec{\mathcal{D}} = \rho \tag{A.2}$$

$$\vec{\nabla} \cdot \vec{\mathcal{B}} = 0 \tag{A.3}$$

$$\vec{\nabla} \times \vec{\mathcal{E}} = -\frac{\partial \vec{\mathcal{B}}}{\partial t} \tag{A.4}$$

$$\vec{\nabla} \times \vec{\mathcal{H}} = \vec{\mathcal{J}} + \frac{\partial \vec{\mathcal{D}}}{\partial t}. \tag{A.5}$$

$\vec{\mathcal{J}}$ is the current density and is related to the electric field by

$$\vec{\mathcal{J}} = \sigma \vec{\mathcal{E}} \tag{A.6}$$

where σ is the conductivity.

$\vec{\mathcal{D}}$ is the electric displacement vector, and is related to the electric intensity vector $\vec{\mathcal{E}}$ by the constitutive relation

$$\vec{\mathcal{D}} = \epsilon_o \vec{\mathcal{E}} + \vec{\mathcal{P}} \tag{A.7}$$

where ϵ_o is the permittivity of free space ($8.854187817 \cdot 10^{-12}$ farad/meter).

$\vec{\mathcal{P}}$ is the electric polarization vector and is given by[5]

$$\vec{\mathcal{P}} = \epsilon_o \chi \vec{\mathcal{E}} \tag{A.8}$$

where χ is the complex susceptibility. (Note that in some references,[6] the permittivity of free space may be included in the definition of χ, giving the slightly altered definition $\vec{\mathcal{P}} = \chi \vec{\mathcal{E}}$.)

The complex susceptibility χ possesses both a real and an imaginary component. Unfortunately, the sign conventions on the imaginary part of χ are not uniformly established in the laser community. Thus, both $\chi = \chi' - j\chi''$ and $\chi = \chi' + j\chi''$ are seen in the literature.[7,8]

The definition for the electric polarization vector $\vec{\mathcal{P}}$, Equation (A.8), may be substituted back into the constitutive relation for $\vec{\mathcal{D}}$, Equation (A.7), as

$$\vec{\mathcal{D}} = \epsilon_o \vec{\mathcal{E}} + \epsilon_o \chi \vec{\mathcal{E}} = \epsilon_o (1 + \chi) \vec{\mathcal{E}} = \epsilon_o \epsilon_r \vec{\mathcal{E}} = \epsilon_m \vec{\mathcal{E}} \tag{A.9}$$

which yields the expression $\epsilon_r = 1 + \chi$ for the complex relative dielectric constant of a material. (Notice that the complex index of refraction $n_c = n + j\kappa$ is also related to χ and ϵ_r as $\epsilon_r = (n_c)^2 = 1 + \chi$.)

In many laser materials there is a large nonresonant dielectric polarization associated with the host crystal lattice ($\chi \approx 2$) and a smaller resonant dielectric polarization associated with the laser atoms ($\chi \ll 1$). So the constitutive relation is actually formed from three terms as

$$\vec{\mathcal{D}} = \epsilon_o \vec{\mathcal{E}} + \vec{\mathcal{P}}_{\text{lattice}} + \vec{\mathcal{P}}_{\text{atoms}} \tag{A.10}$$

---

[5]This definition of the polarizability is used by Amnon Yariv, *Optical Electronics*, 4th ed. (Philadelphia, PA: Saunders College Publishing, 1991), p. 6; and Anthony Siegman, *Lasers* (Mill Valley, CA: University Science Books, 1986), p. 103.

[6]This definition of the polarizability is used by Robert W. Boyd, *Nonlinear Optics* (Boston, MA: Academic Press, 1992), p. 2.

[7]$\chi = \chi' - j\chi''$ in Amnon Yariv, *Optical Electronics*, 4th ed. (Philadelphia, PA: Saunders College Publishing, 1991), p. 6.

[8]$\chi = \chi' + j\chi''$ in Anthony Siegman, *Lasers* (Mill Valley, CA: University Science Books, 1986), p. 107.

$$\vec{\mathcal{D}} = \epsilon_o \vec{\mathcal{E}} + \epsilon_o \chi_{\text{lattice}} \vec{\mathcal{E}} + \epsilon_o \chi_{\text{atoms}} \vec{\mathcal{E}} \tag{A.11}$$

$$\vec{\mathcal{D}} = \epsilon_o (1 + \chi_{\text{lattice}}) \vec{\mathcal{E}} + \epsilon_o \chi_{\text{atoms}} \vec{\mathcal{E}}. \tag{A.12}$$

Siegman developed a simple way around this lengthy notation by expressing the constitutive relation in the form[9]

$$\vec{\mathcal{D}} = \epsilon_h \vec{\mathcal{E}} + \vec{\mathcal{P}}_{\text{atoms}} \tag{A.13}$$

$$\vec{\mathcal{D}} = \epsilon_h \vec{\mathcal{E}} + \epsilon_h \chi_{\text{atoms}} \vec{\mathcal{E}} = \epsilon_h (1 + \chi_{\text{atoms}}) \vec{\mathcal{E}} \tag{A.14}$$

where ϵ_h (the dielectric permittivity of the host crystal alone) replaces ϵ_o (the dielectric permittivity of free space).

$\vec{\mathcal{B}}$ is the magnetic induction vector and is related to the magnetic intensity vector $\vec{\mathcal{H}}$ by the constitutive relation

$$\vec{\mathcal{B}} = \mu_o \vec{\mathcal{H}} + \mu_o \vec{\mathcal{M}} \tag{A.15}$$

where μ_o is the permeability of free space ($4\pi \cdot 10^{-7}$ henry/meter) and $\vec{\mathcal{M}}$ is the magnetization.

It turns out that most laser interactions of interest are those where the electric field (rather than the magnetic field) interacts with the material. Thus, interactions involving the electric polarization vector $\vec{\mathcal{P}}$ are typically of more importance than those involving the magnetization vector $\vec{\mathcal{M}}$. (There are certainly exceptions to this such as the magnetic dipole transition in the iodine laser at 1.3 μm.)

A.3.2 A General Wave Equation for Light Propagation in a Material

Maxwell's equations can be used to derive a general wave equation for light propagation in a material. This (admittedly complex) wave equation can be subjected to a number of simplifying assumptions to model wave propagation in real materials.

Begin with Maxwell's equation for $\vec{\mathcal{H}}$, Equation (A.5)

$$\vec{\nabla} \times \vec{\mathcal{H}} = \vec{\mathcal{J}} + \frac{\partial \vec{\mathcal{D}}}{\partial t}. \tag{A.16}$$

Substitute in Equation (A.6) for $\vec{\mathcal{J}}$

$$\vec{\nabla} \times \vec{\mathcal{H}} = \sigma \vec{\mathcal{E}} + \frac{\partial \vec{\mathcal{D}}}{\partial t}. \tag{A.17}$$

Multiply by μ_o and differentiate by t, yielding

$$\vec{\nabla} \times \mu_o \frac{\partial \vec{\mathcal{H}}}{\partial t} = \mu_o \sigma \frac{\partial \vec{\mathcal{E}}}{\partial t} + \mu_o \frac{\partial^2 \vec{\mathcal{D}}}{\partial t^2}. \tag{A.18}$$

Consider the constitutive relation for $\vec{\mathcal{B}}$, Equation (A.15)

$$\vec{\mathcal{B}} = \mu_o \vec{\mathcal{H}} + \mu_o \vec{\mathcal{M}}. \tag{A.19}$$

[9] Anthony Siegman, *Lasers* (Mill Valley, CA: University Science Books, 1986), p. 105.

In most laser materials, $\vec{\mathcal{M}}$ is negligible, so

$$\vec{\mathcal{B}} = \mu_o \vec{\mathcal{H}} \tag{A.20}$$

and differentiating with respect to time gives

$$\frac{\partial \vec{\mathcal{B}}}{\partial t} = \mu_o \frac{\partial \vec{\mathcal{H}}}{\partial t}. \tag{A.21}$$

Now, considering Maxwell's equation for $\vec{\mathcal{E}}$, Equation (A.4)

$$\vec{\nabla} \times \vec{\mathcal{E}} = -\frac{\partial \vec{\mathcal{B}}}{\partial t} \tag{A.22}$$

and reorganizing and substituting from Equation (A.21) gives

$$\mu_o \frac{\partial \vec{\mathcal{H}}}{\partial t} = -\vec{\nabla} \times \vec{\mathcal{E}}. \tag{A.23}$$

Substituting Equation (A.23) into Equation (A.18),

$$\vec{\nabla} \times \left(\vec{\nabla} \times \vec{\mathcal{E}}\right) = -\left(\mu_o \sigma \frac{\partial \vec{\mathcal{E}}}{\partial t} + \mu_o \frac{\partial^2 \vec{\mathcal{D}}}{\partial t^2}\right) \tag{A.24}$$

and using the vector identity

$$\vec{\nabla} \times \vec{\nabla} \times \vec{\mathcal{E}} = \vec{\nabla}(\vec{\nabla} \cdot \vec{\mathcal{E}}) - \nabla^2 \vec{\mathcal{E}} \tag{A.25}$$

on Equation (A.24) gives the final answer of

$$\vec{\nabla}(\vec{\nabla} \cdot \vec{\mathcal{E}}) - \nabla^2 \vec{\mathcal{E}} = -\left(\mu_o \sigma \frac{\partial \vec{\mathcal{E}}}{\partial t} + \mu_o \frac{\partial^2 \vec{\mathcal{D}}}{\partial t^2}\right). \tag{A.26}$$

Other than the assumption that $\vec{\mathcal{M}}$ is negligible, this is a general wave equation that can be applied to a variety of laser problems.

A.3.3 Light Propagation in a Vacuum

In a vacuum with no net static charge, $\vec{\nabla} \cdot \vec{\mathcal{E}} = 0$, $\sigma = 0$, and $\chi = 0$. Using the simplification, Equation (A.7), that the constitutive relation is

$$\vec{\mathcal{D}} = \epsilon_o \vec{\mathcal{E}} + \epsilon_o \chi \vec{\mathcal{E}} \tag{A.27}$$

this can be substituted into Equation (A.26), yielding the familiar wave equation for free space

$$\nabla^2 \vec{\mathcal{E}} = \mu_o \epsilon_o \frac{\partial^2 \vec{\mathcal{E}}}{\partial t^2}. \tag{A.28}$$

For a one-dimensional analysis, a solution to Equation (A.28) is a plane wave of the form $\mathcal{E}_o e^{j(\omega t - kz)}$, where the wavevector $k = \omega / c_o = 2\pi / \lambda_o$ and the wave velocity in free space is $c_o = 1/\sqrt{\mu_o \epsilon_o}$.

A.3.4 Light Propagation in a Simple Isotropic Material with No Net Static Charge

In a simple isotropic material with no net static charge, where $\vec{\nabla} \cdot \vec{\mathcal{E}} = 0$ and $\sigma = 0$ but $\chi \neq 0$, and using the simplification Equation (A.7) that the constitutive relation is

$$\vec{\mathcal{D}} = \epsilon_o \vec{\mathcal{E}} + \epsilon_o \chi \vec{\mathcal{E}} = \epsilon_o (1 + \chi) \vec{\mathcal{E}} = \epsilon_o \epsilon_r \vec{\mathcal{E}} = \epsilon_m \vec{\mathcal{E}} \tag{A.29}$$

the wave equation takes the form

$$\nabla^2 \vec{\mathcal{E}} = \mu_o \epsilon_m \frac{\partial^2 \vec{\mathcal{E}}}{\partial t^2}. \tag{A.30}$$

For a one-dimensional analysis, the solution Equation (A.30) is a plane wave of the form $\mathcal{E}_o e^{j(\omega t - kz)}$, where the wavevector $k = \omega\sqrt{\mu_o \epsilon_m} = \omega / c = 2\pi / \lambda_m$ (where λ_m is the wavelength in the material equal to λ_o / n) and where the wave velocity in the material is $c = 1/\sqrt{\mu_o \epsilon_m} = c_o / n$. The refractive index n is defined by $\sqrt{\mu_o \epsilon_m / \mu_o \epsilon_o}$.

In other words, in this simple material the velocity of light and the wavelength have both been reduced by a factor of n and the wavevector k has been increased by a factor of n.

A.3.5 Light Propagation in a Simple Laser Material with No Net Static Charge

Now, for a more realistic approach, consider an isotropic laser material with no net static charge, where $\vec{\nabla} \cdot \vec{\mathcal{E}} = 0$ but $\sigma \neq 0$, and $\chi \neq 0$. Using the formalism Equation (A.13) that the constitutive relation is

$$\vec{\mathcal{D}} = \epsilon_h \vec{\mathcal{E}} + \vec{\mathcal{P}}_{\text{atoms}} \tag{A.31}$$

and taking $\vec{\mathcal{P}} = \vec{\mathcal{P}}_{\text{atoms}}$, the wave equation takes the form,

$$\nabla^2 \vec{\mathcal{E}} = \left(\mu_o \epsilon_h \frac{\partial^2 \vec{\mathcal{E}}}{\partial t^2} + \mu_o \frac{\partial^2 \vec{\mathcal{P}}}{\partial t^2} + \mu_o \sigma \frac{\partial \vec{\mathcal{E}}}{\partial t} \right). \tag{A.32}$$

This is the original simple wave equation, but with two additional terms to model the increasing complexity of the material.

A.3.6 A One-Dimensional Wave Equation for a Less Simple Isotropic Material

Assume that a typical laser beam propagates in the z direction. If the x and y dimensions of the laser beam are many wavelengths (as is often the case) then $\partial/\partial x$ and $\partial/\partial y$ are much less than $\partial/\partial z$. Under these circumstances,

$$\nabla^2 \approx \frac{\partial^2}{\partial z^2}. \tag{A.33}$$

Substituting this back into Equation (A.32) gives the general one-dimensional wave equation

$$\frac{\partial^2 \vec{\mathcal{E}}}{\partial z^2} - \left(\mu_o \epsilon_h \frac{\partial^2 \vec{\mathcal{E}}}{\partial t^2} + \mu_o \frac{\partial^2 \vec{\mathcal{P}}}{\partial t^2} + \mu_o \sigma \frac{\partial \vec{\mathcal{E}}}{\partial t} \right) = 0 \tag{A.34}$$

or, making the substitution $\vec{\mathcal{P}} = \epsilon_m \chi \vec{\mathcal{E}}$, Equation (A.14),

$$\frac{\partial^2 \vec{\mathcal{E}}}{\partial z^2} - \left(\mu_o \epsilon_h \frac{\partial^2 \vec{\mathcal{E}}}{\partial t^2} + \mu_o \epsilon_h \chi \frac{\partial^2 \vec{\mathcal{E}}}{\partial t^2} + \mu_o \sigma \frac{\partial \vec{\mathcal{E}}}{\partial t} \right) = 0 \tag{A.35}$$

and reorganizing,

$$\frac{\partial^2 \vec{\mathcal{E}}}{\partial z^2} - \mu_o \epsilon_h \left(\frac{\partial^2 \vec{\mathcal{E}}}{\partial t^2} + \chi \frac{\partial^2 \vec{\mathcal{E}}}{\partial t^2} + \frac{\sigma}{\epsilon_h} \cdot \frac{\partial \vec{\mathcal{E}}}{\partial t} \right) = 0. \tag{A.36}$$

For a one-dimensional analysis, the solution to Equation (A.36) is a plane wave of the form

$$\psi(z) = \mathcal{E}_o e^{j(\omega t - k_c z)} \tag{A.37}$$

where the complex wavevector k_c is given by

$$k_c = \omega \sqrt{\mu_o \epsilon_h} \left(1 + \chi - j \frac{\sigma}{\omega \epsilon_h} \right)^{1/2}. \tag{A.38}$$

Now, $\chi - j\frac{\sigma}{\omega \epsilon_h}$ is much less than 1, so the first term of the binomial expansion can be used as an approximation,

$$(1+x)^{1/2} = 1 + \frac{x}{2} - \frac{x^2}{8} + \cdots \approx \left(1 + \frac{x}{2}\right) \quad \text{for } x \ll 1 \tag{A.39}$$

yielding

$$k_c \approx \omega \sqrt{\mu_o \epsilon_h} \left(1 + \frac{\chi}{2} - j \frac{\sigma}{2\omega \epsilon_h} \right). \tag{A.40}$$

Now χ can be expanded into real and imaginary parts as

$$k_c = \omega \sqrt{\mu_o \epsilon_h} \left(1 + \frac{\chi'}{2} + j \frac{\chi''}{2} - j \frac{\sigma}{2\omega \epsilon_h} \right) \tag{A.41}$$

following the definition that $\chi = \chi' + j\chi''$.

Now consider how Equation (A.41) fits back into the general plane wave solution Equation (A.37),

$$\mathcal{E}_o e^{j(\omega t - k_c z)} = \mathcal{E}_o \exp\left[j \left(\omega t - \left[\omega \sqrt{\mu_o \epsilon_h} \left(1 + \frac{\chi'}{2} + j \frac{\chi''}{2} - j \frac{\sigma}{2\omega \epsilon_h} \right) \right] z \right) \right]. \tag{A.42}$$

Notice that real terms in the expression for k_c are multiplied by j, yielding imaginary exponentials and thus generating sinusoidal behavior. Imaginary terms in the expression for k_c are also multiplied by j, but the product of the two imaginary numbers will generate a real number and thus a real exponential. Thus, imaginary terms in the expression for k_c will yield either gain or loss terms.

Therefore, real terms in the expression for k_c are given by

$$\mathrm{Re}(k_c) = \omega \sqrt{\mu_o \epsilon_h} + \omega \sqrt{\mu_o \epsilon_h} \left(\frac{\chi'}{2} \right) \tag{A.43}$$

where the basic sinusoidal wave propagation in the material is expressed as

$$T_{\text{sinusoid}} = e^{j\omega\sqrt{\mu_o\epsilon_h}\cdot z} \tag{A.44}$$

and the additional phase shift factor is given by

$$T_{\text{phase}} = e^{j\omega\sqrt{\mu_o\epsilon_h}(\chi'/2)\cdot z}. \tag{A.45}$$

Imaginary terms in the expression for k_c will be multiplied by j and will generate gain or loss. Thus, the imaginary part of k_c given as

$$\text{Im}(k_c) = \omega\sqrt{\mu_o\epsilon_h}\left(\frac{\chi''}{2}\right) - \frac{\sigma}{2\omega\epsilon_h} \tag{A.46}$$

where the generated gain or loss in the material is given as

$$T_{\text{gain/loss}} = e^{\omega\sqrt{\mu_o\epsilon_h}(\chi''/2)\cdot z} \tag{A.47}$$

and the ohmic (or background) loss is given by

$$T_{\text{ohmic-loss}} = e^{-\sigma/2\omega\epsilon_h\cdot z}. \tag{A.48}$$

A.4 LENSES AND TELESCOPES

A.4.1 Lenses

The lens is the most common optical component used in the laser laboratory. Lenses can be used to focus (convex lenses), diverge (concave lenses), focus or defocus only in one transverse direction (cylindrical lenses), or collimate (telescopes). There are many references on lenses available to the interested reader.[10, 11, 12, 13, 14, 15, 16, 17]

A number of different types of lenses are used in the laser laboratory. The more common can be listed as follows:[18]

Thin lens: A lens whose thickness does not appreciably alter its optical properties.

Thick lens: A lens whose focusing properties are a function of its thickness.

---

[10] Eugene Hecht, *Optics,* 2d ed. (Reading, MA: Addison-Wesley, 1987).

[11] Melles Griot, 1995–6 Catalog, Irvine, CA 92714.

[12] Michael Bass, ed, *Handbook of Optics*, Vol. I and II, Optical Society of America, (New York: McGraw Hill, 1995).

[13] Max Born and Emil Wolf, *Principles of Optics,* 6th ed. (New York: Pergamon Press, 1980).

[14] B. B. Rossi, *Optics* (Reading, MA: Addison-Wesley, 1957).

[15] M. V. Klein and T. E. Furtak, *Optics*, 2d ed. (New York: John Wiley and Sons, 1986).

[16] Jurgen R. Meyer-Arendt, *Introduction to Classical and Modern Optics,* 4th ed. (Englewood Cliffs, NJ: Prentice-Hall, 1995).

[17] Frank Pedrotti and Leno Pedrotti, *Introduction to Optics*, 2d ed. (Upper Saddle River, NJ: Prentice-Hall, 1993).

[18] Eugene Hecht, *Optics,* 2d ed. (Reading, MA: Addison-Wesley, 1987), p. 136. Notice that Hecht uses the sign convention that $\mathbf{R}_2 < 0$ for bi-convex lenses.

Simple lens: A lens formed from a single piece of optical material.

Compound lens: A lens formed from several pieces of optical material cemented together.

Spherical lens: A lens whose curved surface(s) may be described by the equation of a sphere whose origin lies on the centerline of the lens.

Cylindrical lens: A lens with a surface curved in only one direction.

Aspheric lens: A lens whose curved surface(s) take on arbitrary shapes.

The most common lenses seen in the laser laboratory are simple spherical lenses. The configuration of a spherical lens is determined by the curvature of the two surfaces. If the radii of the two surfaces are called $\mathbf{R}_1$ and $\mathbf{R}_2$, and if the sign convention is that $\mathbf{R}_1$ and $\mathbf{R}_2$ are greater than zero for a bi-convex lens (see Figure A.1), then there are six possible configurations:

Bi-convex lens: A lens where $\mathbf{R}_1 > 0$ and $\mathbf{R}_2 > 0$

Bi-concave lens: A lens where $\mathbf{R}_1 < 0$ and $\mathbf{R}_2 < 0$

Plano-convex: A lens where $\mathbf{R}_1$ is flat and $\mathbf{R}_2 > 0$

Plano-concave lens: A lens where $\mathbf{R}_1$ is flat and $\mathbf{R}_2 < 0$

Meniscus convex: A lens where $\mathbf{R}_1 > 0$ and $\mathbf{R}_2 < 0$ and where $|\mathbf{R}_1| < |\mathbf{R}_2|$

Meniscus concave: A lens where $\mathbf{R}_1 > 0$ and $\mathbf{R}_2 < 0$ and where $|\mathbf{R}_1| > |\mathbf{R}_2|$

These six spherical lens configurations are illustrated in Figure A.2.

A.4.2 Classical Lens Equations

A lens is frequently used to focus a source object to an image object. Classical optics formulas are often used to compute the locations of the source and image. To calculate the

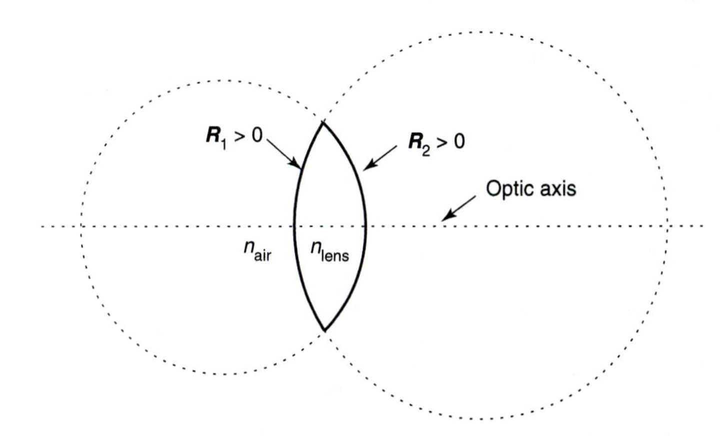

Figure A.1 The configuration of a spherical lens is determined by the curvature of its two surfaces. The sign convention is that $\mathbf{R}_1$ and $\mathbf{R}_2$ are greater than zero for a bi-convex lens.

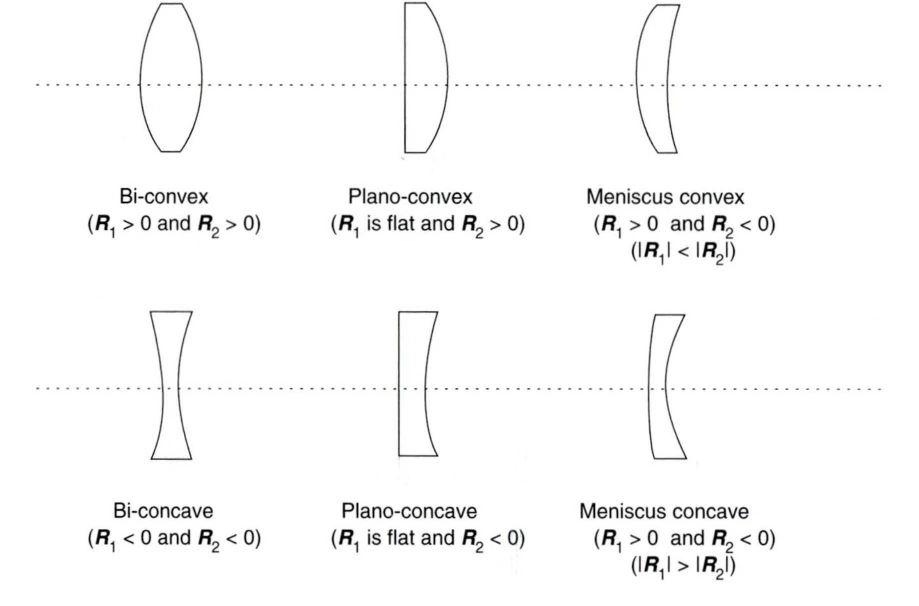

Figure A.2 Common lenses used in the laser laboratory.

location of the image object, consider a simple spherical thin lens with an index of refraction n_{lens} (see Figure A.3). This lens is immersed in a material of index of refraction n_{air} and has a curvature of $\mathbf{R}_1$ on the first surface and $\mathbf{R}_2$ on the second surface. A point source is assumed to be located a distance s_o away from the lens and the point source is focused by the lens to an image location s_i. The general classical optics formula relating s_o to s_i (for a thin lens) is given as

$$\frac{n_{\text{air}}}{s_o} + \frac{n_{\text{air}}}{s_i} = (n_{\text{lens}} - n_{\text{air}}) \left(\frac{1}{\mathbf{R}_1} + \frac{1}{\mathbf{R}_2} \right). \tag{A.49}$$

For a generic convex lens in air, where $n_{\text{air}} = 1$, this reduces to

$$\frac{1}{s_o} + \frac{1}{s_i} = (n_{\text{lens}} - 1) \left(\frac{1}{\mathbf{R}_1} + \frac{1}{\mathbf{R}_2} \right) \tag{A.50}$$

a form that is often called the *lensmaker's equation.*[19]

In laser engineering, the input beam into a lens is often collimated. Under these conditions, the object distance can be taken to infinity and the distance to the image s_i is the focal length of the lens f as

$$\frac{n_{\text{air}}}{f} = (n_{\text{lens}} - n_{\text{air}}) \left(\frac{1}{\mathbf{R}_1} + \frac{1}{\mathbf{R}_2} \right) = \frac{n_{\text{air}}}{s_o} + \frac{n_{\text{air}}}{s_i}. \tag{A.51}$$

[19]Eugene Hecht, *Optics*, 2d ed. (Reading Massachusetts: Addison-Wesley, 1987), p. 138. Notice that Hecht uses the sign convention that $\mathbf{R}_2 < 0$ for bi-convex lenses.

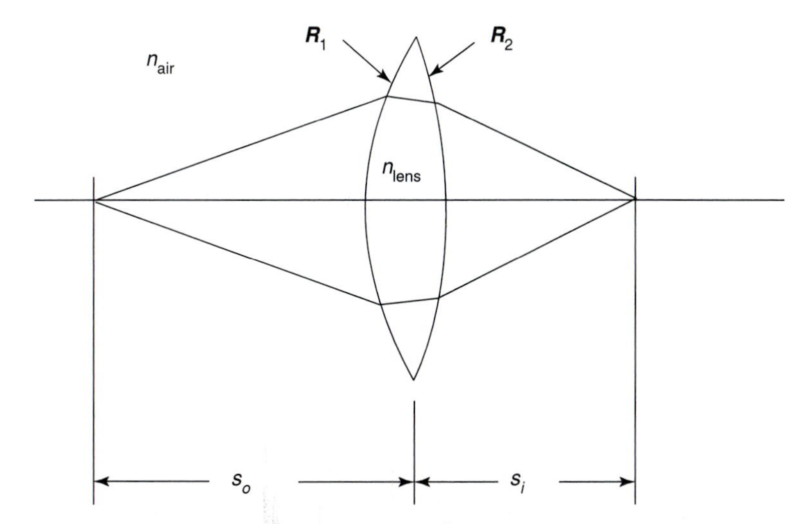

Figure A.3 A lens is frequently used to focus a source object to an image object. This lens is immersed in a material of index of refraction n_{air} and has a curvature of $\mathbf{R}_1$ on the first surface and $\mathbf{R}_2$ on the second surface. A point source is assumed to be located a distance s_o away from the lens and the point source is focused by the lens to an image location s_i.

For a generic convex lens in air, this is often written as

$$\frac{1}{f} = \frac{1}{s_o} + \frac{1}{s_i} \tag{A.52}$$

a form that is often called the *Gaussian lens formula*.[20]

A.4.3 Telescopes

The single most common use of lens systems in laser engineering is for the expansion or reduction in the size of a collimated laser beam. This function is typically performed using a simple telescope system. There are two types of telescopes commonly used in laser systems.

Keplerian telescopes (see Figure A.4) are constructed from two positive lenses and have a real focal point located in the center of the telescope. Such telescopes are useful if spatial filtering is desired because a small aperture (Lyot stop) can be placed at the central focal point in order to remove high-frequency optical noise.

However, many high-power lasers are sufficiently intense to ionize air at the focus of a Keplerian telescope. Although this is certainly dramatic to observe, it does have a negative effect on beam quality. Thus telescopes used with high-power lasers are constructed with one positive and one negative lens to avoid internal focal points. Such telescopes are called *Galilean telescopes* (see Figure A.5).

To examine the telescope more closely, consider a simple optical system formed from two thin lenses of front focal length f_1 and f_2 and separated by a distance d (see Figure A.6).

[20]Eugene Hecht, *Optics*, 2d ed. (Reading, MA: Addison-Wesley, 1987), p. 138.

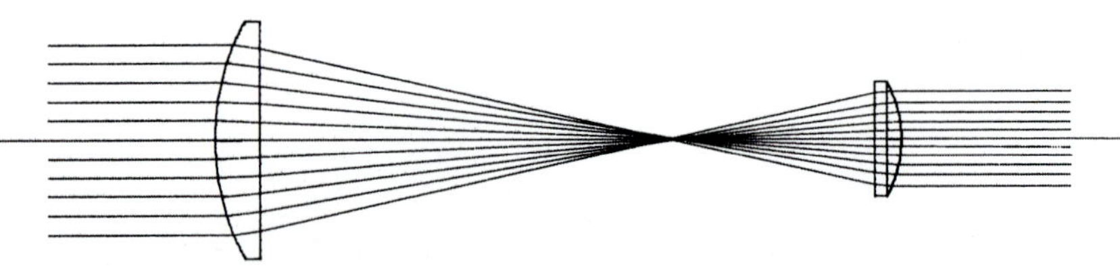

Figure A.4 Keplerian telescopes are constructed from two positive lenses and have a real focal point located in the center of the telescope.

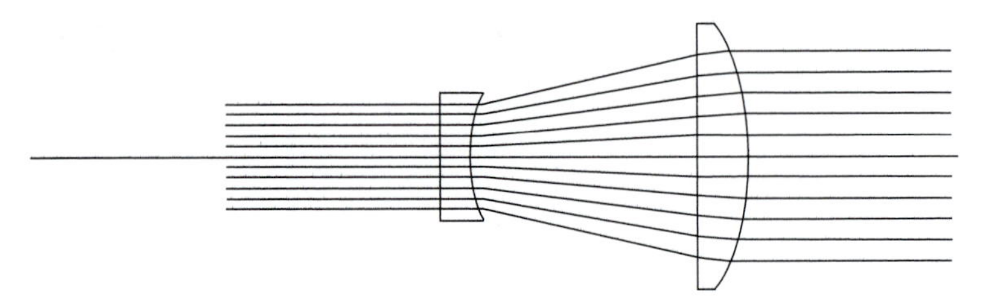

Figure A.5 Galilean telescopes are constructed with one positive and one negative lens to avoid internal focal points.

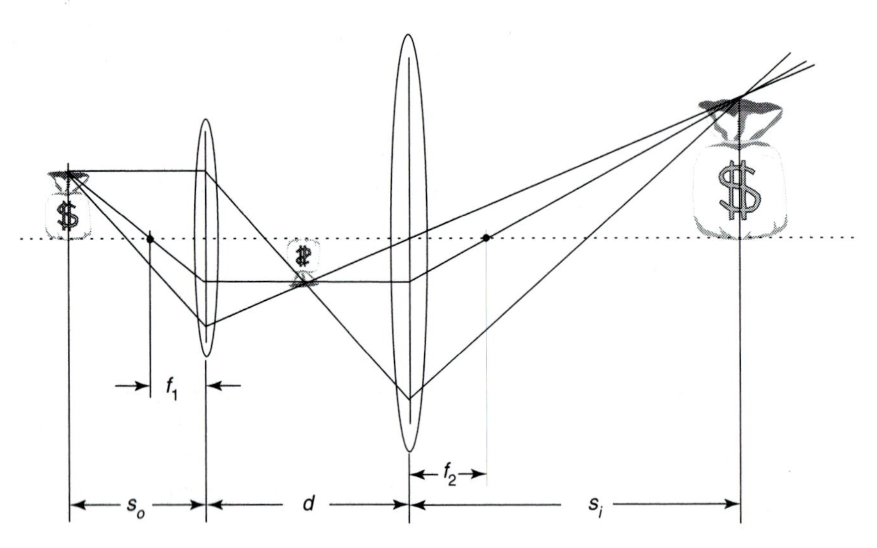

Figure A.6 Two lenses separated by a distance greater than the sum of their focal lengths. (Adapted from E. Hecht, *Optics*, 2d ed., Figure 5.30, p. 147. ©1987, 1974 Addison-Wesley Publishing Company, Inc. Reprinted by permission of Addison-Wesley Longman, Inc.)

The relationship between the image and object distance for such a system is given by[21]

$$s_i = \frac{f_2 d - f_1 f_2 s_o / (s_o - f_1)}{d - f_2 - f_1 s_o / (s_o - f_1)}. \tag{A.53}$$

[21] Eugene Hecht, *Optics,* 2d ed. (Reading, MA: Addison-Wesley, 1987), p. 147.

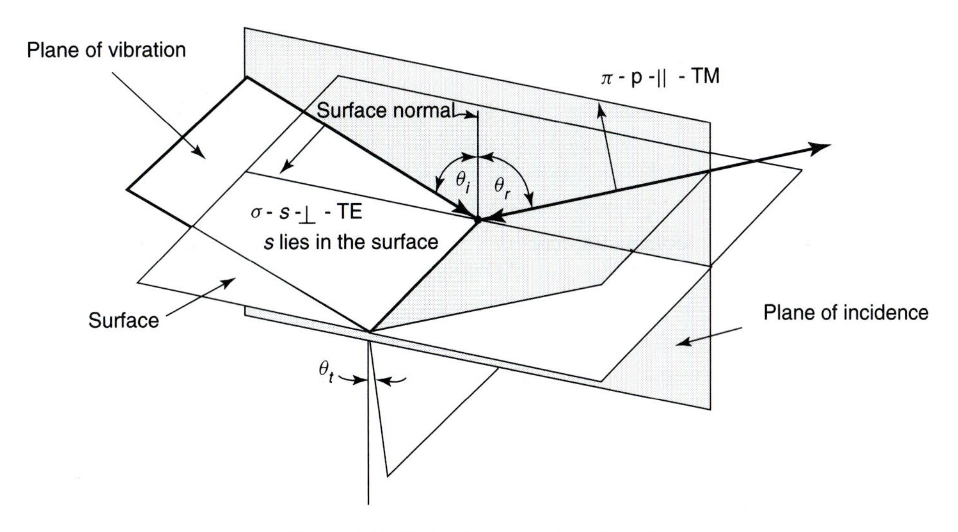

Figure A.7 Basic sign conventions in optics.

The system will act as a beam expander if $d = f_1 + f_2$ and will display a magnification given by[22]

$$M_{\text{lens}} = -\frac{f_2}{f_1} \tag{A.54}$$

where the larger size beam is associated with the side of the telescope with the longer focal length lens and where the negative sign indicates that the image is inverted.

A.5 REFLECTION AND REFRACTION

A.5.1 Nomenclature

A light ray incident upon an object will produce both a reflected ray and a transmitted (refracted) ray. Unfortunately, there is no one standard convention for describing reflection and refraction. Instead, a number of different conventions have been established in various branches of optics. Figure A.7 illustrates the most common sign conventions used for reflection and refraction.

Plane of incidence[23]: The path of a light beam reflecting from a surface defines a *plane of incidence*. The path of the light beam lies entirely in this plane. The plane is perpendicular to the surface.

---

[22]David Halliday and Robert Resnick, *Physics*, Part II, 3d ed. (New York: John Wiley and Sons, 1978), p. 982.

[23]There is some use in the literature of the term *plane of polarization*. The plane of polarization was originally defined (by Malus) as being perpendicular to the direction of electric field vector. Since this is more than slightly confusing, the term *plane of polarization* is carefully not used in modern terminology. The term *plane of vibration* is used instead to describe the plane containing the electric field vector.

θ_i, θ_r, and θ_t: The incident, reflected, and transmitted angles. These angles lie in the plane of incidence and are measured from the surface normal.

σ, $\perp$, s-component, or TE: The electric field (or polarization) component that is perpendicular to the plane of incidence and parallel to the incident surface. (The s-component is the one reflected at Brewster's angle.) A mnemonic is: "s lies in the surface" of the plane of vibration.

π, $\|$, p-component, or TM: The electric field (or polarization) component that lies in the plane of incidence and is orthogonal to the σ or s-component. (The p-component is the one transmitted at Brewster's angle.)

$r_\perp$, $r_\|$, $t_\perp$, $t_\|$: The electric field reflection and transmission coefficients.

$R_\perp$, $R_\|$, $T_\perp$, $T_\|$: The intensity reflection and transmission coefficients.

A.5.2 Snell's Law

Light incident on a surface will be reflected and transmitted. The angle of incidence will equal the angle of reflection $\theta_i = \theta_r$. The refracted angle θ_t for the transmitted ray will be governed by Snell's law as[24]

$$n_i \sin\theta_i = n_t \sin\theta_t. \tag{A.55}$$

(See also Section A.6.)

A.5.3 Total Internal Reflection

If light is traveling from a material of high index ($n_i = n_{\text{high}}$) to a material of low index ($n_t = n_{\text{low}}$), there is an incident angle for which no light is transmitted into the material of lower index. For all angles greater than this angle, the light will be totally reflected off the interface and will remain in the material of higher index. The phenomena is called total internal reflection and the critical angle θ_{TIR}, at which total internal reflectance first occurs, is given by[25]

$$\theta_{\text{TIR}} = \sin^{-1}\left(\frac{n_{\text{low}}}{n_{\text{high}}}\right) = \sin^{-1}\left(\frac{n_t}{n_i}\right). \tag{A.56}$$

A.5.4 Brewster's Angle

For dielectric materials, there is a particular angle of reflection called Brewster's angle, θ_B, for which the reflection coefficient of the π or p-component is zero. Brewster's angle is given as[26]

$$\theta_B = \tan^{-1}\left(\frac{n_t}{n_i}\right). \tag{A.57}$$

---

[24]David Halliday and Robert Resnick, *Physics*, Part II (New York: John Wiley and Sons, 1978), p. 946, eq. 43-11.

[25]David Halliday and Robert Resnick, *Physics*, Part II (New York: John Wiley and Sons, 1978), p. 947, eq. 43-13.

[26]David Halliday and Robert Resnick, *Physics*, Part II (New York: John Wiley and Sons, 1978), p. 1075, eq. 48-2.

At Brewster's angle, the reflected light is all σ or s-component. However, this weakly reflected light is only a small percentage of the total σ or s-component light in the incident beam. Thus, the transmitted light still consists of a significant amount of s-component as well as p-component light.

A.6 FRESNEL EQUATIONS

Snell's law is a subset of a more general set of expressions governing reflection and refraction. These expressions (called the Fresnel equations) govern the magnitude of the reflected and transmitted components. The general forms of the Fresnel equations for the electric field reflection and transmission (assuming well-behaved dielectric materials for which $\mu_i = \mu_t$) are given here.

Field (amplitude) reflection[27]:

$$r_\perp = \left(\frac{E_r}{E_i}\right)_\perp = \frac{n_i \cos\theta_i - n_t \cos\theta_t}{n_i \cos\theta_i + n_t \cos\theta_t} \tag{A.58}$$

$$r_\parallel = \left(\frac{E_r}{E_i}\right)_\parallel = \frac{n_t \cos\theta_i - n_i \cos\theta_t}{n_i \cos\theta_t + n_t \cos\theta_i} \tag{A.59}$$

Field (amplitude) transmission[28]:

$$t_\perp = \left(\frac{E_t}{E_i}\right)_\perp = \frac{2n_i \cos\theta_i}{n_i \cos\theta_i + n_t \cos\theta_t} \tag{A.60}$$

$$t_\parallel = \left(\frac{E_t}{E_i}\right)_\parallel = \frac{2n_i \cos\theta_i}{n_i \cos\theta_t + n_t \cos\theta_i} \tag{A.61}$$

Generally, these equations are simplified by expressing the equations only in terms of the incident (θ_i) and transmitted (θ_t) angles.

Field (amplitude) reflection[29]:

$$\left(\frac{E_r}{E_i}\right)_\perp = \frac{-\sin(\theta_i - \theta_t)}{\sin(\theta_i + \theta_t)} \tag{A.62}$$

$$\left(\frac{E_r}{E_i}\right)_\parallel = \frac{\tan(\theta_i - \theta_t)}{\tan(\theta_i + \theta_t)} \tag{A.63}$$

Field (amplitude) transmission[30]:

$$\left(\frac{E_t}{E_i}\right)_\perp = \frac{2\sin\theta_t \cos\theta_i}{\sin(\theta_i + \theta_t)} \tag{A.64}$$

---

[27]Eugene Hecht, *Optics*, 2d ed. (Reading, MA: Addison Wesley, 1987), p. 95, 96.

[28]Eugene Hecht, *Optics*, 2d ed. (Reading, MA: Addison Wesley, 1987), p. 95, 96.

[29]Eugene Hecht, Optics, 2d ed. (Reading, MA: Addison Wesley, 1987), p. 96.

[30]Eugene Hecht, *Optics*, 2d ed. (Reading, MA: Addison Wesley, 1987), p. 96.

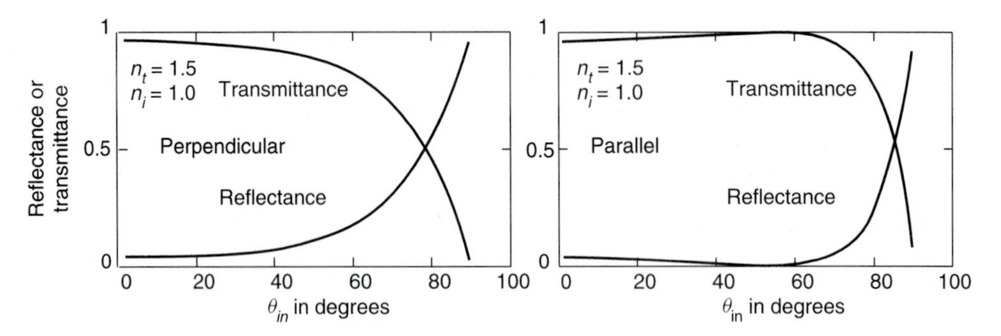

Figure A.8 Reflectance and transmittance versus incident angle.

$$\left(\frac{E_t}{E_i}\right)_{\parallel} = \frac{2\sin\theta_t\cos\theta_i}{\sin(\theta_i+\theta_t)\cos(\theta_i-\theta_t)} \tag{A.65}$$

Notice that these are field reflections and transmissions, not intensity reflections and transmissions. Most photodetectors (and the human eye) do not detect electric fields but, rather, detect intensities. In practice, it is the *intensity* reflection coefficients that are really of interest as they are the ones that govern reflection and transmission coefficients as measured by photodetectors or the human eye. For well-behaved nonabsorbing materials (at $\theta_i \neq 0$) these are given by (see Figure A.8):

Intensity reflection[31]:

$$R_{\perp} = r_{\perp}^2 = \left(\frac{E_r}{E_i}\right)_{\perp}^2 = \frac{\sin^2(\theta_i-\theta_t)}{\sin^2(\theta_i+\theta_t)} \tag{A.66}$$

$$R_{\parallel} = r_{\parallel}^2 = \left(\frac{E_r}{E_i}\right)_{\parallel}^2 = \frac{\tan^2(\theta_i-\theta_t)}{\tan^2(\theta_i+\theta_t)} \tag{A.67}$$

Intensity transmission[32]:

$$T_{\perp} = \frac{n_t\cos\theta_t}{n_i\cos\theta_i}t_{\perp}^2 = \frac{\sin 2\theta_i\sin 2\theta_t}{\sin^2(\theta_i+\theta_t)} \tag{A.68}$$

$$T_{\parallel} = \frac{n_t\cos\theta_t}{n_i\cos\theta_i}t_{\parallel}^2 = \frac{\sin 2\theta_i\sin 2\theta_t}{\sin^2(\theta_i+\theta_t)\cos^2(\theta_i-\theta_t)} \tag{A.69}$$

$$R_{\perp} + T_{\perp} = 1 \tag{A.70}$$

$$R_{\parallel} + T_{\parallel} = 1 \tag{A.71}$$

[31] Eugene Hecht, *Optics*, 2d ed. (Reading, MA: Addison Wesley, 1987), p. 102.
[32] Eugene Hecht, *Optics*, 2d ed. (Reading, MA: Addison Wesley, 1987), p. 102.

A.7 THE EFFECTIVE VALUE OF THE NONLINEAR TENSOR

The term d_{eff} is a materials parameter used in the calculation of nonlinear optical properties. The value of d_{eff} depends on the crystal structure and the type of phase matching. For several common crystals, d_{eff} is given by[33]

	ooe, oeo, eoo	eeo, eoe, oee	
$\bar{4}$2m	$d_{36} \sin\Theta \sin 2\Phi$	$d_{36} \sin 2\Theta \cos 2\Phi$	KDP, KD*P, CDA, CD*A, ADP
3m	$d_{15} \sin\Theta - d_{22} \cos\Theta \sin 3\Phi$	$d_{22} \cos^2\Theta \cos 3\Phi$	$LiNbO_3$, BBO
6, 6m	$d_{15} \sin\Theta$	0	$LiIO_3$

Type I phase matching is when $n_2^e(\Theta) = n_1^o$. Type I phase matching is most efficient when Θ is near 90 degrees and when Φ is near 45 degrees. In Type I phase matching, the fundamental beam is polarized perpendicular with the optic axis and the second harmonic beam is polarized parallel with the optic axis. For the popular crystal KDP, d_{eff} is given by

$$d_{\text{eff}} = d_{36} \sin\Theta \sin 2\Phi. \tag{A.72}$$

(As an aside, notice that for $\bar{4}$2m materials such as KDP, ADP, etc., only three nonzero tensor coefficients exist and they are all equal. Thus $d_{14} = d_{25} = d_{36}$.)

Type II phase matching is when $n_2^e(\Theta) = (1/2)(n_1^e(\Theta)+n_1^o)$ Type II phase matching is most efficient when Θ is near 45 degrees and when Φ is near 90 degrees. In Type II phase matching, the fundamental beam is polarized at 45 degrees to the optic axis and the second harmonic beam is polarized parallel with the optic axis. Again, for KDP d_{eff} is given by

$$d_{\text{eff}} = d_{36} \sin 2\Theta \cos 2\Phi. \tag{A.73}$$

The d components in the above chart are the tensor components for the reduced nonlinear tensor d. The full form of d is given by

$$d_{ijk} = \frac{1}{2}\chi_{ijk}^{(2)} \tag{A.74}$$

so it would have 27 components (with the indices, 111, 112, 113, 121, 122, and so on through all the combinations of the Cartesian axes 1, 2, and 3). However, for second harmonic frequency generation, the two waves are the same and so d_{ijk} is symmetric in its last two indices. Thus, the jk indices are typically mapped to a single number following the pattern

jk	11	22	33	23,32	31,13	12,21
l	1	2	3	4	5	6

[33] V. G. Dmitriev, G. G. Gurzadyan, and D. N. Nikogosyan, *Handbook of Nonlinear Optical Crystals* (Berlin: Springer-Verlag, 1991), p. 24.

This has the subtle side advantage of permitting d_{ijk} to be written as a matrix. Thus, the nonlinear polarization can be written in terms of the modified tensor d_{il} as

$$\begin{bmatrix} P_x(2\omega) \\ P_y(2\omega) \\ P_z(2\omega) \end{bmatrix} = 2 \begin{bmatrix} d_{11} & d_{12} & d_{13} & d_{14} & d_{15} & d_{16} \\ d_{21} & d_{22} & d_{23} & d_{24} & d_{24} & d_{26} \\ d_{31} & d_{32} & d_{33} & d_{34} & d_{35} & d_{36} \end{bmatrix} \begin{bmatrix} E_x(\omega)^2 \\ E_y(\omega)^2 \\ E_z(\omega)^2 \\ 2E_y(\omega)E_z(\omega) \\ 2E_x(\omega)E_z(\omega) \\ 2E_x(\omega)E_y(\omega) \end{bmatrix}. \tag{A.75}$$

Luckily, for most crystals of interest, many of these coefficients are 0. For example, for a crystal of class 3m, the modified tensor d_{il} is

$$d_{il} = \begin{bmatrix} 0 & 0 & 0 & 0 & d_{31} & -d_{22} \\ -d_{22} & d_{22} & 0 & d_{31} & 0 & 0 \\ d_{31} & d_{31} & d_{33} & 0 & 0 & 0 \end{bmatrix} \tag{A.76}$$

and similarly for the $\bar{4}$2m point group,

$$d_{il} = \begin{bmatrix} 0 & 0 & 0 & d_{36} & 0 & 0 \\ 0 & 0 & 0 & 0 & d_{36} & 0 \\ 0 & 0 & 0 & 0 & 0 & d_{36} \end{bmatrix} \tag{A.77}$$

and for the 6 point group,

$$d_{il} = \begin{bmatrix} 0 & 0 & 0 & 0 & d_{31} & 0 \\ 0 & 0 & 0 & d_{31} & 0 & 0 \\ d_{31} & d_{31} & d_{33} & 0 & 0 & 0 \end{bmatrix}. \tag{A.78}$$

A.8 PROJECTS AND DESIGN ACTIVITIES

A.8.1 Gas Laser Activities

There are a number of interesting hands-on and design activities that can be performed with a HeNe or argon-ion laser. Good references are the collection of Stong's classic *Amateur Scientist* articles,[34] the recent text by Zare,[35] and Newport Corporation's project series.[36]

Designing and building a HeNe laser (from scratch). The idea of designing and building a HeNe laser by hand (fabricating, baking out, and filling the tube) is very compelling. However, this is not a trivial project and should not be attempted unless good

[34]These are C. L. Stong's articles from *The Amateur Scientist*, reprinted as *Light and Its Uses*, ed Jearl Walker (San Francisco, CA: W. H. Freeman and Company, 1952).

[35]R. N. Zare, B. H. Spencer, D. S. Springer and M. P. Jacobson, *Laser Experiments for Beginners* (Sausalito, CA: University Science Books, 1995).

[36]Newport has a series called *Projects in Optics* and *Projects in Holography* that are associated with some of their teaching kits.

resources are available to support glass-blowing, vacuum bake-out, and gas handling. If such resources are available, Malacara et al.[37] and Stahlberg et al.[38] may prove useful in addition to the above references.

Designing and building a HeNe laser (from an existing plasma tube). Designing and building a HeNe laser using an existing plasma tube as a starting base is readily achievable in an undergraduate class.

For a variation on this, surplus HeNe tubes can be obtained and a HeNe laser power supply designed and built to drive the tubes. Maxson et al.[39] and Posakony[40] may prove helpful in the design of HeNe power supplies.

Hints:

1. This project requires a high-quality laser plasma tube. Options include tubes with a window on one end and a high-reflecting curved mirror on the other[41] or two windows.[42]
2. Building a HeNe power supply is not a trivial task. Since building a power supply is *not* a laser design project (but rather an electronics one), groups may wish to consider purchasing a power supply.[43]

Single Pass Gain. A HeNe laser is *not* a high-gain laser! However, it is possible to (roughly!) determine the gain of a HeNe tube. This can either be done in a single pass configuration (requiring a double-window tube) or by inserting a variable loss element inside a HeNe cavity (requiring a single-window tube).

Hints:

1. The critical components are a laser tube with two windows[44] (or a laser tube with a window on one end and a high-reflecting curved mirror on the other[45]) and an appropriate power supply.[46]

---

[37] D. Malacara, L. R. Berriel, and I. Rizo, *Am. J. Phys.* 37:276 (1969).

[38] B. Stahlberg, P. Jungner and T. Fellman, *Am. J. Phys.* 58:878 (1990).

[39] D. Maxon, D. G. Seller, and L. Tipton, *Rev. Sci. Instrum.* 46:1110 (1975).

[40] M. Posakony, *Rev. Sci. Instrum.* 43:270 (1972).

[41] The Melles Griot 05-WHR-570, 4 mW plasma tube (with a 45 cm radius, high-reflecting end mirror) works well for student projects.

[42] The Melles Griot 05-LHB-290, 1 mW double Brewster tube works well for student projects.

[43] The Melles Griot 05-LPL-939-065 matches the 05-LHB-290, 1 mW double Brewster tube; and the 05-WHR-570, 4 mW plasma tube.

[44] The Melles Griot 05-LHB-290, 1 mW double Brewster tube works well for student projects.

[45] The Melles Griot 05-WHR-570, 4 mW plasma tube (with a 45 cm radius, high-reflecting end mirror) works well for student projects.

[46] The Melles Griot 05-LPL-939-065 matches the 05-LHB-290, 1 mW double Brewster tube; and the 05-WHR-570, 4 mW plasma tube.

2. For single-pass gain, an alignment laser can be directed through a double-window plasma tube. The intensity can then be measured with the double-window tube off (losses only) and with the tube on (gain and losses).
3. Another option for measuring the gain is to introduce a variable loss element inside a laser cavity. For example, a Brewster surface inserted in the cavity can be used as a variable loss element. As the Brewster surface is rotated, the loss will change slightly. The output laser performance can then be measured as a function of the calculated loss (from the Fresnel equations). (Notice that this does not work if the mirror is curved or if the Brewster plate is too thick, because the small spatial deviation of the beam crossing the Brewster surface is greater than the alignment tolerance of the cavity.)
4. Alternatively, a variable loss element can be designed to be used in a cavity with a curved front mirror. (The challenge here is to not displace the beam sideways when the element loss is changed!)

Transverse modes. Viewing the multitude of possible transverse modes in a laser system is one of the high points of a laser class. Generally, commercial lasers are sold to run single-$TEM_{0,0}$ mode. Thus, it is difficult to obtain a commercial HeNe that will display transverse modes. However, constructing a simple HeNe from a large diameter plasma tube (as described previously) permits the observation of transverse modes. Hint: Again, the critical components are a laser tube with a large diameter aperture,[47] and an appropriate power supply.[48]

Some interesting activities to try:

- Adjust the external mirror alignment and observe the transverse-mode behavior.
- Adjust the external mirror position and observe the transverse-mode behavior.
- Insert an aperture into the cavity and observe the transverse-mode behavior as the aperture diameter is changed.
- Insert a thin wire or needle into the cavity and observe the transverse-mode behavior as the wire position is changed.
- Adjust the alignment until a pure transverse mode (for example, $TEM_{1,4}$) lies with its horizontal axis along the table. Then install a photodetector and measure the intensity of the transverse mode at various horizontal positions. Plot the data points and compare with the calculated intensity for the mode.

Gaussian beams. Observing Gaussian beam propagation in an aquarium is an excellent way to experimentally confirm the Gaussian beam equations.

The critical components for the experiment are a standard commercial HeNe or argon-ion laser, a small aquarium, and an assortment of lenses.

[47]The Melles Griot 05-WHR-570, 4 mW plasma tube (with a 45 cm radius, high-reflecting end mirror) works well for displaying a wide selection of transverse modes.

[48]The Melles Griot 05-LPL-939-065 matches the Melles Griot 05-WHR-570, 4 mW plasma tube.

Some interesting basic activities to try:

- Place various lenses in the beam and observe the characteristic shape of the beam as it focuses down to the beam waist.
- Measure the location of the beam waist and calculate the focal length of the lenses.
- Measure the Rayleigh range and compare with theory.
- Measure the beam waist at two locations far away from the minimum beam waist and calculate the location and size of the minimum beam waist. Compare with the actual location and size of the minimum beam waist.
- Design lens combinations and calculate the location and size of the minimum beam waist. Compare with the measured location and size of the minimum beam waist.

More advanced activity: Design and construct a simple resonator structure using the argon laser as an external source and locating the fishtank in the middle of the resonator. (This will allow you to see the path the beam takes in the resonator.) Vary the resonator from stable to unstable and observe the changes in the beam.

Stability. An external mirror laser cavity provides an excellent way of testing the stability equation (see Section 4.5). The critical components for the experiment are a laser tube, an appropriate power supply, and whatever additional mirrors are required for the tube. Once the laser is aligned and lasing, the external mirror can be moved and the stability equation and stability diagram (see Section 4.3) mapped.

Some variations to try:

- Use external mirrors with different radii of curvature.
- Use a double-window tube (thus gaining control of both mirrors) and explore the stability diagram with various combinations of flat-curved and curved-curved mirrors.
- Consider a cavity with an included lens. Use the lens matrix method to derive a new set of stability equations. Build cavities with included lenses and test them.

Longitudinal modes and etalons. Viewing the longitudinal modes of a laser as a function of temperature and transverse mode structure is one of the classical laser activities.

Hints:

1. Viewing longitudinal modes requires some type of spectrum analyzer. Using a scanning confocal interferometer as a spectrum analyzer is delightful, because the interferometer is a Fabry-Perot cavity and the instrument itself becomes part of the activity.
2. The critical components are a scanning confocal interferometer[49] and a commercial oscilloscope. (A standard commercial HeNe or argon-ion laser, an optical rail, and appropriate mounting hardware are also required.)

[49]The Melles Griot 13-SAE-024, 480-600 nm, with the 2 GHz head and the 13 SAD-001 electronics module, work well for this laboratory.

3. If a scanning confocal interferometer is *not* available, it is also possible to "roll your own" and design a variable length Fabry-Perot from mirrors and a good set of adjustable mirror mounts.

Some interesting basic activities to try:

- Observe the longitudinal modes immediately after the laser is turned on and compare them with the modes after the laser is fully warmed up.
- Spray hot or cold air on the laser and watch the longitudinal mode location change with changes in temperature.
- Use a laser that can lase on several transverse modes and observe the changes in the longitudinal mode structure for various combinations of transverse modes.

More advanced activities:

- Turn off the piezoelectric scanner on the optical spectrum analyzer and use a white light source and a monochromator to measure the transmittance of the spectrum analyzer as a function of wavelength. As a variation, apply a DC voltage to the piezoelectric scanner (to controllably change the cavity length) and measure the transmittance of the spectrum analyzer as a function of wavelength and applied voltage. (This requires a white light source and a monochromator or second spectrum analyzer.)
- Design and fabricate some simple etalons by depositing chromium or aluminum on glass substrates. Measure the transmittance of the simple etalons using a monochromator and a white light source. (This requires access to an evaporator, a white light source, and a monochromator or second spectrum analyzer.)

Holography. Basic transmission and reflection holography, as well as some more sophisticated techniques such as interference holography, are very showy and popular projects.

Hints:

1. Use high quality plates or films. If economics is an issue, it is better to cut large plates into small plates than to compromise on resolution. Newport Corporation (1-800-222-6440) is a good source of plates and films. (8E75HD-1 plates are a good starting point.)
2. Stability is a key to successful holography. Set up a Michaelson interferometer first. If a Michaelson interferometer doesn't yield stable fringes, then successful holography is not likely.
3. Black and white photography processing knowledge is important for student groups. Holography project success can be dramatically improved by encouraging students to pick up some basic information on black and white photography early in the quarter/semester before the design activity begins.

Raster scanning (laser TV) projects. A number of interesting projects are possible using various scanning technologies. Laser TV and laser light show projects are also crowd pleasers at engineering fairs.

Hints:

1. Laser scanning projects use either resonant scanners, or motors plus multisided mirrors. A common problem is speed stability of the motor.
2. The simplest way to control the intensity of the laser in a scanning project is to use a laser diode and to modulate the operating current. However, the high elliptical output mode of laser diodes usually means that a significant part of the project time is spent collimating the diode!
3. Laser scanning projects tend to be very electronically-oriented because of the complexity of the video scanning. Successful student teams usually have at least one computer engineer on the team.

Laser microphone projects. Optical microphone projects are popular, as the idea of being James Bond and spying on conversations by reflecting a laser beam from a window seems to have a perennial appeal.

Most laser microphone projects begin as an attempt to measure the laser Doppler signal in an interferometer configuration. This turns out to be quite difficult in the typical undergraduate laboratory. However, most interferometers built to show Doppler signals usually do operate effectively as optical microphones! (If a Michaelson interferometer is constructed where one mirror is vibrated by an audio source and where a photodetector plus amplifier arrangement is placed on the output of the interferometer, the system will give a recognizable output signal corresponding to the input audio signal.) However, the effect is a spatial one, not a Doppler shift. The laser beam tends to be modulated spatially and this spatial modulation is picked up by the photodetector and amplified.

Hints:

1. Build the simplest possible optical set-up to display the microphone effect.
2. Consider using a HeNe laser for debugging the electronics, then move to a laser diode if the project requires it.
3. Remember that the Doppler effect is a frequency-dependent effect. If the output of an interferometer system is AM modulated, it is probably spatial modulation rather than Doppler.

Gain saturation. Any particular laser system has an optimal value for the transmittance of the output coupler. (See Chapter 5.)

The critical components for the experiment are a commercial argon-ion or Nd:YAG laser with external mirrors and an adjustable power supply and a set of output mirrors (two is minimum, four is better) at one wavelength with different transmittances. (An optical rail and appropriate mounting hardware are also required.)

The output laser power as a function of the input electrical power can then be measured for each of the available output mirrors. The resulting data can be used to compute the small signal gain, loss, saturation intensity, and optimal output coupler value for the laser.

A.8.2 Nd:YAG Laser Activities

It is more difficult to develop teaching activities using conventional solid-state lasers than simple gas lasers, primarily due to the high starting cost of the equipment. Small HeNe lasers can be purchased for tens or hundreds of dollars and small argon-ion lasers can be purchased (used) for a few thousand dollars. However, lasers such as the Spectra-Physics Quanta-Ray GCR are significantly out of the price range of a typical teaching laboratory!

Additionally, a number of commercial conventional solid-state lasers are constructed in such a way as to optimize turn-key operation. In many cases, this means the laser is constructed as a sealed box with little ability to alter the components in a teaching environment.

However, there are some extremely viable and cost-effective alternatives. First, there are several small companies that sell low-energy conventionally pumped Nd:YAG lasers. These are typically 10 to 50 mJ per pulse, can be operated Q-switched, and provide an optical assembly that permits manipulation of the cavity components.[50] Second, it is sometimes possible to purchase small used Nd:YAG laser systems for a few thousand dollars. Finally, small speciality laser companies sometime build little in-house Nd:YAG lasers as pumps for other processes. It is sometimes possible to convince them to either donate one or build one at cost as part of a cooperative educational exercise with a local educational institute.

Given a small Nd:YAG system, there are a number of possible activities that can be developed. Ideas are listed below.

Gain saturation. The output coupler experiment presented in Section 5.3.3 and described in Section A.8.1 can be performed with Nd:YAG. Since Nd:YAG is a high-gain laser, the reflectance values for the output coupling mirror cover a broader range than with an argon-ion laser. This permits a very nice experiment where output power is measured versus input power for several mirrors, and the small signal gain and saturation intensity of the laser calculated from the experimental results.

Unstable resonator. The high gain of Nd:YAG permits the construction of unstable resonators similar to those discussed in Section 4.5. Interesting activities include: building an unstable resonator with a dot front coating (like a DCR laser), altering resonator properties by changing length and dot size (and comparing observations with theory), and observing the mode structure while passing a razor blade through the unstable resonator.

Relaxation oscillations and Q-switching. Many of the smaller Nd:YAG lasers can be operated Q-switched (Section 6.2). Interesting activities include: observing the relaxation oscillations (Section 6.1) as a function of pumping (and comparing obser-

[50]Representative companies include CASIX (China) and New Wave (Mountain View, CA). Additional companies can be found in *The Photonics Buyers Guide to Products and Manufacturers, 1996 Book 2* of the *Photonics Directory*, 42d ed. (Pittsfield, MA: Laurin Publishing Co., Inc., 1996).

vations with theory) as well as observing the Q-switched pulse as a function of timing and hold-off voltage (and comparing observations with theory).

Making a dye laser. A small Nd:YAG laser can pump a small dye cell to create a dye laser. Interesting activities include: building and aligning the dye laser, using prisms or gratings to tune the dye laser, and using the dye laser to perform spectroscopy activities.

Injection seeding. On a more advanced level, a small Nd:YAG can be injection seeded (not discussed in this text) with a single mode diode laser source.[51] This permits activities such as observing the longitudinal mode linewidth as the laser is injection seeded.

A.8.3 Transition Metal Laser Activities

As with Nd:YAG, it is more difficult to develop teaching activities using transition-metal solid-state lasers than HeNe lasers because of the high starting cost of the equipment. The problem is further complicated by the fact that typically two lasers are needed, both the primary laser (such as Ti:sapphire) and the pump laser (such as an argon-ion, Nd:YAG, or diode laser).

Laboratory suggestions for the lab with neither an Nd:YAG nor a Ti:sapphire laser include:

Ring lasers. A simple ring laser (Section 11.4.1) can be constructed using a Brewster window HeNe laser plasma tube (see Section A.8.1 for HeNe tube information). Specific activities include investigating the directionality of the ring laser and measuring the longitudinal mode properties of the ring in various configurations.

Birefringent filters. Birefringent filters (Section 11.4.2) can be constructed and their transmission spectra measured on a spectrophotometer (most Chemistry departments have spectrophotometers). Options include activities comparing theoretical and experimental transmission properties of birefringent filters as a function of number of elements and thickness.

Laboratory suggestions for the lab with only an Nd:YAG include:

Autocorrelator. A simple autocorrelator (Section 11.5) can be constructed using an interferometer geometry, a speaker (with an attached mirror), and a specially cut nonlinear crystal. Options include activities comparing the pulse width for Nd:YAG pulses measured on the autocorrelator versus those measured by various photodetectors.

Nonlinear optics. Q-switched Nd:YAG lasers (even small ones) are likely to be able to demonstrate external second harmonic generation (Section 7.2). Additionally (depending on the energy of the laser), Raman shifting (Section 7.4) and self-focusing (Section 7.5) demonstrations may be possible. Options include: second harmonic generation activities where the doubling efficiency is measured versus various parameters

---

[51] A good introductory treatment of injection locking and seeding can be found in Anthony Siegman, *Lasers* (Mill Valley, CA: University Science Books, 1986), Chapter 29.

(such as mode quality, focus, crystal length, etc.) and compared with theory, Raman-shifting activities where a prism or grating is used to separate the wavelengths, and self-focusing activities where highly nonlinear materials are intentionally damaged.

OPO. Depending on the power and energy of the Nd:YAG laser, it may be possible to construct a small singly resonant OPO (Section 7.3). Some companies sell OPO kits for this type of experiment.[52] Options include tuning the OPO and measuring its output wavelength as a function of temperature or angle. If the student lab has a possible pump laser (an argon-ion, Nd:YAG, or diode laser) and also can afford to purchase optics for a Ti:sapphire laser, then a number of additional interesting activities are possible.

Building and tuning a Ti:sapphire laser. The process of constructing a Ti:sapphire ring geometry resonator can be used as a laboratory exercise. Once the resonator is built, additional options include installing birefringent filters in the resonator, building a grating tuned version, self-mode-locking the resonator, and so on.

Diode pumped Ti:sapphire laser. An alternative path for the laboratory that has Ti:sapphire laser components (but no pump laser) is to construct a Ti:sapphire laser with a diode laser pump.

A.8.4 Successful Student Projects

- Optimizing the collimation of a diode laser
- Transmission and reflection holography
- Characterization of a pulsed Nd:YAG
- Scattering in random media
- Fiber-optic communication link
- Characterization of a carbon dioxide laser
- Two-color fiber-optic communication link
- Design and construction of ND filters
- A laser raster display system
- Laser unequal path interferometry
- Scattering in the oceanic environment
- Hole drilling using a pulsed Nd:YAG
- Photorefractive mirrors
- Design and fabrication of a HeNe laser
- Design of an in situ absorption meter for ocean optical sensing
- Laser doppler radar
- Low-temperature photoluminescence
- Lateral shearing interferometry

[52]Representative companies include CASIX (China) and New Wave (Mountain View, CA). Additional companies can be found in *The Photonics Buyers Guide to Products and Manufacturers, 1996 Book 2* of the *Photonics Directory*, 42d ed. (Pittsfield, MA: Laurin Publishing Co., Inc., 1996).

- Design and fabrication of a nitrogen laser
- Characterization of an argon-ion laser
- Photorefractive keratectomy
- Optimizing a carbon dioxide laser
- Fiber optic interferometer
- Operation of an audio CD system
- Noncontact measuring of skin blood flow rate using speckle phenomenon
- Laser graph
- Laser microphone experiment
- Laser seismograph project
- Laser doppler anemometry
- Spatial phase modulation using liquid crystal displays
- Laser photobleaching
- Results of laser scattering on water, ice, snow, and pavement
- A fiber-optic hydrophone
- Holographic interferometry of mono- and bimetallic pennies
- Where is the doppler shift?
- Laser doppler velocimeter
- Laser rastering
- Laser marking
- Holographic neural network
- Holography
- Laser display system
- Holography: vibration measurements of a piezoelectric device
- Droplet velocity and size from laser doppler velocimetry
- Free-space laser communication
- Character recognition
- Turbulence

A.9 LASER ALIGNMENT

There are a number of strategies for aligning a laser system. However, the majority rely on using one laser (termed the *alignment laser*) to align a second laser. This strategy is to orient the alignment laser so that it is parallel to the optical rail (or similar mounting system) and then to align the plasma tube and mirrors with respect to the alignment laser.

In the case of a plasma tube with a high-reflecting end mirror and a window, the steps are as follows:

1. Orient the alignment laser parallel to the optical rail and centered on it.

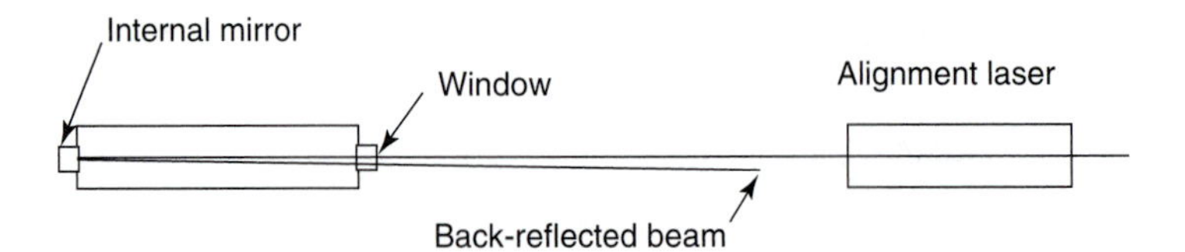

Figure A.9 The plasma tube is adjusted until the alignment beam enters the center of the front window, travels down the center of the laser bore, bounces off the end mirror, and reflects back out the front window.

2. Install the window plasma tube on the optical rail and center the alignment laser beam on the front window of the tube. (Flat window tubes are preferable to Brewster window tubes for this experiment, because it is significantly easier to center the alignment beam on a flat window than on a slanted Brewster window.)
3. Adjust the plasma tube until the alignment beam enters the center of the front window, travels down the center of the laser bore, bounces off the end mirror, and reflects back out the front window (see Figure A.9).

This situation can be recognized as follows:

- A small dot (generated by side-scattering from the alignment beam) can be seen centered on the front window.
- A small dot (generated by transmitted light leaking through the rear side of the back mirror) can be seen centered on the back reflector.
- There is virtually no scattering inside the laser bore.
- The back-reflected beam emerges from the front of the laser and is nearly coincident with the incoming alignment laser.

Once the back-reflected beam can be seen emerging from the laser, then the plasma tube should be adjusted so that the back-reflected beam travels exactly down the path of the alignment laser beam. Since the aperture at the front of the alignment HeNe is typically much larger in diameter than the HeNe beam itself, then a good trick is to use a small piece of paper with a pinhole in it as a temporary aperture (see Figure A.10). When the alignment laser beam passes through the pinhole, the back-reflected beam can be easily seen on the paper. The mechanical assembly is then adjusted until the back-reflected beam enters the pinhole.

Once the plasma tube is aligned with respect to the alignment laser, then the front mirror can be inserted into the cavity. The front mirror mount is then adjusted until the back-reflected beam from the mirror is aligned with the beam from the alignment laser (see Figure A.11).

At this point, the laser is aligned and the high-voltage power supply can be connected. The alignment laser can be removed and replaced with a photodetector. In most cases, the laser will lase as soon as the high voltage is applied. Once the laser is lasing, the plasma tube alignment and external mirror alignment should be adjusted until the beam profile is both symmetric and of maximum power.

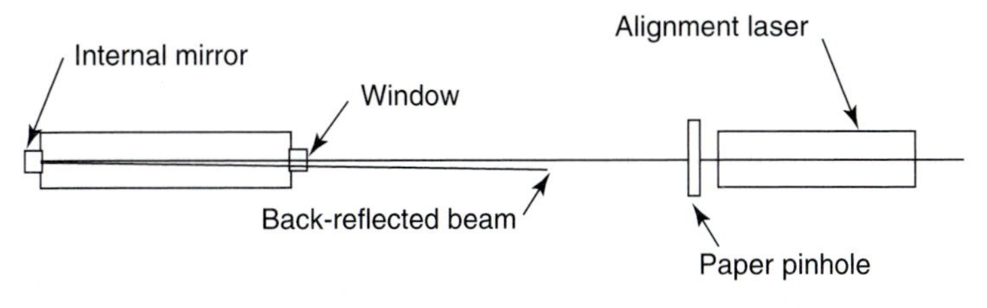

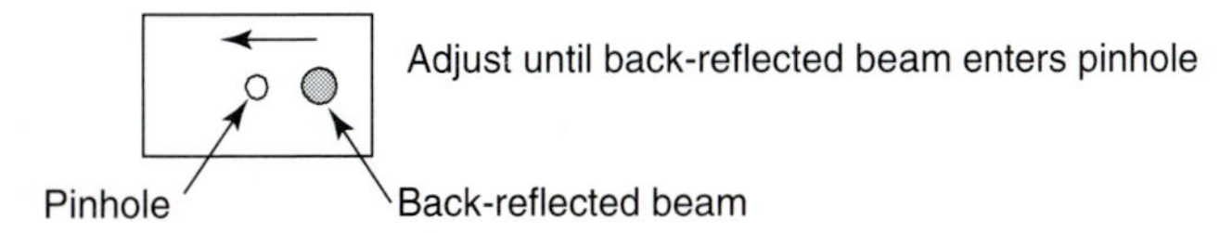

Figure A.10 A good trick in laser alignment is to use a small piece of paper with a pinhole in it as a temporary aperture. When the alignment laser beam passes through the pinhole, the back-reflected beam can be easily seen on the paper. The mechanical assembly is then adjusted until the back-reflected beam enters the pinhole.

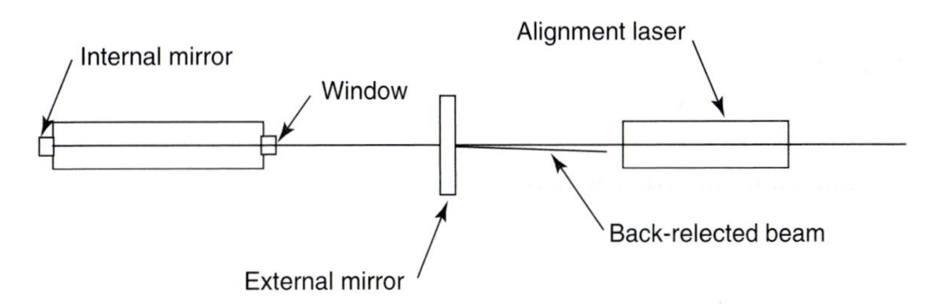

Figure A.11 The front mirror mount is adjusted until the back-reflected beam from the mirror is aligned with the beam from the alignment laser.

If the laser does not lase immediately, adjust the external mirror adjustment screws slightly in each direction. (For example, turn one-quarter turn to the right; then one-quarter turn to the left.) If the laser does not lase after about ten minutes of adjustment, then it is best to begin the alignment process again.

A.10 GLOSSARY OF BASIC LASER TERMS

ABCD law: A general expression that describes how the complex beam parameter is modified by transformation by an optical element.

Acousto-optic: Exhibiting an optical effect with acoustical waves.

Active mode-locking or Q-switching: Using an electrooptic or acousto-optic device to mode-lock or Q-switch.

Beam expander: An optical element used to expand a collimated beam into a collimated beam of a different diameter.

Birefringent materials: Materials with different indices of refraction in different directions.

Blackbody radiation spectrum: The energy (or power or energy density or power density) emitted from a blackbody source, typically plotted as a function of wavelength.

Broadening: An increase in the frequency linewidth of a longitudinal mode or of a gain profile.

Coherence length: The length over which a set of photons remains in phase.

Coherence time: The time over which a set of photons remains in phase.

Coherent (or photon coherent): The property that characterizes a set of photons that are precisely in phase.

Collisional broadening: An increase in the linewidth of a gain profile due to collisions between gas atoms.

Complex beam parameter: A complex (real + imaginary) expression containing both the radius and waist parameters.

Concentric cavity: A laser resonant cavity where $R_1 = R_2 = d/2$.

Confocal cavity: A laser resonant cavity where $R_1 = R_2 = d$.

Constructive interference: The superposition of two optical fields to create a larger field.

Continuous wave (cw): A laser that runs continuously (i.e., a laser that is not pulsed).

Degeneracy: The number of states at the same energy or a material doped so that the Fermi level is above (below) the conduction (valence) band edge.

Degrees of freedom: The set of basis vectors for a system.

Depth of field: The distance over which an image remains focused.

Destructive interference: The superposition of two optical fields of the opposite sign that cancel each other out.

Detailed balance: An accounting process that describes the populations of various states in a closed system.

Einstein A and B coefficients: The coefficients describing the spontaneous and stimulated transition rates between two states.

Electromagnetic mode density: The number of electromagnetic modes in a small volume per unit frequency.

Electronic states: States formed by various configurations of the electrons around an atom.

Electrooptic: Exhibiting an optical effect under an applied field.

Energy and energy density: Energy is measured in joules and energy density is energy per unit area or joules/cm^2.

Energy states: Various discrete configurations of a system that possess distinct energies.

Etalon: An optical element consisting of two plane parallel reflective surfaces.

Extraordinary index: n_e of a material.

F number: Focal length of a lens divided by aperture size.

Fabry-Perot etalon: A device with a selective transmission function similar to that of a comb filter.

Far-field: A distance far from a source, a distance much greater than the Rayleigh range and where a wave is spherical.

Far-field diffraction angle: The divergence angle of an expanding beam. Can be measured as a half or full angle.

Finesse (F): The free spectral range divided by the frequency linewidth. More generally used to describe the property of resonance, as in high-finesse cavity.

Four-state laser: A laser possessing a ground state, a pump state, and an upper and lower laser state.

Free spectral range: The spacing between the modes of a Fabry-Perot etalon. Sometimes used to describe the spacing between longitudinal modes of a laser cavity.

Full-width half-maximum (FWHM): The linewidth of a function as measured at half the total amplitude of the function. Usually applied to Lorentzian and Gaussian lineshape functions.

Gain material: The material containing the energy states that display stimulated emission and produce optical gain.

Gain or optical gain: The stimulated emission process results in more photons coming out of the laser than entered the laser. Thus, the process exhibits gain.

Gain profile: A material does not have gain over all frequencies. The function that describes the gain as a function of frequency is termed the gain profile.

Gain saturation: The reduction in the optical gain as the cavity intensity increases.

Gaussian: A function, often used to describe an inhomogeneous gain profile.

Gaussian mode: The $TEM_{0,0}$ mode of a laser. (Not the same thing as a Gaussian gain profile.)

Getter: A material that absorbs or otherwise attracts contaminants.

Ground state: The lowest energy state in a system.

Half-wave plate: An optical element that converts linearly polarized light into orthogonally oriented linearly polarized light.

Hermite polynomial: A polynomial used in the solution for the tranverse modes for a rectangular laser.

Homogeneous: Broadening processes that operate on all the atoms in a system equally.

Hygroscopic: Water-loving, often applied to crystals that react with water.

Induced absorption rate: The change in excited states per unit time resulting from a radiation field (pump light).

Inhomogeneous: Broadening processes that operate on some groups of atoms in the system differently than other groups.

Isotope broadening: An inhomogeneous broadening process caused by different isotopes in a laser material.

Laguerre polynomial: A polynomial used in the solution for the tranverse modes for a cylindrical laser.

Laser speckle: The characteristic bright/dark speckle pattern exhibited when a laser is directed against a diffuse reflector.

Lensmaker's equation: An equation for determining the image and object distances of a lens from the radii of curvature.

Linewidth, frequency linewidth, or wavelength linewidth: The linewidth is the width of the longitudinal mode or the gain profile. The linewidth is usually measured at full-width half-maximum.

Longitudinal electromagnetic mode: A mode cavity formed by the laser end mirrors.

Longitudinal mode spacing: The frequency spacing between the longitudinal modes, often called the free spectral range of the laser resonant cavity.

Lorentzian: A function, often used to describe a homogeneous gain profile.

Lower laser state: The lower-energy transition in a laser system.

Marginally stable: A resonator that meets the equality condition on the stability equation. Unstable in practice.

Metastable state: An upper laser state is termed metastable when its spontaneous lifetime is significantly longer than the spontaneous lifetime of the lower laser state.

Mode-locking: The process of setting all the longitudinal modes to the same phase in order to create short pulses separated by the round trip travel time of the cavity.

Monochromatic: Light of one color.

Multiplicative gain: The multiplicative coefficient describing the increase in laser intensity from one end of a laser system to the other end.

Near field: A distance close to a source, a distance much less than the Rayleigh range and where a wave is near-planar.

Nonlinear optics: A field of optics that exploits the nonlinear relation between the polarizability and the electric field.

Normalized function: A function whose integral is 1. (The word normalized can also be used to describe a function whose peak value is 1.)

Optical parametric oscillators: A device that generates light at the difference optical frequency between a pump and idler wave.

Ordinary index: n_o of a material.

Output coupler: The output mirror of a laser resonant cavity.

Parasitic transition: An unwanted radiative transition in a laser system. Usually a transition that steals gain from another transition.

Paraxial approximation: A mathematical approximation where the variation in the tranverse mode structure with z is assumed small.

Passive mode-locking or Q-switching: Using a saturable absorber dye or similar mechanism to mode-lock or Q-switch.

Perturbative: A mathematical technique for linearization where a term is assumed to be composed of the sum of a large constant term and a small oscillating term.

Phase matching: Setting the value of the index of refraction at the fundamental frequency equal to the value of the index of refraction at the second harmonic.

Photoelastic: A change in the index of refraction generated by acoustical excitation.

Photon density: Photons per unit volume in a cavity.

Piezoelectric: A crystal where the application of an electric field causes a change in crystal length.

Population inversion: Having a greater population in the upper state of a laser system than in the lower state. A nonthermal equilibrium condition that is necessary for optical gain.

Power and power density: Energy per unit time. Power can be described as average power (such as energy times repetition rate) or peak power (energy divided by temporal pulsewidth).

Pulse width: The temporal width of a laser pulse.

Pulsed: A laser that is not continuous.

Pump state: A state (or band of states) higher in energy than the upper laser state.

Pumping source: The device used to create a population inversion.

Q or quality factor: The resonant frequency of a cavity divided by the linewidth. More generally used to describe the property of resonance, as in high-Q cavity.

Q-switching: The process of blocking the transmission of the cavity to permit a greater upper state population than would exist in steady state.

Quantum efficiency: The percentage of pump band population that transfers to the upper laser state.

Quarter-wave plate: An optical element that converts linearly polarized light into circularly polarized light.

Raman scattering: A process which generates light at the sum and difference frequencies between a pump source and a Raman state in a material.

Rate equations: A system of equations that describes the changes in populations of various states.

Ray matrix or lens matrix technique: A matrix technique for analyzing lens systems based on tracking the slope and position of a ray.

Rayleigh range: A characteristic distance that marks the difference between plane and spherical wave behavior for a Gaussian beam.

Regenerative: A fancy word meaning a closed path.

Relaxation lifetime: The statistical lifetime of an electron in a state.

Relaxation oscillations: Transient oscillatory behavior of the upper state population and photon density.

Resonant cavity or resonator: The front and back mirrors placed around a laser gain material to create a regenerative optical path.

Round trip: Twice the length of the laser. (Front mirror to front mirror.)

Round-trip gain: The multiplicative gain for a round trip of the laser.

Saturable absorber: A dye with an intensity dependent absorption profile. Usually the dye bleaches at high intensities.

Saturation intensity: The intensity of light in a laser cavity for which the gain has dropped to one-half of its small signal value.

Second harmonic generation: The creation of light at twice the optical frequency (or half the wavelength).

Self-focusing: A nonlinear effect where an increase in intensity creates a change in index of refraction and the focusing generated by this change further increases the intensity. The increase in the intensity, further increases the change in the index. The process usually results in optical damage.

Self-trapping: A nonlinear effect where an increase in intensity creates a change in index of refraction and the focusing generated by this change compensates for diffraction.

Single pass: The length of the laser.

Single-pass gain: The multiplicative gain for a single pass of the laser.

Slope efficiency: The slope of the P_{out} versus P_{in} line.

Snell's law: A general law describing the reflection, refraction, and transmission of light from an interface.

Spatial coherence: A property occuring when two points on an electromagnetic wavefront possess the same relative phase.

Spherical wave: A wave with spherical equiphase surfaces. A wave generated by a point source.

Spontaneous emission: The decay of a higher-energy state to a lower-energy state, releasing a photon.

Spontaneous emission rate: The change in excited states per unit time resulting from spontaneous emission.

Spontaneous lifetime: The statistical lifetime for the spontaneous emission decay process.

Stable: A resonator where the beam is permanently trapped.

Stimulated emission: A process producing a photon of the same wavelength, in phase with, of the same polarization as, and in the same direction as the stimulating photon.

Stimulated emission cross-section: A constant used to characterize the material-dependent factors contributing to gain.

Stimulated emission rate: The change in excited states per unit time resulting from a stimulating radiation field (laser light).

Three-state laser: A laser possessing a ground state, a pump state, and an upper laser state.

Threshold power: The threshold input power where a laser begins to generate significant output power.

Transverse electromagnetic (TE) mode: The transverse spot pattern of a laser beam.

Tunability: The ability of the laser wavelength to be adjusted to suit the user.

Unsaturated or small signal gain: The gain coefficient when the upper state population is not affected by the gain process.

Unstable: A resonator where the beam can escape after a certain number of passes.

Upper laser state: The higher energy transition in a laser system.

Walk-off: The distance (or angle) between noncolinear beams.

Wallplug efficiency: The output power of the laser divided by the input power into the power supply.

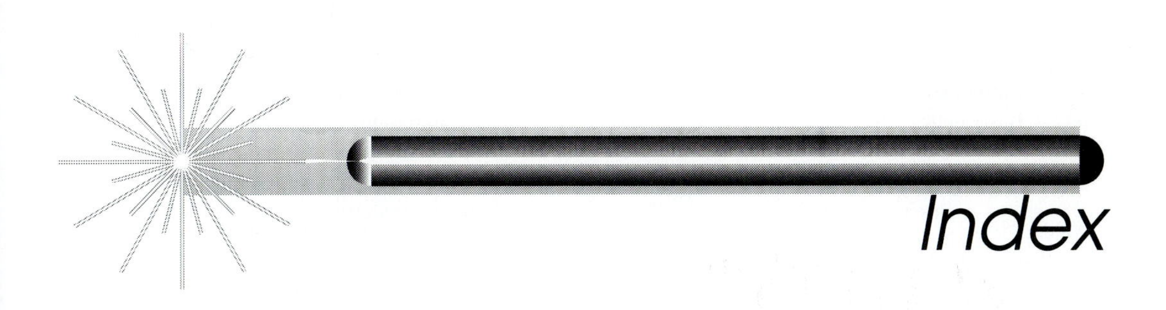

Index

Constants Used in Book

h: Planck's constant: $6.6260755 \cdot 10^{-34}$ J-sec

$\hbar = h/2\pi$: $1.05457267 \cdot 10^{-34}$ J-sec

k_B: Boltzman's constant: $8.61738573 \cdot 10^{-5}$ eV/K, $1.380658 \cdot 10^{-23}$ J/K

c_o: speed of light in a vacuum: $299,792,458$ m/sec

ϵ_o: the permittivity of free space: $8.854187817 \cdot 10^{-12}$ farad/meter)

μ_o: the permeability of free space: $4\pi \cdot 10^{-7}$ henry/meter

q: electron charge: $1.60217733 \cdot 10^{-19}$ C

m_o: rest mass of the electron: $9.109389 \cdot 10^{-31}$ kg